L. Wang (Ed.)

Soft Computing in Communications

Springer-Verlag Berlin Heidelberg GmbH

Studies in Fuzziness and Soft Computing, Volume 136
http://www.springer.de/cgi-bin/search_book.pl?series=2941

Editor-in-chief
Prof. Janusz Kacprzyk
Systems Research Institute
Polish Academy of Sciences
ul. Newelska 6
01-447 Warsaw
Poland
E-mail: kacprzyk@ibspan.waw.pl

Further volumes of this series can be found on our homepage

Vol 117. O. Castillo, P. Melin
Soft Computing and Fractal Theory for Intelligent Manufacturing, 2003
ISBN 978-3-642-53623-6

Vol. 118. M. Wygralak
Cardinalities of Fuzzy Sets, 2003
ISBN 978-3-642-53623-6

Vol. 119. Karmeshu (Ed.)
Entropy Measures, Maximum Entropy Principle and Emerging Applications, 2003
ISBN 978-3-642-53623-6

Vol. 120. H.M. Cartwright, L.M. Sztandera (Eds.)
Soft Computing Approaches in Chemistry, 2003
ISBN 978-3-642-53623-6

Vol. 121. J. Lee (Ed.)
Software Engineering with Computational Intelligence, 2003
ISBN 978-3-642-53623-6

Vol. 122. M. Nachtegael, D. Van der Weken, D. Van de Ville and E.E. Kerre (Eds.)
Fuzzy Filters for Image Processing, 2003
ISBN 978-3-642-53623-6

Vol. 123. V. Torra (Ed.)
Information Fusion in Data Mining, 2003
ISBN 978-3-642-53623-6

Vol. 124. X. Yu, J. Kacprzyk (Eds.)
Applied Decision Support with Soft Computing, 2003
ISBN 978-3-642-53623-6

Vol. 125. M. Inuiguchi, S. Hirano and S. Tsumoto (Eds.)
Rough Set Theory and Granular Computing, 2003
ISBN 978-3-642-53623-6

Vol. 126. J.-L. Verdegay (Ed.)
Fuzzy Sets Based Heuristics for Optimization, 2003
ISBN 978-3-642-53623-6

Vol 127. L. Reznik, V. Kreinovich (Eds.)
Soft Computing in Measurement and Information Acquisition, 2003
ISBN 978-3-642-53623-6

Vol 128. J. Casillas, O. Cordón, F. Herrera, L. Magdalena (Eds.)
Interpretability Issues in Fuzzy Modeling, 2003
ISBN 978-3-642-53623-6

Vol 129. J. Casillas, O. Cordón, F. Herrera, L. Magdalena (Eds.)
Accuracy Improvements in Linguistic Fuzzy Modeling, 2003
ISBN 978-3-642-53623-6

Vol 130. P.S. Nair
Uncertainty in Multi-Source Databases, 2003
ISBN 978-3-642-53623-6

Vol 131. J.N. Mordeson, D.S. Malik, N. Kuroki
Fuzzy Semigroups, 2003
ISBN 978-3-642-53623-6

Vol 132. Y. Xu, D. Ruan, K. Qin, J. Liu
Lattice-Valued Logic, 2003
ISBN 978-3-642-53623-6

Vol. 133. Z.-Q. Liu, J. Cai, R. Buse
Handwriting Recognition, 2003
ISBN 978-3-642-53623-6

Vol 134. V.A. Niskanen
Soft Computing Methods in Human Sciences, 2004
ISBN 978-3-642-53623-6

Vol. 135. J.J. Buckley
Fuzzy Probabilities and Fuzzy Sets for Web Planning, 2004
ISBN 978-3-642-53623-6

Lipo Wang

Soft Computing in Communications

Springer-Verlag Berlin Heidelberg GmbH

Prof. Lipo Wang
School of Electrical and Electronic Engineering
Nanyang Technological University
Block S2, 50 Nanyang Avenue
Singapore 639798
E-mail: elpwang@ntu.edu.sg

ISSN 1434-9922

DOI 10.1007/978-3-540-45090-0

Library of Congress Cataloging-in-Publication-Data

Soft computing in communications / Lipo Wang.
p. cm. -- (Studies in fuzziness and soft computing)
Includes bibliographical references and index.

1. Soft computing. 2. Telecommunication systems. I. Wang, Lipo. II Series.
QA76.9.S63S628 2004
006.3--dc22

http://www.springer.de

Originally published by Springer-Verlag Berlin Heidelberg 2004
MyCopy version of the original edition 2004

Typesetting: Camera-ready by author
Cover design: E. Kirchner, Springer-Verlag, Heidelberg
Printed on acid free paper 62/3020/M - 5 4 3 2 1 0
www.springer.com/mycopy

Preface

Soft computing, as opposed to conventional "hard" computing, tolerates imprecision and uncertainty, in a way very much similar to the human mind. Soft computing techniques include neural networks, evolutionary computation, fuzzy logic, and chaos. The recent years have witnessed tremendous success of these powerful methods in virtually all areas of science and technology, as evidenced by the large numbers of research results published in a variety of journals, conferences, as well as many excellent books in this book series on Studies in Fuzziness and Soft Computing.

This volume is dedicated to recent novel applications of soft computing in communications. The book is organized in four Parts, i.e., (1) neural networks, (2) evolutionary computation, (3) fuzzy logic and neurofuzzy systems, and (4) kernel methods.

Artificial neural networks consist of simple processing elements called neurons, which are connected by weights that may be adjusted during learning. Part 1 of the book has seven chapters, demonstrating some of the capabilities of two major types of neural networks, i.e., multiplayer perceptron (MLP) neural networks and Hopfield-type neural networks.

MLPs can be trained to respond with desired output when presented with certain input to arbitrary precision, which is why MLPs are said to be universal approximators. Hence MLPs are very useful in controlling non-linear dynamic systems. In the chapter written by Davoli and Maryni, MLPs are used to provide optimal admission control for bandwidth allocation in multiservice access multiplexers. Zerguine and Shafi study performance of an MLP-based decision feedback equalizer in non-linear channels. Ibnkahla investigates adaptive identification of nonlinear channels using natural gradient descent.

Hopfield-type neural networks (HNNs) have recurrent connections between neurons and minimize energy functions while the states of the neurons are updated. Because of this intriguing property, HNNs have been used in solving numerous optimization problems. Compared with other optimization approaches, HNNs are highly parallel and can be easily implemented with hardware. In this book, AboElFotoh applies the HNN in maximizing topology connectivity. Funabiki describes binary HNNs for combinatorial optimization problems in communication networks. Sharif, Chuah, and Hinton survey applications of neural networks for robust multiuser detection in direct-sequence code-division multiple-access systems. Wang, Li, Wan, and Soong employ a chaotic HNN for minimizing interference in cellular mobile communications by optimal channel assignment.

Evolutionary computation (EC) techniques, e.g., genetic algorithms (GA), are motivated from natural processes in which individuals evolve and improve themselves for the purpose of survival. Part 2 of this book includes three chapters on applications of GA in various optimization problems in communications.

In particular, Leung proposes orthogonal genetic algorithms with applications to multimedia multicast routing. Gu and Chu solve QoS multicast routing problems using GA. Krishnamachari and Wicker investigate base station location optimization in cellular wireless networks using GA and other heuristic search algorithms. Werner, Bray, Allard, and Werner study the synthesis and design of communication antennas using GA.

Fuzzy logic (FL) was invented to mimic human's vague concepts and approximate reasoning. Neurofuzzy systems, also known as fuzzy neural networks, combine FL with adaptive weights found in neural networks and are also universal approximators, like MLPs. Part 3 of this book contains five chapters on applications of FL and neurofuzzy systems in communications.

Koliver, Farines, and Nahrstedt report their work on QoS adaptation based on FL. Lian, Liu, and Chiu discuss chaotic signal synchronization and secure communications using FL design methods. Lo, Chang, and Chen study FL and neurofuzzy channel assignment schemes for hierarchical cellular systems. Gil and Park describe a neurofuzzy mobility prediction system and its application to restoration of mobility databases. Yuang and Tien present a framework for transporting voice over wireless LAN using neurofuzzy control.

Support vector machines and kernel methods have recently enjoyed great success in many areas. Part 4 of this book consists of two chapters on applications of kernel methods in various communications problems. Kuh focuses on the theory and applications of a newly developed kernel algorithm called the least squares support vector machine. Chen investigates least bit error rate adaptive multiuser detection.

I would like to sincerely thank all authors who have spent time and effort to make important contributions to this book. My gratitude also goes to Professor Janusz Kacprzyk and Dr. Thomas Ditzinger for their most kind support and help for this book.

Lipo Wang

Contents

Part 1: Neural Networks

Neural Approximation of Resource Allocation Functions in Multiservice Access Multiplexers 1
Franco Davoli and Piergiulio Maryni

Performance of the Multilayer Perceptron-based Decision Feedback Equalizer with Lattice Structure in Non-Linear Channels 33
Azzedine Zerguine and Ahmar Shafi

Nonlinear Channel Identification Using Natural Gradient Descent: Application to Modeling and Tracking 56
Mohamed Ibnkahla

Maximizing Topology Connectivity Using a Hopfield Neural Network 72
Hosam M.F. AboElFotoh

Binary Neural Network Approaches to Combinatorial Optimization Problems in Communication Networks 88
Nobuo Funabiki

Robust Multiuser Detection Using Artificial Neural Networks 106
B. S. Sharif, T. C. Chuah, and O. R. Hinton

Minimizing Interference in Cellular Mobile Communications by Optimal Channel Assignment Using Chaotic Simulated Annealing 131
Lipo Wang, Sa Li, Chunru Wan, and Boon Hee Soong

Part 2: Evolutionary Computation

Orthogonal Genetic Algorithms with Application to Multimedia Multicast Routing 147
Yiu-Wing Leung

Solving the QoS Multicast Routing Problems Using Genetic Algorithms 175
Qijun Gu and Chao-Hsien Chu

Base Station Location Optimization in Cellular Wireless Networks using Heuristic Search Algorithms 201
Bhaskar Krishnamachari and Stephen Wicker

The Synthesis and Design of Communication Antennas Using Genetic Algorithms 220
Douglas H. Werner, Matthew G. Bray, Rene J. Allard, and Pingjuan L. Werner

Part 3: Fuzzy Logic and Neurofuzzy Systems

QoS Adaptation Based on Fuzzy Theory 245
Cristian Koliver, Jean-Marie Farines, and Klara Nahrstedt

Fuzzy Chaotic Synchronization and Communication --- Signal Masking Encryption 268
Kuang-Yow Lian, Peter Liu, and Chian-Song Chiu

Intelligent Channel Assignment Schemes for Hierarchical Cellular Systems 291
Kuen-Rong Lo, Chung-Ju Chang, and Yih-Shen Chen

A New Mobility Prediction System using Neuro-Fuzzy Model and Its Application to Restoration of Mobility Databases 320
Joon-Min Gil and Chan Yeol Park

Voice Over Wireless LAN Using Intelligent Control 341
Maria C. Yuang and Po-Lung Tien

Part 4: Kernel Methods

Least Squares Kernel Methods and Applications 361
Anthony Kuh

Least Bit Error Rate Adaptive Multiuser Detection 384
Sheng Chen

Neural Approximation of Resource Allocation Functions in Multiservice Access Multiplexers

Franco Davoli[1] and Piergiulio Maryni[2]

[1] Department of Communications, Computer and Systems Science, DIST–University of Genova
Via Opera Pia 13, 16145 Genova, Italy
[2] Datasiel S.p.A., Via Merano 22, 16154 Genova, Italy

Abstract. In an access node to a multiservice network (e.g., a base station in an integrated services cellular wireless network the hub station in a double hop satellite system, or the Optical Line Terminal (OLT) in a broadband Passive Optical Network (PON)), the output link bandwidth is adaptively assigned to different users, and dynamically shared between guaranteed bandwidth traffic and asynchronous traffic types. The bandwidth allocation is effected by an admission controller, whose goal is to minimize the refusal rate of bandwidth reservation requests as well as the loss probability of cells queued in a finite buffer. Optimal admission control strategies are approximated by means of backpropagation feedforward neural networks, acting on the embedded Markov chain of the guaranteed bandwidth traffic dynamics; the neural networks operate in conjunction with a higher level bandwidth allocation controller, which performs a stochastic optimization algorithm. The case of unknown, slowly varying input rates is explicitly considered. Numerical results are presented that evaluate the approximation and the ability to adapt to parameter variations.

Keywords: Neural Networks, Multiservice Networks, Bandwidth Allocation.

1 Introduction

A peculiar aspect of high speed multiservice networks is the necessity of carrying a great variety of traffic classes, characterized by different statistical behavior and performance requirements (as expressed by Quality of Service (QoS) indices). A great deal of attention is being paid to satisfy all such requirements (and at the same time to ensure efficiency in the use of resources) within environments such as the Asynchronous Transfer Mode (ATM) [1], or within IP networks with integrated or differentiated services [2–5]. Since a mix of different transfer modes is very likely to remain within the existing networks, at least in the medium term, multiplexing devices will be used to provide a physical and/or MAC layer structure, capable of carrying traffic of asynchronous nature and non-stringent QoS requirements (like best-effort IP packets or ATM cells generated by Available Bit Rate (ABR) services), as well as guaranteed bandwidth, connection-oriented and real-time-constrained streams. In particular, even "hybrid" Time Division Multiplexing (TDM), where slots in a frame can be dynamically assigned to different traffic types, can maintain a position in the user access area. Actually, TDM still plays a major role in application areas such as cellular mobile radio networks [6,7], and is likely to be used, though in combination with Code Division Multiple Access (CDMA), also in the forthcoming third generation wireless networks [8,9]. A further example is constituted by Passive Optical Networks (PONs), used in the distribution of multimedia services (see, e.g., [10], [11]), where a TDM frame carries ATM cells in the uplink and downlink between Optical Network Units (ONU) in the users' premises and the Optical Line Termination (OLT) network interface. Still another instance is given by a satellite system (typically a VSAT one), where the hub station multiplexes downstream traffic destined to the terminals. These situations may be considered as possible applications of the model and of the control techniques developed in the paper. However, more generally, we do not explicitly need a framed channel in our model: we can consider the presence of a scheduler, which dynamically assigns the link capacity among different flows, according to some scheduling algorithm that ensures servicing of the real-time flows within the required time limits (see, e.g. [12,13]). In general, in order to allocate the available link capacity among different traffic classes and ensure QoS, while at the same time obtaining an efficient use of the bandwidth resources, some form of control must be exerted on the multiplexer. In this respect, several attempts have been made to allocate the available bandwidth between different service classes (e.g., isochronous and asynchronous), and to find the "optimal" allocation policies in order to minimize a cost related to the basic QoS indicators (e.g., blocking probability for isochronous connections and packet loss rate or delay for asynchronous packets) [14–17].

The purpose of the present chapter is to investigate the use of neural networks and stochastic approximation algorithms that allow the practical implementation of optimal admission control and bandwidth allocation strategies in a hybrid multiplexer of the type mentioned above, serving multiple users, with different traffic types. In general, neural networks have been widely used in access control and bandwidth allocation problems in the literature, especially in the context of ATM networks (e.g., [18–21]). A first investigation of the application of neural networks in hybrid access multiplexing problems has been made in [22], though in the case of a single user, and a hierarchical neural network controller in the multiuser case was considered in [23]. However, the latter was characterized by a non-stationary control and a finite horizon optimization, whereas a quite different (and more realistic) setting was considered in [24], by seeking stationary control strategies over an infinite horizon, and by explicitly considering time-varying traffic parameters. In particular, in [24] we considered the inclusion of the neural network controller in a two-level hierarchical scheme over an infinite time horizon, where bandwidth is allocated among a number of user sites, which in turn independently perform admission control. As will be apparent in the following, the use of neural approximations in this kind of problems, whose dynamics is characterized by finite state Markov chains, presents a practical interest mainly in the adaptive control case (i.e., when the controller has to adjust its parameters to follow slowly varying traffic statistics). The purpose of the chapter is to put the models and results of [22] and [24] in a unified framework.

An important advantage of neural networks over other techniques is their ability to approximate complex control functions. Moreover, with neural networks, most of the complexity is handled off-line in the "training phase" and the result is that in the on-line phase the value of the approximated functions can be retrieved in a very simple and efficient. The main non neural network approaches, in this kind of problems, belong to the family of Dynamic Programming techniques, that often require to recompute the control function each time some parameter of the system to be controlled changes (even slightly). On the other hand, neural networks can be trained, instead, to approximate the control function against a set of variations and, hence, to be ready to promptly react to changes while the system is running.

The chapter is organized as follows. In the next section, we formally define the system model. The approximation of the optimal strategies for a single user by means of a neural controller is outlined in Section 3, where non-constant arrival rates in both guaranteed bandwidth and best effort traffic are considered. In Section 4 we illustrate in more detail how the bandwidth can be divided among multiple users and in Section 5 some numerical results are reported. Section 6 draws the final conclusions.

2 The System Model

We consider an access multiplexer, handling a slotted channel shared among M users. Each user m ($1 \leq m \leq M$) generates two types of traffic: a real-time guaranteed bandwidth (e.g., isochronous) and connection-oriented one, with average connection request arrival rate λ_1^m requests/s and average holding time $1/\mu_1^m$ s, and an asynchronous connectionless one (cell-switched), with average arrival rate λ_2^m cells/frame, with fixed length (one slot). The "frame" time in our case is just the time T required by an isochronous source to generate b cells. The cells originating from the asynchronous connectionless traffic can be stored in a buffer, which has a finite capacity K_m. As a matter of fact, we can think in terms of an ATM multiplexer handling Continuous Bit Rate (CBR) traffic on a guaranteed rate basis, as well as ABR traffic; we consider the former only at the call-level, whereas we are interested in the cell-level behavior of the latter. The ABR cells may be originated, in turn, by the segmentation of variable length packets; in the case, for instance, of TCP/IP data, limiting the cell loss is desirable to reduce retransmission of lost packets. The arrival rates of both types of traffic are not supposed to be constant, but slowly varying within a given range (slowly with respect to connection dynamics). The dependence on time of λ_1^m and λ_2^m will be always supposed to hold, even if, for the sake of notational simplicity, it is not explicitly annotated. Guaranteed bandwidth connection requests are subject to a probabilistic admission control scheme. More specifically, given that user m has already r connections in progress, a new connection request is accepted with probability $\beta_t^m(r)$. The acceptance probability values for all possible realizations of r are collected in the vector $\boldsymbol{\beta}^m$. Figure 1 depicts a skeleton of the system model. CBR traffic requires continuation of service (b cells every

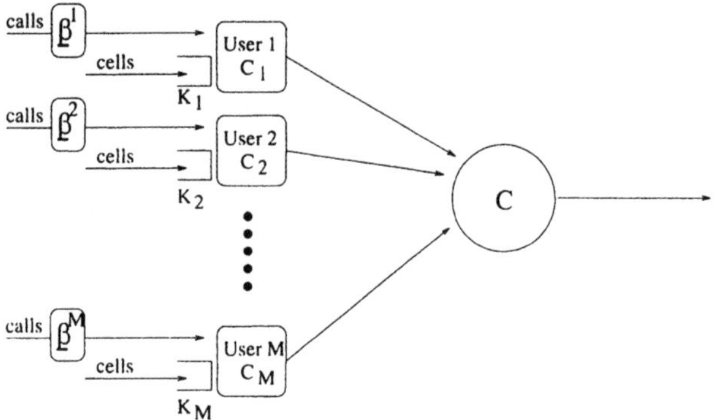

Fig. 1. The skeleton of the system model.

T s) until the end of the connection and, in order to guarantee this rate, pre-emption is performed with respect to the asynchronous traffic. Arrivals in the queue of the latter will be modeled as the superposition of geometric distributions, a widely used approach in modeling cell arrivals at ATM switches (see, e.g., [25]). Connection request inter-arrival and holding times will be supposed to have exponential distribution; we can then represent the connection dynamics as a continuous-time birth-death process. The goal of the assignment strategy should be that of deciding how to share the channel among the m users and how to share the amount of transmission resources assigned to each user between the two traffic types. It can be seen, therefore, as a hierarchical control task, where the resources of a large system are first partitioned into m subsystems, one per user, and then further partitioned among two traffic types. We call the first partitioning task *multiuser partitioning*, and the second one *multiservice partitioning*. Multiservice partitioning, for user m, is achieved by means of the acceptance probability vector $\boldsymbol{\beta}_t^m$. Figure 2 illustrates this double partitioning (i.e., hierarchical) scheme. In the following, the transfer capacity of the system will not be described in terms of bits/s, but rather in terms of the maximum number C of CBR connections that can be simultaneously in progress with all the system bandwidth

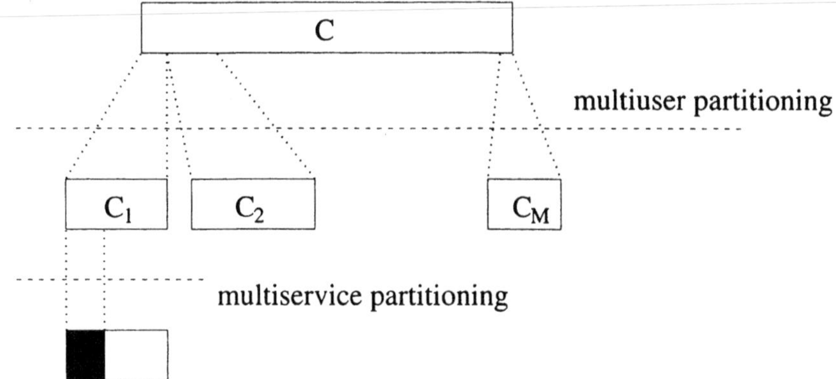

Fig. 2. Graphical description of the two-level partitioning.

and of the service rate seen by the asynchronous traffic, given the number of connections in progress. More in particular, if at time instant t there are r_t connections in progress (for a given subsystem m), then the service rate seen by the asynchronous traffic (for the same subsystem) is modeled as being constant with rate $\mu_2^m(r_t)$. Figure 3 gives a pictorial representation of this way of describing the system capacity. In the simplest case, $\mu_2^m(r_t)$ can be equal to $(C_m - r_t)b$, if b is the number of bandwidth units (cells per reference time T) used by a guaranteed bandwidth connection and C_m is the capacity assigned to the m-th user subsystem. In other situations (e.g., those where the asynchronous flows from different users contend for the available bandwidth), $\mu_2^m(r_t)$ can capture the effect of a given channel access control mechanism, by representing the maximum service rate achievable. Thus, the

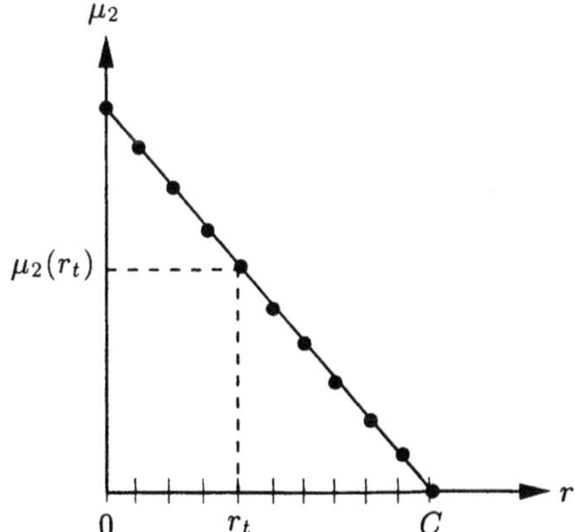

Fig. 3. Graphical description of the system load (μ_2 refers to the rate before partitioning).

result of the multiuser partitioning task is to subdivide the total "capacity" of C simultaneous guaranteed bandwidth flows among the M users; the generic user m will be, therefore, assigned a capacity equivalent to C_m connections, with $\sum_m^M C_m = C$. Figure 4 shows a pictorial example of such a resource partitioning between two subsystems (i.e., $M = 2$). Since in multiuser parti-

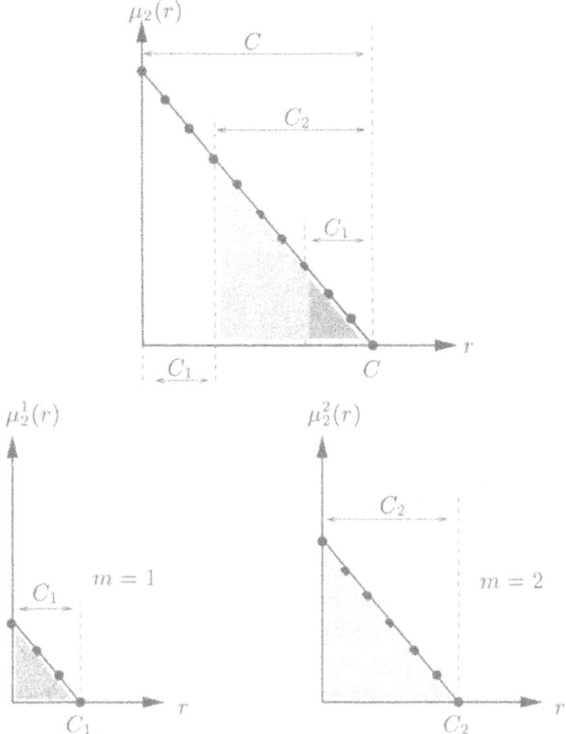

Fig. 4. Graphical description of the partitioning among subsystems ($M = 2$).

tioning we will need to use results which pertain to the multiservice one, we start first by looking at the latter in more detail.

3 Multiservice Partitioning

Let us now address the multiservice partitioning, the control mechanism that, for each subsystem (i.e., for each user), allows to share the networking resources between the two traffic types. In the following, we will assume that the generic user m ($1 \leq m \leq M$) is assigned a capacity of C_m connections as a result of the multiuser partitioning. The guaranteed bandwidth traffic is therefore represented by a controlled M/M/C_m/C_m queueing system; we consider the embedded Markov chain that arises from looking only at the instants at which connections are set–up and torn–down. We can do so, since we suppose our allocation strategy to act explicitly only on the acceptance of connection attempts of this traffic type. Cell traffic is simply assigned to slots that the guaranteed bandwidth connections do not utilize: no explicit control is done on the acceptance of cells. Moreover, we assume that cell dynamics runs on a faster time scale with respect to the other traffic (as seen at the connection level): this implies that we can think of the queue process in terms of a stationary one. In other words, we suppose the time between two successive events of the birth–death process characterizing the real-time traffic to be large enough to make the transient behavior of the queue process negligible. A similar way of reasoning has been followed in [26]: the approximation is applied to the joint stationary distribution of the two traffic types. When in the system there are, say, r calls in progress, we will model the cell queue as a $\mathrm{Geom}(bC)/D/\mu_2^m(r)/(K_m + \mu_2^m(r))$ one (see the Appendix). Therefore, in our context, the state of the whole subsystem at a certain time instant is represented only by the number r of connections in progress ($r \in \{0, 1, \ldots, C_m\}$). Let $t = 0, 1, \ldots,$ represent a discrete–time variable, counting the occurrence of an event in the real-time traffic process for the generic user m (i.e., a connection has been set–up or it has ended), and let r_t be the number of connections in progress at time t (actually, the notation should be t_m and r_{t_m}; however, for the sake of notational simplicity, we have dropped the subscript m). Whenever an incoming connection request for a "circuit" appears, it is either accepted with probability $\beta_t^m(r_t)$ or it is blocked (refused) with probability $1 - \beta_t^m(r_t)$. Since the channel is only able to carry C_m connections, we will have $\beta_t^m(C_m) \equiv 0$. Thus $\beta_t^m(r_t)$ represents a randomized control strategy that maps r_t into a probability of call acceptance, and whose choice will be one of the main goals of our treatment. Since we suppose that asynchronous traffic cells occupy slots unused by the real-time traffic, this strategy plays the role of a "probabilistic movable boundary", and realizes a partition of the available capacity. Notice that, since we consider time varying input rates, the functions β_t^m (i.e., the vectors with components $\beta_t^m(s)$, $s = 0, 1, \ldots, C_m$) turn out to be also time varying.

The following sections 3.1 and 3.2 will present possible solutions to the multiservice partition problem under two different points of view. Both of them minimize a given cost function but, while in the first the focus is on a finite horizon problem, the second considers an infinite horizon one. Since at this level of partitioning we are working on the subsystem of the generic m-th user, whenever possible, for the sake of notational simplicity, the dependence on m is not made explicit.

3.1 Controlling the System within a "Receding Horizon" Problem

In such a framework the state of the whole system is represented by the number r_t of calls in progress ($r_t \in \{0, 1, \ldots, C\}$), and by the two arrival rates $\lambda_t^{(1)}$

and $\lambda_t^{(2)}$. For the sake of notational simplicity, let us define $\boldsymbol{\Lambda}_t \triangleq \text{col}\,[\lambda_t^{(1)}, \lambda_t^{(2)}]$. We will also indicate the control function as $\hat{\beta}_t(i, \boldsymbol{\Lambda}_t)$ in the following.

In order to simplify the computation involved in the minimization problem we will present, it turns out to be more convenient to transform the continuous time process into an equivalent Markov chain by means of a "uniformization" procedure [27], that works as follows. Let $\lambda_{max}^{(1)}$ be defined as $\sup_t(\lambda_t^{(1)})$, which we suppose to be finite and known. Then $\mu_1 C + \lambda_{max}^{(1)}$ becomes the maximum rate at which connection/disconnection events can occur (this comes from the maximization of $\mu_1 i + \lambda_t^{(1)} \hat{\beta}_t(i, \boldsymbol{\Lambda}_t)\,, i = 0, \ldots, C$), and this rate is independent of the state of the system. In this new framework, the process becomes a discrete time Markov chain with the following transition probabilities:

$$P\{r_{t+1} = j \,|\, r_t = i, \boldsymbol{\Lambda}_t\} = \begin{cases} \dfrac{\lambda_t^{(1)} \hat{\beta}_t(i, \boldsymbol{\Lambda}_t)}{\mu_1 C + \lambda_{max}^{(1)}} & j = i + 1 \\ \dfrac{\mu_1 i}{\mu_1 C + \lambda_{max}^{(1)}} & j = i - 1 \\ 1 - \dfrac{\lambda_t^{(1)} \hat{\beta}_t(i, \boldsymbol{\Lambda}_t)}{\mu_1 C + \lambda_{max}^{(1)}} - \dfrac{\mu_1 i}{\mu_1 C + \lambda_{max}^{(1)}} & j = i \\ 0 & \text{otherwise} \end{cases} \tag{1}$$

We have therefore introduced another event in the process (i.e., the case $j = i$ in (1)), in which nothing happens, but we have greatly simplified the model, since now the process rate is the same independently of the state of the system. Let $P_{\hat{\beta}_t(r_t, \boldsymbol{\Lambda}_t)}$ be the transition probability matrix, which, for the sake of notational simplicity, will be denoted as P_t in the following. Moreover, let

$$\pi_t(s) = P\,\{r_t = s\} \tag{2}$$

and

$$\boldsymbol{\pi}_t \triangleq \text{col}\,[\pi_t(0), \pi_t(1), \ldots, \pi_t(C)] \tag{3}$$

Then

$$\boldsymbol{\pi}_{t+1} = [P_t]' \boldsymbol{\pi}_t \tag{4}$$

where the superscript $'$ denotes transpose. Let us define $\xi(t)$ as the time instant (in continuous time) at which the t-th event occurs. Given a time horizon of T events, we can then introduce the following cost function:

$$J_t = E \int_{\xi(t)}^{\xi(t+T)} h(r_\eta, \boldsymbol{\Lambda}_t, \beta_\eta(r_\eta, \boldsymbol{\Lambda}_t)) d\eta \ , \tag{5}$$

where we have chosen

$$h(s,\boldsymbol{\Lambda}_t,\hat{\beta}_\tau(s,\boldsymbol{\Lambda}_t)) = \sigma_1 \left[\alpha - \hat{\beta}_\tau(s,\boldsymbol{\Lambda}_t)\right]^2 \lambda_t^{(1)} b + \sigma_2 P_{\text{loss}}(s,\lambda_t^{(2)}) \tag{6}$$

$P_{\text{loss}}(s,\lambda_t^{(2)})$ is the loss probability of a Geom$(bC)/D/\mu_2(s)/(K_m+\mu_2(s))$ system with finite buffer K_m and input rate $\lambda_t^{(2)}$ (see Appendix, where $\lambda_2^m = \lambda_t^{(2)}$); σ_1 and σ_2 are weighting coefficients. The term in brackets in the r.h.s. of (6) represents the square deviation between a desired acceptance probability α, $0<\alpha\leq 1$, for connection requests and the one provided by the access control strategy in state s. In our equivalent model with uniform rate the state of the system does not change between two subsequent events and, therefore, $h(s,\beta_\eta(s,\boldsymbol{\Lambda}_t))$ is constant. Hence, we can rewrite (5) as follows (the dependence on r_t has been indicated explicitly and we have defined $\delta(\tau)\triangleq\xi(\tau+1)-\xi(\tau)$):

$$J_t(r_t,\boldsymbol{\Lambda}_t) = E\left\{\sum_{\tau=t}^{t+T-1} h(r_\tau,\boldsymbol{\Lambda}_\tau,\beta_\tau(r_\tau,\boldsymbol{\Lambda}_\tau))(\xi(\tau+1)-\xi(\tau))\right\} =$$
$$= \sum_{\tau=t}^{t+T-1} E_{r_\tau}\{h(r_\tau,\boldsymbol{\Lambda}_\tau,\beta_\tau(r_\tau,\boldsymbol{\Lambda}_\tau))\}E\{\delta(\tau)\}$$

Notice that, by means of (1), the random variables $\xi(\tau)$, which represent the occurrence of an event, do not depend on the state of the system.

Moreover, by means of (2), and by considering that the time $\delta(\tau)$ between two subsequent events is expressed by a random variable with exponential distribution and average $1/(\lambda_{max}^{(1)}+\mu_1 C)$, we can write:

$$\hat{J}_t(r_t,\boldsymbol{\Lambda}_t) = \hat{h}(r_t,\boldsymbol{\Lambda}_t,\hat{\beta}_t(r_t,\boldsymbol{\Lambda}_t)) + \sum_{\tau=t+1}^{t+T-1}\sum_{s=0}^{C} \hat{h}(s,\boldsymbol{\Lambda}_\tau,\hat{\beta}_\tau(s,\boldsymbol{\Lambda}_\tau))\,\pi_\tau(s,\boldsymbol{\Lambda}_\tau) \tag{8}$$

where

$$\hat{h}(s,\boldsymbol{\Lambda}_t,\hat{\beta}_\tau(s,\boldsymbol{\Lambda}_t)) =$$
$$= \frac{1}{\mu_1 C+\lambda_{max}^{(1)}}\left\{\sigma_1\left[\alpha-\hat{\beta}_\tau(s,\boldsymbol{\Lambda}_t)\right]^2 \lambda_t^{(1)} b + \sigma_2 P_{\text{loss}}(s,\lambda_t^{(2)})\right\} \tag{9}$$

Finally, it has to be considered that, in practical cases, variations of $\boldsymbol{\Lambda}_t$ turn out to be very slow compared to variations of r_t. Indeed, during a "short" period, input rates may change so slowly that they can be considered constant. If, in particular, this is true for the time horizon $(t,t+T-1)$, we can assume $\boldsymbol{\Lambda}_\tau = \boldsymbol{\Lambda}_t$ for any $\tau = t,t+1,\ldots,t+T-1$. The cost function (8) turns then into the following:

$$\hat{J}_t(r_t,\boldsymbol{\Lambda}_t) = \hat{h}(r_t,\boldsymbol{\Lambda}_t,\hat{\beta}_t(r_t,\boldsymbol{\Lambda}_t)) + \sum_{\tau=t+1}^{t+T-1}\sum_{s=0}^{C} \hat{h}(s,\boldsymbol{\Lambda}_t,\hat{\beta}_\tau(s,\boldsymbol{\Lambda}_t))\,\pi_\tau(s,\boldsymbol{\Lambda}_t)$$

(10)

Notice that the cost function (10) embraces a finite horizon period characterized by T events.

We can then define the following repetitive (or "receding horizon") optimal control problem (see, for instance, [28]).

Problem 1: *at each discrete event t, find the strategies $\beta_\tau(r_\tau, \boldsymbol{\Lambda}_t)$, $\tau = t, t+1, \ldots, t+T-1$, that minimize cost (10), for any $r_\tau \in \{0, 1, \ldots, C\}$, and for any π_t, defined on a suitable probability space.*

In the repetitive control scheme, only the first control function, namely $\beta_t(r_t, \boldsymbol{\Lambda}_t)$, will be actually applied. In fact, the controller "sees" a time horizon that shifts forward at each t, and reconsiders the problem as starting again anew. Obviously, as neither P_t nor J_t depend explicitly on the discrete time instant t, this gives rise to a stationary strategy, whose off-line computation is effected only once. Hence, no on-line computational effort is requested from the controller, which is clearly a notable advantage of the proposed methodology.

As regards QoS constraints, the required operating conditions (in terms of connection blocking and packet loss rate) should be met as closely as desired, by means of a suitable choice of the weighting coefficients. This is true, in particular, for the connection blocking probability, which is often specified as a performance objective that should not be violated (i.e., should remain below a fixed upper bound).

The Neural Controller The functional optimization Problem 1 cannot be solved analytically, so we shall move toward an approximate technique, which consists in assigning the control strategies a given structure, in which the values of a certain number of parameters have to be determined, in order to minimize the cost function. This method is not new in control theory, but it does not seem that it met with great success, as the selected structures were characterized by too small a number of free parameters to attain satisfactory approximation properties; moreover, they required rather complex computational procedures to determine the optimal values of the unknown parameters.

Multilayer feedforward neural networks are not affected by such drawbacks; on the contrary, they are very attractive candidates for the design of nonlinear control strategies. As will be shown, this kind of neural networks can involve an arbitrarily large number of parameters (synaptic weights), but they are well suited for the use of distributed optimizing algorithms that keep the complexity of calculations substantially independent of the number of parameters to be tuned. However, it should be stated beforehand that the above considerations cannot lead us to deduce that, in general, the neural algorithms presented in the following should perform better than the dynamic programming technique or other classical methods for the solution of optimal control problems (this is especially true in a finite-state MDP). As will be shown by numerical results, the proposed algorithms involve some specific computational difficulties (typically, arising from the considerable length of their convergence time). On the other hand, our neural controller is a very promising candidate for the effective implementation of adaptive strategies, capable of responding to variations in the input rates, with very limited computational effort, whereas Dynamic Programming would require the entire re-calculation of the optimal strategy.

Now, we focus on the approximate solution of the functional optimization Problem 1. In this respect, let us constrain the control strategies to take on a fixed structure of the form

$$\tilde{\beta}_\tau(r_\tau, \lambda_t^{(1)}, \lambda_t^{(2)}) = f(\hat{\gamma}(r_\tau, \lambda_t^{(1)}, \lambda_t^{(2)}, \boldsymbol{w}_\tau), r_\tau) \tag{11}$$

where $\hat{\gamma}$ is the input/output mapping of a multilayer feedforward neural network and $\boldsymbol{w}_\tau$ is the vector of the synaptic weights to be determined. We have indicated explicitly the dependence of the acceptance strategy on the input rates. The function $f(\cdot)$ takes into account constraints on the output of the neural network, that is

$$f(\hat{\gamma}(r_\tau, \lambda_t^{(1)}, \lambda_t^{(2)}, \boldsymbol{w}_\tau), r_\tau) = \begin{cases} \hat{\gamma}(r_\tau, \lambda_t^{(1)}, \lambda_t^{(2)}, \boldsymbol{w}_\tau), & r_\tau < C \\ 0 & r_\tau = C \end{cases} \tag{12}$$

which means that we cannot accept more connections than the number of slots we have in each frame.

The control scheme gives rise to a chain of T neural networks, each followed by the dynamic system [29].

If we now substitute (11) into (1) and (10), and (1) into (10), the cost function takes on the form $\tilde{J}_t(\boldsymbol{w}, \boldsymbol{\Lambda}_t, \boldsymbol{\pi}_t)$, where $\boldsymbol{w} \triangleq \text{col}\,[\boldsymbol{w}_t, \boldsymbol{w}_{t+1}, \ldots, \boldsymbol{w}_{t+T-1}]$ (we have dropped the subscript t in the parameter vector $\boldsymbol{w}$, which is independent of time). Since the cost function depends both on the vector $\boldsymbol{w}$ to be determined and on vectors $\boldsymbol{\Lambda}_t$, $\boldsymbol{\pi}_t$, we eliminate the dependence on $\boldsymbol{\Lambda}_t$, $\boldsymbol{\pi}_t$ by averaging $\tilde{J}_t(\boldsymbol{w}, \boldsymbol{\Lambda}_t, \boldsymbol{\pi}_t)$ with respect to the above vectors (we assume $\boldsymbol{\Lambda}_t$, $\boldsymbol{\pi}_t$ uniformly distributed). It is worth noting that, though this procedure is rather arbitrary, it is not unusual. For example, it has been applied to solve parametric LQ optimal control problems (see, for instance, [30], where the gain matrix of the controller is determined after averaging the cost function with respect to the initial state, considered as a random vector). It is certainly true that another way of eliminating the dependence of $\tilde{J}_t(\boldsymbol{w}, \boldsymbol{\Lambda}_t, \boldsymbol{\pi}_t)$ on $\boldsymbol{\Lambda}_t$, $\boldsymbol{\pi}_t$ may consist in adopting a min–max approach, which means to maximize the cost function with respect to $\boldsymbol{\Lambda}_t$, $\boldsymbol{\pi}_t$. Our choice in favour of the expectation procedure is essentially motivated by the reassuring experimental results given in the following and confirmed, in general, in the literature for similar optimization problems. The min–max approach, however, appears more appropriate in some practical cases, for example, when there is the danger of incurring unacceptably high costs. The possibility of using such an alternative approach should be examined carefully. We have then to solve the following optimization problem.

Problem 2: *find the vector* w° *that minimizes* $\underset{\Lambda_t,\pi_t}{\mathrm{E}}\left[\tilde{J}_t(w,\Lambda_t,\pi_t)\right]$.

□

It follows that the functional Problem 1 has been reduced to an unconstrained nonlinear programming problem. Let us now describe in some detail the neural networks that implement our control strategies.

Each of the T neural networks implementing the strategies $\hat{\gamma}_\tau$ is composed of L layers, and, in the generic layer l, n_l neural units are active. The input/output mapping of the q-th neural unit of the l-th layer is given by

$$y_q^\tau(l) = g\left[z_q^\tau(l)\right], \quad q = 1,\ldots,n_l\,;\, l = 1,\ldots,L \tag{13}$$

$$z_q^\tau(l) = \sum_{p=1}^{n_{l-1}} w_{pq}^\tau(l) y_p^\tau(l-1) + w_{0q}^\tau(l) \tag{14}$$

where $y_q^\tau(l)$ is the output variable of the neural unit, $g(x) = 1/(1+e^{-x})$ is a sigmoidal activation function, and $w_{pq}^\tau(l)$, $w_{0q}^\tau(l)$ are the weight and bias coefficients, respectively. All these coefficients are the components of the vector w_τ appearing in the control law (11); the variables $y_1^\tau(0), y_2^\tau(0)$, and $y_3^\tau(0)$ coincide, respectively, with r_τ, $\lambda_t^{(1)}$, and $\lambda_t^{(2)}$, and the variable $y_1^\tau(L)$ yields the output generating the control strategies $\tilde{\beta}_\tau$. The structure of the neural network is presented in Fig. 5. Clearly, the solution of Problem 2 constitutes an approximation for Problem 1.

The unconstrained nonlinear programming Problem 2 can be solved by means of some descent algorithm. We focus our attention on methods of the gradient type, as, when applied to neural networks, they are well suited to distributed computation. The gradient algorithm can be written as follows

$$\begin{aligned} w(i+1) = w(i) - \theta_i \nabla_w \Big\{ &\hat{h}(r_t,\Lambda_t,\tilde{\beta}_t(r_t,\Lambda_t)) + \\ &\sum_{\tau=t+1}^{t+T-1} \sum_{s=0}^{C} \hat{h}(s,\Lambda_t,\tilde{\beta}_\tau(s,\Lambda_t))\,\pi_\tau(s)\Big\} + \bar{\theta}[w(i) - w(i-1)] \end{aligned} \tag{15}$$

where θ_i is the descent step–size, which is generally made dependent from the iteration number. More specifically, we have chosen $\theta_i = K_1/(i+K_2)$, where K_1 and K_2 are two constant values. Moreover, the last term in (15) represents the so–called "momentum term", which is usually introduced in training neural networks in order to accelerate the convergence ($\bar{\theta}$ is a suitable positive scalar).

Let us refer to the notation reported in Fig. 5 as regards a generic layer of the neural network acting at stage τ, and consider that, at each stage, the input pattern has to go through the entire state space $\{s, s = 0,1,\ldots,C\}$,

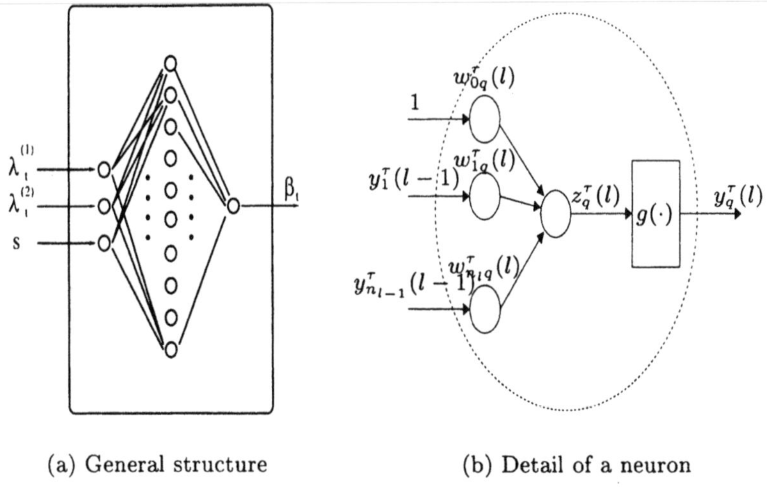

(a) General structure (b) Detail of a neuron

Fig. 5. The neural network structure

in order to generate all values of $\beta_\tau(.)$ that are, in turn, needed to write the matrix P_τ (one for each row). The model of the control scheme over the whole horizon is as represented in Fig. 6.

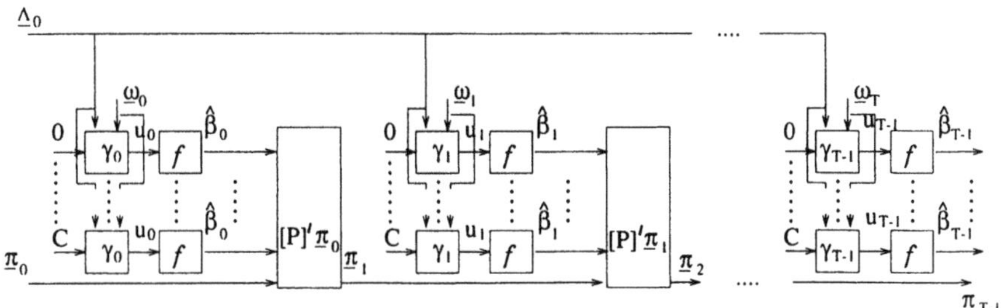

Fig. 6. The neural network chain model

For simplicity, the time index t has been omitted in this figure, and the stages are labelled from 0 to $T-1$. The additional input $\boldsymbol{\Lambda}_0$ represents the vector of arrival rates to be considered for the current control horizon. For greater clarity, we have also made explicit the fact that, at each stage, the same network is fed with all the possible values of the state (i.e., $0 \ldots C_m$).

The learning algorithm will therefore consist of the iteration up to convergence of the following procedure:

1. Consider stage $\tau = t$.
2. Pick randomly an initial probability vector π_t and a vector Λ_t.
3. Perform the forward step for every input pattern $s \in 0, 1, \ldots, C$, and compute $\pi_{\tau+1}$.
4. Set $\tau \Leftarrow \tau + 1$.
5. If $\tau < t + T$ go to 3.
6. Perform the backward steps and update the weights.

3.2 Controlling the System within an "Infinite Horizon" Problem

In this case we still consider time varying input rates; however, we suppose such variations to be slow enough, with respect to connections dynamics, to let the Markov process describing the connections reach stationarity. Under this assumption, if we look now at the connections dynamics (for a given value of the input rates), we will see that the process r_t is a controlled birth-death one, with upward rate $\lambda^{(1)}\beta(r_t)$ and downward rate $\mu_1 r_t$, as depicted in Figure 7. We compute the strategy β by minimizing a local cost function

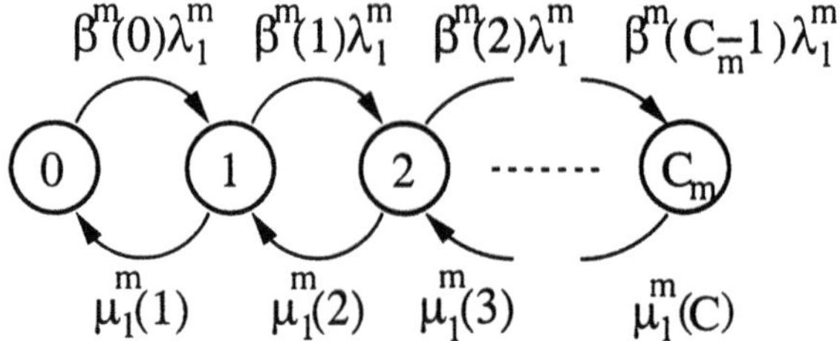

Fig. 7. connection dynamics for user m.

with the following structure:

$$J = \sigma_1 \cdot f_1(P_{\text{block}}) + \sigma_2 \cdot f_2(P_{\text{loss}}) \tag{16}$$

where σ_1 and σ_2 are weighting coefficients, $f_1(\cdot)$ and $f_2(\cdot)$ are two generic functions, whose choice will be specified, P_{block} is the connection blocking probability and P_{loss} represents the cell loss probability for asynchronous traffic (all this is related to the mth user). In the following we will consider

$$f_2(P_{\text{loss}}) \equiv \bar{P}_{\text{loss}} = \sum_{s=0}^{C} P_{\text{loss}}(s) \cdot \pi(s) \tag{17}$$

where $P_{\text{loss}}(s)$ is the packet loss probability given that s $(0 \le s \le C)$ connections are in progress (see the Appendix); as regards $f_1(\cdot)$ can now easily consider two different forms, namely

$$1)\ \sum_{s=0}^{C}(\alpha-\beta(s))^2\cdot\pi(s) \tag{18a}$$

$$2)\ \max^2\left\{0\,;\left[\sum_{s=0}^{C}(1-\beta(s))\cdot\pi(s)\right]-\alpha\right\}. \tag{18b}$$

In the above expressions, $\pi(s)$ represents the probability of having s connections in progress (i.e., the probability for user m of being in state s). The form of (18a), analogously to the first term in the r.s.h. of (6), "enforces" a probabilistic control strategy, owing to the way the control function enters the cost, and its rationale is to keep the blocking probability at each state below a given threshold value. As a matter of fact, the cost function (16) with (18a) may be regarded as the infinite horizon version of the finite stage cost previously considered in Section 3.1. On the other hand, (18b) contains the average blocking probability

$$\bar{P}^m_{\text{block}}=\left[\sum_{s=0}^{C}(1-\beta(s))\cdot\pi(s)\right] \tag{19}$$

as a whole, and comes into effect only when this quantity exceeds a given threshold α. In practice, this form represents a penalty function, trying to enforce a (probabilistic) constraint on the average blocking probability; notice that, whenever $\alpha=0$, (18b) would lead to a deterministic control strategy. In the following we will denote the two cost functions as J^1 and J^2.

By looking at Figure 7, with some trivial calculations, it is straightforward to come up with the following recursive way of computing $\pi=(\pi(0),\ldots,\pi(C))$ as the stationary distribution of a birth-death process:

$$\begin{cases}\pi(n+1)=\pi(n)\cdot\dfrac{\lambda_1\cdot\beta(n)}{\mu_1\cdot(n+1)}\\[2ex] \pi(0)=\left[1+\displaystyle\sum_{n=1}^{C}\frac{\left(\prod_{i=0}^{n-1}\beta(i)\right)\cdot(\lambda_1)^n}{(\mu_1)^n\cdot n!}\right]^{-1}\end{cases} \tag{20}$$

If we had fixed input rates, we would have now all the information for computing the optimum control law β^* by using, for example, a gradient descent algorithm. Since we have variable input rates (and therefore variable β^*), it is reasonable to employ a more complex framework, by introducing a neural network controller. Our aim is to keep the local admission controller as simple as possible by approximating our probabilistic admission control strategies over a range of input rates, to avoid solving a mathematical programming problem for each new value of the rates that is observed. The latter may be tracked over time, by means of some estimation mechanism (see, e.g., [31]).

The Neural Controller Let us again constrain the control strategies to take on a fixed structure of the form

$$\tilde{\beta}(r, \lambda_t^{(1)}, \lambda_t^{(2)}) = f(\gamma(r, \lambda_t^{(1)}, \lambda_t^{(2)}, \boldsymbol{w}_m), r) \tag{21}$$

where γ is the input/output mapping of a multilayer feedforward neural network and $\boldsymbol{w}_m$ is the vector of the synaptic weights to be determined. We have indicated explicitly the dependence of the acceptance strategy on the input rates. The function $f(\cdot)$ takes into account constraints on the output of the neural network, that is

$$f(\gamma(r, \lambda_t^{(1)}, \lambda_t^{(2)}, \boldsymbol{w}), r) = \begin{cases} \gamma(r, \lambda_t^{(1)}, \lambda_t^{(2)}, \boldsymbol{w}), & r < C \\ 0 & r = C \end{cases} \tag{22}$$

which means that we cannot accept more calls than the number of slots we have in each frame. If we now substitute (21) in (20), (17) and (18a/18b) in (16), the cost function takes on again the form $\hat{J}(\boldsymbol{w})$, as in Section 3.1.

The structure of a single hidden layer of the neural network is the same as the one presented in Fig. 5 with the only difference that, in this case, the dependence on τ (and on t) should not be considered. Again, as in Section 3.1, the variable $y(0)$ coincides with r, $y(1)$ coincides with $\lambda_t^{(1)}$, $y(2)$ coincides with $\lambda_t^{(2)}$, and the variable $y_1(L)$ yields the output generating the control strategies $\tilde{\beta}$.

The gradient algorithm can now be written as follows

$$\boldsymbol{w}(i+1) = \boldsymbol{w}(i) - \theta_i \nabla_{\boldsymbol{w}} J|_{(\lambda^{(1)}(i), \lambda^{(2)}(i))} + \bar{\theta}[\boldsymbol{w}(i) - \boldsymbol{w}(i-1)] \tag{23}$$

where θ_i and $\bar{\theta}$ are the same as in Section 3.2. It is worth noting that the algorithm in (23) actually implements a stochastic approximation with respect to the input rates (see [32,33]), which consists in iterating a classical gradient algorithm by picking randomly, at each descent iteration, the values of the input rates; the dependence of the gradient on the realization of $\lambda^{(1)}(i)$ and $\lambda^{(2)}(i)$ at the i-th iteration has been therefore indicated.

The neural network is then trained by following the classical backpropagation algorithm [34].

To conclude this section, it is worth making the following general remarks on the use of neural networks, which apply to both approaches considered in this and in the previous subsection (and to the numerical results to be discussed in Section 5).

- It is "almost always" possible to say that the same level of approximation obtained with more than one hidden layer can be achieved with a single hidden layer, as is the case of the net proposed in the paper. This is a consequence of the Weierstrass property possessed by nets with a single hidden layer and linear output function (see, among others, Leshno et al. [35]). The term "almost always" stems from the fact that, in certain particular optimization problems, it is possible to show that nets with two hidden layers can perform tasks that cannot be achieved by nets with a single hidden layer [36].

- As concerns the number of neural units present in the single hidden layer, we have deemed of little significance (given the net's simplicity) the use of "pruning" or other, anyhow heuristic, techniques. We have therefore chosen only to perform various experiments on the number of neurons. In practice, we have kept increasing the number of neurons, as far as the cost value became practically insensitive to the number used. This is clearly a "trial and error" rule, but we considered its application to be reasonable in this case.

4 Multiuser Partitioning

The goal of the multiuser partitioning task is to decide how M users have to share a common resource. More in particular, we need to identify how to subdivide a total bandwidth equivalent to C isochronous connections. In doing this, we will not take into account the exact values of the input rates, but only their range of variation (or their probabilistic distribution). Our aim is to find a simple "average" partitioning over an infinite horizon (in practice, a partitioning that can last for a relatively long time, with respect to fluctuations in the input rate). We now define a global cost as the sum of all the local ones,

$$J = \sum_{m}^{M} J_m(\beta^m, C_m) \tag{24}$$

where the dependence on $\lambda^{(1)}$ and $\lambda^{(2)}$ has not been explicitly indicated. Cost (24) should be averaged with respect to the input rates, and then minimized with respect to $\beta^m$and C_m, $m = 1, \ldots, M$. In order to avoid averaging over the input rates, we prefer to derive the optimal values of β^m for each given value of C_m by means of a stochastic approximation technique (see again [32]).

Let us further define the minimum cost, with respect to the control strategy β^m

$$J_m^*(C_m) = J_m(\beta^{*\,m}, C_m) \tag{25}$$

In this framework, we can use a classic *dynamic programming* algorithm in a resource partitioning context [37]. Let us define the cost matrix $\mathcal{C}$, each element of which represents the minimum average local cost $J_m^*(C_m)$ of user m, if capacity C_m is assigned. In order to compute the minimum average local costs $J_m(\beta^{*m}, C_m)$, we can use a fast technique such as the stochastic gradient descent one, as in the previous section, namely

$$\beta^m(i+1) = \beta^m(i) - \epsilon_i \nabla_{\beta^m} J_m|_{(\lambda^{(1)}(i),\lambda^{(2)}(i))} \tag{26}$$

where ϵ_i is an iteration dependent descent step-size, that also takes on the form $\epsilon_i = \tilde{K}_1/(i + \tilde{K}_2)$, and $\lambda^{(1)}(i)$, $\lambda^{(2)}(i)$ are the input rate values at iteration i (see again [32]).

The matrix $\mathcal{C}$ has dimensions $(C+1) \times M$, where C is the total capacity (in terms of number of CBR connections) and M is the total number of users. Let us further define a cost-to-go function in this way

$$\begin{aligned} I_M(d) &= J_M^*(d) \\ I_m(d) &= \min_{0 \le c \le d} \{J_m^*(c) + I_{m+1}(d-c)\} \end{aligned} \tag{27}$$

The cost-to-go has to be interpreted as follows. If for users $m, \ldots, M$ there are d resource units (i.e, bandwidth units) left (i.e., $I_m(d)$), and user m uses, say, c resource units, then for users $m+1, \ldots, M$ there will be $d-c$ resource units left. Thus, the optimum combination of c and $d-c$ is the one shown in (27), which has to be computed for each d, $0 \le d \le C$. Going backward from user M to user 1, it is possible, by means of (27) to compute the cost-to-go matrix $\mathcal{I}$. Once this matrix has been computed, the following forward step allows to find the optimum allocation:

$$\begin{cases} C_1 &= \arg\min_{0 \le c \le C} (J_1^*(c) + I_2(C-c)) \\ C_m &= \arg\min_{0 \le c \le \sum_{i=1}^{m-1} C_i} (J_m^*(c) + I_{m+1}(C - \sum_{i=1}^{m-1} C_i)) \\ & \qquad 2 \le m \le M-1 \\ C_M &= C - \sum_{i=1}^{M-1} C_i \end{cases} \tag{28}$$

The result of the dynamic programming algorithm is the optimal set $C_1, C_2, \ldots, C_M$, such that $\sum_m^M C_m = C$, which minimizes (24).

5 Numerical Results

In this section we will give some numerical results relative to the allocation strategy described in the previous sections. We limit ourselves to the infinite horizon case, where results from both cost functions will be presented. The parameters which characterize the system are the global capacity C and, for the generic user m ($m = 1, \ldots, M$), the buffer size for asynchronous traffic

K_m, the range of variation of the input rates $\lambda^m_{1,\min}$, $\lambda^m_{1,\max}$, $\lambda^m_{2,\min}$, $\lambda^m_{2,\max}$, their distribution within such range, the two cost weighting factors σ^m_1 and σ^m_2, the structure of the neural networks (number of levels L and number of neurons per level $N_0, \ldots, N_L$), the learning rate factors K^m_1 and K^m_2, and, as regards J^m_2, the parameter α_m. For both cases we considered:

Network:	$C = 50$	$M = 10$
	$K_m = 20$	$\mu^m_1 = 0.0315$
Neural Net:	$L = 2$	$N_0 = 3$
	$N_1 = 10$	$N_2 = 1$
Cost:	$\sigma^m_1 = 1;$	$\sigma^m_2 = 1;$
	$K^m_1 = 100;$	$K^m_2 = 100;$
	$\alpha_m = 0.01$	

for all m, $m = 1, \ldots, M$; as regards μ_2, we considered it as a linear function of the number of CBR connections: each CBR connection corresponds to $b = 4$ asynchronous servers (i.e., $\mu^m_2(s) = 4 \cdot (C_m - s)$, $m = 1, \ldots, M$). Input rates are supposed to vary with uniform distribution within given ranges. More in particular, with respect to input rates, we present results from three main cases (for both costs J^m_1 and J^m_2). In the first case (case I), the range of input rates is the same for all users (i.e., $\lambda^m_{1,\min}, \lambda^m_{1,\max}, \lambda^m_{2,\min}$, and $\lambda^m_{2,\max}$, are the same for all $m = 1, \ldots, M$); in the second case (case II), all users have the same range for isochronous arrivals only; in the third case (case III) half of the users use a range while the other half uses another range, different with respect to both types of traffic. These three cases, along with the actual values adopted, are depicted in Figure 8, where the superscripts m have been dropped in the input rates on the axes.

With such figures, the multiuser partitioning algorithm gives, for both costs (J^m_1 and J^m_2), the following bandwidth unit assignments:

	$C_{m,\ m=1\ldots5}$	$C_{m,\ m=6\ldots10}$
Case I	5	5
Case II	6	4
Case III	7	3

It is interesting to notice that, with both cost forms, minimizing the sum of the average costs seems to always lead to give more resources to those users that have the lower input rates. In other words, the cost reduction obtained by giving more resources to the first set of users (Users $1 \div 5$) is larger than the cost increase obtained by giving less resources to the second set of users (Users $6 \div 10$). This is mainly because of the role played by the cell loss probability (the one given in (17)): this behavior can, hence, be partially driven through the weighting parameters σ^m_1 and σ^m_2.

As regards the multiservice partitioning, let us consider first the results obtained by using the cost function J^m_1 and then the ones relative to J^m_2.

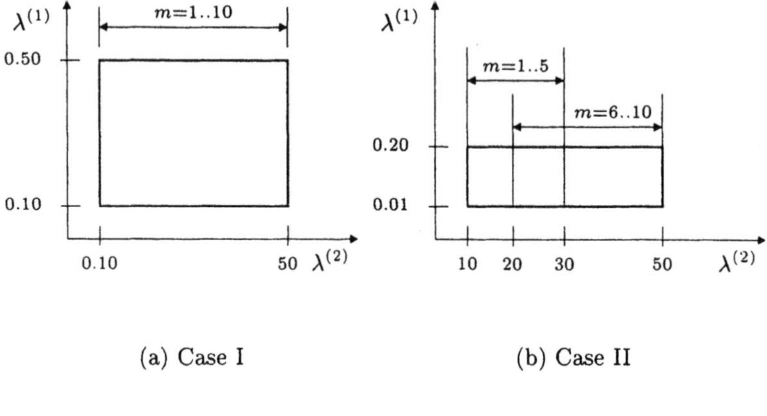

(a) Case I

(b) Case II

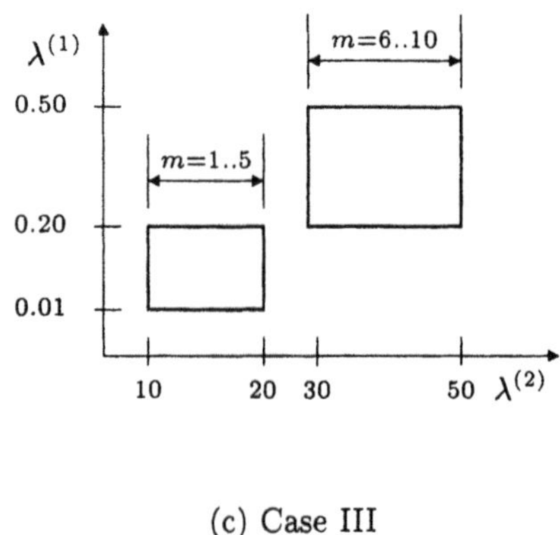

(c) Case III

Fig. 8. The input rates configurations (the numbers within the graph represent the users over which the range of values of $\lambda^{(1)}$ and $\lambda^{(2)}$ is applicable).

5.1 Cost function J_1^m

Figures 9 to 11 show the behavior of the neural network approximation for a generic user within each of the groups that have the same capacity assignment, for the three cases of input rate ranges (we have dropped the index m). The output of the neural networks (one per user) have been compared with the ones obtained by the application of an exact gradient method, where the descent is based on the given values of the input rates. The values of the components of the acceptance probability $\boldsymbol{\beta}$ are plotted versus the state s of connections in progress and the connection requests input rate $\lambda^{(1)}$. For

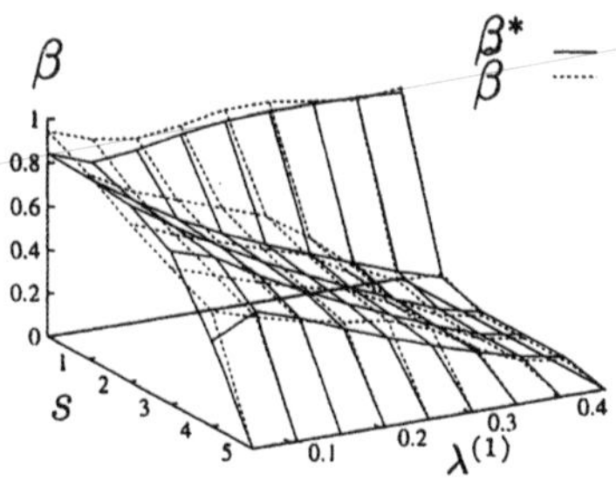

Fig. 9. Gradient descent (β^*) vs Neural Net. approx ($\tilde{\beta}$). Case I

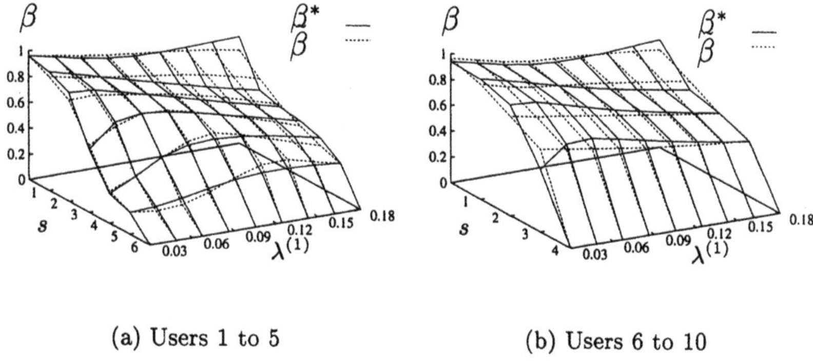

(a) Users 1 to 5

(b) Users 6 to 10

Fig. 10. Gradient descent (β^*) vs Neural Net. approx ($\tilde{\beta}$). Case II

the sake of plot readability, $\lambda^{(2)}$ has been kept fixed to its average value (i.e., $\lambda_{2,\min} + (\lambda_{2,\max} - \lambda_{2,\min})/2$), even if the neural network has been trained within the range given above. It is worth highlighting that the neural network has been trained only once for the whole set of CBR input rates (i.e., $\lambda^{(1)}$) and therefore the values of $\tilde{\beta}$ have been simply obtained as outputs of the neural network (giving the values of $\lambda^{(1)}$, $\lambda^{(2)}$ and s as input). As regards the gradient method, in order to compute β^*, the descent algorithm has been performed for each value of $\lambda^{(1)}$ (and $\lambda^{(2)}$). The plots, and many others not reported here for lack of space, show that the neural network approximation is very good. On the other hand, the neural network has needed about 10,000 (off-line) steps to be trained, while the gradient method needs about 500 (on-line) descent steps to give good results. Under this point of view, choosing between the two mechanisms is mainly a matter of available (on-line) processing power. The quantity of memory needed by the neural network is, in this case, very little since very few neurons are needed. Experiments with larger

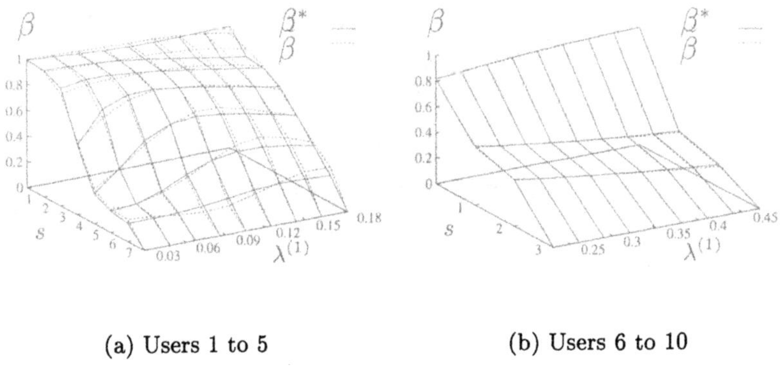

(a) Users 1 to 5 (b) Users 6 to 10

Fig. 11. Gradient descent (β^*) vs Neural Net. approx ($\tilde{\beta}$). Case III

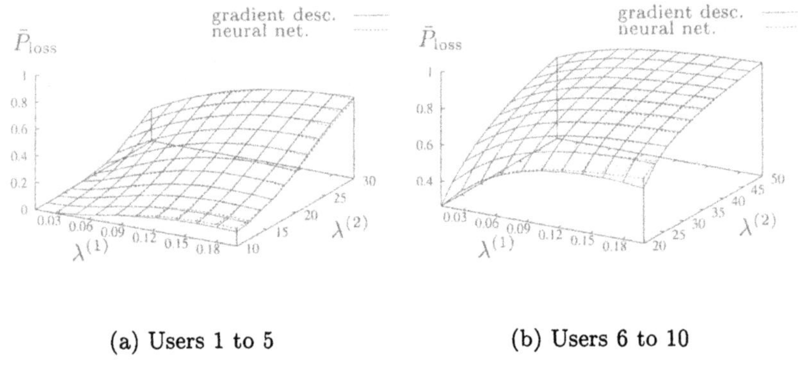

(a) Users 1 to 5 (b) Users 6 to 10

Fig. 12. Average cell loss probabilities (optimal vs. approx.). Case II

scale systems have shown that the increase in memory size needed is reasonable (i.e., smaller than linear). In order to show that the small differences between the "optimal" functions obtained with the exact gradient descent and those obtained with the neural networks give rise to small differences also in the cost functions and in their single elements, we show in Figure 12 the difference in cell loss probabilities in the two situations; the differences between the complete cost functions is shown in Figure 13 (we have limited the comparison to case II; all other cases give a similar behavior). Notice that in this last figure the average cost is evaluated over the entire range of values of both input rates. Moreover, Figure 14 highlights the difference in cell loss probabilities between the controlled and the uncontrolled ($\beta(s) = 1$, $\forall s$) case (always with respect to case II).

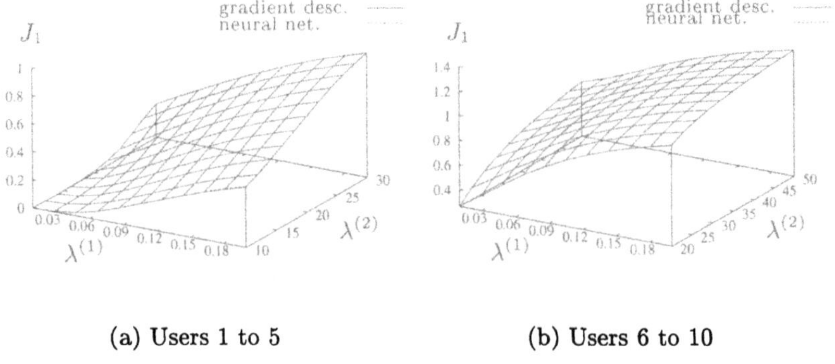

(a) Users 1 to 5 (b) Users 6 to 10

Fig. 13. Cost computed with optimal vs. approx values of β. Case II

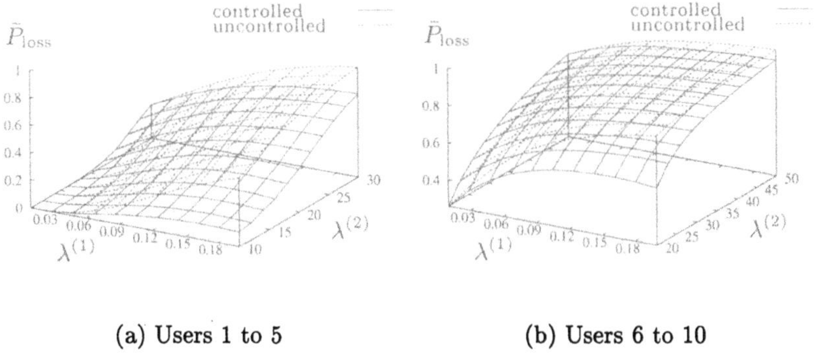

(a) Users 1 to 5 (b) Users 6 to 10

Fig. 14. Average cell loss probabilities. Controlled (optimal) vs. uncontrolled ($\beta(s) = 1,\ \forall s$) situation. Case II

5.2 Cost function J_2^m

Within this cost we have $\alpha = 0.01,\ \forall m$. This means that, as regards the guaranteed bandwidth part, this cost "comes into being" only when the average blocking probability exceeds 1% of the total incoming calls. Figures 15 to 17 shows the behavior of the neural network approximations for the same cases considered above. The plots have to be interpreted exactly as the ones obtained with cost J_1^m. The "tendency" towards a deterministic behavior of the acceptance strategies is apparent. As regards the evaluation of the performance in terms of the cost functions, in Figures 18 to 20 the set of comparisons is carried out as above, showing a good level of accuracy.

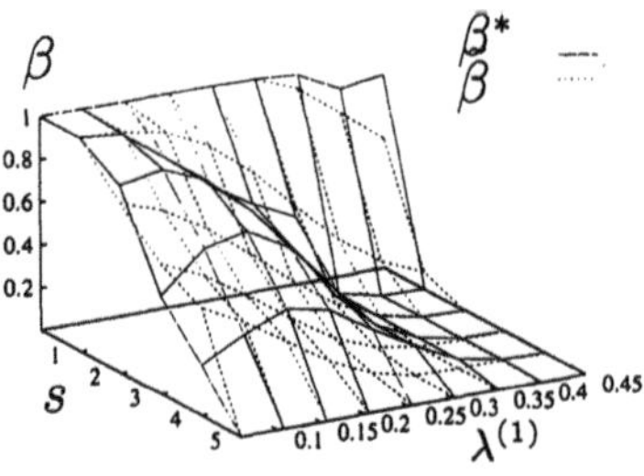

Fig. 15. Gradient descent (β^*) vs Neural Net. approx ($\tilde{\beta}$). Case I

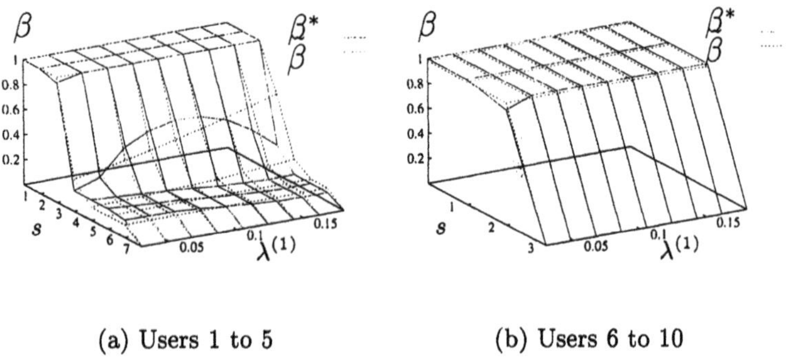

(a) Users 1 to 5

(b) Users 6 to 10

Fig. 16. Gradient descent (β^*) vs Neural Net. approx ($\tilde{\beta}$). Case II

6 Conclusions

A two-level bandwidth allocation and admission control strategy has been defined in the paper, in the context of an access multiplexer to a multiuser, multiservice broadband telecommunication network. This task is accomplished by partitioning the bandwidth among a set of users and, for each user, by approximating optimal admission control strategies for two traffic types, by means of multilayer feedforward neural networks. Numerical results have shown the capacity to follow variations in traffic parameters by means of a fast and simple adaptation mechanism.

Appendix

The Geom(N)/D/R/B queue

In a synchronous Geom(N)/D/R/B queue (see the Appendix in [25]) R servers are available to N mutually independent customers, each requesting

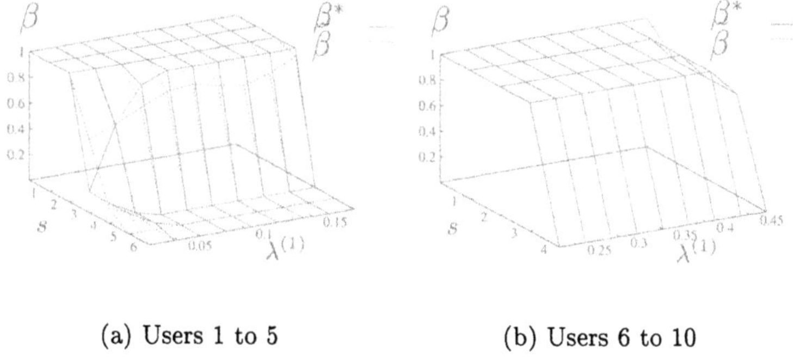

(a) Users 1 to 5 (b) Users 6 to 10

Fig. 17. Gradient descent (β^*) vs Neural Net. approx ($\tilde{\beta}$). Case III

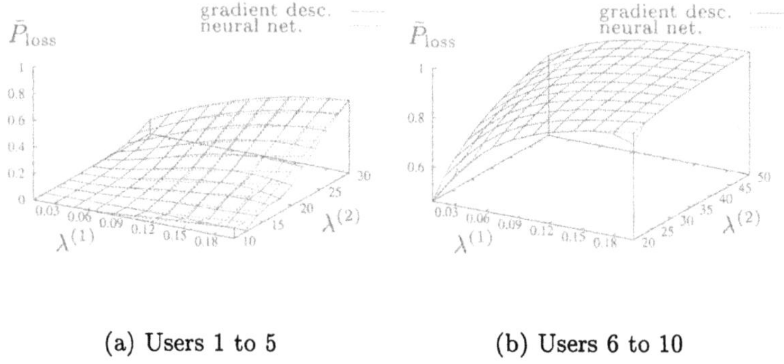

(a) Users 1 to 5 (b) Users 6 to 10

Fig. 18. Average cell loss probabilities (optimal vs. approx.). Case II

service with probability p $(0 \leq p \leq 1)$, and B positions (including the R places for the customers in service) are available to hold the customers in the system. The system is synchronous in the sense that the time axis is slotted, each slot lasting one (deterministic) service time of a customer. All arrivals and departures take place at slot boundaries. The probability distribution of customer requests A in a generic slot is binomial:

$$a_i = \Pr[A = i] = \binom{N}{i} p^i (1-p)^{N-i} \qquad (0 \leq i \leq N) \tag{29}$$

with mean value pN.

An assumption which simplifies computations is that within the queue, *first* customers in service (at most R) are removed and *then* new customers are stored. If t denotes the (discrete) slot time, the system evolution is described

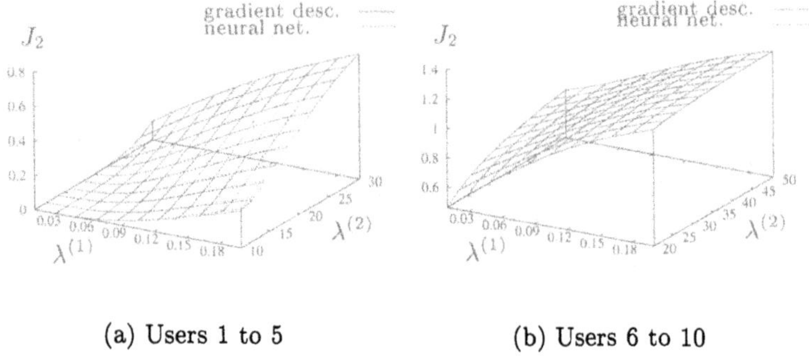

(a) Users 1 to 5 (b) Users 6 to 10

Fig. 19. Cost computed with optimal vs. approx values of β. Case II

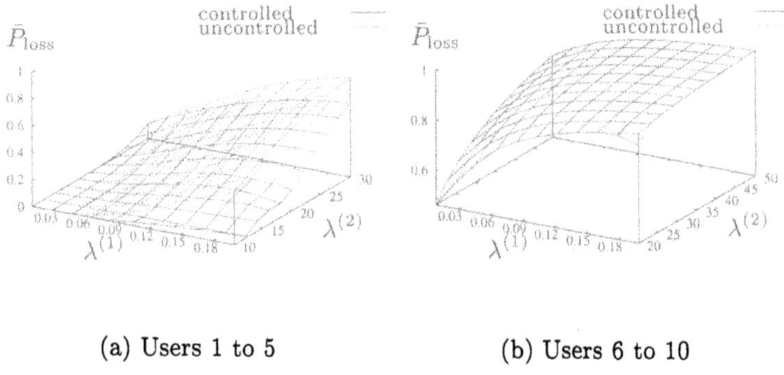

(a) Users 1 to 5 (b) Users 6 to 10

Fig. 20. Average cell loss probabilities. Controlled (optimal) vs. uncontrolled ($\beta(s) = 1, \ \forall s$) situation. Case II

by:

$$Q_t = \min \{\max\{0, Q_{t+1} - R\} + A_t, B\} \tag{30}$$

in which Q_t and A_t represent the customers in the queue and the new customers requesting service, respectively, at the beginning of slot t.

Let us denote with q_i the probability of having i customers in the system at a generic time t (i.e., $q_i = \Pr[Q_t = i]$); then the balance equations (from which the q_i can be obtained) are derived by considering the following three cases:

Case I: $N \leq B - R$:

$$\begin{cases} q_0 = \sum_{i=0}^{R} q_i \, a_0 \\ q_n = \sum_{i=0}^{R} q_i \, a_n + \sum_{i=1}^{n} q_{R+i} \, a_{n-i} & (0 < n \leq N) \\ q_n = \sum_{i=n-N}^{n} q_{R+i} \, a_{n-i} & (N < n \leq B - R) \\ q_n = \sum_{i=n-N}^{B-R} q_{R+i} \, a_{n-i} & (B - R < n \leq B - 1) \\ q_B = 1 - \sum_{i=0}^{B-1} q_i \end{cases} \tag{31}$$

Case IIa: $N > B - R$ and $N \geq B$:

$$\begin{cases} q_0 = \sum_{i=0}^{R} q_i a_0 \\ q_n = \sum_{i=0}^{R} q_i \, a_n + \sum_{i=0}^{n} q_{R+i} \, a_{n-i} & (0 < n \leq B - R) \\ q_n = \sum_{i=0}^{R} q_i \, a_n + \sum_{i=1}^{B-R} q_{R+i} \, a_{n-i} & (B - R < n \leq B - 1) \\ q_B = 1 - \sum_{i=0}^{B-1} q_i \end{cases} \tag{32}$$

Case IIb: $N > B - R$ and $N < B$:

$$\begin{cases} q_0 = \sum_{i=0}^{R} q_i \, a_0 \\ q_n = \sum_{i=0}^{R} q_i \, a_n + \sum_{i=1}^{n} q_{R+i} \, a_{n-i} & (0 < n \leq B - R) \\ q_n = \sum_{i=0}^{R} q_i \, a_n + \sum_{i=1}^{B-R} q_{R+i} \, a_{n-i} & (B - R < n \leq N) \\ q_n = \sum_{i=n-N}^{B-R} q_{R+i} \, a_{n-i} & (N < n \leq B - 1) \\ q_B = 1 - \sum_{i=0}^{B-1} q_i \end{cases} \tag{33}$$

Once the q_i are known, it is possible to compute the average carried load of the R servers in the following way.

$$\rho R = \sum_{i=0}^{R} i q_i + R \sum_{i=R+1}^{B} q_i \tag{34}$$

The loss probability P_{loss} is then computed by considering that the average load offered to the queue is pN:

$$P_{\text{loss}} = 1 - \frac{\rho R}{pN} \tag{35}$$

The model fits in our system, for the generic user m, $(m = 1, \ldots, M)$ by setting $\lambda_2^m = pN$ (i.e., $p = \lambda_2^m / N$), $R = \mu_2^m(r)$ (hence $B = \mu_2^m(r) + K_m$), and $N = b \cdot C$ (see Figures 3 and 4) representing the maximum possible value (before resource partitioning) for the cell output rate (in our case, $b = 4$).

The computation of P_{loss} is performed for all the values that $\mu_2^m(r)$ can assume. If $\mu_2^m(r) = 0$, then $P_{\text{loss}} = 1$.

References

1. Schwartz M., (1996) Broadband Integrated Networks. Prentice Hall, Upper Saddle River, NJ
2. White P. P., Crowcroft J., December (1997) The Integrated Services in the Internet: state of the art. Proc IEEE, **85**, no. 12, 1934–1946
3. Braden R., Clark D., Shenker S., June (1994) Integrated Services in the Internet architecture: an overview. Internet Engineering Task Force, Request for Comments, RFC 1633. [Online]. Available FTP: ds.internic.net/rfc/rfc1633.txt
4. Blake S., Black D., Carlson M., Davies E., Wang Z., Weiss W., December (1998) An architecture for differentiated services. Internet Engineering Task Force, Request for Comments, RFC 2475. [Online]. Available FTP: ds.internic.net/rfc/rfc2475.txt
5. Li T., Rekhter Y., October (1998) A provider architecture for differentiated services and traffic engineering (PASTE). Internet Engineering Task Force, Request for Comments, RFC 2430. [Online]. Available FTP: ds.internic.net/rfc/rfc2430.txt
6. Padgett J. E., Günther C. G., Hattori T., (1995) Overview of wireless personal communications. IEEE Commun. Mag., **33**, 28–41
7. Falconer D. D., Adachi F., Gudmundson B., (1995) Time division multiple access methods for wireless personal communications. IEEE Commun. Mag., **33**, 50–57
8. Ortigoza-Guerrero L., Aghvami A. H., November (1998) A distributed dynamic resource allocation for a hybrid TDMA/CDMA system. IEEE Trans. Vehic. Technol., **47**, no. 4, 1162–1178
9. Samukic A., November (1998) UMTS Universal Mobile Telecommunications System: development of standards for the third generation. IEEE Trans. Vehic. Technol., **47**, no. 4, 1099–1104

10. Van de Voorde I., Van der Plas G., April (1997) Full service optical access networks: ATM transport on Passive Optical Networks. IEEE Commun. Mag., **35**, no. 4, 70–75
11. Frigo N.J., November/December (1996) Local access optical networks. IEEE Network, **10**, no. 6, 32–36
12. Parek A., Gallager R., (1993) A generalized processor sharing approach to flow control – the single node case. IEEE/ACM Trans. Networking, **1**, no. 3, 366–375
13. Parek A., Gallager R., (1993) A generalized processor sharing approach to flow control – the multiple node case. IEEE/ACM Trans. Networking, **2**, no. 2, 137–150
14. Maglaris B. S., Schwartz M., (1982) Optimal fixed frame multiplexing in integrated line- and packet-switched communication networks. IEEE Trans. Info. Theory, **IT-28**, 273–283
15. Ross K. W., Tsang D. H. K., (1989) Optimal circuit access policies in an ISDN environment: a Markov decision approach. IEEE Trans. Commun., **37**, 934–939
16. Meempat G., Sundareshan M., (1993) Optimal channel allocation policies for access control of circuit-switched traffic in ISDN environments. IEEE Trans. Commun., **COM-41**, 338–350
17. Bolla R., Davoli F., April (1997) Control of multirate synchronous streams in hybrid TDM access networks. IEEE/ACM Trans. Networking, **5**, no. 2, 291–304
18. October (1995) Special issue on: Neurocomputing in High-Speed Networks. IEEE Commun. Mag., **33**, no. 10
19. Hiramatsu A., September (1991) Integration of ATM call admission control and link capacity control by distributed neural networks. IEEE J. Select. Areas Commun., **9**, 1131–1138
20. Nordstrom E., Carlstrom J., Gallino O., Asplund L., October (1995) Neural networks for adaptive traffic control in ATM networks. IEEE Commun. Mag., **33**, 43–49
21. Tham C. K., Soh W. S., March/April (1998) Multi-service connection admission control using modular neural networks. Proc. INFOCOM 98, S. Francisco, CA
22. Bolla R., Davoli F., Maryni P., Parisini T., August (1998) An adaptive neural network admission controller for dynamic bandwidth allocation. IEEE Trans. on Systems, Man and Cybernetics, Part B, Special Issue on Artificial Neural Networks, **28**, no. 4, 592–601
23. Karbowski A., Małowidzki M., Bolla R., Davoli F., Parisini T., July (1995) A hierarchical neural optimization structure for access control and resource sharing in integrated communication networks. Proc. IFAC Symp. on Large Scale Syst. Theory and Appl., London, UK, 889–894
24. Davoli F., Maryni P., February (2000) A two-level stochastic approximation for admission control and bandwidth allocation. IEEE Journal on Selected Areas in Communications, Special Issue on Intelligent Techniques in High Speed Networks, **18**, no. 2, 222–233
25. Pattavina A., (1998) Architecture and Performance in Broadband ATM Networks. Wiley, Chichester, UK
26. Ghani S., Schwartz M., (1994) A decomposition approximation for the analysis of voice/data integration. IEEE Trans. Commun., **42**, 2441–2452
27. Serfozo R., (1979) An equivalence between continuous and discrete time Markov decision processes. Op. Res., **27**, 616–620

28. Mayne D. Q., Michalska H., (1990) Receding horizon control of nonlinear systems. IEEE Trans. Automat. Contr., **35**, 814–824
29. Parisini T., Zoppoli R., (1994) Neural networks for feedback feedforward nonlinear control systems. IEEE Trans. Neural Networks, **5**, 436–449
30. Kleinman D. L., Athans M., (1968) The design of suboptimal linear time-varying systems. IEEE Trans. Automat. Contr., **AC-13**, 150–159
31. Warfield R., Chan S., Konheim A., Guillaume A., (1994) Real-time traffic estimation in ATM networks. Proc. ITC'94, 907–916
32. Kushner H. J., Yin G. G., (1997) Stochastic Approximation Algorithms and Applications. Springer-Verlag, New York
33. Parisini T., Zoppoli R., (1998) Neural approximations for infinite–horizon optimal control of nonlinear stochastic systems. IEEE Trans. Neural Networks, **9**, 1388–1408
34. Rumelhart D. E., McClelland J. L., (1986) Parallel Distributed Processing. MIT Press, Cambridge, MA
35. Leshno M., Ya V., Pinkus A., Schocken S., (1993) Multilayer feedforward networks with a nonpolynomial activation function can approximate any function. Neural Networks, **6**, 861–867
36. Sontag E. D., (1992) Feedback stabilization using two-hidden-layer nets. IEEE Trans. Neural Networks, **3**, 981–990
37. Ross K., (1995) Multiservice Loss Models in Broadband Telecommunication Networks. Springer-Verlag, London

Performance of the Multilayer Perceptron-based Decision Feedback Equalizer with Lattice Structure in Non-Linear Channels

Azzedine Zerguine and Ahmar Shafi

Electrical Engineering Department, King Fahd University of Petroleum and Minerals, Dhahran 31261, Saudi Arabia.
E-mail: {azzedine, ahmar}@kfupm.edu.sa.

Abstract. The severely distorting channels limit the use of linear equalizers and the use of nonlinear equalizers then becomes justifiable. Neural networks-based equalizers are computationally efficient alternative to currently used nonlinear filter realizations , e.g. the Voltera type.

In this chapter, the effect of whitening the input data in a multilayer perceptron-based decision feedback equalizer (DFE) is evaluated. It is shown from computer simulations that whitening the received data employing adaptive lattice channel equalization algorithms improves the convergence rate and bit error rate performances of multilayer perceptron-based decision feedback equalizers. The adaptive lattice algorithm is a modification to the one developed by Ling and Proakis.

Keywords: Multi Layer Perceptron (MLP), Decision Feedback Equalizer (DFE), Lattice Filters, Non-Linear Channels.

1 Introduction

Equalization techniques for combating inter-symbol interference (ISI) on bandlimited time dispersive channels may be subdivided into two sub-groups, linear and non-linear equalization.

A decision feedback equalizer is a non-linear equalizer that is widely used in situations where ISI is very large. It has been proved theoretically and experimentally that the DFE performs significantly better than a linear equalizer of equivalent complexity [1].

To further enhance the performance of the DFE, the multilayer perceptron (MLP) has been incorporated in the DFE. It is shown that the MLP-based DFE (MLP DFE) [2], using the BP algorithm [3], gives a significantly improved performance over the LMS DFE [4].

Since the BP algorithm is no more than a generalized LMS algorithm [5], it then suffers from the same problems that the LMS algorithm exhibits, particularly the relatively low rate of convergence when applied to channels with spectral nulls in their frequency responses. These channels are known to yield a large eigenvalue spread of the autocorrelation matrix of the signal at their outputs. This kind of channel characteristics is often encountered in time-varying channels.

A mutlilayer perceptron-based equalizer with a recursive least squares (RLS) [5] adaptation algorithm has already been presented and its performance compared to that of the LMS DFE in [6]-[7] and it was shown to perform better than the LMS DFE over a frequency-selective Rayleigh fading channel at the expense of higher complexity.

A powerful means of improving the performance of the MLP DFE equalizer, without much increase in complexity, is through the use of a whitening process. This can be achieved using a lattice filter as an input stage to the MLP DFE equalizer. In this case, the effect of the eigenvalue spread will be reduced substantially in both time-invariant and time-varying channels.

In [8]-[9], the performance of the MLP DFE using a whitening scheme in its input was evaluated. It was shown that great improvement in performance was obtained through the use of this technique over both the LMS DFE and the MLP DFE for time-invariant and time-varying channels. The whitening process is carried out through a simple modification of the adaptive lattice DFE originally proposed in [10]. As an extension to this work, the present material extends this idea to the case of no-linear channels. Similarly here, great improvement in performance is obtained through the use of this technique over both the LMS DFE and the MLP DFE.

The rest of the paper is organized as follows. In Section 2, a brief review of artificial neural networks is given where the proposed new structure of the MLP-based DFE is presented along with the modification to the adaptive lattice DFE [10]. Non-linear system models is treated in Section 3, while the performance of the proposed algorithm is demonstrated by the simulation results of Section 4. Finally, conclusions as well as suggestions for further work are outlined in Section 5.

2 Artificial Neural Networks

Because of the capabilities of artificial neural networks in efficiently modeling arbitrary nonlinearities, there has been recent interest in employing them in adaptive equalization for data communication channels [2], [11]-[12]. In this case, the linear adaptive filter is replaced by a neural network. Different artificial neural network architectures such as multilayer perceptron, radial basis functions, and recurrent neural networks have all been proposed in the literature for channel equalization [13]. Among all these structures, the most commonly and widely-used is the MLP structure. The popularity of MLP-based equalizers is due in part to their computational simplicity, finite parameterization, stability, and smaller structure size for a particular problem as compared to other structures.

A multilayer perceptron consists of several hidden layers of neurons that are capable of performing complex, non-linear mappings between the input and output layer. The hidden layers provide the capability to use the nonlinear sigmoid function (to be defined later) to create intricately-curved partitions of space with complex nonlinear decision boundaries [14]. Furthermore, it has been shown that only three layers are needed by the MLP to generate these boundaries [14].

The basic element of the multilayer perceptron is the neuron as depicted in Figure 1. Each neuron in the layer has primary local connections and is characterized by a set of real weights $[w_{1j}, w_{2j}, ...w_{Nj}]$ applied to the previous layer to which it is connected and a real threshold level I_j. The jth neuron in the pth layer accepts real inputs $v_h^{(p-1)}(h = 1, 2, ..., N)$ from the $(p-1)^{th}$ layer and produces an output $v_j^{(p)}$, which is also a real scalar, expressed in the following way:

$$v_j^{(p)} = f_j\left(\sum_{h=1}^{N} w_{hj} v_h^{(p-1)} + I_j^{(p)}\right). \tag{1}$$

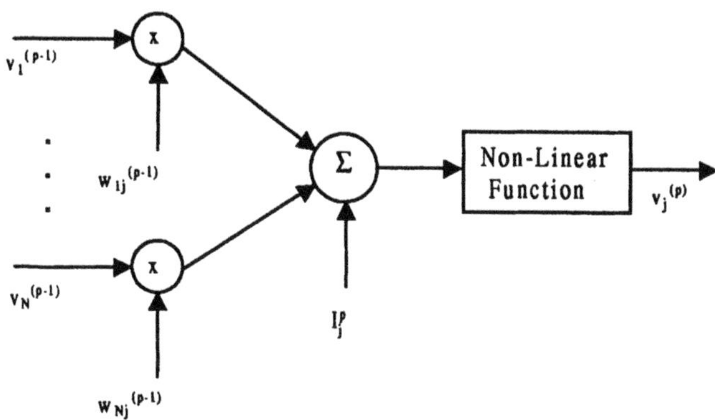

Figure 1: j^{th}-Neuron in the pth layer

This output value $v_j^{(p)}$ serves also as input to the $(p+1)^{th}$ layer (next layer) to which the neuron is connected. In the above expression, $f_j(.)$ represents the nonlinearity function. The most commonly used one in the perceptron is of the sigmoid type, defined as [15]:

$$f_j(x) = \frac{1 - e^{-x}}{1 + e^{-x}}, \tag{2}$$

where $f_j(x)$ is always in the range [-1,1], $\forall x \in \mathbf{R}$ (the set of real numbers). The weights $\{w_{hj}\}$ and thresholds levels $\{I_j\}$ are updated during training [2].

The MLP did not receive much attention in applications until the introduction of the BP algorithm [16]. The BP algorithm was used in both linear equalizers [17] and nonlinear equalizers (DFE) [2], and it was found that in both cases, the neural network-based configuration outperformed its non-neural network-based counterpart. Other structures of neural networks, e.g., radial basis functions [18]-[19] and recurrent neural networks [20], have been tested on channel equalization. In this work, however, only the MLP DFE [2] will be considered as it is more advantageous than its linear counterpart. Figure 2 depicts the block diagram of MLP-Based DFE.

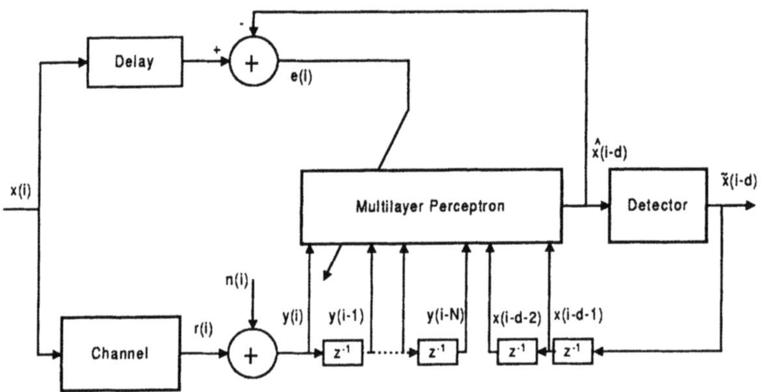

Figure 2: Block diagram of the MLP-based DFE.

2.1 The Learning Phase

In the BP algorithm, the output value is compared with the desired output, resulting in an error signal. This error signal is fed back through the network whose weights are adjusted to minimize this error. The increments used in updating the weights, Δw_{hj}, and threshold levels, ΔI_j, of the pth ($p \in [1, 2, ..., P]$) layer are updated, respectively, according to the following relations:

$$\Delta w_{hj}^{(p)}(i+1) = \eta \delta_j^{(p)} v_j^{(p-1)} + \alpha \Delta w_{hj}^{(p)}(i) \tag{3}$$

and

$$\Delta I_j^{(p)}(i+1) = \beta \delta_j^{(p)} \tag{4}$$

where η is the learning gain, α is the momentum parameter, and β is the threshold level adaptation gain.

The error signal $\delta_j^{(p)}$ for layer p is calculated starting from the output layer P, as:

$$\delta_j^{(P)} = \frac{(z_j - v_j^{(P)})(1 - v_j^{2(P)})}{2} \tag{5}$$

and is then recursively back propagated to lower layers ($p \in [1, 2, , P - 1]$) according to

$$\delta_j^{(p)} = (1 - v_j^{2(p)}) \sum_l \frac{\delta_l^{(p+1)} w_{jl}^{(p+1)}}{2}, \tag{6}$$

where l is over all neurons above the neuron j in the $(p + 1) - st$ layer and z_j is the desired output. To allow for a rapid learning, a momentum term, $\Delta w_{hj}^{(p)}(i)$, scaled by α, is used to filter out high frequency variations of the weight vector. Consequently, the convergence rate is much faster and the fast weight changes are smoothed out.

2.2 Proposed Scheme

Because the BP algorithm is no more than a generalized least-mean squares (LMS) algorithm [21], it then suffers from the same problems as the LMS algorithm, particularly the slow rate of convergence when applied to channels with spectral nulls in their frequency responses. These channels are known to yield a large eigenvalue spread of the autocorrelation matrix of the signal at their outputs. This kind of channel characteristics is often encountered in time-variant channels. Moreover, as seen later, the BP-based MLP DFE has an equal performance to that of the LMS DFE in time varying channels. The MLP DFE equalizer can then hardly be justified for the equalization of such channels. Eventually, a whitening procedure must be used to improve the performance of the MLP DFE with the BP algorithm as an updating scheme.

Since lattice equalizers are known to be insensitive to the eigenvalue spread of the channel autocorrelation matrix [22]-[23], adaptive lattice (AL) algorithms can generate a set of orthogonal signal components that can be used as inputs to the equalizer lending themselves to fast convergence. Ling and Proakis [10] derived a generalized least squares multi-channel lattice DFE (L-DFE) algorithm, implemented and investigated it on both time-invariant and time-varying channels. The results showed that the L-DFE has a much better performance than the LMS DFE.

If this scheme is used as a pre-whitening process to the MLP DFE, the performance of the MLP DFE in terms of convergence rate, BER and steady-state MSE will therefore be expected to improve on both time-invariant and time-varying channels. Figure 3 gives the details of this new configuration where the L-DFE [10] is used as a whitening scheme for the MLP DFE [2].

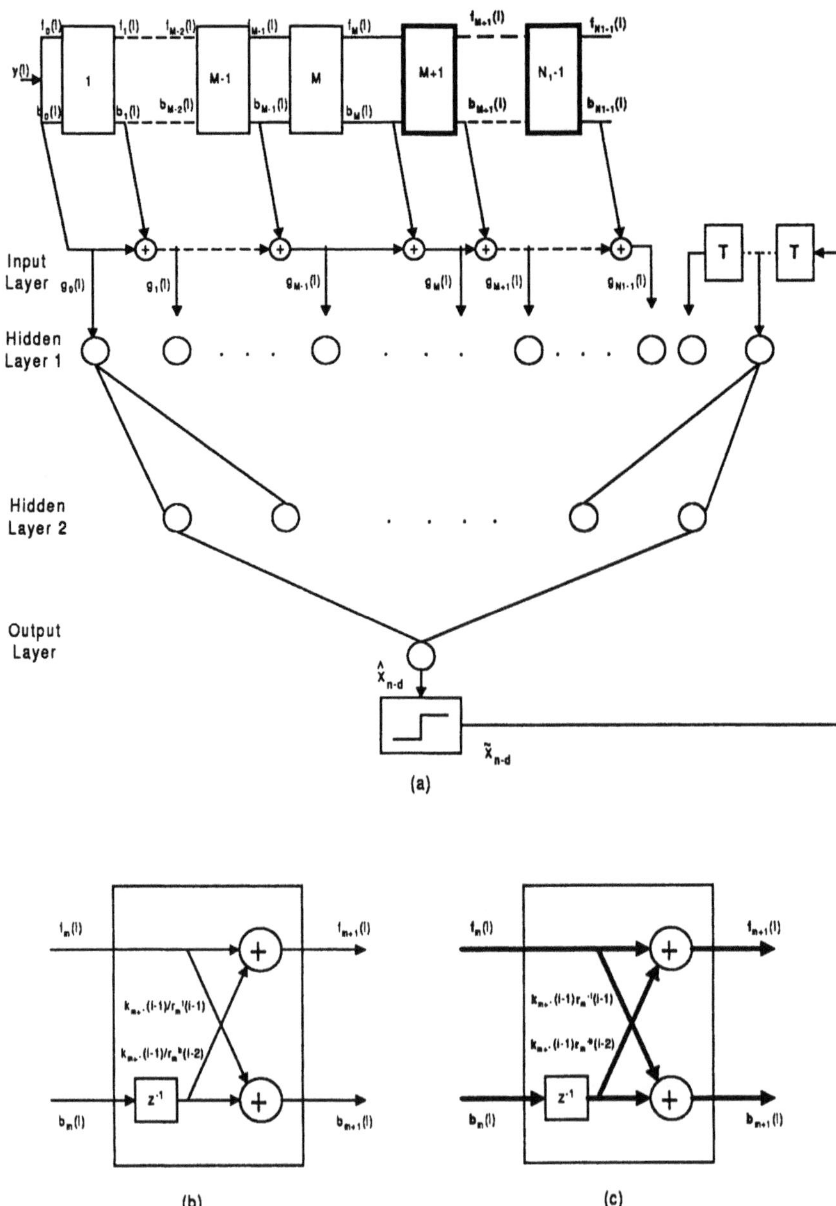

Figure 3: Block diagram of overall system lattice filter MLP DFE

The L-DFE consists of a lattice predictor part and a joint estimator part. The predictor has $M = N_1 - N_2$ scalar stages followed by $N_2 - 1$ two-dimensional stages. The L-DFE input is the received signal $y(i)$, while its outputs $\{g_m(i)\}$ and delayed versions of the decision device output are used as inputs to the MLP DFE which has one input layer, two hidden layers and one output layer. As shown in Figure 3, the outputs of the L-DFE are used as inputs to the the feed-forward section of the lattice-based MLP DFE, and depending on the values of N_1 and N_2, the L-DFE will then consist of either a scalar stage or a two-dimensional one. However, it should be noted that the neural network carries out solely the estimation of the signal.

The outputs $\{g_m(i)\}$ that account for the modification to L-DFE [10] are detailed as follows. During the initialization process, it is defined as:

$$g_o(i) = 0. \tag{7}$$

As for the scalar stages ($0 < m \leq N_1 - N_2$), they are given by:

$$g_m(i) = g_{m-1}(i) + k_m^g(i-1)b_{m-1}(i)/r_{m-1}^b(i-1). \tag{8}$$

Finally, in the two-dimensional stages ($N_1 - N_2 < m < N_1$), they are expressed as :

$$g_m(i) = g_{m-1}(i) + \mathbf{k}_m^{gT}(i-1)\mathbf{r}_{m-1}^{-b}(i-1)\mathbf{b}_{m-1}(i) \qquad (m \leq N_1), \tag{9}$$

where all the terms in the above expressions are defined in Appendix A which gives all the details of the algorithm.

3 Non-linear System Models

A non-linear dynamic single-input single-output channel is a mapping of an input sequence $\{x_i\}$ to an output sequence $\{y_i\}$,

$$y_i = \theta(x_i, x_{i-1}, \ldots, x_{i-N}), \tag{10}$$

where $\theta(.)$ is an arbitrary (nonlinear) multivariable function, i denotes the time instant, and N is the memory length of the system, which need not be finite. Though, the implementation of a system with infinite memory requires in general infinite complexity, by introducing a state space model

$$\begin{cases} S_{i+1} = f(s_i, x_i) \\ y_i = h(s_i, x_i), \end{cases} \tag{11}$$

where, $s_i = (s_i^{(1)}, s_i^{(2)}, \ldots, s_i^{(M)})$ is the state vector, a finite hardware implementation is made possible. The generic block diagram of such a nonlinear dynamic system is shown in Figure 4. It consists of two memoryless blocks f and h with memory elements in between. The input-to-state map f has

$M+1$ inputs (x_i and the M elements of s_i), M outputs (the M elements of s_i), while the state-to-output map h has $M+1$ inputs but only one output. Consequently, the cost of implementing f is much higher than that of implementing h. This has motivated several attempts (the best known by Wiener [24] to approximate nonlinear systems by models of the form (11), but with linear input-to-state mapping f. In the following, we discuss this approximation.

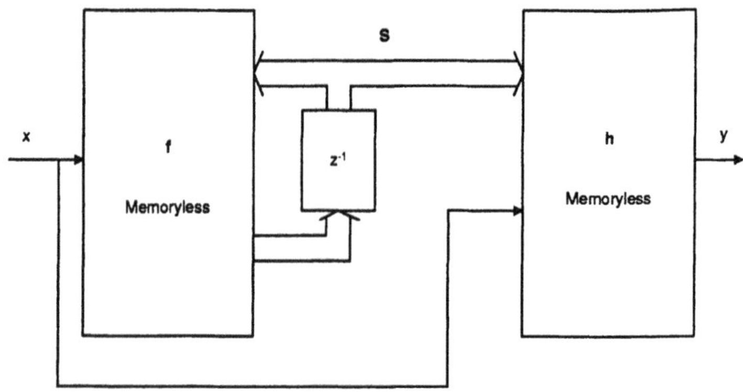

Figure 4: A nonlinear system model.

3.1 Linear Dynamic Polynomial Nonlinearity Approximation

Figure 5 shows the block diagram of one kind of approximation which consists of single-input multi-output linear time invariant (LTI) operator followed by a multi-input single-output memoryless polynomial nonlinearity. The LTI part can be realized as a finite dimensional linear system

$$s_{i+1} = \mathbf{A}s_i + bx_i, \qquad |eig(\mathbf{A})| < 1 \tag{12}$$

which replaces the nonlinear mapping f of (11) is constrained to be a multivariable polynomial of degree L, i.e.

$$y_i = P(s_i). \tag{13}$$

Boyd [25] has shown that such a structure can be used as an approximation, and by letting L and M grow to infinity, one obtains better and better approximation of an arbitrary nonlinear systems with fading memory. It is also demonstrated that an even simpler configuration in Figure 6 is sufficient to approximate discrete time systems. It corresponds to the popular finite

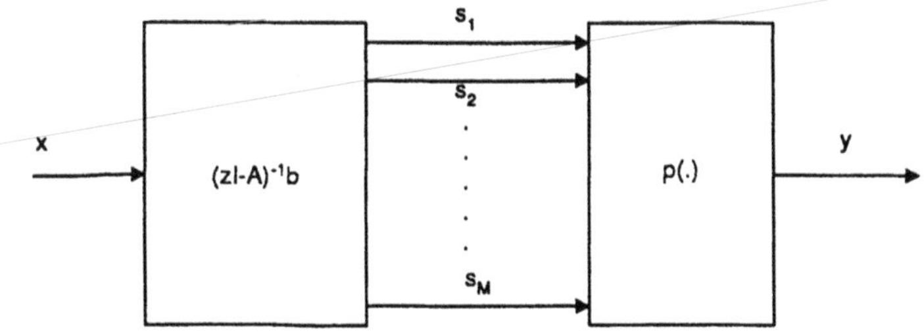

Figure 5: An approximation consisting of linear dynamic system with polynomial nonlinearity.

Voltera model [26]:

$$\begin{aligned} y_i &= h^{(0)} + \sum_{m1=0}^{N} h^{(1)}_{m1} x_{i-m1} + \sum_{m1=0}^{N} \sum_{m2=0}^{N} h_{m1m2} x_{i-m1} x_{i-m2} + \ldots \\ &+ \sum_{m1=0}^{N} \cdots \sum_{mL=0}^{N} h^{(L)}_{m1m2\ldots mL} x_{i-m1} x_{i-m2} \ldots x_{i-mL} + \ldots, \end{aligned} \tag{14}$$

where, $h^{(l)}$ $(0 \leq l \leq L)$ is called the discrete-time Voltera Kernel associated with an input data product of order l, and N is the length of the system's memory. For a detailed discussion of the Voltera and Wiener theory the reader may refer to [27].

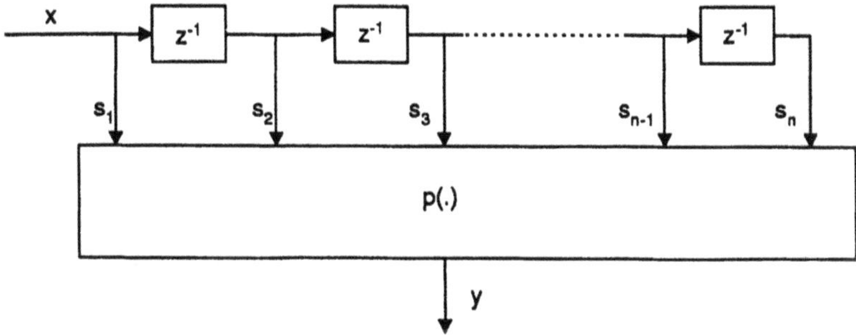

Figure 6: A simple approximation structure for discrete time nonlinear system.

3.2 Non-linear Channels

In satellite communications, the traveling wave tube amplifier in satellite usually operates in the saturated region because of the several signal attenuation and the limited power sources of the satellite. Such operation results in severely nonlinear channels. The nonlinear effect of the channel not only spreads the signal spectrum, but also introduces nonlinear amplitude and phase distortions to the channel in-band signal. Figure 7 shows the nonlin-

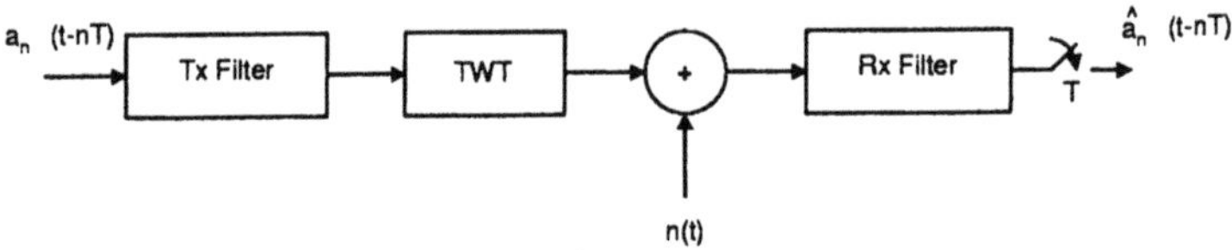

Figure 7: A satellite channel model.

ear satellite channel model. Besides satellite links, this model has also been applied to voice-band telephone channels. The first general simple Voltera model with order three for the voice-band channel was proposed by Falconer [28]. A more simplified version of Falconer's model was used by Biglieri et al [26] as shown in Figure 8, where the output $y(n)$ is related to the input $x(n)$ through the following relation:

$$y(n) = a_1 x(n) + a_2 x^2(n) + a_3 x^3(n). \tag{15}$$

The coefficients a_1, a_2, and a_3 in (15) are scalars, which control the degree of nonlinearity. Therefore, Figure 8 results in a highly simplified version of the Voltera model shown in Figure 6. Hence, a significant reduction in the number of parameters of the model is obtained.

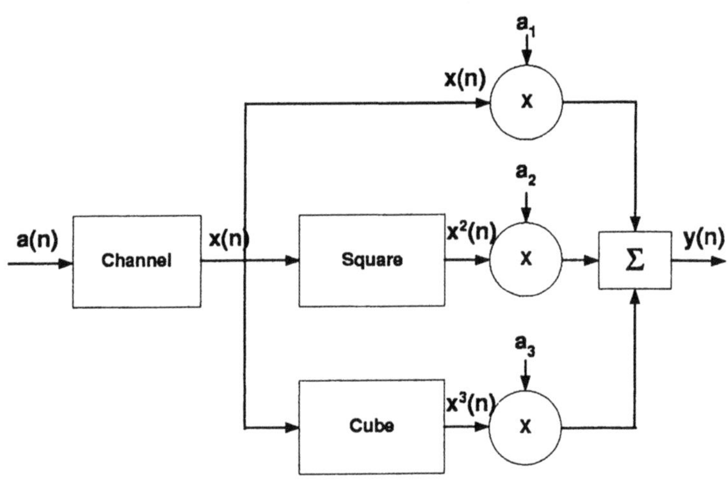

Figure 8: A simplified nonlinear channel model.

4 Performance Analysis

The performance of the new structure (MLP DFE with lattice structure) is compared to that of the LMS DFE and MLP DFE. Both the LMS DFE and the MLP DFE structures use nine samples in the feedforward section and two samples in the feedback section. For the latter structure, this results in eleven input samples in its input layer. The number of neurons in the first and second hidden layer, and the output layer are 9, 3, and 1, respectively, and is denoted by the triplet (9,3,1) for ease of reference. Finally, the L-DFE was initially set up with $N_1 = 4$ and $N_2 = 1$ giving rise to four outputs $(g_m(i), 0 \leq m \leq 3)$. These four outputs and the delayed output from the decision device will provide the five inputs to the MLP DFE. In this case, a (6,3,1) configuration for the lattice-based MLP DFE is used such that the two MLP configurations have similar complexity. The back propagation algorithm is used to update the MLP DFE where the learning gain parameter η, momentum parameter α, and threshold level adaptation gain β, have been chosen as in [2], namely 0.07, 0.3 and 0.05, respectively. The weighting factor for the lattice algorithm is 0.99 and the step size for the LMS algorithm used in the LMS DFE is the one that gives the fastest convergence. The digital message applied to the channel is made of uniformly distributed bipolar random numbers (-1,1). The channel noise is considered to be as additive white Gaussian noise (AWGN).

During this part of the simulations, the performance measure is obtained through the use of learning curves and BER curves.

4.1 Channel Models

Two time-invariant channel models are used in the simulation and are described by their transfer functions $H_1(z) = 0.3482 + 0.8704z^{-1} + 0.3482z^{-2}$, and $H_2(z) = 0.408 + 0.816z^{-1} + 0.408z^{-2}$. The eigenvalue spreads for the first and second channel are 25 and 81, respectively. Also, it should be pointed out here that the first channel model matches the the one used in [2], while the second channel matches one of the time-invariant channel models used in [10].

Two nonlinear channels with different degrees of nonlinearity are used. The first channel, denoted as non-linear channel 1, uses the linear channel $H_1(z)$ as the FIR filter in the nonlinear channel. Whereas, for the second nonlinear channel, denoted as non-linear channel 2, the linear channel $H_2(z)$ is used as the FIR filter. The system is tested for two levels of nonlinearity, first one of mild nature and the other one of higher degree. The two degrees of nonlinearities are controlled by setting the first, second and third-order coefficients, a_1, a_2, a_3 to 1, 0.1 and 0.05 as given in [29], respectively, and in the second case, the three coefficients are set equal to 1, thus increasing the nonlinearity content.

4.2 Learning Characteristics

The learning curves for the lattice-based MLP DFE and the MLP DFE over the two nonlinear channels are shown in Figures 9-12. The signal-to-noise ratio (SNR) is set to 20 dB. Due to its poor performance, the LMS DFE is not considered in the comparison. It will be considered only in the comparison ofbit error rate (BER). As can be seen from Figure 9, the learning performance over channel 1 with $a_1 = 1$, $a_2 = 0.1$, $a_3 = 0.05$, there is little deterioration in the convergence time as compared to that of channel 1 in [9]. The steady state mean squared error for both of the cases is 1 dB greater than that of the corresponding linear channel.

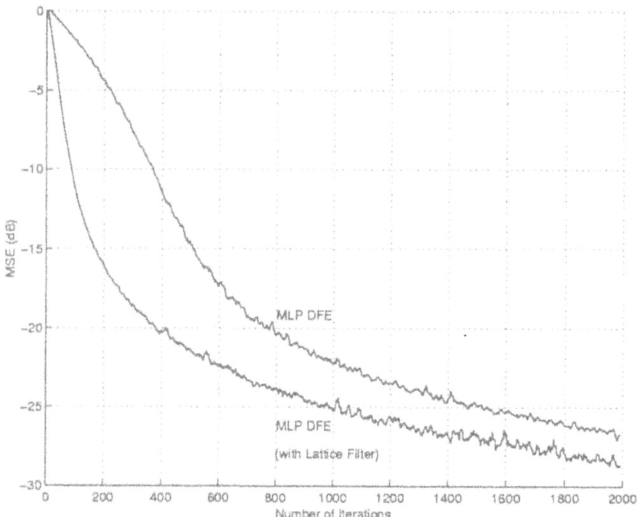

Figure 9: Learning curves for channel 1 with $a_1 = 1$, $a_2 = 0.1$, and $a_3 = 0.05$.

If the nonlinearity content of the channel is increased by taking all coefficients equal to 1, the performance is expected to deteriorate. This is supported by the learning curves for the lattice-based MLP DFE and the MLP DFE as depicted in Figure 10. Notice here that the difference in the convergence time and the MSE have increased between the two configurations, showing the strength of the Lattice filter for the equalization of nonlinear channels. The MLP-DFE converges in about 1700 iterations to an MSE value of -10 dB, while the decorrelation of the input data causes the MSE to drop to around -15 dB which is attained in about 1500 iterations, thus a gain of about 4 dB in MSE and 200 iterations.

When channel 2 will be used, the performance is expected to degrade due to the increase in eigen-value spread. Figures 11-12 show the learning curves for this channel. Figure 11 corresponds to the case when $a_1 = 1$, $a_2 = 0.1$, $a_3 = 0.05$ and Figure 12 corresponds to the case when a_1=a_2= a_3=1. There

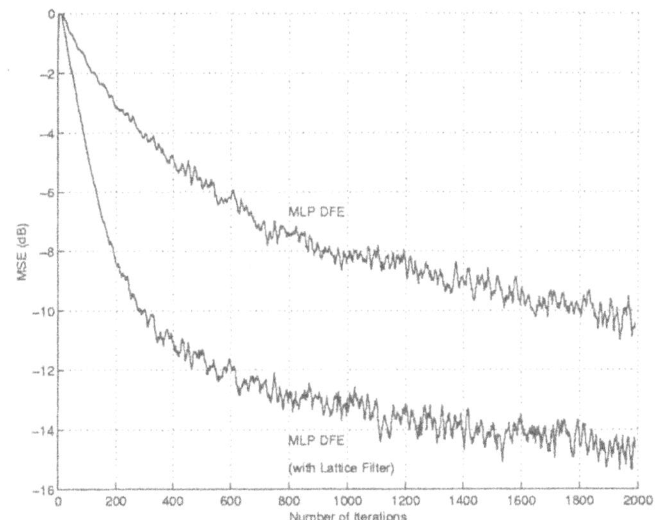

Figure 10: Learning curves for channel 1 with $a_1 = a_2 = a_3 = 1$.

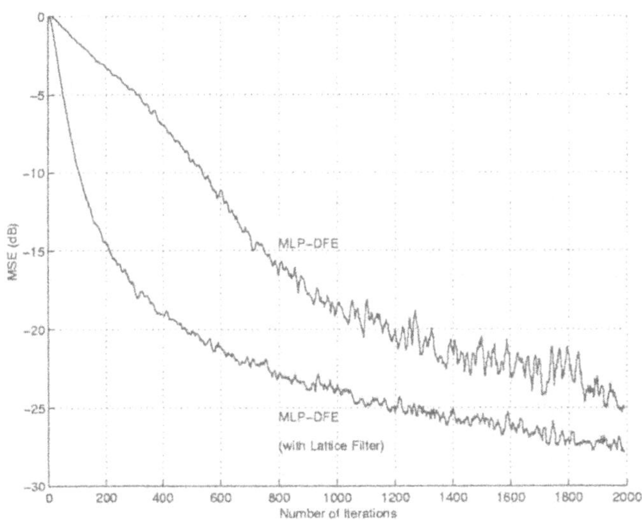

Figure 11: Learning curves for channel 2 with $a_1 = 1$, $a_2 = 0.1$, and $a_3 = 0.05$.

is a clear degradation in the convergence time for the MLP-DFE as it could not converge even after 2000 iterations. On the other hand, the MLP-DFE with pre-orthogonalization converges to some extent, to an MSE value of -27 dB which is 2.5 dB lower than that of the MSE attained by the MLP-DFE after 2000 iterations. The performance improvement increases for the case of the highly nonlinear channel where we find an MSE difference of 4 dB after 2000 iterations.

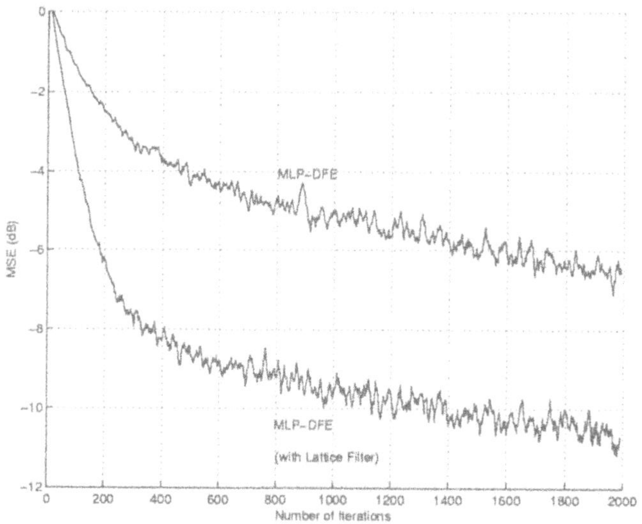

Figure 12: Learning curves for channel 2 with $a_1 = a_2 = a_3 = 1$.

4.2.1 Comparison between (9,3,1) and (3,2,1) structures

Here, we investigate the comparative loss due to the reduction in size of the MLP structure. The trend remains the same for both nonlinear channels. The learning curves for the two nonlinear channels with a_1=1, a_2=0.1 and a_3=0.05, and a SNR = 10 dB are shown in Figures 13 and 14 . As shown by these Figures, the MSE is unaffected by the decrease in network size, but there is an increase in the convergence time of the respective equalizer configurations. The loss in convergence rate is found to be smaller for the lattice-based MLP-DFE than for the conventional MLP-DFE. The MLP-DFE suffers from a loss of about 200 iterations, while its lattice-based counterpart has only a loss of about 100 iterations.

4.3 Bit Error Rate Performance

To further investigate the consistency in performance of the lattice-based MLP DFE, BER performances are studied for the three equalizer configu-

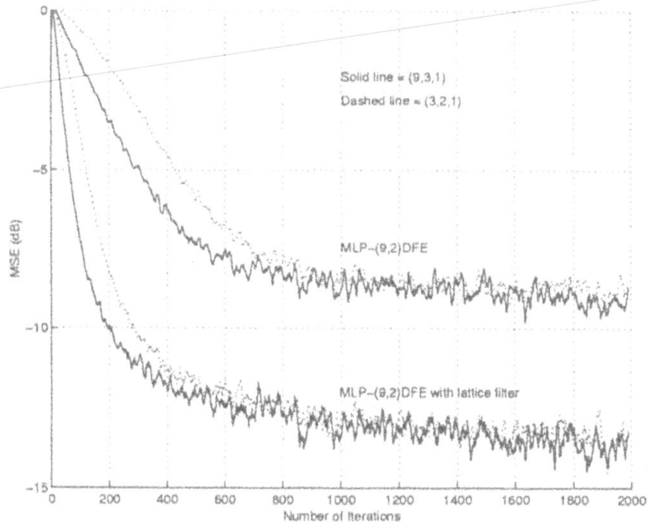

Figure 13: Comparison of (9,3,1) and (3,2,1) MLP structure sizes for channel 1 with $a_1 = 1$, $a_2 = 0.1$, and $a_3 = 0.05$.

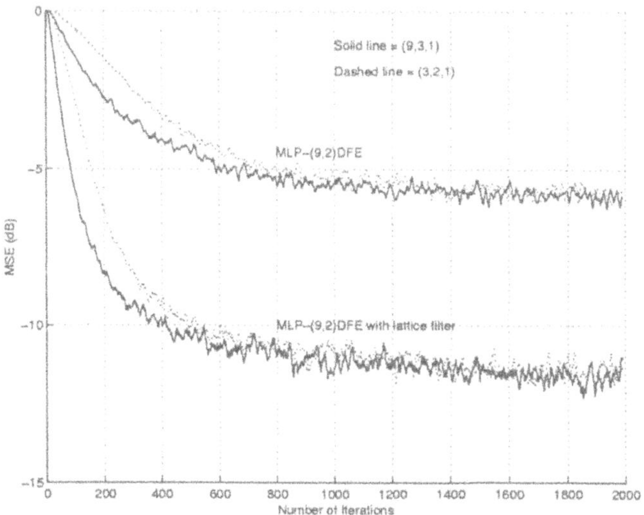

Figure 14: Comparison of (9,3,1) and (3,2,1) MLP structure sizes for channel 2 with $a_1 = 1$, $a_2 = 0.1$, and $a_3 = 0.05$.

rations and are shown in Figures 15-18. As expected, the LMS DFE shows a very poor performance, whereas the MLP DFE equalizer performs better than the LMS DFE. This is so due to the fact the MLP equalizer has the capability to draw nonlinear decision regions. This capability is not shared by the LMS DFE equalizer. This has also been shown in [29] where larger delay lines were used to equalize nonlinear channels. This confirms the fact that the linear equalizer is inherently incapable of dealing with nonlinear distortions, regardless of the equalizer length. However, as shown in Figures 15-18, the lattice-based MLP DFE performs better than the two configurations in all the scenarios presented in this study.

Figure 15, which corresponds to channel 1 with a_1=1, a_2=0.1 and a_3=0.05, shows that a gain of about 2 dB is made by the lattice-based MLP DFE at a BER of 10^{-3}. This shows the impact of pre-orthogonalization of data on non-linear channels. Even the conventional MLP DFE, as depicted in this figure, did not reach this performance.

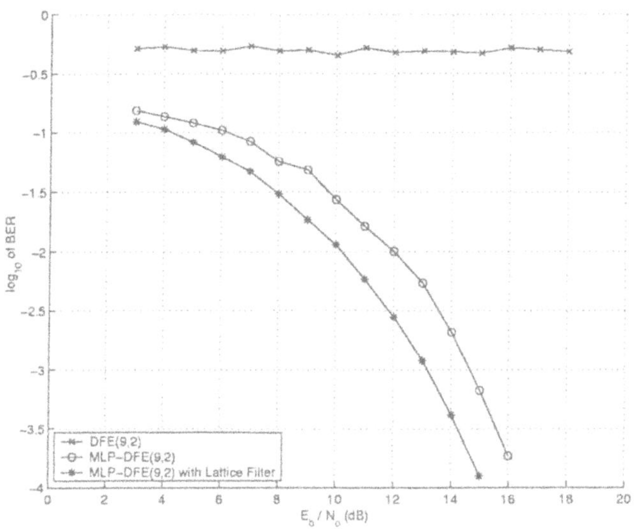

Figure 15: BER curves for channel 1 with $a_1 = 1$, $a_2 = 0.1$, and $a_3 = 0.05$.

The same trend continues for the most severe nonlinearity of $a_1 = a_2 = a_3$=1 for channel 1, as depicted in Figure 16 . The performance has now deteriorated due to the increase in the degree of nonlinearity. The difference in the performance of the two configurations has also increased to around 4 dB at a BER of 10^{-3}.

The BER curves for channel 2 with the two different degrees of nonlinearities are shown in Figures 17 and 18. Figure 17 depicts the results of channel 2 with a_1=1, a_2=0.1 and a_3=0.05, while Figure 18 corresponds to channel 2 with $a_1 = a_2 = a_3 = 1$. For both of these figures, the eigenvalue

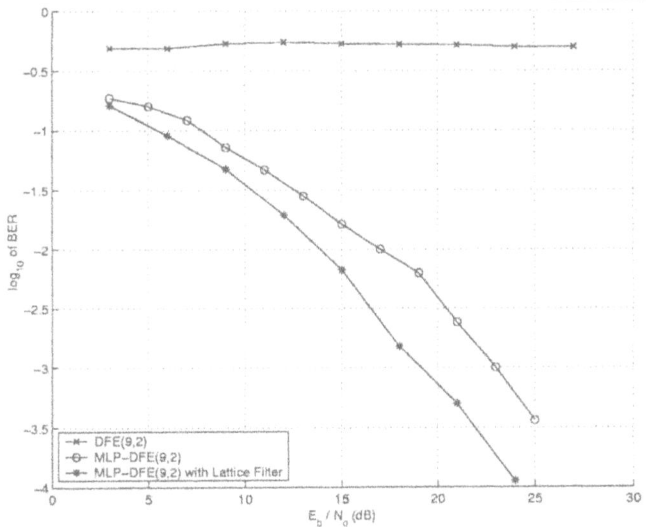

Figure 16: BER curves for channel 1 with $a_1 = a_2 = a_3 = 1$.

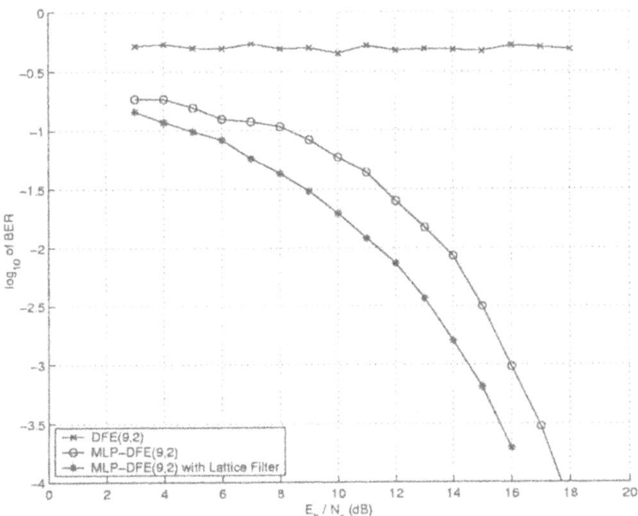

Figure 17: BER curves for channel 2 with $a_1 = 1$, $a_2 = 0.1$, and $a_3 = 0.05$.

spread has increased relatively to that of channel 1. Moreover, in the case of Figure 18, the degree of nonlinearity has increased as well. It is to be expected that the performance of the three equalizers would be affected here. Indeed, at BER=10^{-3}, a 2 dB-increase in performance is obtained for the case of Figure 17 and a more than 5 dB one for the case Figure 18.

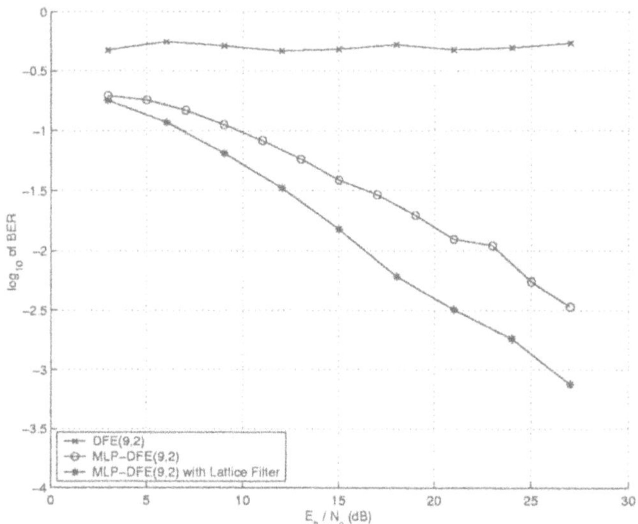

Figure 18: BER curves for channel 2 with $a_1 = a_2 = a_3 = 1$.

If one compares the results of Figures 15-16 to those of Figures 17-18, respectively, one can observe that a BER of 10^{-3} is obtained with a SNR of 13 dB in Figure 15 and with a SNR of 14.5 dB in Figure 17, therefore a 1.5 dB difference between these two scenarios. However, a 7 dB difference in SNR between that of Figure 16 and Figure 18. This is mainly due to the severity of both the channel and its nonlinearity.

In conclusion, the lattice-based MLP DFE outperformed both the MLP DFE and the LMS DFE.

5 Conclusions and Suggestions for Further Work

In conclusion, this work has presented the improvements brought about by whitening the received data samples before they are applied to the MLP DFE. The lattice-based MLP DFE outperformed both the LMS DFE and MLP DFE.

- The results of our study can be summarized as follows:

1. The use of the lattice filter as a whitening scheme for the MLP DFE equalizer's input data results in substantial improvements in terms of convergence rate, steady-state MSE and BER.

2. The convergence rate of the lattice-based MLP DFE is insensitive to the eigenvalue spread of the channel correlation matrix for moderate nonlinearities.
3. The whitening scheme makes the lattice-based MLP DFE less sensitive to the reduction in size of the MLP, down to a tolerable level.
4. Increasing the size of the lattice DFE enhances further the performance of the lattice-based MLP DFE.
5. Increasing the effect of the nonlinearity led to the lattice-based MLP DFE outperforming by far both the MLP DFE and the LMS DFE.
6. For nonlinear channels, the LMS DFE simply fails to perform well. On the contrary, the MLP DFE performs satisfactorily.
7. Finally, it is easy to verify that, computationally the LMS algorithm is the simplest of all and that by using the lattice filter at the input of the MLP DFE, a load of $(20N_1 + 39N_2 - 39)$ computations is added to the lattice-based MLP DFE. Note that, as explained earlier, it was made sure that the two MLP configurations have similar complexity.

- Some suggestions to extend the work are given below:

1. We have used one type of the neural equalizer, namely the multi-layer perceptron for all of our simulations. Other neural-based equalizers, e.g., radial-basis function and recurrent neural networks configurations, are also well worth testing.
2. The proposed equalization scheme discussed here should also be tested for the case of frequency-selective fading channels.
3. In our simulations, bipolar data was used. The system should also be analyzed for some large-constellation modulation schemes, e.g., quadrature amplitude modulation.
4. Finally, all the equalizers in our simulations are T-spaced. The performance of the proposed scheme can also be analyzed for fractionally-spaced equalizers [30].

6 Appendix A

6.1 The Least Squares Lattice DFE algorithm

The algorithm stated below is adapted from [10], and is summarized for convenience. Note that bold faced characters represent matrices or vectors. Specifically, $\mathbf{f}_m(i)$, $\mathbf{b}_m(i)$ and $\mathbf{k}_m^g(i)$ are 2×1 vectors, while $\mathbf{r}_m^f(i)$, $\mathbf{r}_m^b(i)$ and $\mathbf{k}_m(i)$ are 2×2 matrices. We also denote the inverse of $\mathbf{r}_m^f(i)$ and $\mathbf{r}_m^b(i)$ by $\mathbf{r}_m^{-f}(i)$ and $\mathbf{r}_m^{-b}(i)$, respectively. All other quantities are scalars. Finally, $\widetilde{x}_{n-d}$ is the detected sample. In this algorithm γ is chosen to be a very small positive number, while w is a positive number close to 1 (typically 0.95 - 1.0).

6.1.1 Initialization:

$$b_o(i) = f_o(i) = y(i)$$
$$r_o^f(i) = r_o^b(i) = w * r_o^f(i-1) + |y(i)|^2$$
$$e_o(i) = \tilde{x}_{n-d}(i), \quad g_o(i) = 0, \quad \alpha_o(i) = 1$$
$$r_m^f(0) = r_m^b(0) = \gamma, \quad k_m(0) = 0, k_m^g(0) = 0 \qquad (m = 1, ..., N_1 - N_2 - 1)$$
$$k_M^b(0) = 0 \qquad (M = N_1 - N_2)$$
$$\mathbf{k}_m(0) = \mathbf{0}, \quad \mathbf{k}_m^g(0) = \mathbf{0}, \qquad \mathbf{r}_m^f(0) = \mathbf{r}_m^b(0) = \gamma\mathbf{I} \qquad (m = N_1 - N_2, ...,$$

6.1.2 Scalar Stages: $(0 < m \leq N_1 - N_2)$ unless otherwise specified

$$f_m(i) = f_{m-1}(i) - k_m(i-1)b_{m-1}(i-1)/r_{m-1}^b(i-2)$$
$$b_m(i) = b_{m-1}(i-1) - k_m(i-1)f_{m-1}(i)/r_{m-1}^f(i-1)$$
$$k_m(i) = w * k_m(i-1) + \alpha_{m-1}(i-1)f_{m-1}(i)b_{m-1}(i-1)$$
$$r_m^f(i) = r_{m-1}^f(i) - |k_m(i)|^2/r_{m-1}^b(i-1) \qquad (m < N_1 - N_2)$$
$$r_m^b(i) = r_{m-1}^b(i-1) - |k_m(i)|^2/r_{m-1}^f(i) \qquad (m < N_1 - N_2)$$
$$g_m(i) = g_{m-1}(i) + k_m^g(i-1)b_{m-1}(i)/r_{m-1}^b(i-1)$$
$$\alpha_m(i) = \alpha_{m-1}(i) - |b_{m-1}(i)\alpha_{m-1}(i)|^2/r_{m-1}^b(i)$$
$$e_m(i) = \tilde{x}_{n-d}(i) - g_m(i) \qquad (0 \leq m \leq N_1 - N_2)$$
$$k_m^g(i) = w * k_m^g(i-1) + \alpha_{m-1}(i)e_{m-1}(i)b_{m-1}(i)$$

6.1.3 Transitional Stage: $(M = N_1 - N_2)$

$$b_M^*(i) = e_{M-1}(i-1) - k_M^b(i-1)f_{M-1}(i)/r_{M-1}^f(i-1)$$
$$k_M^b(i) = w * k_M^b(i-1) + \alpha_{M-1}(i-1)f_{M-1}(i-1)e_{M-1}(i-1)$$
$$\mathbf{f}_M(i) = [f_M(i) \; e_M(i-1)]^T$$
$$\mathbf{b}_M(i) = [b_M(i) \; b_M^*(i)]^T$$
$$\mathbf{r}_M^f(i) = w * \mathbf{r}_M^f(i-1) + \alpha_M(i-1)\mathbf{f}_M(i)\mathbf{f}_M^T(i)$$
$$\mathbf{r}_M^b(i) = w * \mathbf{r}_M^b(i-1) + \alpha_M(i)\mathbf{b}_M(i)\mathbf{b}_M^T(i)$$

6.1.4 Two Dimensional Stages: $(N_1 - N_2 < m < N_1)$ unless otherwise specified

$$\mathbf{f}_m(i) = \mathbf{f}_{m-1}(i) - \mathbf{k}_m(i-1)\mathbf{r}_{m-1}^{-b}(i-2)\mathbf{b}_{m-1}(i-1)$$
$$\mathbf{b}_m(i) = \mathbf{b}_{m-1}(i-1) - \mathbf{k}_m(i-1)\mathbf{r}_{m-1}^{-f}(i-1)\mathbf{f}_{m-1}(i)$$
$$\mathbf{k}_m(i) = w * \mathbf{k}_m(i-1) + \alpha_{m-1}(i-1)\mathbf{b}_{m-1}(i-1)\mathbf{f}_{m-1}^T(i)$$
$$\mathbf{r}_m^f(i) = w * \mathbf{r}_m^f(i-1) + \alpha_m(i-1)\mathbf{f}_m(i)\mathbf{f}_m^T(i)$$
$$\mathbf{r}_m^b(i) = w * \mathbf{r}_m^b(i-1) + \alpha_m(i)\mathbf{b}_m(i)\mathbf{b}_m^T(i)$$
$$g_m(i) = g_{m-1}(i) + \mathbf{k}_m^{gT}(i-1)\mathbf{r}_{m-1}^{-b}(i-1)\mathbf{b}_{m-1}(i) \qquad (m < N_1)$$
$$\alpha_m(i) = \alpha_{m-1}(i) - \alpha_{m-1}^2(i)\mathbf{b}_{m-1}^T(i)\mathbf{r}_{m-1}^{-b}(i)\mathbf{b}_{m-1}(i)$$
$$e_m(i) = \widetilde{x}_{n-d}(i) - g_m(i)$$
$$\mathbf{k}_m^g(i) = w * \mathbf{k}_m^g(i-1) + \alpha_{m-1}(i)\mathbf{b}_{m-1}(i)e_{m-1}(i) \qquad (m < N_1)$$

References

[1] Qureshi S (1985) Adaptive equalization. IEEE Proc 73(9) : 1349-1387

[2] Siu S, Gibson GJ, Cowan CFN (1990) Decision feedback equalization using neural network structures and performance comparison with standard architecture. IEE Proc 137: 221-225

[3] Lippmann RP (1987) An introduction to computing with neural nets. IEEE ASSP Magazine 4

[4] Proakis JG (1989) Digital Communications. McGraw-Hill, New York

[5] Haykin S (1991) Adaptive Filter Theory. Prentice-Hall, Englewood Cliffs, NJ

[6] Azimi-Sadjadi MR, Citrin S (1989) Fast learning of multilayer neural nets using recursive least squares technique. Proc IEEE Int Conf Neural Networks, Washington DC

[7] Coloma J, Carrasco RA (1994) MLP equaliser for frequency selective time-varying channels. Electronics Letters 30 : 503-504

[8] Shafi A, Zerguine A, Bettayeb M (1999) Neural Network-Based Decision Feedback Equaliser with Lattice Structure. Electronic Letters 35 : 1705-1707

[9] Zerguine A, Shafi A, Bettayeb M (2001) Multilayer Perceptron-based DFE with Lattice Structure. IEEE Trans Neural Networks 12(3) : 532-545

[10] Ling F, Proakis JG (1985) Adaptive lattice decision-feedback equalizers: Their performance and application to time-variant multipath channels. IEEE Trans Comm COM-33 : 348-356

[11] Chen S, Gibson GJ, Cowan CFN, Grant PM (1990) Adaptive equalization of finite nonlinear channels using multilayer perceptrons. Signal Processing 20 : 107-119

[12] Gibson GJ, Siu S, Cowan CFN (1991) The application of nonlinear structures to the reconstruction of binary signals. IEEE Trans Signal Processing 39 : 1877-1884

[13] Mulgrew B (1996) Applying radial basis functions. IEEE ASSP Magazine 13 : 50-65

[14] Wieland A, Leighton R (1987) Geometric analysis of neural network capabilities. First Intl Conf Neural Networks : 385-393

[15] Haykin S (1994) Neural Networks: A Comprehensive Foundation. Macmillan College Publishing Company

[16] Rumelhart DE, McClelland JL (1986) Parallel Distribution Processing: Explorations in the Microstructure of Cognition 1. Cambridge, M.A: MIT Press

[17] Gibson GJ, Siu S, Cowan CFN (1989) Multilayer perceptron structures applied to adaptive equalizers for data communications. IEEE Proc ICASSP Glasgow : 1183-1186

[18] Chen S, Mulgrew B, Grant PM (1993) A Clustering Technique for Digital Communication Channel Equalization using Radial Basis Function Networks. IEEE Trans Neural Networks 4(4) : 570 -590

[19] Kumar PC, Saratchandram P, Sundararajan N (1998) Nonlinear Channel Equalization using Minimal Radial Basis Function Neural Networks. IEEE Proc ICASSP Seattle : 3373-3376

[20] Parisi R, Di Claudio ED, Orlandi B, Rao D (1997) Fast Adaptive Digital Equalization by Recurrent Neural Networks. IEEE Trans Signal Processing 45(11) : 2731 -2739

[21] Widrow B, McCool JM, Larimore MG, Johnson CR (1976) Stationary and nonstationary learning characteristics of the LMS adaptive filter. IEEE Proc 64 : 1151-1162

[22] Satorius EH, Alexandre ST (1979) Channel Equalization Using Adaptive Lattice Algorithms. IEEE Trans Comm COM-27 : 899-905

[23] Makhoul J (1978) A Class of all-zero Lattice Digital Filters: Properties and Applications. IEEE Trans Acoust Speech and Signal Process ASSP-26(4) : 304-314

[24] Wiener N (1958) Nonlinear Problems in Random Theory. John Wiley & Sons, New York

[25] Boyd S, Chua LO (1985) Fading Memory and the Problem of Approximating nonlinear Operators with Voltera Series. IEEE Trans Circuits and Systems CAS-32(11) : 1150-1161

[26] Biglieri E, Gersho A, Gitlin RD, Lim TL (1984) Adaptive Cancellation of nonlinear Intersymbol Interference for Voic-Band Data Transmission. IEEE JSAC SAC-2(5) : 765-777

[27] Schetzen M (1980) The Voltera and Wiener Theories of Nonlinear Systems. John Wiley & Sons, New York

[28] Falconer DD (1978) Adaptive Equalization of Channel nonlinearities in QAM Data Transmission, Systems. Bell Sys Tech J 57(7) : 2589-2611

[29] Peng M (1994) Neural Networks Applications in Linear and Nonlinear Channel Equalization. PhD Thesis, Northeastern University, Boston, Massachusetts

[30] El-Hennawey MS, Zerguine A, Hassan EE (1990) Non-Equally Fractionally-Spaced Equalizers. Electronic Letter 26(16) : 1254-1256

Nonlinear Channel Identification Using Natural Gradient Descent: Application to Modeling and Tracking

Mohamed Ibnkahla

Electrical and Computer Engineering Department,
Queen's University, Kingston, Ontario, K7L 3N6 Canada

Abstract. This chapter applies the natural gradient (NG) descent for adaptive identification of nonlinear channels with memory. The nonlinear channel is comprised of a discrete-time linear filter H followed by a zero-memory nonlinearity $g(.)$. The adaptive system is a neural network which is composed of a linear adaptive filter Q followed by a two-layer memoryless nonlinear neural network (NN). It is shown that the NG learning method significantly outperforms the ordinary gradient descent method in terms of convergence speed and mean squared error (MSE) performance. The chapter studies the sensitivity of the different NN parameters to the natural gradient descent and gives applications to channel tracking and to the modeling of high power amplifiers.

Keywords. Neural networks, backpropagation algorithm, natural gradient descent, channel identification.

1 Introduction

In telecommunications, channel identification allows to estimate and characterize the communication channel. The resulting channel model may be used for several purposes, e.g. receiver design, performance evaluation of the communication channel, power control, adaptive modulation, failure detection, tracking, cellular system design, handoff initiation, etc. Therefore, it is important to have efficient and fast methods for communication channel identification [3].
Several techniques have been proposed in the literature for nonlinear channel identification. Most of these techniques are based on parameterized nonlinear models such as Wiener-Hammerstein models, Volterra series, wavelet networks, neural networks, etc. The parameter estimation can be performed using non-adaptive techniques such as least squares methods and higher order statistics-based methods, and adaptive techniques such as the backpropagation[1] (BP) algorithm and adaptive gradient learning [3, 4, 5, 6, 11, 12, 13, 14, 15, 18].

Recently, neural networks (NNs) [7, 16] have been applied to modeling and identification of nonlinear systems such as memoryless channels and nonlinear dynamical systems [4, 7, 11, 12, 13]. Neural Network approaches have shown excel-

[1] In this chapter, the use of the word 'backpropagation' (BP) without specifying the gradient descent method, refers to the classical BP algorithm trained with the ordinary gradient descent. (section 2.1).

lent performance compared to classical techniques.

Neural Networks trained with the BP algorithm [7, 16] have, however, two major drawbacks: First, their convergence is slow, which can be inadequate for on-line training. Secondly the NN parameters may be trapped in a non-optimal local minimum, leading to sub-optimal approximation of the system [12]. Natural Gradient (NG) learning [1] in the other hand, has been shown to have better convergence capabilities than the classical BP algorithm because it takes into account the geometry of the manifold in which the NN weights evolve. Therefore, NG learning can better avoid the plateau phenomena, which characterize the BP learning curves.

The unknown nonlinear channel studied in this chapter (figure 1) is a nonlinear Wiener system composed of a linear filter $H(z) = \sum_{k=0}^{N_h-1} h_k z^{-k}$ followed by a zero-memory nonlinearity $g(.)$.

This nonlinear channel structure corresponds to a typical uplink satellite communication channel which is composed of a linear filter followed by a traveling wave tube (TWT) amplifier [3, 9, 17].

The same structure can be found in other applications, for example in microwave amplifier design when modeling solid state power amplifiers (SSPA) [11], in adaptive control of nonlinear systems [13], and in biomedical applications when modeling the relationships between cardiovascular signals [4, 5].

The nonlinear system output signal is corrupted by a zero-mean additive white gaussian noise $N_0(n)$, it can be expressed at time n as:

$$d(n) = g\left(\sum_{k=0}^{N_h-1} h_k x(n-k)\right) + N_0(n) \tag{1}$$

The NN model (figure 1) is composed of an adaptive filter $Q(z) = \sum_{k=0}^{N_q-1} q_k z^{-k}$ followed by a two-layer (zero-memory) adaptive neural network. The two-layer NN is composed of a scalar (real-valued) input, M neurons in the input layer, and a scalar output.

Our adaptive structure has two goals:

i) Identify the linear filter H by the adaptive filter Q, and

ii) Model the nonlinearity $g(.)$ by the zero-memory nonlinear NN.

The learning is performed using the input-output signals only (i.e. we consider the unknown nonlinear system as a black box).

The neural network output at time n is expressed as:

$$s(n) = \sum_{k=1}^{M} c_k f\left(w_k \sum_{i=0}^{N_q-1} q_i x(n-i) + b_k\right) = \sum_{k=1}^{M} c_k f\left(w_k Q^t X(n) + b_k\right) \tag{2}$$

where $\{w_k\}$, $\{b_k\}$, and $\{c_k\}$, $k=1,\ldots,M$, are the NN weights, $Q = [q_0\, q_1 \ldots\, q_{N_Q-1}]^t$, and $X(n) = [x(n)\, x(n-1) \ldots\, x(n-N_Q+1)]^t$. $(.)^t$ denotes the transpose.

The network and filter parameters are updated in order to minimize the loss function l (or squared error) between the system output and the NN output

$$l(\theta(n)) = \frac{1}{2}e(n)^2 = \frac{1}{2}(d(n) - s(n))^2 \qquad (3)$$

where θ represents the set of the adaptive parameters, i.e.:
$\theta = [w_1\, w_2 \ldots w_M\, b_1\, b_2 \ldots b_M\, c_1\, c_2 \ldots c_M\, q_0\, q_1\, q_{N_Q-1}]'$.

Different NN-based algorithms are presented and tested in this chapter. The purpose of the chapter is to show that the NG learning overcomes the ordinary gradient descent method in terms of convergence speed and MSE performance. The chapter studies the part of the adaptive system which is the most sensitive to NG learning (in terms of performance improvement).

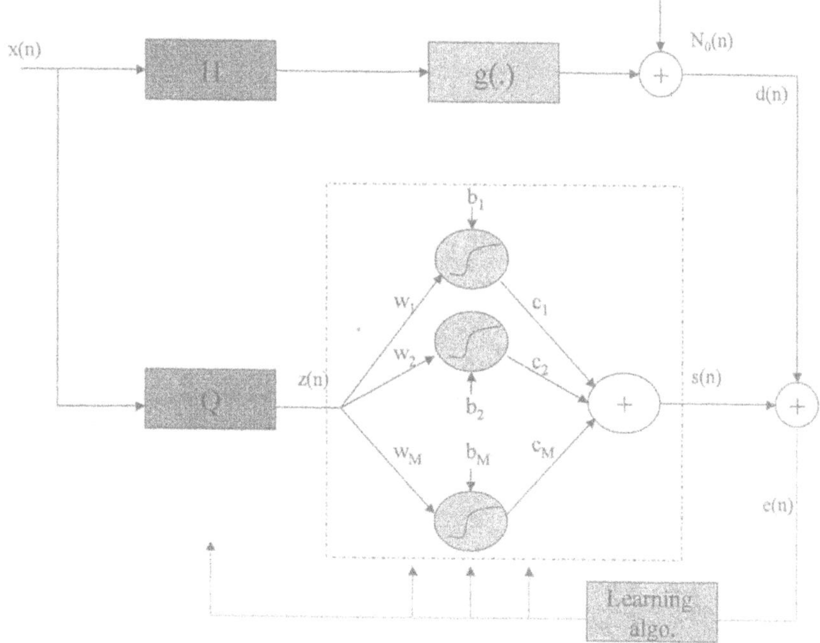

Fig. 1. Neural network system identification structure

Comparisons between classical NN and other adaptive system identification approaches are not within the scope of this chapter. These comparisons have been extensively studied by several authors (see for example [10] for an extended bibliography).

The following section presents the different NN algorithms. Section 3 presents computer simulations, illustrations, and applications to tracking and TWT modeling.

2 Algorithms

2.1 The LMS-Backpropagation (LMS-BP) algorithm

The LMS-BP algorithm updates the weights by following a gradient descent of the error surface according to the ordinary gradient:

$$\theta(n+1) = \theta(n) - \mu\nabla_{\theta}l(\theta(n)) \tag{4}$$

where μ is a small positive constant, and ∇_{θ} represents the ordinary gradient with respect to vector θ, which is expressed as:

$$\nabla_{\theta}l(\theta(n)) = -e(n)\nabla_{\theta}s(n) \tag{5}$$

where:

$$\nabla_{\theta}s(n) = \begin{pmatrix} c_1 Q' X(n) f'(w_1 Q' X(n) + b_1) \\ . \\ . \\ c_M Q' X(n) f'(w_M Q' X(n) + b_M) \\ c_1 f'(w_1 Q' X(n) + b_1) \\ . \\ . \\ c_M f'(w_M Q' X(n) + b_M) \\ f(w_1 Q' X(n) + b_1) \\ . \\ . \\ f(w_M Q' X(n) + b_M) \\ x(n)\sum_{k=1}^{M} c_k w_k f'(w_k Q' X(n) + b_k) \\ . \\ . \\ x(n - N_Q + 1)\sum_{k=1}^{M} c_k w_k f'(w_k Q' X(n) + b_k) \end{pmatrix} \tag{6}$$

where f' denotes the first derivative of f.
(In the RHS of equation 2.3 the weights are at time *n*, index *n* has been removed to make the equation clearer.)

This algorithm will be called LMS-BP, because the update of the NN weights $\{w_1\, w_2 \ldots.\, w_M\, b_1\, b_2 \ldots\, b_N\, c_1\, c_2 \ldots\, c_M\}$ in equation (4) corresponds to the classical BP algorithm [16], and the update of the filter weights $\{q_0\, q_1 \ldots.\, q_{N_Q - 1}\}$ corresponds to the LMS algorithm [8].

2.2 Natural Gradient (NG) learning

The ordinary gradient is the steepest descent direction of a cost function if the space of parameters is an orthonormal coordinate system. It has been shown [1, 2] that in the case of multi-layer NNs, the steepest descent direction (or the natural gradient (NG)) of the loss function is actually given by: $-\tilde{\nabla}_{\theta}l(\theta) = -\Gamma^{-1}\nabla_{\theta}l(\theta)$, where Γ^{-1} is the inverse of the Fisher information matrix (FIM):

$$\Gamma = [\gamma_{i,j}] = [E(\frac{\partial l(\theta)}{\partial \theta_i}\frac{\partial l(\theta)}{\partial \theta_j})] \tag{7}$$

Thus, the natural gradient learning algorithm updates the parameters as:

$$\theta(n+1) = \theta(n) - \mu\,\Gamma^{-1}\,\nabla_\theta\, l(\theta(n)) \tag{8}$$

The calculation of the expectations in the FIM requires the knowledge of the PDFs of x and s, which are not always available. Moreover, the calculation of the inverse of the FIM is computationally very costly. A Kalman filter technique will be used for an on-line estimation of the FIM inverse:

$$\hat{\Gamma}^{-1}(n+1) = (1+\varepsilon_n)\hat{\Gamma}^{-1}(n) - \varepsilon_n\hat{\Gamma}^{-1}(n)\nabla_\theta s(n)(\nabla_\theta s(n))^t\hat{\Gamma}^{-1}(n) \tag{9}$$

where $\nabla_\theta s(n)$ is the (ordinary) gradient of $s(n)$ (see equation 6).

A search-and-converge schedule will be used for ε_n, in order to obtain a good tradeoff between convergence speed and stability:

$$\varepsilon_n = \frac{\varepsilon_0 + \frac{c_\varepsilon n}{\tau}}{1 + \frac{c_\varepsilon n}{\tau\varepsilon_0} + \frac{n^2}{\tau}} \tag{10}$$

such that small n corresponds to a 'search' phase (ε_n is close to ε_0) and large n corresponds to a 'converge' phase (ε_n is equivalent to c_ε / n for large n). ε_0, c_ε and τ are positive real constants.

Using this online Kalman filter technique, the update of the weights (i.e. equation 8) becomes:

$$\theta(n+1) = \theta(n) - \mu\,\hat{\Gamma}^{-1}\,\nabla_\theta\, l(\theta(n)) \tag{11}$$

This algorithm will be called the coupled NGLMS-NGBP, because the filter parameter space and the NN parameter space together are considered as a single space.

2.3 The disconnected NGLMS-NGBP algorithm

Since the filter and the memoryless NN are physically separated, then a choice can be made concerning the parameter space:

- Either we consider a single parameter space for the filter coefficients and neural network weights (as we have done above),

- Or we consider two different parameter spaces, one for the filter, the other for the neural network. In this case, the parameter space of filter Q can be described with a Fisher Information matrix (FIM) $\Gamma_2 = [\gamma_{i,j}(Q)]$ which equals: $\gamma_{i,j}(Q) = E[\frac{\partial l(Q)}{\partial q_i}\frac{\partial l(Q)}{\partial q_j}]$. ($l(Q)$ denotes that the cost function l is considered as a function of Q.)

The parameter space for the NN is described by a new vector: $\theta_{NN} = [w_1\ w_2 \ldots.\ w_M\ b_1\ b_2 \ldots.\ b_M\ \ c_1\ c_2 \ldots..\ c_M]^t$, its FIM will be denoted as Γ_1.

The same Kalman filter technique as in section 2.2 will be used here in order to avoid the explicit calculation of the inverses of the two FIMs, Γ_1^{-1} and Γ_2^{-1}, which will be estimated online by $\hat{\Gamma}_1^{-1}$ and $\hat{\Gamma}_2^{-1}$ as follows:

$$\hat{\Gamma}_1^{-1}(n+1) = (1+\varepsilon_n)\hat{\Gamma}_1^{-1}(n) - \varepsilon_n \hat{\Gamma}_1^{-1}(n)(\nabla_{\theta_{NN}} s)(\nabla_{\theta_{NN}} s)^T \hat{\Gamma}_1^{-1}(n) \tag{12}$$

$$\hat{\Gamma}_2^{-1}(n+1) = (1+\varepsilon_n)\hat{\Gamma}_2^{-1}(n) - \varepsilon_n \hat{\Gamma}_2^{-1}(n)(\nabla_Q s)(\nabla_Q s)^T \hat{\Gamma}_2^{-1}(n) \tag{13}$$

The adaptive system parameters are, therefore, updated as follows:

$$\begin{pmatrix} \theta_{NN}(n+1) \\ Q(n+1) \end{pmatrix} = \begin{pmatrix} \theta_{NN}(n) \\ Q(n) \end{pmatrix} - \mu \begin{pmatrix} \hat{\Gamma}_1^{-1}(n)\nabla_{\theta_{NN}} l(\theta_{NN}) \\ \hat{\Gamma}_2^{-1}(n)\nabla_Q l(Q) \end{pmatrix} \tag{14}$$

where $\hat{\Gamma}_1^{-1}$ is the inverse Fisher matrix for the NN weights (a 3Mx3M matrix), and $\hat{\Gamma}_2^{-1}$ is the inverse Fisher matrix for the filter weights (an $N_Q \times N_Q$ matrix). The other terms are expressed as:

$\nabla_{\theta_{NN}} l(\theta_{NN}) = -e(n)\nabla_{\theta_{NN}} s(n)$, where:

$$\nabla_{\theta_{NN}} s(n) = \begin{pmatrix} c_1 Q' X(n) f'(w_1 Q' X(n) + b_1) \\ \cdot \\ \cdot \\ c_M Q' X(n) f'(w_M Q' X(n) + b_M) \\ c_1 f'(w_1 Q' X(n) + b_1) \\ \cdot \\ \cdot \\ c_M f'(w_M Q' X(n) + b_M) \\ f(w_1 Q' X(n) + b_1) \\ \cdot \\ \cdot \\ f(w_M Q' X(n) + b_M) \end{pmatrix} \tag{15}$$

and $\nabla_Q l(Q) = -e(n)\nabla_Q s(n)$, where:

$$\nabla_Q s(n) = (\sum_{k=1}^{M} c_k w_k f'(w_k Q' X(n) + b_k))\ X(n) \tag{16}$$

This algorithm will be called the separated (or disconnected) NGLMS-NGBP algorithm because the two spaces are treated separately. Note that the computational complexity is lower than the single space NG algorithm because here we deal with two small matrices rather than a large one (i.e. this is equivalent to neglecting the coupling terms between the filter and the NN in matrix Γ). In the simulations below, we will show that these terms are negligible in practice (see figure 2).

Other variants of this algorithm can be derived easily, depending where we would like to apply the NG procedure. For example, if we would like to use the classical LMS algorithm for the adaptive filter and use the NG for the NN, then we keep the upper equations in 14 which concern the update of θ_{NN} and use the LMS algorithm for Q (i.e. the update of filter Q weights in equation 4).

3 Illustrations and Applications

3.1 Identification of a nonlinear channel with memory

3.1.1 Channel model and algorithms

Concerning the unknown structure to be identified, we have taken the nonlinearity as:

$$g(x) = \frac{\alpha x}{1 + \beta x^2}, \ \alpha = 2, \beta = 1 \cdot$$

This function has the same shape as amplitude conversions (AM/AM) of several TWT amplifiers used in satellite communications.

The input signal has been taken as a white gaussian sequence with variance 1 and
filter H weights were taken as:
H=[1.4961 1.3619 1.0659 0.6750 0.2738 -0.0575 -0.2636 -0.3257 -0.2634 -0.125]'. The noise variance was $\sigma = 0.002$ and the learning rate was fixed to $\mu = 0.007$ (for all algorithms), $\varepsilon_0 = 0.005$, $c_\varepsilon = 1$ and $\tau = 70{,}000$ (for the NG algorithms), and $\lambda = 0.99$ (for the RLS algorithm). 50 Monte Carlo runs have been done to estimate the learning curves.

The filter-nonlinearity structure corresponds to a typical model of uplink satellite channels [3]. Filter Q is composed of 10 weights which have been initialized with 0. The NN was composed of $M=5$ neurons which have been initialized with small random values (the same values have been taken for the different algorithms, so that the initial point in the MSE surface is the same for all algorithms). The number of neurons has been chosen equal to 5 because a higher number does not significantly improve the approximation of the nonlinearity, whereas a lower number strongly affects the approximation performance.

In order to study the efficiency of natural gradient learning, and to see which part of the adaptive system is the most sensitive to NG learning (in terms of improvement of the algorithm performance), the following algorithms have been implemented:

- The classical LMS algorithm for the adaptive filter and the Back Propagation for the neural network (LMS-BP).
- The classical LMS for the adaptive filter and the Natural Gradient for the neural network (LMS-NGBP).
- Natural Gradient LMS for the adaptive filter and Back Propagation for the Neural Network (NGLMS-BP).
- Natural Gradient LMS for the adaptive filter and Natural Gradient for the neural network, the parameter space is considered as a single space as explained in section 2.2 (coupled NGLMS-NGBP).
- Natural Gradient LMS for the adaptive filter and Natural Gradient for the neural network and. Both algorithms are separated in the sense of section

2.3 (NGLMS-NGBP, disconnected).

We have also implemented the RLS algorithm [8] for the adaptive filter (instead of LMS) and tested the following algorithms: RLS-BP, RLS-NGBP, NGRLS-BP, NGRLS-NGBP (coupled), and NGRLS-NGBP (disconnected). See the appendix for the NGRLS-NGBP algorithm.

3.1.2 Identification results and sensitivity to Natural Gradient descent

Figure 2 compares the learning curves obtained by coupled and separated versions of the NGLMS-NGBP algorithm as well as the NGRLS-NGBP algorithm. It can be seen that there is no significant difference between the coupled and separated versions. This shows that the coupling terms in the FIM inverse can be neglected. Therefore, in what follows we will keep the separated version which is computationally less complex.

Figure 3 compares the learning curves of 8 algorithms. It can be seen that, in order to obtain an improvement of the convergence speed and the MSE performance, the NG learning should be applied at least to the NN part. If the ordinary gradient is used for the NN, then the algorithm may trapped in a local minimum, whether we apply the NG to the filter or not. When the NG is employed to both parts (i.e. filter Q and the NN), then there is only a slight improvement compared to the case where we apply the NG to the NN part only.

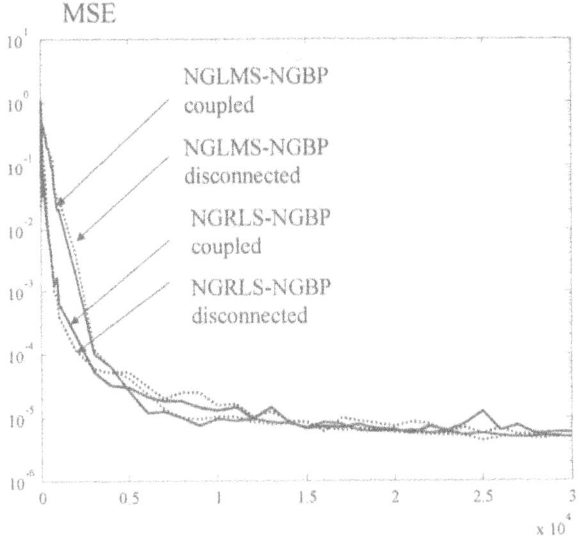

Fig. 2. Comparison of coupled and disconnected NG algorithms

This means that it is more interesting to apply NG to the neural network part only rather than applying it to the whole structure. This considerably reduces the computational complexity while keeping a good overall performance. For example, figure 3 shows that, in order to achieve an MSE of 10^{-4}, the LMS-NGBP needs

6,000 iterations, whereas the LMS-BP needs more than 17,000 iterations.

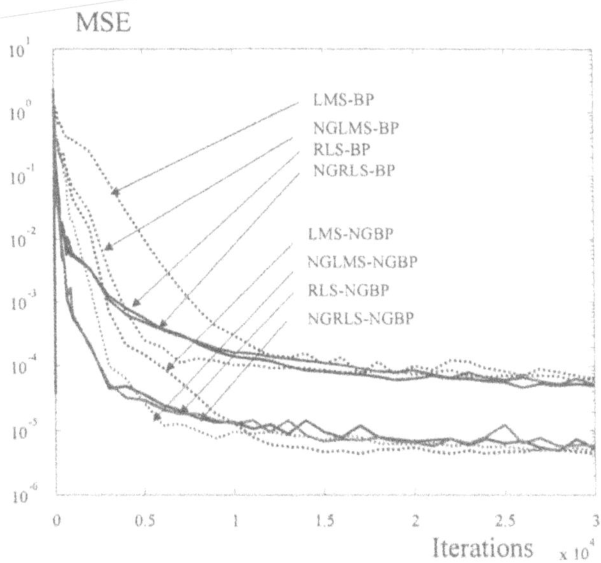

Fig. 3. Learning curves of the different algorithms

Computer simulations show that Q converges to H (to within a scale factor), for all algorithms. In figure 4 we have superimposed the impulse response of filter H and that of the converged filter Q for each of the above algorithms (after normalization with the inverse of the scale factors).

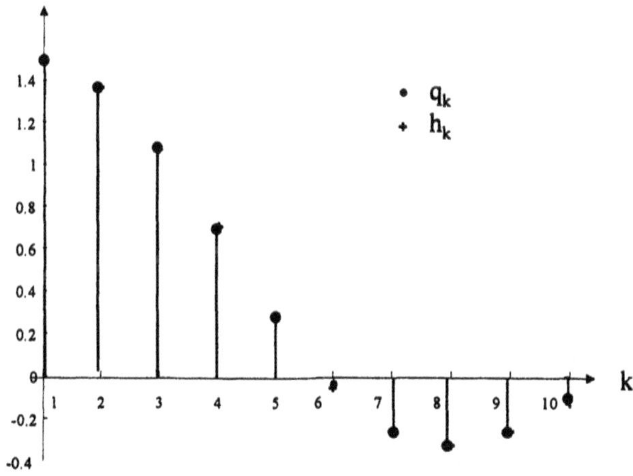

Fig. 4. Impulse response of the converged filters of the different algorithms (after normalization). The impulse responses are superimposed with H

Concerning the nonlinearity, it also has been successfully identified by the memoryless NN. Figure 5 superimposes function *g(z)* and the NN transfer function obtained by each of the algorithms.

Table 1 gives the generalization MSE (i.e. the MSE obtained by the different converged structures for an input that was not used in the learning process). It can be seen that the NG approaches yield better MSE approximation performance than ordinary gradient-based approaches (the improvement is of the order of 10).

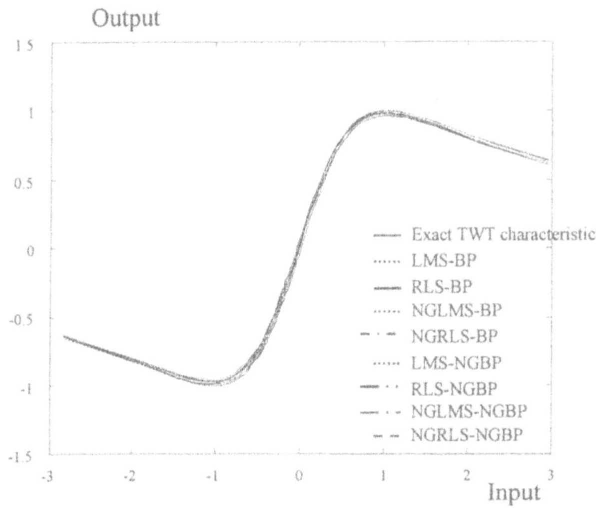

Fig. 5. Input-output conversions of the normalized nonlinear memoryless part of the different algorithms and comparison with the exact TWT characteristic

Table 1. Generalization MSE for the different algorithms

Algorithm	Generalization MSE
LMS-BP	$5.2\ 10^{-5}$
NGLMS-BP	$5.1\ 10^{-5}$
RLS-BP	$4.9\ 10^{-5}$
NGRLS-BP	$4.8\ 10^{-5}$
LMS-NGBP	$4.8\ 10^{-6}$
RLS-NGBP	$4.9\ 10^{-6}$
NGLMS-NGBP	$5\ 10^{-6}$
NGRLS-NGBP	$4.9\ 10^{-6}$

3.2 Tracking capabilities

In order to illustrate the tracking capabilities of the algorithms, we simulated a change in the nonlinearity *g(.)* occurring during on-line learning at the 25,000th iteration. This may happen for example in satellite communications where TWT amplifier characteristics are subject to change because of thermo-dynamical perturbations

It can be seen from figure 6 that, when the change occurs, the MSE considerably increases, then it is decreased by the algorithm. The NG approach again is much faster to track the change than the ordinary gradient. The final MSE is also better (the ordinary gradient algorithms stayed in a local minima until the end of the learning process).

It should be noted that the NG variants of the RLS-BP algorithm have in general a slightly better convergence speed than the NG variants of the LMS-BP algorithm, except for the tracking problem (figure 6), where the NGLMS-NGBP algorithm was faster to track the change than the NGRLS-NGBP algorithm.

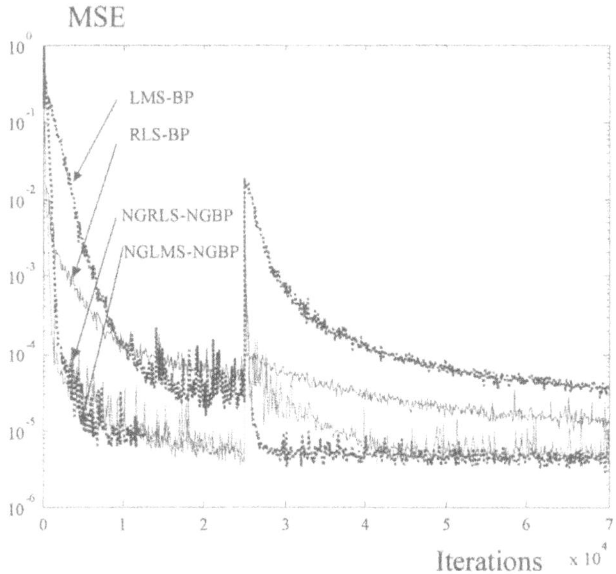

Fig. 6. Tracking of change in the nonlinear system

3.3 Case of memoryless channels

In this section, the nonlinear system is only composed of the memoryless nonlinearity. This corresponds, for example, to an important class of TWT amplifiers used in satellite communications, where the TWT acts as a memoryless nonlinearity.

The adaptive system is composed of a memoryless perceptron (figure 7). The NG learning algorithm can be derived easily from the updating equations of θ_{NN} presented in section 2.3, by fixing Q equal to 1.

For illustration, figure 8 (a and b) gives the results for the modeling of an Intelsat IV TWT amplifier using the classical BP and the NG approach. The TWT input-output measured data was used to train the neural network. It can be seen that the NG algorithm provides faster convergence speed and lower MSE performance than the BP algorithm.

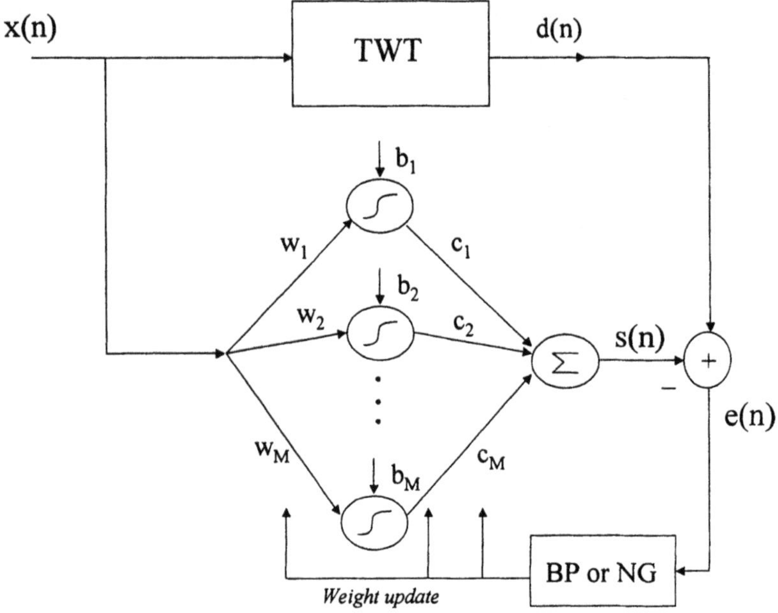

Fig. 7. NN adaptive modeling of a TWT amplifier

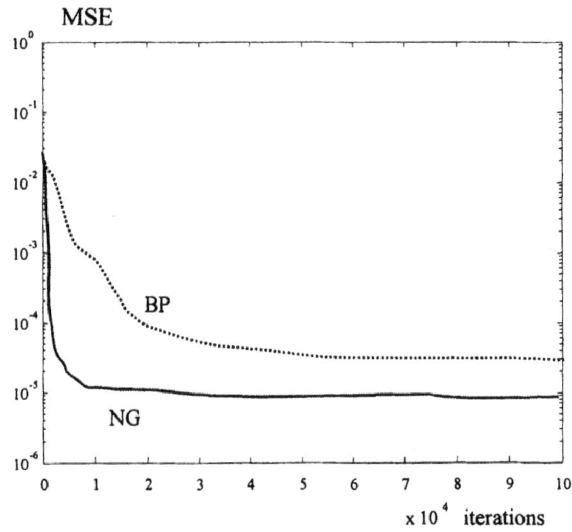

Fig. 8-a. Modeling the AM/AM conversion of Intelsat IV TWT amplifier: Learning curves of the BP and NG algorithms

Other results on TWT modeling are summarized in Table 2. The modeling MSE of neural network models and classical TWT models are given in the table for different TWT amplifiers. It can be seen that the NG outperforms all other models.

In order to illustrate the capabilities of the BP and NG algorithms for on-line tracking, we introduced a new data set simulating a change in the TWT amplitude conversion (figure 9-b). In this simulation, the change has occurred at the 30,000th iteration of the learning process. At the time of change, the learning error increases suddenly, it then decreases until it reaches a minimum (figure 9-a). Here again, the NG has faster tracking speed and lower MSE performance: It converged in 2,000 iterations after the change, whereas the BP algorithm converged in 10,000 iterations after the change. The transfer functions after the change are illustrated in figure 9-b.

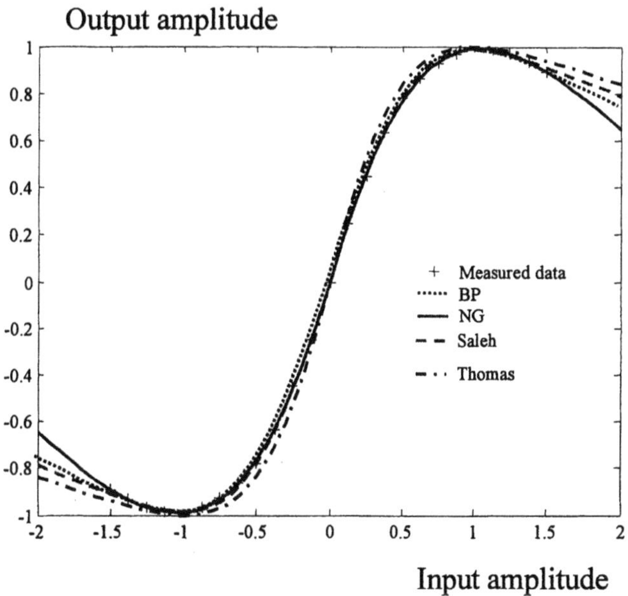

Fig. 8-b. Modeling the AM/AM conversion of Intelsat IV TWT amplifier: Comparison of the resulting transfer function with other models

Table 2. MSE Performance of the different Models for the approximation of the Inphase-Quadrature and Amplitude-Phase TWT characteristics

Model [17]:	**BP NN**	**NG NN**	**Saleh [17]**	**Hetrakul-Taylor [9]**	**Thomas *et al.* [19]**	**Berman-Mahle [20]**
Inphase	$6.1\ 10^{-3}$	$\mathbf{2.6\ 10^{-4}}$	$3.20\ 10^{-3}$	$3.30\ 10^{-3}$	-	-
Quadrature	$6.2\ 10^{-3}$	$\mathbf{2.7\ 10^{-4}}$	$5.20\ 10^{-4}$	$1.60\ 10^{-3}$	-	-
AM/AM	$3.0\ 10^{-5}$	$\mathbf{9.0\ 10^{-6}}$	$1.40\ 10^{-4}$	-	$2.00\ 10^{-4}$	-
AM/PM	$8.1\ 10^{-5}$	$\mathbf{3.2\ 10^{-6}}$	$1.38\ 10^{-4}$	-	-	$8.00\ 10^{-4}$

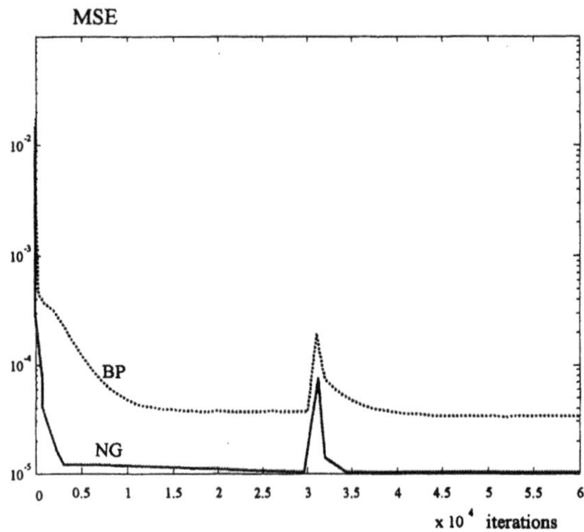

Fig. 9-a. Tracking of change in a TWT AM/AM conversion: Learning curves

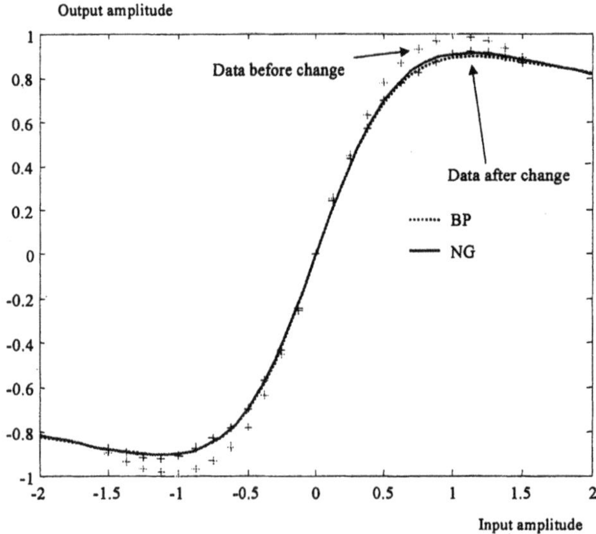

Fig. 9-b. Tracking of change in a TWT AM/AM conversion: NN transfer functions after tracking

4 Conclusion

This chapter proposed different NN algorithms for the identification of nonlinear channels. The unknown system was comprised of a linear filter followed by a memoryless nonlinearity. The NN structure was composed of a linear adaptive filter followed by a memoryless nonlinear NN. Computer simulations have shown that the NG descent gives faster convergence speed and smaller MSE than the ordinary gradient descent. This is due to the capability of the NG descent to better escape from the MSE surface local minima. The NG approach was also applied to modeling memoryless TWT amplifiers and was shown to outperform classical TWT models. Finally, the NG approach has shown better tracking capabilities than the classical BP algorithm.

Acknowledgements:

This work has been supported in part by the Canadian Space Agency / the Canadian Institute for Telecommunications Research (CSA/CITR), by the Natural Sciences and Engineering Research Council of Canada (NSERC), and by the Ontario Premier's Research Excellence Award (PREA).

Appendix: NGRLS-NGBP algorithm

We use here the estimation of the steepest descent to calculate the gain and the estimated covariance matrix at each step, in the recursive least square (RLS [8]) algorithm. We have again two estimations of the inverse of Fisher matrixes, one for the perceptron parameters and one for the filter coefficients.

We simply replace the classic gradient by the steepest descent estimation in the classic RLS equations. This gives:

$$\begin{pmatrix} \theta_{NN}(n+1) \\ Q(n+1) \end{pmatrix} = \begin{pmatrix} \theta_{NN}(n) \\ Q(n) \end{pmatrix} - e(n) \begin{pmatrix} \mu \hat{\Gamma}_1^{-1}(n) \nabla_{\theta_{NN}} s \\ K(n) \end{pmatrix} \tag{A.1}$$

$$K(n+1) = \frac{\hat{P}(n)\hat{\Gamma}_2^{-1}(n)\nabla_Q s}{\lambda + (\nabla_Q s)^T \hat{\Gamma}_2^{-1}(n)\hat{P}_n \hat{\Gamma}_2^{-1}(n)\nabla_Q s} \tag{A.2}$$

$$\hat{P}(n+1) = \frac{1}{\lambda}(\hat{P}(n) - K(n)(\nabla_Q s)^T \hat{\Gamma}_2^{-1}(n)\hat{P}(n)) \tag{A.3}$$

$$\hat{\Gamma}_1^{-1}(n+1) = (1+\varepsilon_n)\hat{\Gamma}_1^{-1}(n) - \varepsilon_n \hat{\Gamma}_1^{-1}(n)(\nabla_{\theta_{NN}} s)(\nabla_{\theta_{NN}} s)^T \hat{\Gamma}_1^{-1}(n) \tag{A.4}$$

$$\hat{\Gamma}_2^{-1}(n+1) = (1+\varepsilon_n)\hat{\Gamma}_2^{-1}(n) - \varepsilon_n \hat{\Gamma}_2^{-1}(n)(\nabla_\theta s)(\nabla_\theta s)^T \hat{\Gamma}_2^{-1}(n) \tag{A.5}$$

Note that we can consider also the whole space $\theta = [w_1\, w_2\;\; w_M\, b_1\, b_2\;\; b_M\;\; c_1\, c_2\;\; c_M\;\; q_0\, q_1\;\; q_{N_Q-1}]^t$, and derive the single space NGRLS-NGBP algorithm as we did in section 2.1.

References

1. Amari S I (1998) Natural Gradient Works Efficiently in Learning. Neural Computation **10**, 251-276
2. Amari S I, Park H, and Fukumizu K (2000) Adaptive method for realizing natu-

ral gradient learning for multi-layer perceptrons. Neural Computation **12**, 1399-1409

3. Benedetto S, Biglieri E (1999) Digital Transmission With Wireless Applications, Kluwer Academic Publishers
4. Bershad N, Celka P, and Vesin J M (1999) Stochastic analysis of gradient adaptive identification of nonlinear systems with memory for gaussian data and noisy input and output measurements. IEEE Trans. Signal Processing **47**, 675-689
5. Bershad N, Celka P, and Vesin J M (2000) Analysis of stochastic gradient tracking of time-varying polynomial Wiener systems. IEEE Trans. Signal Processing **48**, 1676-1686
6. Ghogho M, Meddeb S, and Bakkoury J (1997) Identification of time-varying nonlinear channels using polynomial filters. Proc. IEEE Workshop on Non Linear Signal and Image Processing, Michigan, USA
7. Haykin S (1997) Neural Networks: A Comprehensive Foundation. IEEE Press.
8. Haykin S (1996) Adaptive Filter Theory. Prentice Hall
9. Hetrakul P, Taylor D (1976) The effects of transponder nonlinearity on binary CPSK digital transmission. IEEE Trans. Communications **29**, 546-553
10. Ibnkahla M (2000) Applications of neural networks to digital communications: A survey. Signal Processing **80**, 1185-1215
11. Ibnkahla M, Bershad N J, Sombrin J and Castanié F (1998) Neural network modeling and identification of non linear channels with memory: Algorithms, applications and analytic models. IEEE Trans. Signal Processing **46**, 1208-1220
12. Ibnkahla M (2002) Statistical analysis of neural network modeling and identification of nonlinear channels with memory. IEEE Trans. Signal Processing **50**, 1508-1517
13. Narendra KS, Parthasarathy F (1990) Identification and control of dynamical systems using neural networks. IEEE Trans. Neural Networks **1**, 4-27
14. Prakriya S, Hatzinakos D (1995) Blind identification of LTI-ZMNL-LTI nonlinear channel models. IEEE Trans. Signal Processing **43**, 3007-3013
15. Ralston J, Zoubir A and Bouashash B (1997) Identification of a class of nonlinear systems under stationary non Gaussian Excitation. IEEE Trans. Signal Processing **45**, 719-735
16. Rumelhart D, Hinton G and Williams R (1986) Learning internal representations by error propagation. In: Rumelhart D, McClelland J Eds. Parallel Distributed Processing. MIT Press, pp. 318-362
17. Saleh A (1981) Frequency-independent and frequency–dependent nonlinear models of TWT amplifiers. IEEE Trans. Communications **29**
18. Sjoberg J *et al.* (1995) Nonlinear black box modeling in system identification: A unified overview. Automatica **31**, 1691-1724
19. Thomas M, Weidner M, and Durrani S (1974) Digital amplitude-phase keying with M-ary alphabets. IEEE Trans. Communications **22**, 168-180
20. Berman A, Mahle C (1970) Non linear phase shift in traveling-wave tubes as applied to multiple access communications satellites. IEEE Trans. Communications **18**, 37-48

Maximizing Topology Connectivity Using a Hopfield Neural Network

Hosam M.F. AboElFotoh

Mathematics & Computer Science Department, Kuwait University,
P.O. Box 5969 Safat, Kuwait 13060.

Abstract. In this chapter we use a Hopfield-type neural network to solve the following topology optimization problem. Given a set of n sites, a set of possible links, each link has an associated cost and a probability of failure, we are required to choose a subset of links such that the cost is minimized under the constraint that the network connectivity is not less than a given threshold. The network connectivity is defined to be the probability that every pair of nodes can communicate with each other (the all-terminal reliability). The neural network consists of $n \times (n-1)/2$ Hysteresis McCulloch-Pitts neurons. The solution of the problem corresponds to the minimum of a Lyapunov (energy) function that represents the constraints of the problem. (This chapter is based on reference [1].)

Keywords. All-terminal reliability, connectivity, Hopfield neural network, topology optimization, McCulloch-Pitts neurons, energy function

1 Introduction

1.1 Topological Design Problem

Providing reliable communication paths between network sites is an essential goal of any network topology. In particular, in order to cope with link/node failures, a *backbone* network should provide alternate communication paths between all possible communicating pairs of nodes. An alternate path is chosen by the routing algorithm in case of failure of the current path between two nodes. Therefore, given a set of sites to be connected via point-to-point links to form a backbone (or a wide area) network, it is desirable to layout the links in such a way that multiple paths exist between all pairs of nodes. To limit the cost, we need a method for selecting links to be laid out, that achieves high connectivity (reliability) at a minimal cost.

The general definition of the topological design problem for a large hierarchical network was first presented by Boorstyn et. al. [2]. They divided the global design problem into two subproblems: the design of the backbone and the design of the local distribution networks. The objective was to minimize cost under traffic, delay and reliability constraints. Since then, many researchers have dealt with different variations of topological design problems [3-7]. In this chapter we focus on maximizing the connectivity of a backbone network. The measure of connectivity used is the probability that an operating path exists between every pair of nodes in the network. This measure is known as the *all-terminal reliability* of the network [8]. The design problem can be defined as follows.

Given

1. a set of sites (nodes),
2. a set of candidate links; each link has an associated cost and reliability (i.e. probability of operation), and,
3. the objective value of the backbone network's all terminal reliability R_0 $(0<R_o<1)$,

select a set of links such that the sum of link costs is minimized, and, the network's all terminal reliability is $\geq R_0$. We assume that the network nodes do not fail. Therefore, the network failure will be due to failure of links.

Equivalently, the problem is to find a topology x , selected from all possible topologies constructed from the network's nodes and possible links, so that the all terminal reliability of x is not less than a target value R_o, and the total link costs of x is less than any other network topology, that satisfies the reliability constraint. It has been shown that the design of networks considering all-terminal reliability, along with other objectives or constraints, is an NP-hard combinatorial problem [2,9]. Furthermore, the calculation of the overall reliability of a network is by itself an NP-hard problem [8].

As a result, the continuing researches in the area of topological optimization design problem have either been enumerative-based [5-7], or heuristic-based [10-12]. Enumerative-based methods use exhaustive search techniques and guarantee optimality. However, they can be applied to small network sizes only. Heuristic-based methods can be applied to larger networks, but do not guarantee optimality. Nevertheless, a near-optimal solution, where the cost might not be the lowest possible, is usually accepted given that the reliability constraint is satisfied.

Although heuristics do not guarantee optimality, it (especially the genetic search) showed practically their strength and ability to find near-optimal solutions. Kumar et. al. [12] developed a genetic algorithm (GA) for solving three network design problems for optimizing the reliability, diameter and average distance. Deeter et. al. [10] presented a GA to optimize the design of networks when considering all-terminal reliability. Dengiz et. al. [11] presented a GA with knowledge-based steps, to solve the problem.

In Section 2, we present a method for finding an optimal or a near-optimal solution to the problem using a Hopfield-type neural network, named after the first inventor of optimization neural networks (ONN) [13]. In the following section we give a brief introduction to ONN.

1.2 Neural Network Approach For Solving Optimization Problems

The first ONN was introduced by Hopfield and Tank [13] in 1985, and referred to by *Hopfield neural network.* Since then, tremendous efforts have been done to study and improve ONNs. As a result, new models were introduced, and large number of optimization problems have been efficiently solved [14].

1.2.1 Advantages of Hopfield neural network

Hopfield neural networks have several advantages over traditional techniques for certain types of optimization problems [14]. First, they can find optimal or near optimal solutions quickly for large problems where efficient algorithms do not exist. Second, because the neurons of the network operate in parallel (in the actual circuit), computation time is minimized. Furthermore, Hopfield neural networks can also handle situations in which some constraints are weak (desirable, but not absolutely required). In fact, they have been applied to solve many problems in the area of communication networks [15-18].

1.2.2 Characteristics of Optimization Neural Networks

An artificial neural network model consists of two components: *neurons* and *synaptic links*. The output of the ith neuron is given by $V_i = f(U_i)$, where U_i is the input signal of the ith neuron , and f is called the neuron's input/output function. The input signal of a neuron is determined by a linear sum of weighted output signals from all neurons. The weight corresponds to the strength of the corresponding synaptic link. Usually ONN are fully interconnected, in the sense that each neuron is connected to every other neuron through synaptic links. In ONN each neuron represents a hypothesis, with the hypothesis is true if the neuron is "on", i.e. its output equals 1, false if the neuron is "off" (output equals 0). In addition, the weights of links between the neurons are fixed to represent the constraints of the problem and the cost function to be optimized. The solution of the problem corresponds to the minimum of a *Lyapunov (energy)* function, as illustrated in the following example.

Example [20]. Consider the map-coloring problem. We want to color the regions of a map using only four colors, such that no two adjacent regions have the same color. Let *n* be the number of regions, and the four colors be numbered 1 through 4. We use *n* x 4 array of neurons. The output of neuron (X,i) V_{Xi} is set to 1 if and only if region *X* is colored using color *i*. We form the following energy function:

$$E = \frac{A}{2} \sum_{X=1}^{n} \left(\sum_{i=1}^{4} V_{Xi} - 1\right)^2 + B \sum_{X=1}^{n} \sum_{\substack{Y=1 \\ Y \neq X}}^{n} \sum_{i=1}^{4} d_{XY} V_{Xi} V_{Yi}$$

where d_{XY} is 1 if regions X and Y are adjacent to each other, 0 otherwise. *A* and *B* are positive constant numbers. The first term represents the constraint that each region is assigned one and only one color. The term is minimized if one and only one of the four neuron's outputs V_{Xi} , i=1,2,3,4, is one. The second term is minimized if for every two adjacent regions X and Y (d_{XY}=1) either V_{Xi} or V_{Yi} is 0 (or both). The objective of the ONN is to minimize the energy function E. The output of the neurons in a stable state corresponding to the minimum of E will represent a solution to the problem.

1.2.3 The Energy Function and Its Convergence

Energy function (E) is a bounded function of the state variables of a dynamical system such that all state changes results in a decrease in the value of the function, which causes the system to reach a stable solution [13,14]. In ONN the state variables are the neuron outputs. The dynamics of the network is determined so that E (therefore the cost function) is minimized as the neurons update their activations (inputs). E is usually composed of two parts. The first ($E_{objective}$) represents the solution cost to be minimized. The second part keeps the state of the neural network in a valid solution (i.e. represents the problem constraints). The sum of these two parts forms the computational energy function:

$$E = E_{objective} + \sum_{c} \lambda_c \cdot E_{constraint_c} \cdot$$

The interactions between the ith neuron and other neurons are determined by the motion equation. The change of the input state of the ith neuron (U_i) is given by the partial derivatives of the computational energy function E with respect to the output of the ith neuron (V_i), where E is an n-variable function: $E(V_1,V_2,\ldots,V_n)$. The *motion* equation of the ith neuron is given by:

$$dU_i/dt = -\partial E(V_1,V_2,...,V_n)/\partial V_i \qquad (1)$$

and, taking dt to be unit time:

$$\Delta U_i = -\partial E(V_1,V_2,...,V_n)/\partial V_i \qquad (2)$$

In order to update the activation of the neuron, the first order Euler method is used. It is the simplest among the existing numerical methods. Based on the first order Euler method, the value of $U_i(t+1)$ is determined by $U_i(t)$ and $\Delta U_i(t)$:

$$U_i(t+1) = U_i(t) + \Delta U_i(t) \qquad (3)$$

Hill climbing. Because only the local minimum convergence is guaranteed in optimization neural networks many heuristics have been used to make it easier to escape the *local minima*, and increase the frequency of the *global minimum* convergence [14]. The idea is to excite some neurons such that the energy is allowed to start at a different point on the function landscape. One of the most important such heuristics is *hill climbing*. A new term is added to the motion equation in order to encourage the neuron to have nonzero output if some constraints of the problem have not been met, yet. Otherwise, the term has no effect (= 0).

The optimal solution of the problem is the absolute minimum of this energy. However, one or more local minima can be considered as acceptable solutions of the problem. If a local minimum is not acceptable, the hill-climbing technique may be used. In this case, neurons are re-excited to start in a different point in the energy space, so that the ONN could settle in a different minimum (possibly the global one).

1.2.4 Hysteresis McCulloch-Pitts neuron model

The output of the neuron is updated from the activation (input) U using a non-linear input-output function. Different models of neurons are used for the ONN depending on the input/output function. A major disadvantage of early models was the oscillatory behavior around the minima points. In order to suppress this oscillatory behavior, the hysteresis McCulloch-Pitts neuron model was introduced and the convergence of the neural networks consisting of such neurons has been proved [19,21]. Because of suppressing the oscillatory behavior, hysteresis McCulloch-Pitts model shortens the convergence time considerably. The input/output function of the ith hysteresis McCulloch-Pitts neuron is given by:

$$V_i = \begin{cases} 1 & \text{if } U_i > \text{UTP}, \quad \text{(upper trip point)} \\ 0 & \text{if } U_i < \text{LTP}, \quad \text{(lower trip point)} \\ \text{unchanged otherwise,} & \end{cases}$$

where UTP is always larger than LTP. The shape of the function is given in Fig.1.

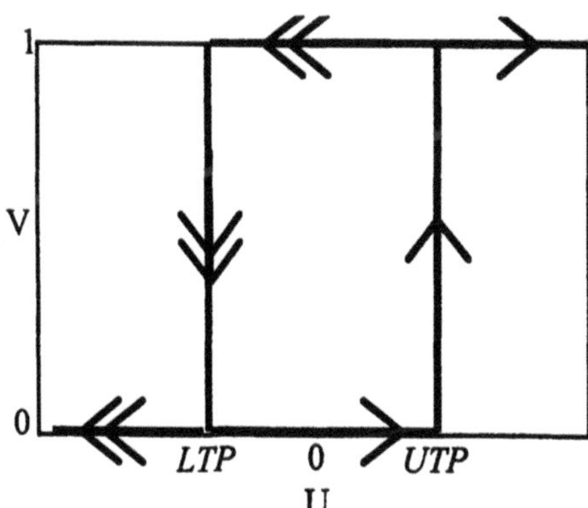

Fig. 1. The Hysteresis McCulloch-Pitts Input/Output Function

2 Network graph model and problem formulation

The network is represented by a probabilistic graph $G = (N, L)$, where N is a set of n *nodes* (*vertices*) which represent the sites of the network. The set L is the set of the *edges* representing communication links which are bi-directional (The extension to the directed graph case is straightforward). A probability of failure (operation) is associated with each edge $e \in L$, denoted q_e. ($p_e = 1 - q_e$). We assume that all link failures are statistically independent, and, that all nodes are fixed and perfectly reliable.

The *all-terminal* reliability measure, denoted $R(G)$, is defined to be the probability that for every pair of nodes $s,t \in N$, there is a path from s to t. It is also defined as the probability that the graph contains at least a spanning tree (a tree subgraph that contains all vertices). In a directed graph case, the all-terminal reliability (*reachability*), denoted $Conn_A(G)$, is the probability that there are paths from a distinguished source s to every other node. Equivalently, this is the probability that the digraph contains at least a *spanning arborescence* (a directed spanning tree) *rooted at s*. (For a complete reference on reliability measures see [8])

2.1 Statement Of The Problem

Notation:

$G(N, L)$	graph G, with the set of nodes = N and the set of links = L.
(i,j)	a bi-directional link between nodes i & j.
p_e, q_e	link (reliability, unreliability) for link e; $q_e + p_e = 1$.
x	Architecture of network design: $x_{ij} = 1$ if (i,j) is selected, else $x_{ij} = 0$.
$R(x)$	Reliability of network design x.
n	number of nodes, $\|N\|$.
n^*	$= n(n - 1)/2$: maximum number of *links*.
l	number of links.
c_{ij}	cost of link (i,j).
$C(x)$	Total cost of network design $= \sum_{i=1}^{N} \sum_{j=i+1}^{N} x_{ij} c_{ij}$.
R_o	Minimum network reliability constraint.

The design problem can then be stated mathematically as follows:

Given $G(N, L)$ and q_e for each e in L , find a design x (selection of links) that minimizes $C(x)$ and satisfies $R(x) \geq R_o$.

The networks we consider can be *fully connected networks, or, non-fully connected networks*. In the fully connected network, all links between any pair of nodes are available for selection, i.e., $|L| = n^*$. On contrast, in the non-fully connected network, only a subset of all possible links is available for selection, i.e., $|L| < n^*$.

3 Neural Network Solution

3.1 Problem Representation

To represent a communication network of n nodes, we use a two-dimensional upper triangular matrix of $n^* = n \times (n-1)/2$ neurons. A neuron n_{ij} represents a link between vertex i and vertex j. We need only the upper triangle since the links are assumed to be bi-directional, and subsequently edge (i, j) and edge (j, i) refer to the same link. The matrix representation has the advantage of simple presentation and addressing of the neurons. To extend the solution to include the directed network case, we should use the full matrix representation. In this case the processing element n_{ij} is no longer identical to n_{ji}, since the link (i, j) is different from the link (j, i).

The output of the ij^{th} neuron represents the selection of link (i,j) in the network design. The output V_{ij} of the neuron is defined as follows:

$$V_{ij} = \begin{cases} 1 & \text{iff link } (i,j) \text{ is selected in the design,} \\ 0 & \text{otherwise.} \end{cases}$$

The output of neurons corresponding to links that are not available for selection in the non-fully connected networks are always set to zero (*stuck_at_0*). We can also set the output to 1 if the corresponding link must be in the design (*stuck_at_1*).

3.2 The Energy Function

In order to solve the design problem, using neural networks, we first have to define an energy function whose minimization process drives the neural network into one of its energy minima states. This stable state corresponds to a solution of the network design problem. The energy function must favor states that correspond to a selection of links with a reliability greater than or equal the threshold value (R_o). Among these states it must also favor the one which has the lowest total cost. An example of such energy function which satisfies such requirements can be formulated as follows:

$$E1 = A\left(R - R_o\right)^2 + B \sum_{i=1}^{n} \sum_{j=i+1}^{n} V_{ij} c_{ij} \qquad (4)$$

where R is a function of V_{ij}'s, $1 \le i \le n$ and $1 \le j \le n$, that represents the reliability. A and B are constants. The representation of R as a function of neuron outputs is dealt with in Section 3.3.

The contribution of the A-term in the energy (Eq. 4) reflects the difference between the network reliability and the target value. When the network reliability is closer to R_0, the A-term will contribute by a less value in the energy function; i.e. the energy favors (has the minimum value for) the state in which it has the minimum difference value. The same argument can be given for the cost term (the B-term). The B-term encourages the neural network to select the links which have the minimum total cost. This is done by summing up the costs of the links selected in the design. A high total cost will result in a high contribution in the energy value, while the lowest total cost will make the B-term in the energy function turns down to its minimum value. A final remark considering the terms of the energy is that both terms are not strong constraints of the problem. Rather, they are desirable ends.

However, if the expression for R contains terms of the form $(Vij)^2$, a neural network, based on the above energy, is not guaranteed to converge all the time. It can be verified that the energy at some points is increased.

To avoid this potential convergence problem, we may redefine the energy function as follows:

$$E_2 = -A \times R + B \sum_{i=1}^{n} \sum_{j=i+1}^{n} V_{ij} c_{ij} \tag{5}$$

The A term encourages the neural network to increase the reliability of the design. The B term, as in Eq. 4, contributes to the energy function by the total cost. Once again, we may be faced with an undesirable behavior. Since the reliability measure contributes directly to the energy, the neural network may prefer states that have reliabilities greater than R_o, regardless of greater total cost [1].

This lead us to design a new energy function (E_3) with three energy terms. This function shares the advantages of the previous ones and, at the same time, overcomes their drawbacks. The first term represents the reliability. The second term represents the cost of the selected links. A new term (C-term) is added to represent the absolute value of the difference of the reliability R from R_o. This term discourages the neural network from adding new links to increase the reliability far beyond the threshold R_o. Note that the third term uses the absolute difference instead of the square value to avoid the convergence problem encountered with E_1. This is represented as follows:

$$E_3 = -A \times R + B \sum_{i=1}^{n} \sum_{j=i+1}^{n} V_{ij} c_{ij} + C|R - R_o| \tag{6}$$

The following theorem [1] formulates sufficient conditions to guarantee convergence of energy functions when minimized using an ONN and the Hysteresis McCulloch-Pitts neuron model:

***Theorem 1*:**

The following conditions are sufficient to guarantee the convergence of the energy function E, using the Hysteresis McCulloch-Pitts neuron model and the motion equation $\Delta U_{ij} = -\frac{\partial E}{\partial V_{ij}}$:

(1) E is quadratic in Vij, $1 \leq i \leq n$ and $1 \leq j \leq n$, i.e., does not include powers of Vij greater than 2.

(2) The second partial derivative $\frac{\partial^2 E}{\partial V_{ij}^2} \leq 0$.

Proof: Consider the change of the energy function E due to a change in Vij:

$$\Delta E = E(Vij(t+1)) - E(Vij(t))$$

(We shall use Vij to denote $Vij(t)$ for simplicity)
Using Taylor's expansion:

$$\Delta E = \frac{\partial E}{\partial Vij}\Delta Vij + \frac{\partial^2 E}{\partial Vij^2}\frac{(\Delta Vij)^2}{2!} + \cdots + \frac{\partial^k E}{\partial Vij^k}\frac{(\Delta Vij)^k}{k!}$$

$$= \Delta_1 + \Delta_2 + HO$$

where $\Delta_1 = \frac{\partial E}{\partial Vij}\Delta Vij$, $\Delta_2 = \frac{\partial^2 E}{\partial Vij^2}\frac{(\Delta Vij)^2}{2!}$

and HO denotes higher order terms.

Substituting for $\frac{\partial E}{\partial Vij}$ in Δ_1 from the motion equation:

$$\Delta 1 = \frac{\partial E}{\partial V_{ij}}\Delta V_{ij}$$

$$= \frac{\partial E}{\partial V_{ij}}\frac{\Delta V_{ij}}{\Delta U_{ij}}\Delta U_{ij}$$

$$= -\left(\frac{\partial E}{\partial V_{ij}}\right)^2 \frac{\Delta V_{ij}}{\Delta U_{ij}}.$$

Let $\frac{\Delta V_{ij}}{\Delta U_{ij}}$ be $\frac{V_{ij}(t+\Delta t) - V_{ij}(t)}{U_{ij}(t+\Delta t) - U_{ij}(t)}$..

It is necessary and sufficient to consider the following four regions in the input/output function of the Hysteresis McCulloch-Pitts neuron model [19,21]:

Region1: $U_{ij}(t) >$ UTP and $V_{ij}(t) = 1$.

Region2: $LTP \leq U_{ij}(t) \leq UTP$ and $V_{ij}(t) = 1$.

Region3: $LTP \leq U_{ij}(t) \leq UTP$ and $V_{ij}(t) = 0$.

Region4: $U_{ij}(t) < LTP$ and $V_{ij}(t) = 0$.

In region 1: we consider the four possible cases for $U_{ij}(t+\Delta t)$:

$(a)\, U_{ij}(t+\Delta t) > U_{ij}(t)$.

$(b)\; LTP < U_{ij}(t+\Delta t) < U_{ij}(t)$.

$(c)\, U_{ij}(t+\Delta t) < LTP < U_{ij}(t)$.

$(d)\, U_{ij}(t+\Delta t) = U_{ij}(t)$.

In (a), (b), and (d), $V_{ij}(t+\Delta t) = V_{ij}(t) = 1 \Rightarrow \dfrac{\Delta V_{ij}}{\Delta U_{ij}} = 0$.

Therefore, $\Delta_1 = 0$.

In (c), $V_{ij}(t+\Delta t) = 0 \Rightarrow \dfrac{\Delta V_{ij}}{\Delta U_{ij}} = (0-1)/(\textit{negative}\text{value}) > 0$.

Therefore, $\Delta_1 < 0$.

Hence, $\Delta_1 \leq 0$ is always satisfied in Region 1. Similarly, in Regions 2-4, $\Delta_1 \leq 0$ is always satisfied. (1)

Now if E is quadratic in V_{ij}, then $\dfrac{\partial^k E}{\partial V_{ij}{}^k} = 0$, for $k > 2$.

Therefore, HO = 0. (2)

So if $\dfrac{\partial^2 E}{\partial V_{ij}{}^2} \leq 0$, then $\Delta_2 \leq 0$. (3)

From (1), (2), and (3) $\Delta E \leq 0$. (Q.E.D.)

Recall that E_1 of Eq. 4 is defined in terms of $(R\text{-}R_0)^2$. If R includes terms of the form $V_{ij}{}^2$, and subsequently R^2 includes terms of the form Vij^4, then, it can be easily shown that $\partial^3 E_1/\partial V_{ij}{}^3 \neq 0$, which may cause the energy to increase at some points.

3.3 Defining The Reliability Term

A major challenge in defining the energy function is expressing the reliability (R) in terms of the neuron outputs, or equivalently, in terms of the link selection of the design. Taking into account the NP-hard complexity of the exact reliability calculation, together with the iterative behavior of the neural networks, bounding, or estimation of the all-terminal reliability must be used. A number of techniques for computing upper and lower bounds of the reliability for large arbitrary networks exist [8].

Among alternative strategies for bounding the reliability, the edge packing strategy seems to be the most effective. Simulation results based on such bounds have been reported in [1]. In the simulator, both upper and lower bounds of system reliability were used to express the reliability in the energy function. We define these bounds in the following theorems.

***Theorem 2**:(An upper Bound on All-Terminal Reliability.)*

Given $G = (N, L)$, p_e for each edge $e \in L$. Let s be a vertex in N. Then

$$R(G) \leq \prod_{\substack{i=1 \\ i \neq s}}^{n} \left[1 - \left(\prod_{j \in Adj(i)} (1 - p_{ji}) \right) \right],$$

where $Adj(x)$ denotes the set of vertices connected to vertex x.

Proof: .see [8].

Using the upper bound does not guarantee the satisfaction of the reliability constraint ($R(G) \geq R_o$). This is so, since the reliability could be less than the target value R_o, although the upper bound is not. However, experimental results [1,8] have shown that for large edge probability of operation values, the upper bound is usually close enough to the actual value of R. This suggests using an all-terminal reliability lower bound for small values of p_e, or incorporating both lower bound and upper bound in the energy function for arbitrary values.

***Theorem3:** (A Lower Bound on All-Terminal Reliability.)*

Given $G = (N, L)$, p_e for each edge $e \in L$. Let s be a vertex in N, and let Π be a random permutation of N, such that $\Pi(s) = 1$. Define $In(i)$ to be the set of nodes $\{j: (j, i) \in L \text{ and } \Pi(j) < \Pi(i)\}$. Then

$$R(G) \geq \prod_{\substack{i=1 \\ i \neq s}}^{n} \left[1 - \left(\prod_{j \in In(i)} (1 - p_{ji}) \right) \right].$$

Proof: see [1].

Now, we can use bounds to obtain an expression for the reliability in terms of neuron outputs. We substitute for R in the energy function of Eq. 6 using the lower bound of Theorem 3.

$$E=-A\prod_{i=1,i\neq s}^{n}\left(1-\prod_{j\in In(i)}(1-p_{ij}V_{ij})\right)+B\sum_{i=1}^{n}\sum_{j=i+1}^{n}c_{ij}V_{ij}+$$

$$C\left|A\prod_{i=1,i\neq s}^{n}\left(1-\prod_{j\in In(i)}(1-p_{ij}V_{ij})\right)-Ro\right| \quad (7)$$

3.4 The Motion Equation

The motion equation for neuron ij is derived from the energy function of Eq. 7 using Eq. 2, and is given by:

if $(R > Ro)$

$$\frac{dU_{ij}}{dt}=-\frac{\partial E}{\partial V_{ij}}=A\frac{\partial R}{\partial V_{ij}}-Bc_{ij}-C\frac{\partial R}{\partial V_{ij}}, \quad (8)$$

if $(R < Ro)$

$$\frac{dU_{ij}}{dt}=-\frac{\partial E}{\partial V_{ij}}=A\frac{\partial R}{\partial V_{ij}}-Bc_{ij}+C\frac{\partial R}{\partial V_{ij}}. \quad (9)$$

3.5 Hill-Climbing

As mentioned earlier, in order to allow the state of the system to escape from the local minima and to converge to the global minimum, a hill-climbing term is introduced to allow the energy to start from a new point (uphill) on the function landscape. We add the term $+D\,h(R - R_o)$ to the motion Eqs (8) and (9), where $h(x)$ is 1 if $x < 0$, and 0 otherwise, and D is a constant. The hill-climbing term performs the excitatory force only when the reliability of the topology constructed so far is less than R_o.

The activation (inputs) of the neurons can be updated asynchronously or synchronously [14]. In order to update the activation of the neuron, the first order Euler method is used. Based on Eqs. 2 and 3, the value of $U_{ij}(t+1)$ is determined by $U_{ij}(t)$ and $\Delta U_{ij}(t)$, $U_{ij}(t+1) = U_{ij}(t) + \Delta U_{ij}(t)$. where $\Delta U_{ij}(t)$ is given by Eqs. 8 and 9, and Δt is set to 1. In an actual circuit implementation the synaptic links weights (from the output of neuron x to the input of neuron y) is the multiplier of Vx in the motion equation of neuron y.

Using the above motion equations and the energy function E (Eq. 7) we have the following convergence theorem.

Theorem 4:

The energy function of Eq. 7 is an energy (Lyapunov) function that always converges (using motion equations 8 and 9) to a stable state.

Proof: see [1] (We need to show that ΔE is always negative unless the climbing-hill term is used).

5 Simulation of the Optimization Neural Network

5.1 Program Outline:

The following procedure describes an implementation of an ONN simulator reported in [1]. The data set of A, B, C, D, UTP, LTP, U_max, and U_min are empirically determined. U_max and U_min are the constant upper limit and the constant lower limit of $U_{ij}(t+1)$ respectively. Instead of starting the simulation with an empty set of links, the simulator initializes the design as a minimum cost spanning tree (a set of links that forms a tree spanning all nodes such that the total link cost is minimum).

Procedure (ONN Simulator):

0. Set t, UTP, LTP, U_max, and U_min.
1. Initialize n, R_o, P, constants A, B, C, and D, and the costs of the links.
2. For (experiment = 1 to max_experimrents) Do

2.1 Construct an MCST.

2.2 Initiate U_{ij}(t) for i= 1,...,n-1 and j= i+1,...,n as follows:
  if link (i,j) is in MCST then U_{ij}(t) = random value > UTP.
  Otherwise U_{ij}(t) = random value > LTP.

2.3 Initiate V_{ij}(t) as follows: if U_{ij}(t) > 0 then V_{ij}(t) = 1, 0 otherwise.

2.4 While (t < max_t) and ($|\Delta E|$ > *epsilon*) loop

  2.4.1 Select a neuron (i,j) (a link) randomly.

  2.4.2 Use the motion equations (8) or (9) to compute $\Delta u_{ij}(t)$.

  2.4.3 Compute $U_{ij}(t+1)$ based on first Euler method (Eq. 3).

  $$U_{ij}(t+1) = U_{ij}(t) + \Delta U_{ij}(t).$$

  2.4.4 If $U_{ij}(t+1)$ > U_max then $U_{ij}(t+1)$ = U_max.

  2.4.5 If $U_{ij}(t+1)$ < U_min then $U_{ij}(t+1)$ = U_min.

  2.4.6 Evaluate Vij(t+1)

  $$V_{ij}(t+1) = 1, \text{ if } U_{ij}(t+1) > \text{UTP}$$
  $$= 0, \text{ if } U_{ij}(t+1) < \text{LTP}$$
  unchanged otherwise.

 {End while loop.}

2.5 If (solution found by the experiment is the best so far)
  best-solution = current-solution.

{End for loop.}

3. Return best-solution.

End (ONN Simulator).

The initialization of problem variables and energy parameters in step 1 are input by the user at the beginning of each run. Each problem for the ONN is run many times (= max_experiments), each with different random selection sequence of neurons. This is done in order to gauge the natural variability of the ONN. Furthermore, each experiment starts with a network which has the characteristics of being highly reliable. As stated in the algorithm, this step is done by constructing the MCST for the network.

5.2 Setting The Parameters A, B, C and D:

The choice of the neural network's parameters can affect its performance and the solution quality. This is so since the parameters of the energy function affect the shape of the energy landscape. Furthermore, input saturation affects the speed of the energy convergence to a solution [22], i.e. the performance of the neural network. As a result, the right choice of the upper and lower limits of the input activation values is also important.

One heuristic that helps in setting the parameter values is the weight of each term in the motion equation. It is clear that the A-term and the *C*-term are much less than 1, while the *B*-term is of $O(n*Cmax)$ where *Cmax* is the maximum link costs. To balance the terms so that they all have comparable weights in the energy, the parameters *A*, *B*, and *C* must be set properly [1]. In addition, the parameters *A* and *C* must not be equal so they do not neutralizes each other, since the *A*-term and the *C*-term are identical in the motion equation.

The results reported in [1] showed that the neural network approach was more promising than an enumerative-based methods, as well as GA, when applied to this design problem. The process of satisfying the design constraints, i.e., minimizing the cost and maximizing the reliability, is achieved efficiently. It is encouraging that the ONN simulator (run on a PC) could identify optimal solutions (or near-optimal solutions), even in search spaces up to 10^{16} or 2^{225}.

5.3 Some Numerical Results

We include here some numerical results that demonstrate the effectiveness of ONN approach, in particular, for large network sizes where no results have been reported by other methods [3,11]. As shown in the following table (Table 1), no results have been reported by method in [3] for number of links greater than 45, and, no results have been reported by method in [11] for number of links greater than 300.

In Table 1, N is the number of nodes, L is the number of links, p is the link probability, and R_0 is the target (threshold) reliability. The last three columns are the costs reported by the branch-and-bound method of [3], the Genetic Algorithm method of [11], and the cost obtained from the simulation of the ONN, respectively. A full set of comparison results can be found in [1].

Table 1. Simulation results

N	L	p	R_0	[3]	[11]	ONN
10	45	0.9	0.9	154	156	154
10	45	0.9	0.95	197	205	197
10	45	0.95	0.95	136	136	136
15	105	0.9	0.95		317	304
20	190	0.95	0.95		926	286
25	300	0.95	0.9		1606	402
30	450	0.95	0.95			604
40	780	0.9	0.9			1084
50	1225	0.95	0.9			2054

6 Conclusions

In this chapter we have presented a solution to the problem of topology optimization under reliability constraints using a Hopfield neural network. An energy function is formulated in terms of the links' probabilities of failures such that its minimization yields a minimum design cost while satisfying the reliability constraint. The neural network contains one neuron for each possible link. An output of 1 indicates the selection of the link. The efficiency of the neural network solution has been demonstrated via simulation. Simulation results show that the neural network solution is capable of obtaining a solution for large network sizes where other techniques could not be applied or requires an extensive CPU time.

References

1. AboElFotoh HMF, Al-Sumait LS (2001) A Neural Approach to Topological Optimization of Communication Networks with Reliability Constraints. IEEE Transactions on Reliability, **50:4**, 397-408
2. Boorstyn RR, Frank H (1977) Large-scale network Topological Optimization. IEEE Trans. Communication, **com 25:1**, 29-47
3. Chopra YC, Sohi BS, Tiwari, RK, Aggarwal KK (1984) Network Topology for Maximizing the Terminal Reliability in a computer Communication Network. Microelectronics and reliability, **24:5**, 911-913
4. Aggarwal KK, YC Chopra, Bajwa JS (1982) Topological Layout of links for Optimizing the s-t reliability in a computer Communication Network. Microelectronics and reliability, **22:3**, 341-345
5. Jan Rong-Hong (1993) Design of reliable networks. Comp Operation Research, **20:1**, 25-34
6. Jan Rong-Hong, Hwang Fung-Jen, Cheng Sheng_Tzong (1993) Topological Optimization of a Communication Network subject to a Reliability Constraint", IEEE Trans Reliability, **42:1**, 63-70

7. Aggarwal KK, Chopra YC, Bajwa JS (1982) Topological Layout of links for optimizing the overall reliability in a computer communication system. Microelectronics and reliability, **22:3**, 347-351
8. Colbourn CJ (1987) The Combimatorics of Network Reliability. Oxford University Press
9. Garey MR, Johnson DS (1979) Computers and Intractability: A Guide to the Theory of NP-Completeness. W. H. Freeman and Co., San Francisco
10. Deeter DL, Smith AE (1997) Heuristic Optimization of Network design Considering All-terminal Reliability. IEEE 1997 Proceedings annual Reliability and Maintainability Symposium, 194-199
11. Dengiz B, Altipamak F, Smith AE (1997) Local Search Genetic Algorithm for Optimal Design of Reliable Networks. IEEE Trans Evolutionary Computation., **1:3**, 179-188
12. Kumar A, Pathak RM, Gupta YP, Parsaei HR(1995) A Genetic algorithm for Distributed System Topology Design. Computers ind Eng, **28:3**, 659-670
13. Hopfield JJ, Tank DW (1985) Neural computation of decisions in optimization problems. Biol Cybern **52**, 141-152
14. Takefuji Y, Wang J (1996) Neural computing for Optimization and Combinatorics. World Scientific Publishing Co
15. Fanabiki N, Takefuji Y (1992) A Neural Network Parallel Algorithm for Channel Assignment Problems in Cellular Radio Networks. IEEE Trans Vehicular Technology, **41:4**, 430-437
16. Ali MKM, Kamoun F (1993) Neural networks for Shortest Path computation and Routing in Computer networks. IEEE Trans Neural Networks, **4:6**, 941-955
17 Funabiki N, Nishikawa S (1996) A binary neural network approach for link activation problems in multihop radio networks. IEOCI Trans Comm, **E79-B:8**, 1086-1093
18. Funabiki N, Nishikawa S (1997) A Binary Hopfield Neural Network Approach for Satellite Scheduling problems. IEEE Trans Neural Networks, **8:2**, 441-445
19. Takefuji Y, Lee KC (1991) An Artificial Hysteresis Binary Neuron: a Model Suppressing the Oscillatory Behaviors of Neural Dynamics. Biol Cybern, **64**, 353-356
20. Takefuji Y, Lee KC (1991) Artificial neural networks for four-coloring problems and k-colorability problems. IEEE Trans on Circuits and Systems, **38:3**, 326-333
21. Wang L, Ross J (1990) Synchronous neural networks of non-linear threshold elements with hysteresis. Proc. National Academy of Sciences (USA), **87**, 988 - 992
22. Wang Lipo (1997) Discrete-Time Convergence Theory and updating Rules for Neural Networks with Energy Function. IEEE Trans Neural Networks, **8:2**, 445-447

Binary Neural Network Approaches to Combinatorial Optimization Problems in Communication Networks

Nobuo Funabiki1

Okayama University, Okayama, 700-8530 Japan

Abstract. For the best use of communication networks, some mathematical problems that can be formulated as *NP-hard* combinatorial optimization problems should be solved efficiently by algorithms. Neural networks provide their elegant solutions as a highly-parallel, hardware-friendly approach. Particularly, binary neural networks that are composed of binary neurons with discrete internal states are much suitable for the current digital technology. This chapter first reviews the framework of binary neural networks for combinatorial optimization problems. Then, it scans through three studies on binary neural networks for typical combinatorial optimization problems in communication networks.

Keywords: Binary neural network, communication network, NP-hard, combinatorial optimization

1 Framework of Binary Neural Network

In a binary neural network to solve a combinatorial optimization problem, each neuron has an integer input U_{ij} and a binary output V_{ij}. Here, without losing generality, we discuss a two-dimensional neural network. For convenience, a neuron with $V_{ij} = 1$ is called *active*, and a neuron with $V_{ij} = 0$ is *non-active*. The output is computed from the input by following a nonlinear, nondecreasing function called the *neuron function*:

$$V_{ij} = f(U_{ij}). \tag{1}$$

The simplest form of this neuron function is the binary model by McCulloch and Pitts [1]:

$$\text{If } U_{ij} > 0 \text{ then } f(U_{ij}) = 1, \text{else } f(U_{ij}) = 0. \tag{2}$$

A binary neural network can be viewed as a nonlinear function to connect multiple inputs and multiple outputs. Then, the outputs are fed back into the inputs through a set of differential equations, which is called the *motion equation*. After initial states for neuron inputs and outputs are given from outside by certain ways, this closed loop between neuron inputs and outputs cause autonomic dynamical state transitions, until the state of the binary neural network reaches a stable one. The stable state may represent a solution of the problem, if it satisfies the constraints of the problem. Here, the binary output of a neuron directly corresponds to the binary variable of the combinatorial optimization problem to be solved, and the state of the binary neural network means the state of neuron outputs. We note that a binary neural network may sometimes converge to a non-solution state where some constraints are not satisfied, or keep the oscillation of states.

The *energy function* should be defined first to solve a combinatorial optimization problem by a binary neural network. The energy function is usually designed such that its output becomes minimum when the state of the binary neural network reaches a solution of the problem. The energy function $E(V)$ is usually given by a linear combination of the term to represent the constraints and the term for the objective function in the problem:

$$E(V) = Ap(V) + Bq(V) \tag{3}$$

where A and B are constant coefficients to balance the weights between two terms, and V is a vector of neuron outputs. The function $p(V)$ represents the constraints of the problem where $p(V) = 0$ must be satisfied in the solution. The function $q(V)$ represents the objective function of the problem where $q(V)$ should be minimized in the solution.

Then, the motion equation should be derived by taking the partial derivative of the energy function E with respect to the neuron output V_{ij}:

$$\begin{aligned} \Delta U_{ij} &= -\frac{\partial E}{\partial V_{ij}} \\ U_{ij} &= U_{ij} + \Delta U_{ij}. \end{aligned} \tag{4}$$

The motion equation determines the direction of state transitions, where it describes all the constraints and the objective function in the combinatorial optimization problem. The motion equation guides the state transitions of the binary neural network leading to the final stable state that should represent a solution. However, the motion equation often leads to a local minimum where some constraints may not be satisfied and/or the objective function may not be minimized, although all the constraints of the problem must be satisfied and the objective function should be minimized in the final state. Therefore, heuristic methods called *omega function* and *hill-climbing term* have been used in our binary neural network, to avoid the local minimum convergence as best as possible. The motion equation usually yields more negative force for neurons to be non-active as seen later. The omega function alleviates it by allowing the negative force only for active neurons at times. The hill-climbing term supplements the positive force for neurons to become active.

Besides, in our binary neural network, two additional schemes, namely *greedy initialization* and *gradual expansion*, have been used together to improve the performance, depending on target combinatorial optimization problems. In the greedy initialization scheme, the initial states of neurons are generated by a greedy heuristic algorithm, so that the search by the binary neural network can be started from more desirable states than completely randomized states. In the gradual expansion scheme, each neuron is weighted depending on their contribution to the final cost of the solution. Less-weighted neurons have more chance to be active than heavily-weighted neurons. As a result, the total cost of the achieved solution becomes smaller than that by the case without the gradual expansion scheme.

In the following sections, we introduce the problem formulations and the binary neural networks for three combinatorial optimization problems in communication networks.

2 Channel Assignment Problem in Cellular Communication Network

In this section, we introduce the binary neural network with the greedy initialization scheme for the *channel assignment problem (CAP)* in the cellular mobile communication network [3]. Over past years, traffic demands for mobile communications have been rapidly increased to receive voice and data services, due to portability and availability in everyplace with no requirement of hard wires. Besides, the successful introduction of packet communications using mobile networks has accelerated this explosive growth of traffic demands. On the other hand, the electromagnetic frequency spectrum allocated for this system has been limited, because of a variety of significant applications using radio waves. Thus, the efficient use of precious frequency band resources has played an important task in the research/development communities in mobile communication networks. The concept of the cellular network has been widely adopted as the efficiency realization of mobile communication networks [2]. This cellular network allows the reuse of the same frequency spectrum or channel in geographically separated regions simultaneously. As a result, CAP has become the critical problem to be solved for its efficient solutions. A solution for CAP is not only requested to avoid the mutual radio interference between close radio channels in proximal regions, but also to maximize the channel utilization.

2.1 Formulation of Channel Assignment Problem (CAP)

The formulation of CAP was defined by Gamst et al. [4]. The servicing region in a cellular network is managed as a set of disjoint hexagonal cells. Each cell occupies a unit area for providing communication services to users that are located in the cell area. When a user requests a call for communication services in the network, a channel must be assigned to this user through which voice or data packets are communicating between the user's mobile terminal and the base station in the cell. This channel assignment must satisfy the constraints to avoid the mutual radio interference between close channels used in proximal cells. In CAP, the following three types of constraints have been considered:

1) *Co-Channel Constraint (CCC)*: the same or its adjacent channels cannot be reused in the cells that are located within a specified distance from each other in the network. This set of channel-reuse prohibited cells is called a *cluster*. In each cluster, any pair of channels assigned to call requests from the cells must have a specified channel distance between them.
2) *Adjacent Channel Constraint (ACC)*: adjacent channels cannot be assigned to adjacent cells in the network simultaneously. In other words, any pair of channels assigned to adjacent cells must have a specified distance. The distance for ACC is usually larger than that for CCC.
3) *Co-Site Constraint (CSC)*: any pair of channels in the same cell must have a specified distance. The distance for CSC is usually larger than that for ACC.

The channel distance defined by constraints is described by the difference on the channel indices in the channel domain. In this paper, the cell cluster size for CCC is denoted by "Nc", the channel distance to satisfy CCC is by "cij", the distance for ACC is by "acc", and the distance for CSC is by "cii". The goal of CAP is to find a channel assignment to every call request with the minimum number of channels or *channel span* subject to the above three constraints.

The three constraints to avoid the channel interference in an n-cell network are altogether described by an $n \times n$ symmetric *compatibility matrix* C. A non-diagonal element c_{ij} $(i \neq j)$ in C represents the minimum distance to be separated between a channel in cell i and a channel in cell j. A diagonal element c_{ii} in C represents the minimum distance between any pair of channels in cell i. Thus, CCC is described by $c_{ij} = cij$, ACC is by $c_{ij} = acc$, and CSC is by $c_{ii} = cii$ respectively.

In addition, a set of call requests in the n-cell network is given by an n-element *demand vector* D. The i-th element d_i in D represents the number of required channels to satisfy the call requests in cell i. Let a binary variable x_{ik} represent whether channel k be assigned to cell i $(x_{ik} = 1)$ or not $(x_{ik} = 0)$ for $i = 1, ..., n$ and $k = 1, ..., m$. Note that m represents the channel span required for the instance. The lower bound on the channel span may be referred in [6][7]. Then, CAP is defined as follows:

$$
\begin{aligned}
&\text{minimize } m \quad \text{such that} \\
&x_{ik} = 0 \text{ or } 1, \quad \text{for } i \in \{1, ..., n\} \text{ and } k \in \{1, ..., m\} \\
&\sum_{k=1}^{m} x_{ik} = d_i, \quad \text{for } i \in \{1, ..., n\} \\
&|k - l| \geq c_{ij}, \quad \text{for } k, l \in \{1, ..., m\}, i, j \in \{1, ..., n\} \\
&\qquad\qquad \text{and } x_{ik} = x_{jl} = 1.
\end{aligned}
\tag{5}
$$

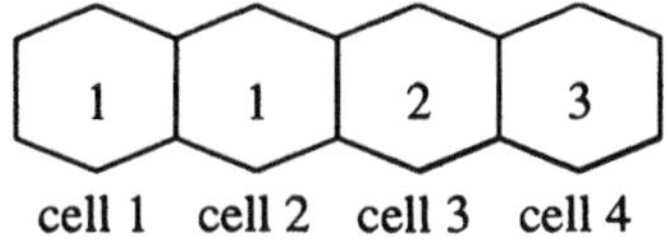

Fig. 1. A four-cell CAP instance.

Let us see a small CAP example of four cells in Figure 1. Each number in a cell represents the number of channels for call requests. This example assumes $Nc = 7$, $cij = 1$, $acc = 2$, and $cii = 5$ for constraints. The cell cluster $Nc = 7$ means that any pair of channels assigned to the seven cells in Figure 2 must be differed by cij channels. Then, the compatibility matrix C is given by:

$$C = \begin{bmatrix} 5\,2\,1\,0 \\ 2\,5\,2\,1 \\ 1\,2\,5\,2 \\ 0\,1\,2\,5 \end{bmatrix} \tag{6}$$

The lower bound on the channel span for this example is 11, because the three channels in cell 4 must occupy $11 (= 5 \times 2 + 1)$ channels to satisfy CSC. One optimum solution can be that $x_{11} = 1$, $x_{25} = 1$, $x_{33} = x_{38} = 1$, and $x_{41} = x_{46} = x_{411} = 1$.

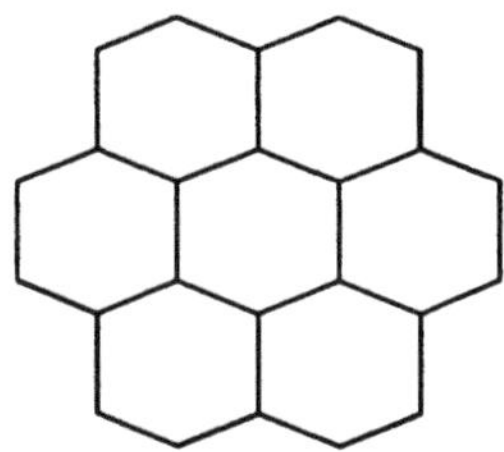

Fig. 2. A seven-cell cluster ($Nc = 7$).

2.2 Binary Neural Network for CAP

In our approach to CAP, the cells in a *greedy region* are assigned channels by the greedy method in [5] as the greedy initialization scheme, before the remaining cells are assigned channels by the binary neural network. The greedy region is initially composed of the cell with the largest degree and its surrounding cells. The degree deg_i for cell i in an n-cell network is given

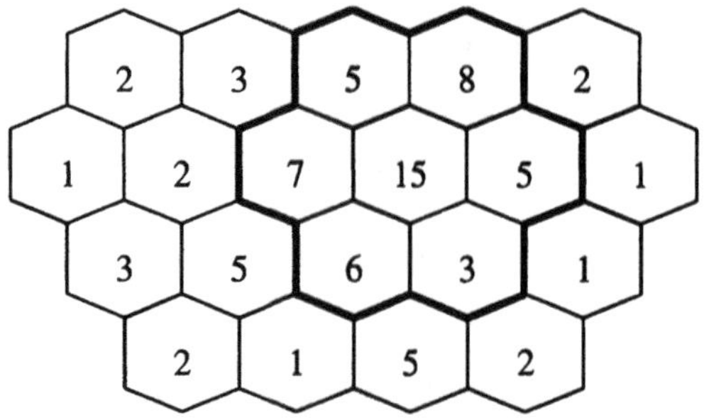

Fig. 3. A 20-cell CAP instance.

by:

$$deg_i = \left(\sum_{j=1}^{n} d_j c_{ij} \right) - c_{ii}. \tag{7}$$

For example, the cell requesting 15 channels in Figure 3 has the largest degree. Note that in this example, each number in a cell represents the number of requesting channels with $Nc = 7$, $cij = acc = 1$, and $cii = 5$. This cell and it surrounding six cells that are highlighted by heavy lines compose the initial greedy region. If the assignment to all the cells by the binary neural network is failed, the greedy region is expanded outside by adding the adjacent cells, and the cells in this expanded greedy region are assigned channels by the greedy method.

Then, the binary neural network composed of $n \times m$ neurons is used to solve the n-cell m-channel CAP. The one-output ($V_{ij} = 1$) of neuron ij represents the assignment of channel j to cell i, and the zero-output ($V_{ij} = 0$) represents no assignment there. Here, the hysteresis binary function is adopted as the neuron function:

$$V_{ij} = \begin{cases} 1 & \text{if } U_{ij} > UTP \\ 0 & \text{if } U_{ij} < LTP \\ \text{unchanged} & \text{otherwise} \end{cases} \tag{8}$$

where UTP and LTP are constants. It has been empirically shown that the hysteresis binary function can suppress the undesirable oscillation that has often been caused in the synchronous state update of binary neurons and improve the convergence property. The theoretical analysis on the convergence can be found in [9][10]. A CMOS design for the hysteresis neuron has been presented in [11].

Then, the energy function E is defined to represent the constraints of CAP:

$$E = \frac{A}{2} \left(\sum_{q=1}^{m} V_{iq} - d_i \right)^2$$

$$+\frac{B}{2}\sum_{i=1}^{n}\sum_{j=1}^{m}V_{ij}\left(\sum_{\substack{q=j-c_{ii}+1\\q\neq j,1\leq q\leq m}}^{j+c_{ii}-1}V_{iq}+\sum_{\substack{p=1\\p\neq i,c_{ip}>0}}^{n}\sum_{\substack{q=j-c_{ip}+1\\1\leq q\leq m}}^{j+c_{ip}-1}V_{pq}\right) \tag{9}$$

where A and B are constant coefficients. The A-term represents the constraint that d_i channels must be assigned to cell i. The B-term represents the constraint that any pair of assigned channels must not cause the interference given by the compatibility matrix.

The motion equation for CAP is derived from the energy function where two heuristic methods are used together for the global convergence:

$$\Delta U_{ij} = -A\left(\sum_{q=1}^{m}V_{iq}-d_i\right)+Ch\left(\sum_{q=1}^{m}V_{iq}-d_i\right)$$
$$-B\begin{cases}\left(\sum_{\substack{q=j-c_{ii}+1\\q\neq j,1\leq q\leq m}}^{j+c_{ii}-1}V_{iq}+\sum_{\substack{p=1\\p\neq i,c_{ip}>0}}^{n}\sum_{\substack{q=j-c_{ip}+1\\1\leq q\leq m}}^{j+c_{ip}-1}V_{pq}\right)V_{ij} & \text{if } t \bmod T<\omega\\ \left(\sum_{\substack{q=j-c_{ii}+1\\q\neq j,1\leq q\leq m}}^{j+c_{ii}-1}V_{iq}+\sum_{\substack{p=1\\p\neq i,c_{ip}>0}}^{n}\sum_{\substack{q=j-c_{ip}+1\\1\leq q\leq m}}^{j+c_{ip}-1}V_{pq}\right) & \text{otherwise}\end{cases} \tag{10}$$

where C is a constant coefficient, t is the number of iteration steps, T and ω are constant parameters, and the function $h(x)$ returns 1 if $x < 0$, 0 otherwise. The B-term describes the omega function, and the C-term does the hill-climbing term.

2.3 Simulation Results

The performance of the binary neural network for CAP is verified through solving benchmarks in [8]. Table 1 shows the channel spans m in solutions that were obtained by our algorithm and by existing algorithms. Our binary neural network found best solutions among them in any instance. Thus, the effectiveness of this binary neural network has been confirmed.

3 Broadcast Scheduling Problem in Packet Radio Network

In this section, we introduce the binary neural network with the gradual expansion scheme for the *broadcast scheduling problem (BSP)* in the packet radio network [17]. The packet radio network provides data communication services to geographically distributed stations or nodes through a shared radio channel [18]. Each node is equipped with a transmitter/receiver unit to either

Table 1. Simulation results for CAP benchmarks.

Instance No.	Constraint			Ours	Existing results					
	Nc	acc	cii	m	[8]	[12]	[13]	[14]	[15]	[16]
1	12	2	5	427	460	-	440	427	440	-
2	7	2	5	427	447	433	436	427	436	427
3	12	2	7	533	536	-	533	-	533	-
4	7	2	7	533	533	533	533	-	533	533
5	12	1	5	381	381	-	381	-	381	-
6	7	1	5	381	381	381	381	-	381	381
7	12	1	7	533	533	-	533	-	533	-
8	7	1	7	533	533	533	533	-	533	533
9	12	2	5	258	283	-	273	258	287	-
10	7	2	5	253	270	263	268	253	269	253
11	12	2	7	309	310	309	309	-	309	-
12	7	2	7	309	310	309	309	-	309	309
13	12	2	12	529	529	529	529	-	529	-

transmit or receive packets via a single channel at one time, and a control unit to perform the channel access control. A node can directly transmit packets only to its neighboring nodes, where a pair of nodes within the direct connectivity is called *adjacent*. When a node needs to send packets to distant nodes, their intermediate nodes perform as repeaters to relay packets.

A *time-division multiple-access (TDMA)* protocol is adopted for packet transmissions to avoid the interference on the single shared radio channel. The whole system is controlled by a single clock, and time is divided into *time-slot (slots)*. At each slot, any node may either transmit or receive one packet, and may receive at most one packet from its adjacent nodes. If a node is scheduled to perform both the transmitter and receiver at the same slot, a *primary conflict* occurs there. If two or more packets reach one node at the same slot, they become annulled by a *secondary conflict*. A *TDMA cycle* is a sequence of slots where every node is scheduled to broadcast packets at least once without conflicts.

Since Baker and Ephremides formulated a packet transmission scheduling problem in a packet radio network, it has been widely studied with a couple of variations, namely the *link scheduling problem* and the *broadcast scheduling problem* [20]. A link is composed of a pair of adjacent nodes to directly transmit packets through a radio channel. The link scheduling problem aims to find the shortest schedule of activating every requested link without conflicts. The broadcast scheduling problem requires finding the shortest TDMA cycle for every node to broadcast a packet to all of its adjacent nodes without conflicts. This section deals with the broadcast scheduling problem.

3.1 Formulation of Broadcast Scheduling Problem (BSP)

The topology of an n-node packet radio network is described by an $n \times n$ connectivity matrix C, where the ij-th element c_{ij} represents the adjacency between nodes i and j. The broadcast scheduling problem requests to find a TDMA cycle for each node to broadcast packets by satisfying the two constraints:

1) every node transmits a packet at least once, and
2) neither the primary conflict nor the secondary conflict occurs in any transmission from any node,

and optimizing the two objective functions:

1) the cycle length or the number of slots is minimized, and
2) the total number of transmissions or slots assigned to nodes is maximized.

The binary neural network seeks to optimize the first objective function for the delay minimization. Then, it seeks the optimization of the second one.

A compatibility matrix D is derived from the connectivity matrix C to describe the primary and secondary conflicts at once. The primary conflict occurs when adjacent nodes or one-hop away nodes are activated at the same slot. The secondary conflict occurs when a pair of two-hop away nodes, which is not within direct transmission but has an intermediate node adjacent to both of them, is scheduled at the same slot, because this intermediate node needs to receive two packets simultaneously.

Figure 4 shows a BSP example of a network with six nodes and five links. The corresponding connectivity and compatibility matrices are descried in (11).

$$C = \begin{bmatrix} 0\,1\,0\,1\,0\,0 \\ 1\,0\,1\,0\,0\,1 \\ 0\,1\,0\,0\,0\,0 \\ 1\,0\,0\,0\,1\,0 \\ 0\,0\,0\,1\,0\,0 \\ 0\,1\,0\,0\,0\,0 \end{bmatrix}, D = \begin{bmatrix} 0\,1\,1\,1\,1\,1 \\ 1\,0\,1\,1\,0\,1 \\ 1\,1\,0\,0\,0\,1 \\ 1\,1\,0\,0\,1\,0 \\ 1\,0\,0\,1\,0\,0 \\ 1\,1\,1\,0\,0\,0 \end{bmatrix} \quad (11)$$

In this example, four nodes of 1, 2, 3, and 6 must be scheduled at different slots to avoid the secondary conflict, because they are within two-hop away from each other. Thus, at least four slots are necessary for the TDMA cycle. An optimum solution can be that node 1 is scheduled at slot 1, nodes 2 and 5 are at slot 2, nodes 3 and 5 are at slot 3, and nodes 4 and 6 are at slot 4.

3.2 Binary Neural Network for BSP

The binary neural network is composed of $n \times m$ neurons for finding an m-slot TDMA cycle in an n-node network. The one output ($V_{ij} = 1$) of neuron ij represents the transmission schedule of node i at slot j, and the zero output

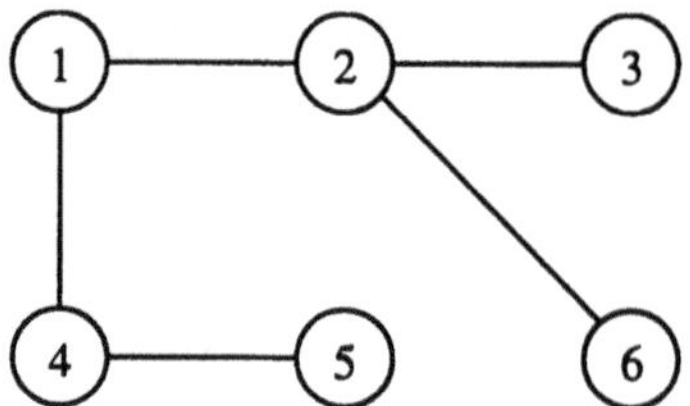

Fig. 4. A packet radio network with six nodes and five links.

($V_{ij} = 0$) does no schedule there. Here, the hysteresis binary function is used as the neuron model.

The binary neural network approach to BSP is composed of two phases to optimize the two objective functions in BSP one by one. In the first phase, the binary neural network seeks a TDMA cycle to satisfy two constraints with the minimum number of time-slots. The energy function E_1 is given by:

$$E_1 = \frac{A_1}{2}\sum_{i=1}^{n}\left(\sum_{k=1}^{m} V_{ik} - 1\right)^2 + \frac{B_1}{2}\sum_{i=1}^{n}\sum_{j=1}^{m}\sum_{k=1}^{n} d_{ik}V_{ij}V_{kj} \tag{12}$$

where A_1 and B_1 are coefficients, and d_{ij} is the ij-th element in D. The A_1-term represents the constraint that each of n nodes must be scheduled once in m slots. The B_1-term represents the constraint that any pair of one-hop or two-hop away nodes must not be scheduled at the same slot. When both constraints are satisfied, E_1 becomes zero.

As the gradual expansion scheme, the number of slots m in a TDMA cycle is gradually increased from an initial value during the iterative computation of the binary neural network, until both constraints are satisfied. The initial value of m is given by the number of vertices in a maximum clique of the graph whose adjacency matrix is given by D. This clique is approximated here because of the NP-hardness of the maximum clique problem. The clique is initially composed of a node with the maximum degree and its adjacent nodes in C, where any pair of them are within two hops. Then, it is expanded by adding the nodes that are adjacent to every node in the currently selected clique. After a maximal clique is extracted, every node in this clique is assigned a different slot from each other to shorten the computation time in the binary neural network.

In the second phase, the number of transmissions is maximized by including as many conflict-free transmissions as possible into the TDMA cycle while the assigned slots in the first phase are fixed. The energy function for this phase is given by:

$$E_2 = \frac{A_2}{2}\sum_{i=1}^{n}\sum_{j=1}^{m}(1 - V_{ij})^2 + \frac{B_2}{2}\sum_{i=1}^{n}\sum_{j=1}^{m}\sum_{k=1}^{n} d_{ik}V_{ij}V_{kj} \quad (13)$$

where A_2 and B_2 are coefficients. The A_2-term represents the objective function of maximizing the total number of neurons with one outputs. The B_2-term is the same as the B_1-term in (12).

3.3 Simulation Results

For the performance evaluation of the binary neural network for BSP, three small benchmark instances in [21], and 600 geometric graph instances with 20 different sizes between 100 nodes and 1000 nodes are solved as in Table 2. In a geometric graph, the locations of n vertices are randomly generated on a unit square, and any pair of two vertices whose distance is less than the threshold r is connected by an edge. In addition to our binary neural network, a greedy algorithm in [19], and a simple graph coloring algorithm are simulated for comparisons. The solution quality of each algorithm is evaluated by two indices. The first and main index is the average delay time η to broadcast packets, which represents the average availability of a packet radio network:

$$\eta = \frac{m}{n}\sum_{i=1}^{n}\left(\frac{1}{\sum_{j=1}^{m} V_{ij}}\right). \quad (14)$$

The second and auxiliary index is the TDMA cycle length m, which represents the least availability of the network.

In our simulations, one run is executed for the binary neural network, 10 runs for the greedy algorithm in [19] (*Greedy*1), and 100 runs for the simple greedy graph coloring algorithm (*Greedy*2) in each instance to normalize the computation speed. Among plural solutions, the best one is selected. Table 2 shows the results. Because *Greedy*1 cannot minimize the TDMA cycle length, it exhibits the worst performance. In any instance, the binary neural network finds better solutions than other algorithms. Thus, the binary neural network can better contribute to the optimal protocol design for packet radio networks.

Table 2. Simulation results for BSP instances.

Instance	Node	Threshold	Ours		*Greedy*1		*Greedy*2	
	n	r	η	m	η	m	η	m
[21]-1	15		7.1	8	10.1	15	7.2	8
[21]-2	30		9.5	10	19.6	30	10.2	11
[21]-3	40		6.2	8	23.2	40	6.2	8
1	100		6.3	8.1	70.3	100	6.7	8.1
2	300		6.7	9.6	210.2	300	7.8	9.7
3	500	$1/\sqrt{n}$	6.8	10.0	350.9	500	8.1	10.1
4	750		6.9	10.2	526.0	750	8.3	10.3
5	1000		7.0	10.6	700.0	1000	8.6	10.7
6	100		17.4	20.2	82.3	100	18.6	20.6
7	300		19.7	23.3	240.4	300	22.2	24.5
8	500	$2/\sqrt{n}$	20.3	24.2	401.3	500	22.9	25.2
9	750		20.9	25.4	602.8	750	24.0	26.4
10	1000		21.2	25.5	805.1	1000	24.3	26.8
11	100		34.1	37.4	89.9	100	36.4	38.7
12	300		39.1	43.2	259.3	300	43.7	46.3
13	500	$3/\sqrt{n}$	41.0	45.2	429.0	500	47.3	50.2
14	750		42.9	47.3	639.4	750	49.4	52.5
15	1000		43.7	48.1	854.8	1000	50.7	53.9
16	100		54.5	57.9	92.2	100	56.3	59.1
17	300		63.6	68.4	270.3	300	69.4	72.5
18	500	$4/\sqrt{n}$	67.4	72.2	444.6	500	74.6	77.7
19	750		70.3	75.2	665.1	750	83.7	87.6
20	1000		72.9	78.2	880.5	1000	86.3	90.2

4 Route Assignment Problem in Multihop Radio Network

In this section, we introduce the binary neural network with the greedy initialization scheme for the *route assignment problem (RAP)* in the multihop radio network [22][23]. In a multihop radio network, packets are transmitted from source nodes to destination nodes by activating the links on a route connecting them. Each node can either send a packet to, or receive a packet from, at most one of its adjacent nodes simultaneously [24]. In order to minimize the number of transmission time slots for given requests described by a set of SD-pairs (Source and Destination), the *route assignment problem* must be solved to assign a transmission route for each SD-pair, and then, the *link activation problem* must be solved to find a link activation schedule for each assigned route [25]. Both problems are known to be *NP*-hard. This section deals with the binary neural network for the route assignment problem.

4.1 Formulation of Route Assignment Problem (RAP)

The topology of a multihop radio network is described by a connected graph $G = (V, E)$. Each vertex in V represents a communication station or node and each edge in E does a bi-directional link between two nodes. When a link exists between two nodes, one packet can be directly transmitted from one node to the other or vice versa in a unit time called a time slot. Every node in the network is synchronously operated by a single clock. The set of SD-pairs is given in this problem by the form of $\{(s_1, d_1), ..., (s_m, d_m)\}$, where s_i is the source node and d_i is the destination node for the i-th SD-pair, and m is the number of SD-pairs.

The route assignment problem (RAP) requires finding a route for every SD-pair such that the resulting transmission schedule has the minimum number of time slots, while it satisfies the constraint that each node can either send a packet to, or receive a packet from, at most one of its adjacent nodes in the same time slot. Thus, the following cost function $Cost$ should be minimized in RAP:

$$\begin{aligned} Cost &= \max\left(Cost_{hop}, Cost_{node}\right) \\ Cost_{hop} &= \max_{1 \le i \le m} \left(\sum_{j=1}^{S_i} hop\left(P_{ij}\right) V_{ij} \right) \\ Cost_{node} &= \max_{1 \le k \le n} \left(\sum_{i=1}^{m} \sum_{j=1}^{S_i} w_{ijk} V_{ij} \right) \end{aligned} \tag{15}$$

where n is the number of nodes in the network, S_i is the number of route candidates and P_{ij} is the j-th route candidate for SD-pair i respectively, $hop(P_{ij})$ is the number of hops on P_{ij}, V_{ij} is the binary variable in the problem such that $V_{ij} = 1$ represents the selection of the j-th candidate for SD-pair i and $V_{ij} = 0$ represents no selection. $Cost_{hop}$ represents the maximum number of hops among SD-pairs. $Cost_{node}$ represents the maximum number of slots to process every packet transmission going through the same node. This cost function determines the lower bound on the total number of slots for the link activation problem. w_{ijk} is the number of transmitting or receiving packets at the k-th node on route P_{ij}:

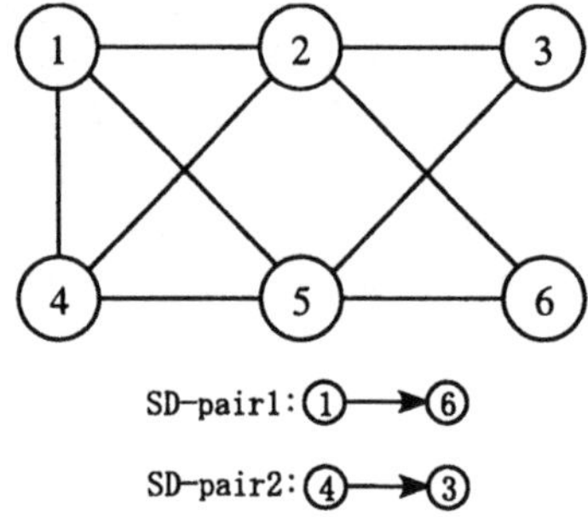

Fig. 5. A RAP example with a 6-node radio network.

$$w_{ijk} = \begin{cases} 2 & \text{if node } k \text{ is intermediate on route } P_{ij} \\ 1 & \text{if node } k \text{ is source or destination of SD-pair } i \\ 0 & \text{otherwise.} \end{cases} \quad (16)$$

In our approach to RAP, plural route candidates are first extracted for each SD-pair by the *k-shortest route extraction method (k-SRE)*. Then, one route is selected for each SD-pair by the binary neural network.

Figure 5 illustrates a RAP example with a 6-node radio network and two SD-pairs. SD-pair 1 has two shortest route candidates, $P_{11} = \{1,2,6\}$ and $P_{12} = \{1,5,6\}$, with length 2 ($hop(P_{11}) = hop(P_{12}) = 2$). SD-pair 2 has two shortest candidates, $P_{21} = \{4,2,3\}$ and $P_{22} = \{4,5,3\}$. For P_{11}, $w_{111} = w_{116} = 1$ (source, destination), $w_{112} = 2$ (intermediate), and $w_{113} = w_{114} = w_{115} = 0$ are given. Among four combinations of route candidate selections, the selection of P_{11} and P_{22}, or P_{12} and P_{21} provides the minimum value of $Cost = 2$ in this example.

4.2 *k*-Shortest Route Extraction Method

The k-shortest route extraction method (k-SRE) extracts plural route candidates first, which is based on the k-shortest path algorithm in [26]. k-SRE extracts only promising route candidates for each SD-pair so as to produce good final solutions, and to reduce the computation time and space in the route selection problem at the same time. Actually, it extracts only routes whose number of hops or length is less than or equal to the largest one, MH, among the shortest route for all the SD-pairs:

$$MH = \max_{1 \leq i \leq m} \left(\min_{1 \leq j \leq S_i} hop(P_{ij}) \right) \quad (17)$$

MH is the lower bound on $Cost_{hop}$ in (15), because otherwise, some SD-pair cannot have any route connecting the source and the destination. Thus, k-

SRE achieves the minimization of $Cost_{hop}$. As a result, only the minimization of $Cost_{node}$ becomes essential in the route selection.

Besides, the number of extracted route candidates K_i is limited by the minimum number of hops for each SD-pair:

$$K_i = \min_{1 \leq j \leq S_i} (hop(P_{ij})) \tag{18}$$

The reason for this limitation is that many redundant route candidates can be extracted when the source node is not far from the destination node.

4.3 Binary Neural Network for RAP

The binary neural network is presented for RAP to minimize $Cost_{node}$. Because one and only one route candidate must be selected for each SD-pair, the maximum neuron function [27] is adopted to always select one route candidate. The binary output V_{ij} of neuron ij in the maximum neuron function is given by:

$$\begin{aligned} &\text{If } U_{ij} = \max(U_{i1}, ..., U_{iS_i}), \\ &\text{then } V_{ij} = 1 \text{ else } V_{ij} = 0. \end{aligned} \tag{19}$$

The greedy initialization scheme for RAP is very simple. It initializes the neuron input by the number of hops for the corresponding route candidate with the minus sign:

$$U_{ij} = -hop(P_{ij}). \tag{20}$$

This simple scheme improves the final solution quality by preferring the selection of shorter route candidates.

The energy function E consists of the term (A-term) to minimize the cost function $Cost_{node}$ in (15) and the term (B-term) to minimize the congestions caused by selecting routes passing through the same nodes:

$$\begin{aligned} E = A\sum_{i=1}^{m}\sum_{j=1}^{S_i}\sum_{q=1}^{n} f\Bigg(&\sum_{k=1,k\neq i}^{m}\sum_{p=1}^{S_k} w_{kpq}V_{kp} \\ &-MH + w_{ijq}\Bigg) V_{ij} \\ &-B\sum_{i=1}^{m}\sum_{j=1}^{S_i}\sum_{p=1,p\neq i}^{m}\sum_{q=1}^{S_p} r_{ijpq}V_{ij}V_{pq} \end{aligned} \tag{21}$$

where A and B are coefficients, and r_{ijpq} is the number of nodes that are shared by two route candidates P_{ij} and P_{pq}. The function $f(x)$ returns x if $x > 0$, and 0 otherwise. The A-term is designed to avoid the selection of route candidates if the total number of transmitting or receiving packets at some node is larger than MH.

4.4 Simulation Results

In order to verify the performance of the binary neural network for RAP, 14 RAP instances in Table 3 are randomly generated, where the network topology follows the geometric graphs on a unit square with the threshold r, and SD-pairs are randomly selected so as not to be overlapped to each other.

Table 3. Simulation results for RAP instances.

Instance	Node n	SD-pair m	Threshold r	Candidates		Max. hops		*Cost*	
				Ours	[24]	Ours	[24]	Ours	[24]
1	50	10	0.2	26	134	5	14	5	5
2	50	20	0.2	66	317	5	14	5	5
3	100	20	0.1	125	417	11	11	11	12
4	100	20	0.2	77	696	5	18	5	6
5	100	20	0.3	135	787	4	14	4	5
6	100	40	0.1	234	806	12	24	12	NA
7	100	40	0.2	212	51	6	18	6	7
8	100	40	0.3	333	1602	5	15	5	7
9	200	40	0.2	253	2135	6	20	6	7
10	200	80	0.2	496	4176	7	20	7	NA
11	300	60	0.2	429	3957	7	24	7	NA
12	300	120	0.2	890	7901	7	24	7	NA
13	400	80	0.2	611	5869	7	30	7	NA
14	500	100	0.2	797	7997	7	28	7	NA

First, the quality of route candidates extracted by the k-shortest route extraction method is evaluated through the comparison with those by the algorithm in [24]. In Table 3, " Candidates " compares the number of route candidates extracted by each algorithm, and "Max. hops" does the maximum number of hops among the route candidates. This result indicates that our method only extracts promising route candidates with smaller numbers of hops, which is critical to obtain high-quality route assignments.

Then, the final solution quality is compared between two algorithms. In Table 3, *Cost* represents the value in (15) of the solution in each instance. " NA " indicates that the corresponding result could not be obtained because it required either too long computation time or too large space. This result supports the superiority of our binary neural network to the existing algorithm for the route assignment problem in the multihop radio network.

5 Conclusions

In this chapter, we briefly review the framework of the binary neural network approach to a combinatorial optimization problem, and show three specific approaches to problems in communication networks. The results in this chapter suggest the great possibility of this approach to further explore the new areas in this field. In addition to these results, a number of works has been done in this field. Some of them can be referred in [28]-[32].

References

1. McCulloch WS, Pitts WH (1943) A logical calculus of ideas immanent in nervous activity. Bulletin Math. Biophysics. **5**, 115
2. MacDonald AH (1979) Advanced mobile phone service: the cellular concept. Bell Syst. Tech. J. **53**, 15-41
3. Funabiki N, Okutani N, Nishikawa S (2000) A three-stage heuristic combined neural-network algorithm for channel assignment in cellular mobile systems. IEEE Trans. Veh. Technol. **49**, 397-403
4. Gamst A, Rave W (1982) On frequency assignment in mobile automatic telephone systems. Proc. GLOBECOM. 309-315
5. Box F (1978) A heuristic technique for assigning frequencies to mobile radio nets. IEEE Trans. Veh. Technol. **VT-27**, 54-64
6. Gamst A (1986) Some lower bounds for a class of frequency assignment problems. IEEE Trans. Veh. Technol. **VT-35**, 8-14
7. Funabiki N, Nakanishi T, Yokohira T, Tajima S, Higashino T (2002) A quasi-solution state evolution algorithm for channel assignment problems in cellular networks. IEICE Trans. Fundamentals. **E85-A**, 977-987
8. Sivarajan KN, McEliece RJ, Letchum JW (1989) Channel assignment in cellular radio. Proc. 39th IEEE Veh. Technol. Conf. 846-850
9. Wang L, Ross J (1990) Synchronous neural networks of nonlinear threshold elements with hysteresis. Proc. Natl. Acad. Sci. USA. **87**, 988-992
10. Wang L (1997) Discrete-time convergence theory and updating rules for neural networks with energy functions. IEEE Trans. Neural Networks. **8**, 445-447
11. Kurokawa T, Lee KC, Cho YB, Takefuji Y (1990) CMOS layout design of the hysteresis McCulloch-Pitts neurons. Electronics Letters. **26**, 2093
12. Wang W, Rushforth CK (1996) An adaptive local-search algorithm for the channel assignment problem (CAP). IEEE Trans. Veh. Technol. **45**, 459-466
13. Sung CW, Wong WS (1997) Sequential packing algorithm for channel assignment under cochannel and adjacent-channel interference constraint. IEEE Trans. Veh. Technol. **46**, 676-686
14. Hurley S, Smith DH, Thiel SU (1997) FASoft: a system for discrete channel frequency assignment. Radio Science. **32**, 1921-1939
15. Rouskas AN, Kazantzakis MG, Anagnostou ME (1999) Minimization of frequency assignment span in cellular networks. IEEE Trans. Veh. Technol. **48**, 873-882
16. Beckmann D, Killat U (1999) A new strategy for the application of genetic algorithms to the channel-assignment problem. IEEE Trans. Veh. Technol. **48**, 1261-1269
17. Funabiki N, Kitamichi J (1999) A gradual neural network algorithm for broadcast scheduling problems in packet radio networks. IEICE Trans. Fundamentals. **E82-A**, 815-824

18. Kahn RE, Gronemeyer SA, Burchfiel J, Kunzelman RC (1978) Advances in packet radio technology. Proc. IEEE. **66**, 1468-1496
19. Ephremides A, Truong TV (1990) Scheduling broadcasts in multihop packet radio networks. IEEE Trans. Commun. **38**, 456-460
20. Baker DJ, Ephremides A (1981) The architectural organization of a mobile radio network via a distributed algorithm. IEEE Trans. Commun. **29**, 1694-1701
21. Wang G, Ansari N (1997) Optimal broadcast scheduling in packet radio networks using mean field annealing. IEEE J. Select. Areas Commun. **15**, 250-260
22. Baba T, Funabiki N, Nishikawa S (1998) A proposal of a greedy neural network for route assignments in multihop radio networks. IEICE Trans. D-I. **J81-D-I**, 700-707 (in Japanese)
23. Baba T, Funabiki N, Nishikawa S (1997) A greedy neural network algorithm for route selection problems in multihop radio networks. Proc. NOLTA ' 97. 245-248
24. Wieselthier JE, Barnhart CM, Ephremides A (1994) A neural network approach to routing without interference in multihop radio networks. IEEE Trans. Commun. **42**, 166-177
25. Barnhart CM, Wieselthier JE, Ephremides A (1994) A neural network approach to solving the link activation problem in multihop radio networks. IEEE Trans. Commun. **43**, 1277-1283
26. Maruyama H (1993) A new kth shortest path algorithm. IEICE Trans. Inf. Syst. **E76-D**, 388-389
27. Takefuji Y (1992) Neural network parallel computing. Kluwer Academic Publishers, 1992
28. Yuhas B, Ansari N (1993) Neural networks in Telecommunications. Kluwer Academic Publishers, 1993
29. Funabiki N, Kitamichi J, Nishikawa S (1997) A gradual neural-network approach for time slot assignment in TDM multicast switching systems. IEICE Trans. Commun. **E80-B**, 939-947
30. Funabiki N, Nishikawa S (1997) A binary Hopfield neural-network approach for satellite broadcast scheduling problems. IEEE Trans. Neural Network. **8**, 441-445
31. Funabiki N, Nishikawa S (1997) A gradual neural-network approach for frequency assignment in satellite communication systems. IEEE Trans. Neural Network. **8**, 1359-1370
32. Funabiki N, Kitamichi J (1998) A gradual neural-network algorithm for jointly time-slot/code assignment problems in packet radio networks. IEEE Trans. Neural Network. **9**, 1523-1528

Robust Multiuser Detection Using Artificial Neural Networks

B. S. Sharif [1,2], T. C. Chuah [1,3], O. R. Hinton [1]

[1] Department of Electrical, Electronic and Computer Engineering, University of Newcastle upon Tyne, Newcastle upon Tyne, NE1 7RU, United Kingdom.

Present addresses: [2] Etisalat College of Engineering, Emirates Telecommunications Corporation, Sharjah, U.A.E.; [3] Faculty of Engineering, Multimedia University, Jalan Multimedia, 63100 Cyberjaya, Selangor, Malaysia.

Abstract

This chapter surveys the applications of artificial neural networks (ANNs) for multiuser detection in direct-sequence code-division multiple-access systems. We aim to stimulate more research on neural network techniques for robust multiuser detection in non-Gaussian channels. Here, we present the *M*-estimation technique for robust decorrelating detection and provide a summary of existing algorithms. We reveal the powerful features of ANNs by focusing on a special recurrent neural network structure for implementing the robust decorrelating detector and highlight its computational saving. Extension of this technique to multicarrier CDMA systems with turbo decoding in Rayleigh fading, impulsive noise channels is then investigated.

Keywords: Artificial neural networks, non-Gaussian noise, CDMA, *M*-estimation, robust detection, turbo-codes.

1 Introduction

The idea of mathematically emulating adaptive learning of biological nervous systems in artificial systems, which was introduced more than four decades ago, has seen tremendous growth in the past few years. These information-processing paradigms are usually grouped under the term artificial neural networks (ANNs), connectionist architectures, or parallel distributed processing. The theory of ANNs is at present so vast and the applications are so numerous in virtually all disciplines that no treatment of the subject can be complete. For recent reviews of the current state of the art, the reader is referred to [1, 2].

In this chapter, we survey the application of ANNs for the design of multiuser detectors in direct-sequence code-division multiple-access (DS-CDMA) Gaussian and non-Gaussian channels. The key issue in designing neural detectors involves identifying the appropriate architectures and algorithms that fit nicely to the underlying models. This issue demands detailed attention, profound understanding and innovative thinking. Our primary aim is to stimulate further research by illustrating the promising aspects of ANNs for multiuser detection in non-Gaussian channels. This problem is at present receiving very little attention by researchers, and its importance has not been widely recognized by the more general neural network research community.

Recently, immense interest has focused on multicarrier transmission as an efficient technique in combating distortions due to multipath propagation in mobile radio communications and for achieving high data rates [3]. In particular, the combination of DS-CDMA and multicarrier transmission techniques, termed multicarrier CDMA (MC-CDMA) [4, 5], is considered a promising alternative to conventional DS-CDMA due to its high spectral efficiency [6] and robustness to channel frequency selectivity [7].

In MC-CDMA the different chips of each information symbol are transmitted in parallel over different narrowband subcarriers using orthogonal frequency division multiplexing (OFDM) modulation. As a result, MC-CDMA is also sometimes called OFDM-CDMA. One reason for the increased popularity of MC-CDMA is the ease of implementation of the modulation and demodulation using an inverse discrete Fourier transform (IDFT) and DFT, respectively. The applications of MC/OFDM-CDMA modulation have been studied for different mobile radio environments [5, 8].

Several detection techniques for MC-CDMA signals have been proposed in the literature. Similar to the detection schemes developed for DS-CDMA [9], these techniques can be classified as simple single-user and improved multiuser detection techniques [10]. In the simple detection techniques, a simple correlator receiver is used. However, these techniques perform poorly in multipath fading channels as a result of the loss of orthogonality, which creates multiuser interference (MUI) and limits the number of active users. Improved detection techniques mitigate the MUI and provide considerably better performance at the expense of a higher receiver complexity. Various iterative detection/decoding strategies have also been proposed recently to improve the MC-CDMA system performances over multipath fading channels, e.g., [11-13].

Thus far, most of the DS/MC-CDMA detection schemes are designed for additive white Gaussian noise (AWGN) channels for reasons of analytical convenience. However, it is well known that many practical noise sources encountered in practice are decidedly non-Gaussian. In this chapter, we concentrate on one particular recurrent neural network (RNN) structure, which has recently been exploited for implementing a robust DS-CDMA detector [14]. By illustrating model equivalence between DS-CDMA and MC-CDMA, we extend the investigation of the RNN technique with joint turbo decoding [15] for MC-CDMA systems in Rayleigh fading impulsive noise channels.

2 Overview of Neural Network Techniques for Multiuser Detection

Aazhang et al. [16] first suggested the use of multilayer perceptrons (MLPs) for multiuser detection. Even though MLPs are capable of drawing arbitrarily complex decision boundaries, the classical back-propagation algorithm trains the detector to classify the symbol bits using the minimum mean squared error (MMSE) criterion, which is clearly distinct from that of minimizing the bit error rate (BER). Though the MLP-based detectors were shown to outperform the matched filter receiver and had a performance that is comparable to that of the optimal detector [9], they suffers from prohibitively long and indecisive training time, large computational complexity, network size uncertainty, and local minima problems. This work was later extended by Mitra and Poor [17], who provided convergence results and performance comparisons of a single-layer perceptron and an adaptive least mean square (LMS) receiver. A related work also appeared in [18].

Mitra and Poor [19] also considered a radial basis function (RBF) approach to multiuser detection. The RBF receiver, however, requires the number of neurons to increase exponentially with the number of users. Kechriotis and Manolakos [20] proposed the optimal detector based on Hopfield neural networks (HNN) [21]. The HNN presents a powerful generalization of the multistage detector [22] and can be implemented using analogue VLSI circuits. A lower-complexity hybrid detector based on the HNN structure was studied in [23]. Other neural detectors can also be found in [24-27].

All the neural net receivers described above are intended for Gaussian noise channels, and would perform poorly in impulsive noise environments. In contrast, the development of robust neural net receivers that can perform satisfactorily in non-Gaussian channels has only been studied by a relatively small number of researchers. For example, robust adaptive receivers based on RBF network were proposed in [28]. Instead of using ANNs as a signal detection unit, one of their interesting applications is to function as a denoising device. For example, Lippmann and Beckman [29] employed a MLP as a pre-processor to assist a matched filter in reducing impulsive noise. This approach has been extended to multiuser DS-CDMA channels in [30].

3 MC-CDMA Channel Model

In this chapter, we consider the uplink transmission model of a synchronous MC-CDMA system shared by K users as shown in Fig. 1. The binary data of the kth user is fed into a channel decoder with code rate R. The encoded bits are binary phase-shift keying (BPSK) mapped yielding data symbols each of duration of T_b. Each coded symbol during the qth interval, $b_k(q)$, is replicated into N parallel copies. Each branch of the parallel stream is multiplied by one chip of a spreading code of length N $\mathbf{s}_k = [s_1^k, \cdots, s_N^k]^T$, where $s_j^k \in \{\pm 1/\sqrt{N}\}$ and then modulated by a subcarrier. Each subcarrier has a unique center frequency $f_c+\Delta f_n$, where the spacing, Δf_n, between the subcarriers is chosen as:

$$\Delta f_j = \frac{j-1}{T_d}, \quad j = 1,\cdots,N \tag{3.1}$$

The choice of Δf_n, given in (3.1) ensures that the overlapping subcarriers are orthogonal which helps in deriving a convenient mathematical model and realizing modulation and demodulation by means of the IDFT and DFT, respectively.

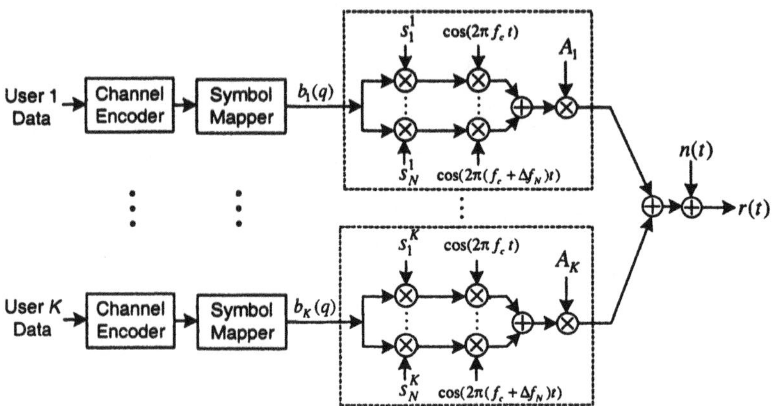

Fig. 1. Continuous-time multiuser MC-CDMA transmission model

The low-pass equivalent continuous-time waveform transmitted by the kth user during the qth symbol is given as [10]:

$$u_k(t) = \frac{1}{\sqrt{N}} \sum_{j=1}^{N} b_k(q) s_j^k \cos(2\pi\Delta f_j)\, p(t - qT_d) \tag{3.2}$$

where $p_{T_d}(t)$ is the unit pulse and satisfies the following:

$$p_{T_d}(t) = \begin{cases} 1 & \text{for } 0 \le t \le t_d \\ 0 & \text{otherwise} \end{cases} \tag{3.3}$$

In this chapter, we consider Rayleigh flat fading channels in which the spectral characteristics of the received signal are preserved at the receiver. However, due to fluctuations in the gain of the channel caused by multipath, the strength of the received signal changes with time according to a Rayleigh distribution. We also assume that the Doppler spread introduces slow fading such that the channel gain remains constant within one symbol interval. The time invariant transfer function for the jth subcarrier of user k is given by

$$g_j^k = a_j^k \exp(i\phi_j^k) \tag{3.4}$$

where a_j^k is a set of mutually independent Rayleigh random variables and the phase shift ϕ_j^k is uniformly distributed on [0, 2π]. Since K users are transmitting simultaneously over an additive channel, the received signal during the qth symbol interval can be written as

$$r_q(t) = \frac{1}{\sqrt{N}} \sum_{k=1}^{K} \sum_{j=1}^{N} A_k b_k(q) s_j^k g_j^k \operatorname{Re}\{\exp(2\pi \Delta f_j)\} p(t - qT_d) + n_q(t) \qquad (3.5)$$

Where A_k denotes the signal amplitude for the kth user, and $n_i(t)$ is the zero mean stationary random process describing the channel noise.

3.1 Discrete-Time MC-CDMA Channel Model

Processing the received signal in the digital domain requires an appropriate discrete-time model. This is accomplished by sampling the continuous-time received signal at the chip rate $1/T_c = N/T_d$. Substituting (3.1) into (3.5) and setting $t=(l-1)T_d/N$, the discrete-time representation of (3.5) becomes

$$\tilde{r}_l(q) = \frac{1}{\sqrt{N}} \sum_{k=1}^{K} A_k b_k(q) \left[\operatorname{Re}\left\{ \sum_{j=1}^{N} s_j^k g_j^k \exp\left(\frac{i2\pi (j-1)(l-1)}{N} \right) \right\} \right] + \tilde{n}_l(q) \qquad (3.6)$$

where $l = 1, \cdots, N$ is the chip sampling index. Recalling that the IDFT x_l of a discrete sequence X_n is related by the following expression [10]:

$$x_l = \frac{1}{\sqrt{N}} \sum_{n=1}^{N} X_n \exp\left\{ i \frac{2\pi}{N} (n-1)(l-1) \right\} \qquad (3.7)$$

Since the inverse fast Fourier transform (IFFT) can be used to efficiently compute the IDFT, (3.7) can be written as

$$x_l = \operatorname{IFFT}\{X_n\} \qquad (3.8)$$

Similarly, the bank of correlators required for (3.6) can be synthesized by the IFFT operation as follows

$$\tilde{r}_l(q) = \operatorname{Re}\left\{ \sum_{k=1}^{K} \operatorname{IFFT}\{A_k b_k(q) s_j^k a_j^k\} \right\} + \tilde{n}_l(q) \qquad (3.9)$$

where the phase shift in (3.4) can be neglected as we assume coherent reception.

For clarity, we drop the symbol index q and define the faded spreading chip as $\tilde{s}_j^k = s_j^k a_j^k$ and place the K Rayleigh faded spreading codes $\tilde{s}_k = [\tilde{s}_1^k, \cdots, \tilde{s}_N^k]^T \in \Re^N$ into the following channel matrix:

$$\boldsymbol{H} = [\tilde{s}_1, \cdots, \tilde{s}_K] \in \Re^{N \times K} \qquad (3.10)$$

Therefore, (3.9) can be expressed using the following matrix notation:

$$\tilde{\boldsymbol{r}} = \mathrm{Re}\left\{\sum_{k=1}^{K} \mathrm{IFFT}\{\boldsymbol{HAb}\}\right\} + \tilde{\boldsymbol{n}}$$

$$= \mathrm{Re}\left\{\sum_{k=1}^{K} \mathrm{IFFT}\{\boldsymbol{Hx}\}\right\} + \tilde{\boldsymbol{n}} \qquad (3.11)$$

where

$$\tilde{\boldsymbol{r}} = [\tilde{r}_1, \cdots, \tilde{r}_N]^T \in \Re^N$$

$$\boldsymbol{A} = \mathrm{diag}(A_1, \cdots, A_K) \in \Re^{K\times K}$$

$$\boldsymbol{b} = [b_1, \cdots, b_K]^T \in \{\pm 1\}^K$$

$$\boldsymbol{x} = [A_1 b_1, \cdots, A_K b_K]^T = [x_1, \cdots, x_K]^T \in \Re^K$$

$$\tilde{\boldsymbol{n}} = [\tilde{n}_1, \cdots, \tilde{n}_N]^T \in \Re^N$$

At the receiver, we perform an N-point FFT operation on the channel output $\tilde{\boldsymbol{r}}$, the demodulated signal can therefore be written as

$$\boldsymbol{r} = \mathrm{Re}\{\mathrm{FFT}\{\mathrm{IFFT}\{\boldsymbol{Hx}\}\}\} + \mathrm{FFT}\{\tilde{\boldsymbol{n}}\}$$

$$= \boldsymbol{Hx} + \boldsymbol{n} \qquad (3.12)$$

A close examination of (3.12) reveals that the MC-CDMA channel output $\boldsymbol{r}$ has the same form as the DS-CDMA channel output from a chip-matched filter [9], except that the channel matrix $\boldsymbol{H}$ is now defined in the frequency domain. A block diagram of this MC-CDMA channel model is shown in Fig. 2. Assuming perfect channel estimation, and let the cross-correlation matrix $\boldsymbol{R} = \boldsymbol{H}^T\boldsymbol{H} \in \Re^{K\times K}$. The matched filter output (matched to $\boldsymbol{H}$) after FFT demodulation can be written as

$$\boldsymbol{y} = \boldsymbol{H}^T\boldsymbol{r} = \boldsymbol{Rx} + \boldsymbol{H}^T\boldsymbol{n} \qquad (3.13)$$

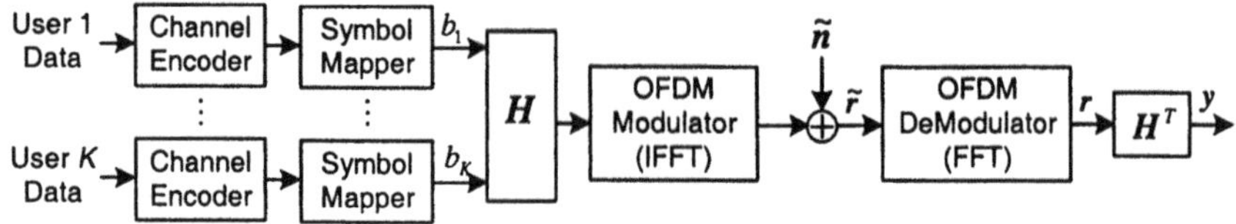

Fig. 2 Discrete-time multiuser MC-CDMA channel model (adapted from [10, pp. 35])

3.2 Impulsive Noise Model

To accommodate the presence of impulsive noise, we model the components of the noise vector $\tilde{n}$ in (3.11) using symmetric α-stable (SαS) distributions. α-stable distributions form a class of algebraic-tailed models which shares several desirable properties with the Gaussian model, such as the *stability property* and *generalized* form of the *central limit theorem*. The class of SαS distributions is best described by the Fourier transform of its probability density function, also called *characteristic function*, as described below [31]

$$\varphi(t) = \exp\left(-\gamma|t|^{\alpha}\right) \tag{3.14}$$

where α $0 < \alpha \leq 2$ is called the *characteristic exponent* and γ $\gamma \geq 0$ is the scale parameter called *dispersion*. The characteristic exponent determines the shape of the distribution. Smaller values of α correspond to heavier tailed distributions and hence more impulsive behavior, and as α increases, the tails become lighter and hence it is less likely to observe outliers. As α approaches 2, the non-Gaussian stable distribution approaches the Gaussian distribution in a continuous fashion. The dispersion is a measure of spread of the samples from the distribution around its mean.

Despite their first introduction into the mainstream of signal processing by Shao and Nikias [31] in 1993, α-stable distributions did not gain widespread acceptance by researchers over the years. There are mainly two reasons for this. First, stable distributions do not have analytic expressions for their densities except for the Gaussian (α=2) and Cauchy (α=1) distributions. This precludes the application of maximum likelihood (ML) estimation and Bayesian inferential techniques, which require an explicit expression for the underlying distributions. However, the advances in computing power has fostered the development and efficient implementation of numerical integrations for power series expansions that had been thought implausible just decades ago. Further, recent results in [32] have also developed the first numerically stable analytical representation for α-stable distributions based on finite mixtures of Gaussians with high accuracy.

The second reason that has caused considerable skepticism and restricted interest in α-stable distributions has been due to the fact that the *p*th moments of a SαS random variable exists only for $p<\alpha$. In particular, the variance (i.e., the second-order moment) of a stable distribution with $\alpha<2$ does not exist, making the use of variance as a measure of noise power meaningless. Therefore, it is widely conceived that noise models with infinite variance are inappropriate in any signal processing context. Although variance has been accepted in the research community as a standard measure associated with the definition of power, its meaning is not universal and could lead to misleading results when the processes exhibit heavy tails [33].

The absence of a finite variance does not, however, preclude other appropriate measures of strength for stable random variables. The variance is one measure of spread of a Gaussian process; the dispersion in a stable process is another, which is more appropriate and meaningful when dealing with outliers. In this chapter, we use the *geometric SNR* (GSNR) [33], which is, to our knowledge, the first approach towards a mathematically and conceptually valid characterization of the relative strength between information-bearing signal and channel noise with infinite variance. The *geometric power* of a SαS random variable is given by [33]

$$S_0 = \frac{(C_g \sigma)^{1/\alpha}}{C_g} \tag{3.15}$$

where $C_g \approx 1.78$ is the exponential of the Euler constant. Let A be the amplitude of a modulated signal in an additive noise channel with geometric noise power S_0. Then the geometric signal-to-noise ratio (GSNR) is defined as [33]

$$\text{GSNR} = \frac{1}{2C_g}\left(\frac{A}{S_0}\right)^2 \tag{3.16}$$

The normalizing constant $2C_g$ ensures that for the Gaussian case (α=2), the definition of the GSNR coincides with that of the standard SNR. Since the limiting case with α=2 corresponds to the Gaussian distribution with variance 2γ [31].

Finally, it is interesting to point out that the use of the GSNR framework based on the SαS noise model also overcomes some of the fundamental limitations of the highly popular Gaussian mixture model, which still adopts variance as a measure of noise power [30].

4 Robust Decorrelating Detection Using the *M*-Estimator

It is well known that the matched filter output $\boldsymbol{y}$ given in (3.13) contains MUI if the off-diagonal elements of $\boldsymbol{R}$ are nonzero. Assuming that $\boldsymbol{R}$ has full rank and is therefore invertible, the linear decorrelating detector (LDD) [34] eliminates MUI by applying the linear transformation $\boldsymbol{R}^{-1}$ to $\boldsymbol{y}$ as follows:

$$\boldsymbol{y}_{\text{LDD}} = \boldsymbol{R}^{-1}\boldsymbol{y} \tag{4.1}$$

Alternatively, another way treating decorrelating detection is to cast the problem into a linear regression model. In particular, it can be shown that the LDD is in fact a least-squares (LS) estimator for $\boldsymbol{x}$.

To proceed, consider the linear regression model from the output of the FFT demodulator as given in (3.12)

$$\boldsymbol{r} = \boldsymbol{H}\boldsymbol{x} + \boldsymbol{n} \tag{4.2}$$

where $\boldsymbol{r}$ is the vector of observed responses, $\boldsymbol{H}$ is a matrix of regressors of full column rank, $\boldsymbol{x}$ is the vector of unknown coefficients of interest, and $\boldsymbol{n}$ is a vector of random errors caused by noise. The estimation of $\boldsymbol{x}$ can be accomplished by minimizing the following energy function

$$E(\boldsymbol{x}) = \sum_{j=1}^{N} \rho(\hat{n}_j) \tag{4.3}$$

where

$$\hat{n}_j = r_j - \sum_{k=1}^{K} \tilde{s}_j^k x_k, \quad j = 1,\ldots,N \tag{4.4}$$

are the residuals, and $\rho(\cdot)$ represents suitably chosen penalty functions (usually symmetric and convex) according to an optimality criterion [35]. Therefore, the estimator of $\boldsymbol{x}$ can be generically expressed as:

$$\hat{\boldsymbol{x}} = \arg \min_{\boldsymbol{x} \in \Re^K} \sum_{j=1}^{N} \rho\left(r_j - \sum_{k=1}^{K} \tilde{s}_j^k x_k \right) \tag{4.5}$$

To obtain (4.5) using a gradient approach, we differentiate (4.3) with respect to x_k

$$\frac{\partial E(\boldsymbol{x})}{\partial x_k} = -\sum_{j=1}^{N} \psi\left(r_j - \sum_{p=1}^{K} \tilde{s}_j^p x_k \right) \tilde{s}_j^k, \quad k = 1,\cdots,K \tag{4.6}$$

where $\psi(\cdot) = \rho'(\cdot)$. The gradient of the energy function in express in vector form as

$$\nabla E(\boldsymbol{x}) = \left[\frac{\partial E(\boldsymbol{x})}{\partial x_1}, \cdots, \frac{\partial E(\boldsymbol{x})}{\partial x_K} \right]^T = -\boldsymbol{H}^T \psi(\boldsymbol{r} - \boldsymbol{H}\boldsymbol{x}) \tag{4.7}$$

Apparently, the solution to (4.5) is obtained by solving the following equation

$$\boldsymbol{H}^T \psi(\boldsymbol{r} - \boldsymbol{H}\boldsymbol{x}) = \boldsymbol{0}_K \tag{4.8}$$

where $\boldsymbol{0}_K$ denotes a K-dimensional zero vector. Any estimator defined by a minimum problem of the form given by (4.5) or (4.8) is called ML type estimator or *M-estimator* [35], since the choice

$$\rho_{\mathrm{ML}}(\hat{n}_j) = -\log f(\hat{n}_j) \tag{4.9}$$

$$\psi_{\mathrm{ML}}(\hat{n}_j) = -\frac{f'(\hat{n}_j)}{f(\hat{n}_j)} \tag{4.10}$$

gives the ordinary ML estimator, where $f(\cdot)$ denotes the density of n_j's.

The M-estimator family includes the classical method of LS, which derives much of its popularity from the existence of the Gauss-Markov theorem and the central limit theorem (CLT). The Gauss-Markov theorem states that if the noise samples n_j's follow a distribution with finite variance, then the LS estimator has the minimum variance of all linear unbiased estimators of $\boldsymbol{H}$. The implication of the CLT is obvious if we recall that n_j's usually represent the perturbation on the r_j's of the sum of a very large number of independent variables, each of which is itself insignificant to include in the regression. If these variables follow a distribution with finite variances and if they arise additively, the n_j's will be Gaussian distributed and LS is the ML estimator in this case [36].

The LS estimator uses the following penalty function and derivative

$$\rho_{\mathrm{LS}}(\hat{n}_j) = (\hat{n}_j)^2 / 2 \tag{4.11}$$

$$\psi_{\mathrm{LS}}(\hat{n}_j) = \hat{n}_j \tag{4.12}$$

Substituting $\psi_{\mathrm{LS}}(\cdot)$ into (4.8), the solution to the LS estimate $\hat{\boldsymbol{x}}_{\mathrm{LS}}$ is given by

$$\begin{aligned} \hat{\boldsymbol{x}}_{\mathrm{LS}} &= \arg \min_{\boldsymbol{x} \in \Re^K} \frac{1}{2} \|\boldsymbol{r} - \boldsymbol{H}\boldsymbol{x}\|_2^2 \\ &= \boldsymbol{R}^{-1} \boldsymbol{H}^T \boldsymbol{r} \\ &= \boldsymbol{R}^{-1} \boldsymbol{y} \end{aligned} \tag{4.13}$$

which yields exactly the output of the LDD as given in (4.1).

Despite the mathematical elegance leading to closed form solutions, the LS method is not appropriate for all environments. In particular, if the n_j's are characterized by an infinite variance distribution, such as those described by α-stable families, the assumptions of the CLT fail to hold and the minimum variance property of LS is no longer applicable. Thus in infinite variance distributions, both the efficiency and optimality of LS are lost. Indeed, it has been widely acknowledged in statistics literature that the performance of LS can be seriously compromised under non-Gaussian errors, especially when a small portion of extreme observations characterized by heavy-tailed densities is present in the data [35]. The lack of robustness of the LS against outliers is mainly due to the penalty function in (4.11), which admits all residuals regardless of their amplitudes to enter quadratically. Thus occurrence of outliers can have a pronounced impact on the estimate.

One approach to reducing the impact of outliers is to replace the LS penalty function with some other criterion that increases less rapidly than the LS. For example, Huber first suggested the following minimax function [35]:

$$\rho_{\mathrm{H}}(\hat{n}_j) = \begin{cases} (\hat{n}_j)^2 / 2, & \text{for } |\hat{n}_j| \le h \\ h|\hat{n}_j| - h^2 / 2, & \text{for } |\hat{n}_j| > h \end{cases} \tag{4.14}$$

with derivative

$$\psi_{\mathrm{H}}(\hat{n}_j) = \begin{cases} \hat{n}_j, & \text{for } |\hat{n}_j| \le h \\ h\,\mathrm{sign}(\hat{n}_j), & \text{for } |\hat{n}_j| > h \end{cases} \tag{4.15}$$

where the cut-off parameter $h>0$ is used to control the robustness and efficiency of the estimator. Apparently, it is seen that $\rho_{\mathrm{H}}(\cdot)$ applies the LS method to residuals smaller that h and penalizes outliers that are greater than h with least absolute value regression.

4.1 Some Known Iterative Algorithms for Robust Regression

With the exception of some elementary forms of penalty functions like the LS, the M-estimator formulated in (4.8) defines a system of nonlinear equations requiring an appropriate iterative procedure. The three most often used methods being [37]:

- **Newton's method**

$$\hat{x}^{n+1} = \hat{x}^n + [H^T < \psi'(r - H\hat{x}^n) > H]^{-1} H^T \psi(r - H\hat{x}^n) \tag{4.16}$$

- **Iteratively reweighted least squares (IRLS)**

$$\begin{aligned} \hat{x}^{n+1} = \hat{x}^n + [H^T < w(r - H\hat{x}^n) > H]^{-1} H^T \\ \times < w(r - H\hat{x}^n) > (r - H\hat{x}^n) \end{aligned} \tag{4.17}$$

- **Huber's method**

$$\hat{x}^{n+1} = \hat{x}^n + [H^T H]^{-1} H^T \psi(r - H\hat{x}^n) \tag{4.18}$$

where $< \hat{n} >$ denotes an N-by-N diagonal matrix with $<>_{ii} = \hat{n}_i$ for any $\hat{n} = [\hat{n}_1, \cdots, \hat{n}_N]^T$, and w is a weight function such that $w(\hat{n}) = \psi(\hat{n}) / \hat{n}$.

The choice between these three methods is not obvious since each of them has its advantages and drawbacks. Among the three, Newton's method has the fastest (quadratic) convergence rate [38]. However, to ensure global convergence, Newton's method must be used with a line search, and for a number of penalty functions, there is also a possibility that the Hessian matrix $H^T < \psi'(r - Hx^n) > H$ may be singular or indefinite even though H has full rank. Besides, it is difficult to implement because it requires that ψ' be calculated at each iteration. Several variants of Newton-like methods have also been proposed and discussed in [37, 39, 40].

The IRLS method, generally attributed to Beaton and Tukey [41], avoids the calculation of ψ' by replacing an approximation $w(r - Hx^n)$ at the expense of a slower, linear convergence rate. In order to use the IRLS algorithm, $\rho(\cdot)$ and $\psi(\cdot)$ must be continuous [42, 43]. Since almost all proposed penalty functions satisfy $\psi(\hat{n}) / \hat{n} > 0$, and as long as H has full rank, the matrix of IRLS algorithm is always positive definite. Two requirements for both Newton and IRLS methods are a good initials estimate, or guess, of the estimator to begin the iteration and a stopping rule to ensure convergence is achieved [37, 42]. More discussions on the IRLS approach can be found in [38, 42-44].

The Huber's method [45] possesses strong global convergence properties despite the fact that it is the slowest of the three methods [38]. In contrast to the other two methods where the updating matrix must be inverted at each iteration, Huber's method enjoys computational convenience since the generalized inverse

$[H^T H]^{-1} H^T$ needs only be computed once. To ensure convergence, Ekblom [46] introduced a step gain μ to all three methods. For example, the modified Huber (MH) method becomes

$$\hat{x}^{n+1} = \hat{x}^n + \mu[H^T H]^{-1} H^T \psi(r - H\hat{x}^n) \tag{4.19}$$

with μ chosen such that the residuals are successively reduced in each iteration step. Despite its slow convergence, the MH method is promising and is worthy of further investigation.

The MH algorithm in (4.19) has been used to implement the robust MDD proposed in [47], where the LS solution is taken as its initial estimate

$$\hat{x}^0 = \mu(H^T H)H^T r \tag{4.20}$$

and the final symbol estimate for the transmitted data is achieved by taking the sign of the converged vector $\hat{x}^n$.

4.2 Recurrent Neural Network Solution to Robust Decorrelating Detection

In this section, we consider the recurrent neural network (RNN) structure originally proposed in [48] for solving systems of linear equations. The RNN has been used in [14] in the context of DS-CDMA for implementing the MDD with reduced complexity without compromising performance.

Using the gradient approach to minimize the energy function $E(x)$, the estimation problem can be mapped directly to an initial value problem described by a set of differential equations expressed in matrix form as [48]:

$$\frac{dx}{dt} = -\eta \nabla E(x) \tag{4.21}$$

where $\eta > 0$ is the gain factor. Substituting (4.7) into (4.21), we get

$$\frac{dx}{dt} = \eta H^T \psi(r - Hx), \quad x(0) \in \Re^K \tag{4.22}$$

The basic estimation process involves iteratively computing the trajectory of $x(t)$ starting at the initial point $x(0)$, and converging to the final estimate $\hat{x}$ as $t \to \infty$. Provided that the penalty function $\rho(\cdot)$ is convex and bounded from below, the following proof of stability condition ensures that (4.22) always has a stable asymptotic solution [48]:

$$\begin{aligned} \frac{dE(x)}{dt} &= \sum_{k=1}^{K} \frac{\partial E(x)}{\partial x_k} \frac{dx_k}{dt} \\ &= (\nabla E(x))^T \frac{dx}{dt} \end{aligned}$$

$$= -(\nabla E(\boldsymbol{x}))^T \eta \nabla E(\boldsymbol{x}) \leq 0 \tag{4.23}$$

In order to realize the RNN digitally, the key step involves converting the differential equations (4.22) into the relevant difference equations through the standard Euler integration method [48]. The resulting recursion realizing the difference equations can be written as

$$\boldsymbol{x}^{n+1} = \boldsymbol{x}^n + \mu \boldsymbol{H}^T \psi(\boldsymbol{r} - \boldsymbol{H}\boldsymbol{x}^n) \tag{4.24}$$

Typically $\boldsymbol{x}^0 = \boldsymbol{0}_K$, and $0<\mu<\mu_{max}$ is the learning gain, where $\mu_{max}=2/\|\boldsymbol{S}\boldsymbol{S}^T\|$ to ensure stability of the algorithm [49]. The discrete-time RNN for implementing the above difference equations is depicted in Fig. 3, where the coefficients of the matrix $\boldsymbol{H}$ and μ form the synaptic weights. The structure is similar to the RNN presented in [14] for DS-CDMA, the only difference being that the spreading codes are attenuated due to fading.

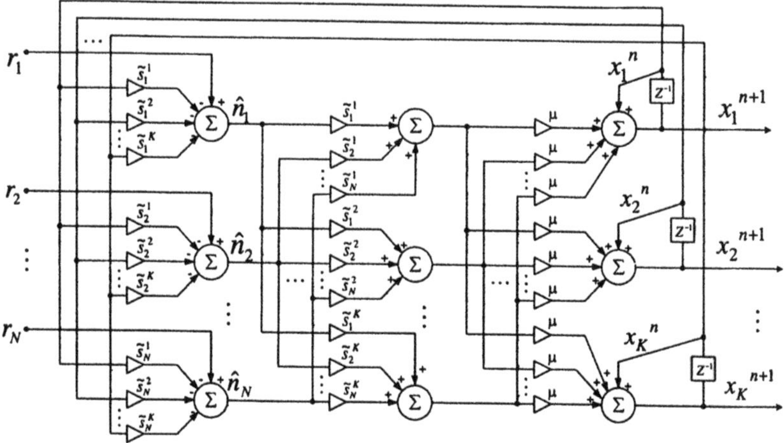

Fig. 3 General structure of the discrete-time RNN for decorrelating detection. The nonlinear neurons in the first layer for realizing the MDD are not shown explicitly

Of particular interest here is the first layer, also called the *sensor layer* because it senses the actual variables x_k and computes the residual defined in (4.4). The structure can be used to implement both the MDD and the iterative version of the LDD. The only distinguishing feature between the two lies in the sensor layer. The iterative LDD is obtained by utilizing linear neurons in this layer [cf. (4.12)]. On the other hand, the MDD can be realized efficiently by incorporating, for example, the nonlinear clipper function $\psi_H(\cdot)$ in (4.15) directly into the neurons of the sensor layer. The nonlinear neurons play an important role of compressing outliers from exceeding the prescribed cut-off parameter h, thereby achieving robustness.

4.3 Numerical Saving of the RNN Approach

It is noted that the RNN solution in (4.24) yields the MH method given in (4.19) except with the matrix inverse $R^{-1} = [H^T H]^{-1}$ removed. Simulation results in [14] have shown that the RNN provides identical performance to the MH approach, regardless of the initial estimates and convergence rates. A rigorous proof of the redundancy of the matrix inverse is not included here for brevity, and will be published elsewhere.

While the removal of this matrix inverse may not attract much interest in the statistical community since the matrix H always remains the same in most statistical problems, the same is not likely to be seen in the case of a wireless multiuser communications network. For example, consider the commercial exploitation of the MDD to a realistic DS/MC-CDMA network. Although the matrix inverse needs only to be computed once for a given number of users and spreading codes, a change in one of these parameters will result in different spreading code combination and a recalculation for the matrix inverse becomes necessary. For DS/MC-CDMA networks accommodating a very large number of users with high user entry/exit rate, the matrix inversion process will be computationally very costly owing to the very large number of possible user combinations and/or permutations resulting in many different cross-correlation matrices. This problem becomes more complicated in fading channels, since the fading parameters also need to be taken into consideration.

The saving in computation of the RNN over the MH method depends on the particular algorithm used to perform the matrix operations and also on the relative sizes of K and N. For example, to implement the MDD using the MH method for a fully loaded system (K=N), the best known algorithm of Coppersmith and Winograd [50] requires $O(N^{2.376})$ to carry out each of the following three operations: the matrix multiplication between H^T and H; the matrix inverse $[H^T H]^{-1}$; and the matrix multiplication between $[H^T H]^{-1}$ and H^T. Therefore in this case, the RNN-based MDD provides a saving of $O(3N^{2.376})$ for a given H over the MH-based MDD.

5 Joint M -Decorrelating Detection and Turbo Decoding for MC-CDMA in Non-Gaussian Rayleigh Fading Channels

Having discussed the robust MDD in the previous section, we now consider joint MDD and turbo decoding for combating MUI, impulsive noise and Rayleigh fading effects. Before presenting the transceiver structure, an introductory treatment of turbo codes is given below.

5.1 Principle of Turbo Coding

Turbo codes came to the attention of researchers with the publication of a seminal paper in 1993 [51]. They are by far the most important breakthrough in coding research, capable of offering reliable communications at a SNR that is within a few tenths of a dB of the Shannon limit. This work was later formalised in [15].

Despite the suddenness with which they burst into the coding community, the innovation of turbo codes involved reviving some previously known concepts and algorithms, namely parallel concatenated encoding and iterative decoding, and combining them using some ingenious ideas.

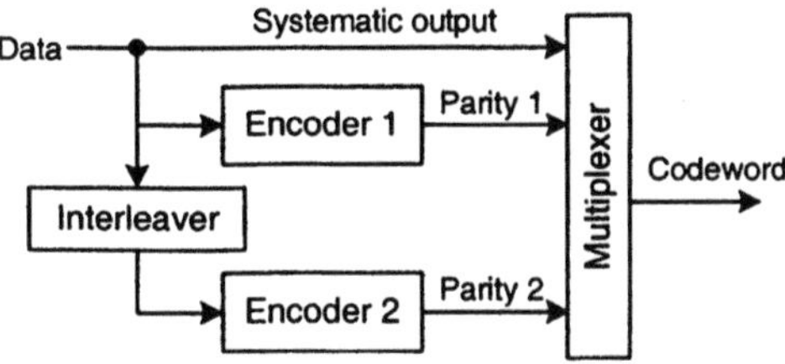

Fig. 4 Block diagram of turbo-encoder

Parallel-concatenated encoders (PCEs) consist of two or more component encoders for block or convolutional codes. In its simplest form, the PCE works as follows. An information block is encoded using the first encoder, generating a parity codeword. Next the original information block is interleaved and fed as the input for the second encoder, producing a second parity codeword. Then the information, the first and second parity codewords are multiplexed and transmitted over the channel, as shown in Fig. 4. The component codes used in the original turbo codes were from a class of convolutional codes known as *recursive systematic convolutional* (RSC) codes; however, encoders for block codes with an efficient iterative decoder may also be used.

The iterative decoder has a soft-input soft-output (SISO) component decoder for each of the constituent codes. These SISO decoders take turns operating on the received data, producing and exchanging soft-outputs in the form of (*a posteriori*) *log-likelihood ratios* (LLRs). The *a posteriori* LLR is the ratio of the probability that a decoded bit is 1 to the probability that it is 0, given the received sequence. The most popular algorithm for doing this is a symbol-by-symbol maximum *a posteriori* (MAP) algorithm [52].

The LLR output from each decoder contains information from several separate sources. Some come from the received codeword itself and is known as the *intrinsic information*. Information is also extracted from the other code and is called *extrinsic information*. It is this information that should be exchanged between decoders, since the intrinsic information is already available to the next decoder, and to pass it on would only dilute the extrinsic information.

Hence the iterative decoder has the structure shown in Fig. 5, in which the intrinsic information is separated from the extrinsic, so that the output of each decoder contains only extrinsic information to be fed back to the next decoder for use as a prior information during the next iteration. The component decoders are connected by interleaver and de-interleaver, denoted by π and π^{-1}, respectively.

The decoding continues in an iterative fashion until the desired performance is attained. It is this feedback action that earns the term "turbo codes", since the decoding process is reminiscent of a turbo engine, in which part of the power at the output is fed back to the input to boost the performance of the whole system. Thus it is the decoding method that gives turbo codes their name, not the codes themselves. The final estimate of the message is found by de-interleaving and hard-limiting the output of the second decoder.

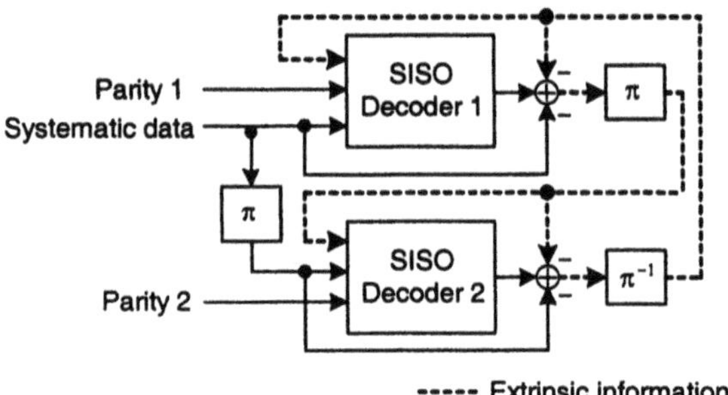

Fig. 5 Block diagram of turbo-decoder

Turbo coding is currently receiving considerable attention for various applications. Review articles on the subject can be found in [53, 54], and specialized monographs are beginning to appear [55, 56].

5.2 The Turbo Transceiver for MC-CDMA

At the transmitter, the binary information data for the kth user, d_k, is encoded by two identical RSC codes separated by a pseudorandom interleaver of length L. The systematic bit stream is retained while the parity bit streams are punctured alternatively to yield rate-1/2 encoding, and the code bits are BPSK symbol mapped to yield $\{b_k^s, b_k^p\}$, where b_k^s and b_k^p denote the systematic and parity bit, respectively, for the kth user. Each coded symbol is then replicated and spread for multicarrier modulation.

The block diagram of the MC-CDMA receiver for joint MDD and turbo decoding is shown in Fig. 6. After downconversion and FFT demodulation, the observation vector $\boldsymbol{r}$ is fed into the RNN for removing MUI and impulsive noise according to (3.24). Next, instead of taking hard decisions immediately on the converged output $\hat{\boldsymbol{x}}$ of the RNN, we can further ameliorate the effects of fading and strive to achieve better performance by feeding each component of the soft output $\hat{\boldsymbol{x}}$ though a demultiplexer into one of the K independent SISO iterative decoders.

Each decoder takes three inputs from the demultiplexer: 1) the systematically encoded channel output $\hat{x}_k^s$ for the respective user; 2) the parity bits $\hat{x}_k^{p1}$ or $\hat{x}_k^{p2}$ transmitted from the associated component encoder; and 3) the *a priori* information provided by the other component decoder. Note that zero parity symbols are reinserted in the parity streams of $\hat{x}_k^{p1}$ and $\hat{x}_k^{p2}$ to replace those bits that were punctured prior to transmission. The component decoders are separated by the interleaver and de-interleaver, denoted by π and π^{-1}, respectively, which facilitate the exchange of extrinsic information.

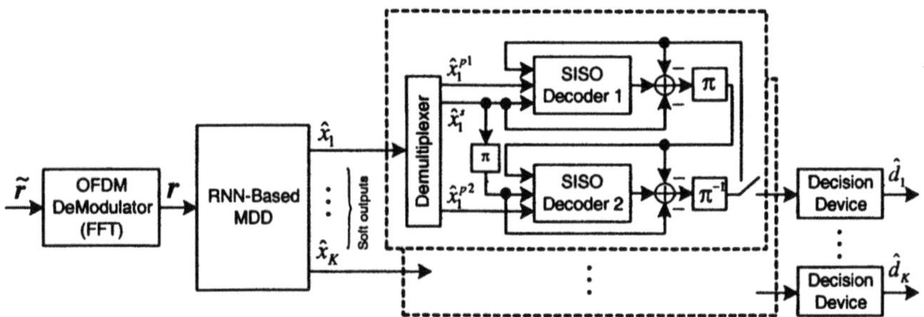

Fig. 6 Block diagram of the joint MDD and turbo decoding structure for multiuser MC-CDMA in impulsive noise Rayleigh fading channels

For a systematic code, the soft output delivered by each component decoder $L(d_k(q)|\{\hat{x}_k(q)\}_{q=1}^{L})$ expressed in the form of *a posteriori* LLR is given as [15]:

$$\Lambda[d_k(q)|\{\hat{x}_k(q)\}_{q=1}^{L}] = \ln\frac{P[d_k(q)=1|\{\hat{x}_k(q)\}_{q=1}^{L}]}{P[d_k(q)=0|\{\hat{x}_k(q)\}_{q=1}^{L}]}$$

$$= \Lambda[d_k(q)] + L_c\hat{x}_k^s(q) + \Lambda_e[d_k(q)] \tag{5.1}$$

where

$$\Lambda[d_k(q)] = \ln\frac{P[d_k(q)=1]}{P[d_k(q)=0]} \tag{5.2}$$

is the *a priori* LLR provided by the other component decoder, and L_c is the channel pre-detection LLR. The final term $\Lambda_e(d_k(q)$ denotes the extrinsic LLR, which represents knowledge gleaned from the decoding process, and is obtained by subtracting the *a priori* information $\Lambda(d_k(q))$ and the received systematic channel output $L_c\hat{x}_k^s(q)$ from the soft output of the decoder. The decoding proceeds by feeding back $\Lambda_e(d_k(q))$ from one component decoder to the input of the other as *a priori* information during the next iteration. This process is repeated several times to enhance the estimate, and the hard decision output is obtained by taking the sign of the de-interleaved *a posteriori* LLR of the second decoder.

6 Simulation Results

In this section, we present some simulation results to assess the performance of the proposed MC-CDMA receiver.

6.1 Convergence Behavior of the RNN

We examine the convergence of the RNN by considering a simple example. Assume that the desired signal vector $\boldsymbol{x}$ in the linear regression model of (4.2) is $\boldsymbol{x} = [1, -4]^T$. The regressor matrix $\boldsymbol{H}$ is a randomly chosen full-rank matrix. Fig. 7 and Fig. 8 plot the trajectories when the noise vector $\boldsymbol{n}$ contains Gaussian and Cauchy noise, respectively. The learning gain for all estimators is set to μ=0.6. Comparisons are made between the MH-algorithm-based MDD, and the RNN-based LDD and MDD. For the RNN-based decorrelators (both LDD and MDD), all initial signal estimates start from zeros, whereas the MH-algorithm-based MDD uses the LS solution in (4.20) as the initial estimates.

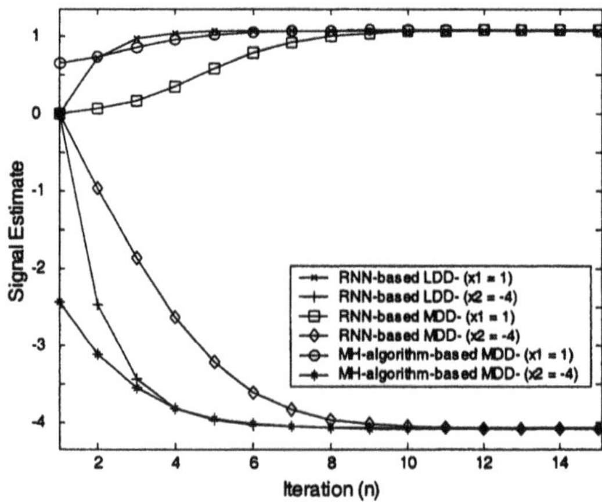

Fig. 7 Computer simulated state trajectories of the RNN-based decorrelating detectors compared with the MH-algorithm-based MDD under Gaussian noise

In the Gaussian channel, it is seen from Fig. 7 that all estimators converge to the same equilibrium states and produce soft-outputs with correct polarities. However, when the channel noise is highly impulsive, Fig. 8 shows that both MDDs have the ability to guard against impulsive noise and produces accurate signal estimates with correct polarities. On the other hand, the RNN-based LDD is adversely affected by the presence of outliers and results in very poor signal estimates with erroneous soft bit decision for the second component with true value of -4.

It is observed from both Fig. 7 and Fig. 8 that the final solutions produced by the RNN-based MDD are identical to those estimated from the MH algorithm, and yet the RNN fully eliminates the computation of $\boldsymbol{H}^{-1}$. Both MDDs have a slower convergence rate as compared with the RNN-based LDD. However, this is not so obvious from the MDD using the MH algorithm in the Gaussian case, since its initial estimates are non-zeros. This sluggish behavior is attributed to the use of less rapidly increasing Huber penalty function, which is, however, more robust against outliers. Nevertheless, this problem can be overcome by increasing the learning gain while satisfying the condition $0<\mu<\mu_{max}$.

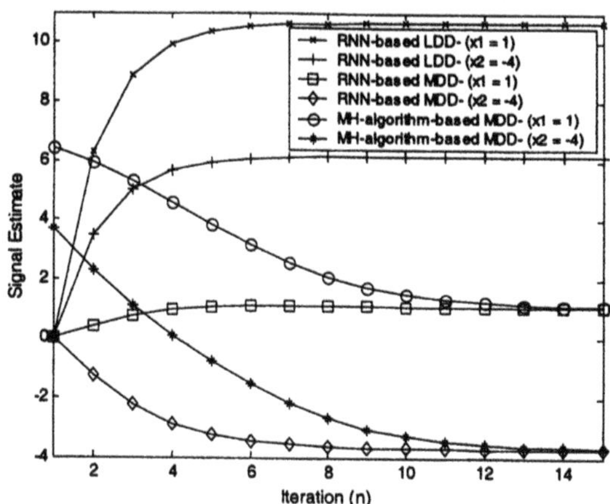

Fig. 8 Computer simulated state trajectories of the RNN-based decorrelating detectors compared with the MH-algorithm-based MDD under Cauchy noise

6.2 BER Performance of the Turbo Receiver for MC-CDMA

Having illustrated the convergence behavior of the RNN, we now study the BER performance of an MC-CDMA system employing the proposed turbo receiver. The system is assumed to be shared by equal-power users (K=15) each with a random spreading code of length N=31. Also, each user employs the same convolutional code generator of (7, 5) in octal notation. The data frame size is L=1024 and the decoders are implemented using the Log-MAP algorithm [57].

Fig. 9 shows the BER of the first user in Gaussian and impulsive noise (α=1.5). The LDD and RNN-based MDD combined with turbo coding are labeled turbo-LDD and turbo-MDD, respectively. For ease of simulation, we do not consider channel estimation problem, and therefore the matrix $\boldsymbol{H}$ simply consists of all original spreading sequences. Turbo coding is not used for non-fading channels. It can be observed that the Rayleigh fading causes dramatic deterioration of the BER performance in both Gaussian and impulsive channels.

For the Gaussian noise channel, Fig. 9(a) shows that both the turbo-LDD and turbo-MDD perform similarly and achieve considerable performance improvements in mitigating distortions due to fading. Nevertheless, the performance after the fifth iteration is still about 3 dB off the fading-free curve. Under impulsive noise, Fig. 9(b) shows that the performance of both the LDD and MDD degrades considerably, though the MDD performs slightly better. However, by incorporating turbo codes, the turbo-MDD quickly regains its performance losses and demonstrates substantial improvement within five iterations.

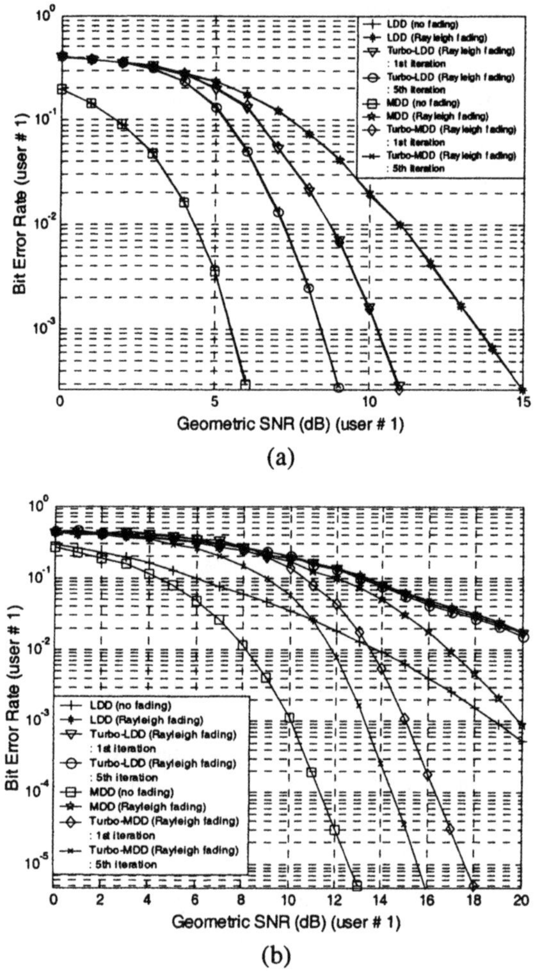

(b)

Fig. 9 BER performance in (**a**) Gaussian noise, (**b**) impulsive noise (alpha=1.5)

This is not the case for the turbo-LDD, as the absence of a bounding mechanism allows entry of impulsive noise to impair the decoding process, since all terms in (5.1) are computed from $\hat{x}$. The saturating function $\psi_H(\cdot)$ in the MDD effectively removes the impact of impulsive noise and allows it to exploit the coding gain. It is seen from Fig. 9(b) that the turbo-MDD at the fifth iteration significantly outperforms the LDD in the fading-free channel after 11dB.

6.3 The Impact of Non-Gaussian Noise on the Turbo Decoder

In this section, we demonstrate the effect of channel noise on the turbo decoder. Fig. 10 and Fig. 11 plot the *a posteriori* LLR over a period of 400 symbols from the second component decoder of the first user in Gaussian and impulsive noise (α=1.5), respectively. The GSNR is 13 dB and the channel conditions are the same as in Fig. 9, except that the first user transmits a string of all 1's. Hence positive LLR values correspond to correct hard decisions.

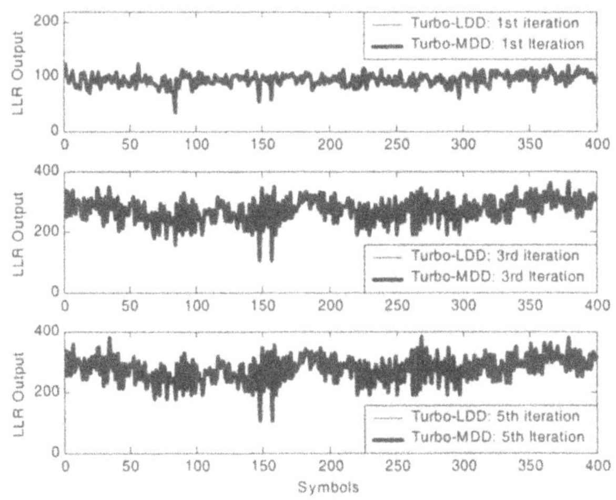

Fig. 10 LLR output from the second component decoder in Gaussian noise

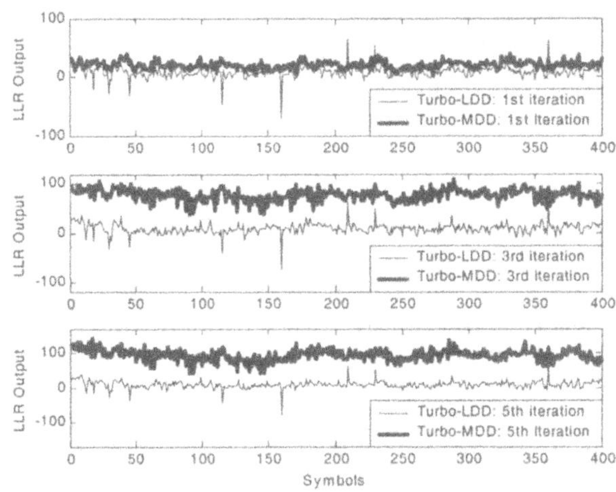

Fig. 11 LLR output from the second component decoder in impulsive noise (α=1.5)

It is seen from Fig. 10 that the LLR outputs from both turbo-LDD and turbo-MDD overlap together and increase in magnitude as the iteration number is increased. However, when the noise is impulsive, the spiky character of the LLR output from the turbo-LDD is obvious from Fig. 11, which does not seem to diminish as the iteration number is increased and causes many erroneous decisions. On the other hand, the LLR outputs from the turbo-MDD improve progressively towards error-free decisions.

7 Conclusions

This chapter provides a survey of the applications of artificial neural networks for multiuser detection. The main motivation here has been to stimulate more research on neural network techniques for robust multiuser detection in non-Gaussian channels. We outlined the M-estimation technique for robust decorrelating detection and discussed existing algorithms. We illustrated the promising features of ANNs by presenting a recurrent neural network (RNN) solution to implement the M-decorrelating detector (MDD) and highlighted the computational saving.

We extended the application of the RNN-based MDD for DS-CDMA systems to MC-CDMA. The MDD is combined with turbo decoders for joint mitigation of MUI, impulsive noise, and Rayleigh fading. The turbo decoder accepts soft outputs from the MDD, which incorporates a bounding mechanism to prevent entry of noise spikes into the decoding stage, thereby achieving substantial performance gains through iterative processing. The turbo decoder, however, does not improve the performance for the linear decorrelating detector as a result of the absence of a bounding mechanism. This results in the presence of outliers in the decoders and thus exhibits irreducible error rates for realistic GSNRs.

Although research on robust multiuser detector using neural network techniques is still in its infancy, we anticipate that in the near future the area will certainly be full of challenges and opportunities.

References

1. Roychowdhury V, Siu KY, Orlitsky A, Eds. (1994) Theoretical Advances in Neural Computation and Learning, Kluwer Academic, Boston
2. Jain LC, Fanelli AM, Eds. (2000) Recent Advances in Artificial Neural Networks: Design and Applications, CRC Press, Boca Raton
3. Proakis JG (1989) Digital Communications, 2nd edn. McGraw-Hill, New York
4. Fazel K (1993) Performance of CDMA/OFDM for mobile communication systems. In Proceedings of the IEEE International Conference on Universal Personal Communications, Ottawa, Ont., Canada, 12-15 Oct., 975-979
5. Yee N, Linnartz JPMG, Fettweis G (1994) Multicarrier CDMA in indoor wireless radio networks. IEICE Trans Commun E77B: 900-904

6. Kaiser S (1995) OFDM-CDMA versus DS-CDMA: performance evaluation for fading channels. In Proceedings of the IEEE International Conference on Communications, Seattle, WA, USA, June, 3:1722-1726
7. Hara S, Prasad R (1999) Design and performance of multicarrier CDMA system in frequency-selective Rayleigh fading channels. IEEE Trans Veh Technol 48: 1584-1595
8. Helard M, Le Gouable R, Helard JF, Baudais JY (2001) Multicarrier CDMA techniques for future wideband wireless networks. Ann Telecommun 56: 260-274
9. Verdu S (1998) Multiuser Detection, Cambridge Univ. Press, Cambridge, U.K.
10. Akhter MS (1998) Signal Processing for MC-CDMA. M.Eng. Dissertation, Faculty of Information Technology, School of Physics and Electronic Systems Engineering, University of South Australia
11. Kafle PL, Sesay AB (2001) Performance of turbo coded multicarrier CDMA with iterative multiuser detection and decoding. In Proceedings of the Canadian Conference in Electrical and Computer Engineering, Toronto, Ontario, Canada, 13-16 May, 105-110
12. Akhter MS, Asenstorfer J, Alexander PD, Reed MC (1998) Performance of multi-carrier CDMA with iterative detection. In Proceedings of the IEEE International Conference on Universal Personal Communications, Florence, Italy, 5-9 Oct., 1:131-135
13. Kaiser S, Papke L (1996) Optimal detection when combining OFDM-CDMA with convolutional and turbo channel coding. In Proceedings of the IEEE International Conference on Communications, Dallas, TX, USA, June, 1:343-348
14. Chuah TC, Sharif BS, Hinton OR (2001) A neural network approach to DS-CDMA multiuser detection in impulsive channels. In Proceedings of the International Conference on Information, Communications, and Signal Processing, Singapore, 15-18 Oct
15. Berrou C, Glavieux A (1996) Near optimum error-correcting coding and decoding: Turbo codes. IEEE Trans Commun 44: 1261-1271
16. Aazhang B, Paris BP, Orsak GC (1992) Neural networks for multiuser detection in code-division multiple-access communications. IEEE Trans Commun 40: 1212-1222
17. Mitra U, Poor HV (1995) Adaptive receiver algorithms for near-far resistant CDMA. IEEE Trans Commun 43: 1713-1724
18. Engel I, Bershad NJ (1997) Statistical convergence analysis of Rosenblatt's perceptron algorithm as a DS-spread spectrum detector. IEEE Trans Signal Process 45: 2843-2846
19. Mitra U, Poor HV (1994) Neural-network techniques for adaptive multiuser demodulation. IEEE J Select Areas Commun 12: 1460-1470
20. Kechriotis GI, Manolakos ES (1996) Hopfield neural network implementation of the optimal CDMA multiuser detector. IEEE Trans Neural Networks 7: 131-141
21. Hopfield JJ (1984) Neurons with graded response have collective computa-

tional properties like those of two-state neurons. Proc Nat Acad Sci 81: 3088-3092
22. Varanasi MK, Aazhang B (1990) Multistage detection in asynchronous code-division multiple-access communications. IEEE Trans Commun 38: 509-519
23. Kechriotis GI, Manolakos ES (1996) A hybrid digital signal processing neural netowrk CDMA multiuser detection scheme. IEEE Trans Circuits Syst II-Analog Digit Signal Process 43: 96-104
24. Chen DC, Sheu B, J. (1998) A compact neural-network based CDMA receiver. IEEE Trans Circuits Syst II-Analog Digit Signal Process 45: 384-387
25. Chen S, Samingan AK, Hanzo L (2001) Support vector machine multiuser receiver for DS-CDMA signals in multipath channels. IEEE Trans Neural Networks 12: 604-611
26. Das K, Morgera SD (1998) Adaptive interference cancellation for DS-CDMA systems using neural network techniques. IEEE J Select Areas Commun 16: 1774-1784
27. Yoon SH, Rao SS (2000) Annealed neural network based multiuser detector in code division multiple access commmunications. IEE Proc-Commun 147: 57-62
28. Sohn I, Gupta SC (1999) Adaptive multiuser detection based on RBF networks in impulsive noise CDMA channels. Int J Wireless Inform Net 6: 59-66
29. Lippmann RP, Beckman P (1989) Adaptive neural net preprocessing for signal detection in non-Gaussian noise. Advances Neural Inform Process Syst 1: 124-132
30. Chuah TC, Sharif BS, Hinton OR (2001) Robust adaptive spread-spectrum receiver with neural-net preprocessing in non-Gaussian noise. IEEE Trans Neural Networks 12: 546-558
31. Shao M, Nikias CL (1993) Signal processing with fractional lower order moments: Stable processes and their applications. Proc IEEE 81: 986-1010
32. Kuruoglu EE (1998) Signal Processing in Alpha-Stable Noise Environments: A Least Lp-Norm Approach. Ph.D. Thesis, Signal Processing and Communications Laboratory, Department of Engineering, University of Cambridge, Cambridge
33. Gonzalez JG (1997) Robust Techniques for Wireless Communications in Non-Gaussian Environments. Ph.D. Dissertation, Department of Electrical and Computer Engineering, University of Delaware, Newark, Delaware
34. Lupas R, Verdu S (1989) Linear multiuser detectors for synchronous code-division multiple-access channels. IEEE Trans Inform Theory 35: 123-136
35. Huber PJ (1981) Robust Statistics, Wiley, New York
36. Blatteberg R, Sargent T (1971) Regression with non-Gaussian stable disturbances: Some sampling results. Econometrica 39: 501-510
37. Shanno DF, Rocke DM (1986) Numerical methods for robust regression: Linear models. SIAM J Sci Stat Comput 7: 86-97
38. Holland PW, Welsch RE (1977) Robust regression using iteratively reweighted least squares. Commun Stat-Theor Meth 6: 813-827
39. Madsen K, Nielsen HB (1990) Finite algorithms for robust linear regression.

BIT 30: 682-699
40. O'Leary DP (1990) Robust regression computation using iteratively reweighted least squares. SIAM J Matrix Anal Appl 11: 466-480
41. Beaton AE, Tukey JW (1974) The fitting of power series, meaning polynomials, illustrated on band-spectroscopic data. Technometrics 16: 147-185
42. Birch JB (1980) Effects of the starting value and stopping rule on robust estimates obtained by iterated weighted least squares. Commun Statist-Simula Comput 9: 141-154
43. Birch JB (1980) Some convergence properties of iterated reweighted least squares in the location model. Commun Statist-Simula Comput 9: 395-369
44. Byrd RH, Pyne DA (1979) Some results on the convergence of the iteratively reweighted least squares algorithm for robust regression. Proc Stat Comp Sec, Am Stat Assoc 87-90
45. Huber PJ (1975) Robust methods of estimation of regresison coefficients. In Proceedings on the 2nd International Summer School on Problems of Model Choice and Regression Analysis, Rheinhardsbrum, G. D. R., 8-18 November
46. Ekblom H (1988) A new algorithm for the Huber estimator in linear models. BIT 28: 123-132
47. Wang X, Poor HV (1999) Robust multiuser detection in non-Gaussian channels. IEEE Trans Signal Process 47: 289-305
48. Cichocki A, Unbehauen R (1992) Neural networks for solving systems of linear equations and related problems. IEEE Trans Circuits Syst I-Fundam Theor Appl 39: 124-138
49. Golub GH, Van Loan CF (1989) Matrix Computation, North Oxford Academic, Oxford
50. Coppersmith D, Winograd S (1990) Matrix multiplication via arithmetic progressions. J Symbolic Comput 9: 251-280
51. Berrou C, Glavieux A, Thitimajshima P (1993) Near Shannon limit error-correcting coding: Turbo codes. In Proceedings of the IEEE International Conference on Communications, Geneva, Switzerland, 1064-1070
52. Bahl LR, Cocke J, Jelinek F, Raviv J (1974) Optimal decoding of linear codes for minimizing symbol error rate. IEEE Trans Inform Theory 20: 284-287
53. Sklar B (1997) A primer on Turbo code concepts IEEE Commun Mag, 94-102
54. Woodard JP, Hanzo L (2000) Comparative study of turbo decoding techniques: An overview. IEEE Trans Veh Technol 49: 2208-2233
55. Heegard C, Wicker SB (1998) Turbo Coding, Kluwer Academic Press, Boston
56. Vucetic B, Yuan J (2000) Turbo Codes- Principles and Applications, Kluwer Academic Press, Boston
57. Robertson P, Hoeher P (1997) Optimal and sub-optimal maximum a posteriori algorithms suitable for turbo decoding. Eur Trans Telecomm 8: 119-125

Minimizing Interference in Cellular Mobile Communications by Optimal Channel Assignment Using Chaotic Simulated Annealing

Lipo Wang, Sa Li, Chunru Wan, and Boon Hee Soong

School of Electrical and Electronic Engineering, Nanyang Technological University
Block S2, 50 Nanyang Avenue, Singapore 639798

Abstract. In this chapter, we deal with the problem of assigning frequency channels to radio cells in a cellular mobile network so that interference between channels is minimized, while demands for channels are satisfied. We solve the channel assignment problem (CAP) using chaotic simulated annealing (CSA) proposed by Chen and Aihara recently. Simulations show that our results are better than existing results found by other algorithms in several benchmarking CAPs.

Keywords: channel assignment, mobile phone, cellular communication, neural network, chaos

1 Introduction

Over recent years, the demand for cellular mobile communication services has been growing rapidly. But electromagnetic spectrum or frequencies allocated for this purpose are limited. Thus optimal assignments of frequency channels are becoming more and more critical. Careful design for a cellular radio network can greatly improve the traffic capacity of the cellular system to accommodate calls, while minimizing interference between calls and guaranteeing the quality of service.

Gamst and Rave defined a general form of channel assignment problems (CAPs) in an arbitrary inhomogeneous cellular radio network [1]: minimizing the span of channels subject to demand and interference-free constraints (denoted as CAP1 in [8]). CAP1 is usually solved by graph coloring algorithms [3] and various techniques have been explored for applying neural networks to CAP1s. Funabiki solved CAP1 by using a parallel algorithm which does not require a rigorous synchronization procedure [9]. Chan et al proposed an approach based on cascaded multilayered feedforward neural networks which showed good performance in dynamic CAP1 [10]. Kim et al proposed a modified Hopfield network to solve CAP1 [11].

Kunz used the Hopfield neural network for solving a different CAP [2]: minimizing the severity of interferences, subject to demand constraints (denoted as CAP2 in [8]). Kunz solved CAP2 by minimizing an energy or cost function representing interference and channel demand constraints. Smith and Palaniswami reformulated CAP2 as a generalized quadratic assignment problem [8] and found good solutions to CAP2 using simulated annealing (SA), a modified Hopfield neural network, and a self-organizing neural network.

In recent years, a large body of work has been carried out on chaotic simulated annealing (CSA) proposed by Chen and Aihara [12]-[22]. Aihara et al proposed a chaotic neural network based on a modified Nagumo and Sato neuron model [7]. Nozawa found [15] that Euler approximation of the continuous-time Hopfield neural network [24] with a negative neuronal self-coupling has chaotic dynamics. Chen and Aihara proposed a neural network model with transient chaos for combinatorial optimization problems [12]. Since this model is similar to simulated annealing, not in a stochastic way but in a deterministically chaotic way, it is regarded as chaotic simulated annealing (CSA). CSA can search efficiently because of its reduced search spaces. Chen and Aihara use CSA to solve the traveling salesman problem (TSP) and showed good performance [12].

In this chapter, we use CSA to solve CAP2 and show that CSA can lead to further improvements on solutions for CAP2 in comparison to solutions obtained by SA, a modified Hopfield neural network, and a self-organizing neural network [8]. We are concerned with only CAP2 because in most of cases, the interference-free lower bound is far greater than the number of available channels.

This chapter is organized as follows. Section 2 reviews CAP2 and its mathematical formulation given by Smith and Palaniswami [8]. In section 3, the CSA algorithm is reviewed. In the section 4, we apply CSA to several benchmarking CAP2s. Finally in section 5, we conclude this chapter.

2 Static Channel Assignment Problems

We assume that a mobile radio network has N cells and the total number of available channels is M. The channel requirements for cell i are given by D_i $(i = 1, 2, \cdots, N)$. $D^T = (D_1, D_2, \cdots, D_N)$ is called the demand matrix. In this chapter, A^T stands for the transposed matrix of A. $C = \{C_{ij}\}$ is the compatibility matrix, where C_{ij} is the minimum frequency separation between cell i and cell j to guarantee an acceptably low signal/interference ratio in each region, $i, j = 1, 2, \cdots N$, and N is the number of cells in the mobile network.

As an example , we consider the following simple data set [3]. The number of cells is $N = 4$ and the demand for channels in each of those base stations is given by

$$D^T = (1, 1, 1, 3) .$$

The compatibility matrix is follows:

$$C = \begin{pmatrix} 5 & 4 & 0 & 0 \\ 4 & 5 & 0 & 1 \\ 0 & 0 & 5 & 2 \\ 0 & 1 & 2 & 5 \end{pmatrix}$$

The diagonal terms $C_{ii} = 5$ means that any two channels assigned to cell i must be at least five frequencies apart in order to satisfy the CSC. $C_{12} = C_{21} = 4$ means that channels assigned to cells 1 and 2 must be four frequencies apart for the CCC and the ACC to be satisfied. We can obtain a interference-free channel assignment for $M = 11$ as follows. Since there are 3 call demands in cell 4, these calls can be assigned to channels 1, 6, and 11, respectively, so that the CSC is satisfied. Since $C_{34} = C_{43} = 2$, we can assign the only call in cell 3 to channel 3 without causing any interference between the calls in cells 3 and 4. Since $C_{24} = C_{42} = 1$, there will be no interference between the calls in cell 2 and cell 4 if we assign the only call in cell 2 to channel 4. The call in cell 1 can be assigned to channel 9 without causing any interference between the two calls in cell 1 and cell 2, since $C_{12} = C_{21} = 4$. We can also see that in this case, other channel assignments may also be interference-free.

The channel assignment problems, as defined by the compatibility matrix C and the demand matrix D in this example, belongs to CAP1 class, if we are interested in the following:

minimize span of channels
subject to demand and noninterference constraints

The solution to this CAP1 is that the minimum span for interference-free channel assignment to exist is $M = 11$ (the lower bound), since we will be unable to find any interference-free assignments if M is less than 11.

Now suppose that there are only ten channels available for the above compatibility matrix C and demand matrix D. It is impossible to find an interference-free channel assignment. In this case, we are interested in the following CAP2:

minimize severity of interference
subject to demand constraint

The CAP2 is useful in practical situations, where an unlimited number of channels is not available, to obtain the best assignments possible given the number of channels available and the call demand in the cellular network.

The solution to CAP2 can be mapped onto a neural network with $N \times M$ neurons [8]. The output of each neuron x_{jk}:

$$x_{jk} = \begin{cases} 1, & \text{if cell j is assigned to channel } k, \\ 0, & \text{otherwise ,} \end{cases} \tag{1}$$

for $j = 1, \cdots, N$ and $k = 1, \cdots, M$. Potential interferences considered here come from the co-channel constraint (CCC), the adjacent channel constraint

(ACC), and the co-site constraint (SCC) [1]. A cost tensor $P_{ji(m+1)}$ is used to measure the degree of interference between cells j and i caused by such assignments that $x_{jk} = x_{il} = 1$ [8], where $m = |k - l|$ is the distance in the channel domain between channels k and l. The cost tensor P can be calculated recursively as follows:

$$P_{ji(m+1)} = \max(0, P_{jim} - 1) \quad , \quad for \;\; m = 1, \cdots, M-1 \quad , \tag{2}$$

$$P_{ji1} = C_{ji} \quad , \qquad \forall j, i \neq j \quad , \tag{3}$$

$$P_{jj1} = 0 \quad , \qquad \forall j \quad . \tag{4}$$

CAP2 can then be formulated as:
minimize

$$F(x) = \sum_{j=1}^{N} \sum_{k=1}^{M} x_{jk} \sum_{i=1}^{N} \sum_{l=1}^{M} P_{ji(|k-l|+1)} x_{il} \quad , \tag{5}$$

subject to

$$\sum_{k=1}^{M} x_{jk} = D_j \quad , \qquad \forall j = 1, \cdots, N \quad , \tag{6}$$

where $F(x)$ is the total interference in the mobile network.

3 Chaotic Simulated Annealing

Chen and Aihara's chaotic simulated annealing (CSA) is described by the following equations [12]:

$$x_{jk}(t) = \frac{1}{1 + e^{-y_{jk}(t)/\varepsilon}} \quad , \tag{7}$$

$$y_{jk}(t+1) = k y_{jk}(t) + \alpha\left(\sum_{i=1, i\neq j}^{N} \sum_{l=1, l\neq k}^{M} w_{jkil} x_{jk}(t) + I_{ij} \right) - z(t)(x_{jk}(t) - I_0) \quad , \tag{8}$$

$$z(t+1) = (1 - \beta) z(t) \quad , \tag{9}$$

where
x_{jk} : output of neuron jk ;
y_{jk} : input of neuron jk ;
w_{jkil}: connection weight from neuron jk to neuron il, with $w_{jkil} = w_{iljk}$ and $w_{jkjk} = 0$;

$$\sum_{i=1, i\neq j}^{N} \sum_{l=1, l\neq k}^{M} w_{jkil} x_{jk} + I_{ij} = -\partial E / \partial x_{jk} : \;\; \text{input to neuron } jk \tag{10}$$

I_{jk} : input bias of neuron jk ;
k : damping factor of nerve membrane ($0 \leq k \leq 1$);
α : positive scaling parameter for inputs ;
β : damping factor ($0 \leq \beta \leq 1$);
$z(t)$: self-feedback connection weight or refractory strength ($z(t) \geq 0$) ;
I_0 : positive parameter;
ε : steepness parameter of the output function ($\varepsilon > 0$) ;
E : energy function.

For any given optimization problem, once we know the energy function E to be minimized, a chaotic neural network can be designed using eqs. (7) (10) to effectively minimize the energy function E [16][12].

The corresponding energy function E for CAP2 can be obtained by combining a constraint term and an interference term suggested by eqs. (5) and (6), respectively :

$$E = \frac{W_1}{2}\sum_{j=1}^{N}(\sum_{k=1}^{M} x_{jk} - D_j)^2 + \frac{W_2}{2}\sum_{j=1}^{N}\sum_{k=1}^{M} x_{jk} \sum_{i=1}^{N}\sum_{l=1}^{M} P_{ji(|k-l|+1)} x_{il}, \quad (11)$$

where W_1 and W_2 represent the relative strength (or importance) of the constraint and the interference, respectively. With eq. (11), the input to neuron jk given in eq. (10) becomes :

$$y_{jk}(t+1) = ky_{jk}(t) - z(t)(x_{jk}(t) - I_0)$$
$$+\alpha\{-W_1 \sum_{q\neq k}^{M} x_{jq} + W_1 D_j - W_2 \sum_{p=1,p\neq j}^{N} \sum_{q=1,q\neq k}^{M} P_{jp(|k-q|+1)} x_{pq}\} \quad . \quad (12)$$

In eq. (8), the term $z(t)(x_{jk}(t) - I_0)$ is related to inhibitory self-feedback with a bias I_0. This term gives the neural network chaotic dynamics when this self-feedback is sufficiently strong. Chaos disappear and the network dynamics stablizes like the Hopfield network when this self-feedback is weak. Eq. (9) represents an exponential cooling schedule for annealing, which leads to transiently chaotic dynamics. β is similar to the temperature in usual stochastic annealing processes.

4 Benchmarking Problem Description

The first benchmarking CAP2 was suggested by Sivarajan [3], denoted as EX1. The number of cells is $N = 4$. The number of channels available is $M = 11$. The demand of channels is given by $D^T = (1, 1, 1, 3)$. We also use a slightly larger extension of EX1, denoted as EX2 [8]:
$N = 5, M = 17, D^T = (2, 2, 2, 4, 3)$.

The Second benchmarking CAP2 used in our simulations is the 21-cell cellular system (HEX1-HEX4) found in [4] (Fig.1). Two different demands are used for HEX as follows.

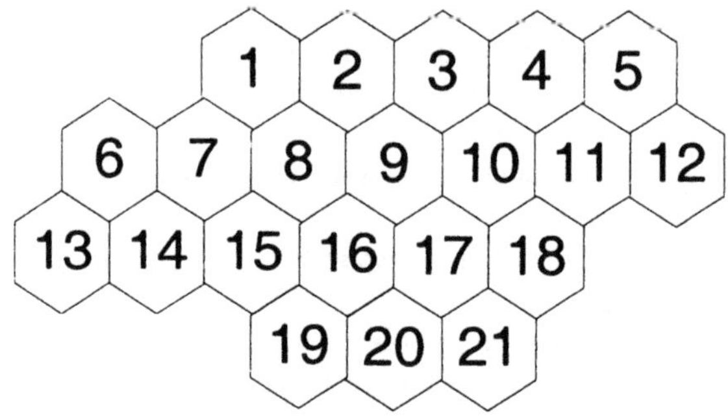

Fig. 1. A 21-cell hexagonal network used in our simulations.

$D_1^T = (2,6,2,2,2,4,4,13,19,7,4,4,7,4,9,14,7,2,2,4,2)$;
$D_2^T = (1,1,1,2,3,6,7,6,10,10,11,5,7,6,4,4,7,5,5,5,6)$.

Two compatability matrices for HEX problems are generated by considering the first two rings of cells around a particular cell as interfering cells. Because HEX1 and HEX3 contain only CCC and CSC, they have the same matrix $C^{(1)}$. HEX2 and HEX4 include CCC, CSC and ACC, so they have the same matrix $C^{(2)}$. We produce the matrices using the details of HEX1-HEX4 listed in table 1 [8] and show the two matrices below (not given in [8]).

Table 1. The descriptions for HEX problems.

Problem	N	M	D	co-channel	adjacent	C_{ii}	C
HEX1	21	37	D_1	yes	no	2	$C^{(1)}$
HEX2	21	91	D_1	yes	yes	3	$C^{(2)}$
HEX3	21	21	D_2	yes	no	2	$C^{(1)}$
HEX4	21	56	D_2	yes	yes	3	$C^{(2)}$

There are two different compatability matrices used for the four HEX problems. The matrices are calculated as follows by considering the first two rings of cells around a particular cell as the sources of interference (Figure 1). The two matrices are both 21×21 in dimension, and they are symmetric matrices according to Gamst and Rave's definition [1]. Each diagonal term C_{ii} represents the minimum separation distance between any two frequencies assigned to cell i, which corresponds to CSC.

Table 1 denotes that diagonal terms $C_{ii}^{(1)}$ in $C^{(1)}$ are 2. The CCC is represented by off-diagonal element C_{ij}=1, and ACC is represented by C_{ij}=2. C_{ij}=0 means that cells i and j are allowed to use the same frequency. From Table 1, $C^{(1)}$ used in HEX1 and HEX3 includes only CCC and CSC but ACC, thus the off-diagonal terms $C_{ij}^{(1)}$ are 1 and 0 corresponding to CCC and no interference, respectively.

$$
C^{(1)} = \begin{pmatrix}
2&1&1&0&0&1&1&1&1&0&0&0&0&1&1&1&0&0&0&0&0\\
1&2&1&1&0&0&1&1&1&1&0&0&0&0&1&1&1&0&0&0&0\\
1&1&2&1&1&0&0&1&1&1&1&0&0&0&0&1&1&1&0&0&0\\
0&1&1&2&1&0&0&0&1&1&1&1&0&0&0&0&1&1&0&0&0\\
0&0&1&1&2&0&0&0&0&1&1&1&0&0&0&0&0&1&0&0&0\\
1&0&0&0&0&2&1&1&0&0&0&0&1&1&1&0&0&0&0&0&0\\
1&1&0&0&0&1&2&1&1&0&0&0&1&1&1&1&0&0&1&0&0\\
1&1&1&0&0&1&1&2&1&1&0&0&0&1&1&1&1&0&1&1&0\\
1&1&1&1&0&0&1&1&2&1&1&0&0&0&1&1&1&1&1&1&1\\
0&1&1&1&1&0&0&1&1&2&1&1&0&0&0&1&1&1&0&1&1\\
0&0&1&1&1&0&0&0&1&1&2&1&0&0&0&0&1&1&0&0&1\\
0&0&0&1&1&0&0&0&0&1&1&2&0&0&1&0&0&1&0&0&0\\
0&0&0&0&0&1&1&0&0&0&0&0&2&1&1&0&0&0&0&0&0\\
1&0&0&0&0&1&1&1&0&0&0&0&1&2&1&1&0&0&1&0&0\\
1&1&0&0&0&1&1&1&1&0&0&0&1&1&2&1&1&0&1&1&0\\
1&1&1&0&0&0&1&1&1&1&0&0&0&1&1&2&1&1&1&1&1\\
0&1&1&1&0&0&0&1&1&1&1&0&0&0&1&1&2&1&1&1&1\\
0&0&1&1&1&0&0&0&1&1&1&1&0&0&0&1&1&2&0&1&1\\
0&0&0&0&0&0&1&1&1&0&0&0&0&1&1&1&1&0&2&1&1\\
0&0&0&0&0&0&0&1&1&1&0&0&0&0&1&1&1&1&1&2&1\\
0&0&0&0&0&0&0&0&1&1&1&0&0&0&0&1&1&1&1&1&2
\end{pmatrix}
$$

$$
C^{(2)} = \begin{pmatrix}
3&2&1&0&0&1&2&2&1&0&0&0&0&1&1&1&0&0&0&0&0\\
2&3&2&1&0&0&1&2&2&1&0&0&0&0&1&1&1&0&0&0&0\\
1&2&3&2&1&0&0&1&2&2&1&0&0&0&0&1&1&1&0&0&0\\
0&1&2&3&2&0&0&0&1&2&2&1&0&0&0&0&1&1&0&0&0\\
0&0&1&2&3&0&0&0&0&1&2&2&0&0&0&0&0&1&0&0&0\\
1&0&0&0&0&3&2&1&0&0&0&0&2&2&1&0&0&0&0&0&0\\
2&1&0&0&0&2&3&2&1&0&0&0&1&2&2&1&0&0&1&0&0\\
2&2&1&0&0&1&2&3&2&1&0&0&0&1&2&2&1&0&1&1&0\\
1&2&2&1&0&0&1&2&3&2&1&0&0&0&1&2&2&1&1&1&1\\
0&1&2&2&1&0&0&1&2&3&2&1&0&0&0&1&2&2&0&1&1\\
0&0&1&2&2&0&0&0&1&2&3&2&0&0&0&0&1&2&0&0&1\\
0&0&0&2&2&0&0&0&0&1&2&3&0&0&1&0&0&1&0&0&0\\
0&0&0&0&0&2&1&0&0&0&0&0&3&2&1&0&0&0&0&0&0\\
1&0&0&0&0&2&2&1&0&0&0&0&2&3&2&1&0&0&1&0&0\\
1&1&0&0&0&1&2&2&1&0&0&0&1&2&3&2&1&0&2&1&0\\
1&1&1&0&0&0&1&2&2&1&0&0&0&1&2&3&2&1&2&2&1\\
0&1&1&1&0&0&0&1&2&2&1&0&0&0&1&2&3&2&1&2&2\\
0&0&1&1&1&0&0&0&1&2&2&1&0&0&0&1&2&3&0&1&2\\
0&0&0&0&0&0&1&1&1&0&0&0&0&1&2&2&1&0&3&1&1\\
0&0&0&0&0&0&0&1&1&1&0&0&0&0&1&2&2&1&2&3&2\\
0&0&0&0&0&0&0&0&1&1&1&0&0&0&0&1&2&2&1&2&3
\end{pmatrix}
$$

For Example, in Figure 1, cell 2 and cell 3 are in the first two rings of cells around cell 1, so CCC exists between cell 1 and cell 2 as well as cell 1 and cell 3. Thus $C^{(1)}_{12}=C^{(1)}_{21}=1$ and $C^{(1)}_{13}=C^{(1)}_{31}=1$. Cell 4 and cell 5 are not among the first two rings of cells around cell 1, so CCC does not exist between cell 1 and cell 4 or between cell 1 and cell 5. Thus $C_{14}^{(1)}=C_{41}^{(1)}=0$ and $C_{15}^{(1)}=C_{51}^{(1)}=0$.

The matrix $C^{(2)}$ used in HEX2 and HEX4 includes CCC, CSC and ACC. The diagonal terms $C_{ii}^{(2)}$ in $C^{(2)}$ are all 3 as shown in Table 1, and the off-diagonal terms of $C_{ij}^{(2)}$ are 1 or 2 corresponding to CCC and ACC, respectively.

We chose the last CAP2 generated from the topographical and morphostructure data from the area of 24×21 km around Helsinki, Finland [23]. Twenty five base station locations were distributed unequally over the area. Kunz used these data to calculate the traffic demand and interference relationships between the 25 base stations [2]. The compatability matrix $C^{(3)}$ is obtained from Kunz data [8][9]. The demand vector is:

$D_3^T = (10, 11, 9, 5, 9, 4, 5, 7, 4, 8, 8, 9, 10, 7, 7, 6, 4, 5, 5, 7, 6, 4, 5, 7, 5)$.

Smith and Palaniswami divided this benchmarking CAP2 into four classes by considering only the first 10 regions (KUNZ1), 15 regions (KUNZ2), 20 regions (KUNZ3), and the entire area (KUNZ4)[8]. The detail is listed in Table 2.

Table 2. The descriptions for KUNZ problems.

Problem	N	M	C	D
KUNZ1	10	30	$[C^{(3)}]_{10}$	$[D_3]_{10}$
KUNZ2	15	44	$[C^{(3)}]_{15}$	$[D_3]_{15}$
KUNZ3	20	60	$[C^{(3)}]_{20}$	$[D_3]_{20}$
KUNZ4	25	73	$C^{(3)}$	D_3

5 Simulation Results

Our simulation results are shown in Table 3. For comparison, Table 3 also includes the simulation results given in [8], i.e., the performances of GAMS/MINOS-5 (labeled GAMS), the traditional heuristics of steepest descent (SD), stochastic simulated annealing (SSA), the original Hopfield network (HN) (with no hill-climbing), the hill-climbing Hopfield network (HCHN), and the self-organizing neural network (SONN). Each of the techniques (except GAMS/MINOS-5) is run from ten different random initial conditions. In Table 3, "Min" means the minimum total interference (eq. (5)) found during these ten times, and "Ave" is the average total interference for the ten runs [8]. The results in Table 3 show that CSA is able to further improve on results obtained by other approaches. In addition, the actual channel assignment results of HEX1 and HEX2 CAP2s with minimum interference are shown in Tables 5 and 6, in case the reader wishes to verify and compare with our results.

Table 3. The simulation results of CSA and other heuristics.

	GAMS	SD		SSA		HN		HCHN		SONN		**CSA**	
problem	Min	Ave	Min	Ave	Min	Ave	Min	Ave	Min	Ave	Min	Ave	Min
EX1	2	0.6	0	0.0	0	0.2	0	0.0	0	0.4	0	**0.0**	**0**
EX2	3	1.1	0	0.1	0	1.8	0	0.8	0	2.4	0	**0.0**	**0**
HEX1	54	56.8	55	50.7	49	49.0	48	48.7	48	53.0	52	**48.1**	**47**
HEX2	27	28.9	25	20.4	19	21.2	19	19.8	19	28.5	24	**18.9**	**18**
HEX3	89	88.6	84	82.9	79	81.6	79	80.3	78	87.2	84	**77.1**	**76**
HEX4	31	28.2	26	21.0	17	21.6	20	18.9	17	29.1	22	**17.7**	**17**
KUNZ1	28	24.4	22	21.6	21	22.1	21	21.1	20	22.0	21	**21.0**	**21**
KUNZ2	39	38.1	36	33.2	32	32.8	32	31.5	30	33.4	33	**31.3**	**31**
KUNZ3	13	17.9	15	13.9	13	13.2	13	13.0	13	14.4	14	**13.0**	**13**
KUNZ4	7	5.5	3	1.8	1	0.4	0	0.1	0	2.2	1	**0.0**	**0**

Table 4. The parameters used in CSA for various CAP2s.

Problem	K	ε	I_0	α	β	$z(0)$	W_1	W_2
EX1	0.9	1/250	0.65	0.0045	0.0005	0.1	1.0	0.02
EX2	0.9	1/250	0.65	0.0045	0.0005	0.1	1.0	0.02
HEX1	0.9	1/150	0.05	0.05	0.0005	0.08	1.0	0.25
HEX2	0.9	1/150	0.05	0.05	0.0005	0.08	1.0	0.25
HEX3	0.9	1/250	0.05	0.05	0.0005	0.08	1.0	0.2
HEX4	0.9	1/250	0.05	0.05	0.0005	0.08	1.0	0.3
KUNZ1	0.9	1/150	0.05	0.05	0.0004	0.08	1.0	0.45
KUNZ2	0.9	1/150	0.05	0.05	0.0005	0.08	1.0	0.45
KUNZ3	0.9	1/150	0.05	0.05	0.0005	0.08	1.0	0.45
KUNZ4	0.9	1/150	0.05	0.05	0.0005	0.08	1.0	0.45

To show the dynamics of the system, the total energy function E (eq. (11)) in HEX2 is plotted as a function of time in Figure 2. We also plot the constraint energy term (eq. (13)) in Figure 3 and the optimization (interference) energy term (eq. (14)) in Figure 4. Figures 5-7 show the three neuronal input terms (eqs. (15), (16) and (17)) for HEX2, respectively.

The constraint energy term in eq. (11) enforces the demand constraint:

$$\frac{W_1}{2}\sum_{j=1}^{N}\left(\sum_{k=1}^{M} x_{jk} - D_j\right)^2 \quad . \tag{13}$$

The optimization (interference) energy term in eq. (11) minimizes the interference:

$$\frac{W_2}{2}\sum_{j=1}^{N}\sum_{k=1}^{M} x_{jk}\sum_{i=1}^{N}\sum_{l=1}^{M} P_{ji(|k-l|+1)}x_{il} \quad . \tag{14}$$

Table 5. Channel assignment for HEX1 problem with interference 47.

Base Station No.	# Channels	Assigned channels
1	2	15,21
2	6	6,13,17,22,26,29
3	2	24,26
4	2	11,32
5	2	20,29
6	4	6,20,22,35
7	4	4,18,21,28
8	13	3,5,7,9,11,14,19,23,25,27,30,32,37
9	19	1,2,4,6,8,10,12,14,16,18,20,22,25,27,29,31,33,35,37
10	7	7,15,17,21,28,34,36
11	4	5,13,23,30
12	4	2,4,14,24
13	7	8,10,12,15,17,23,27
14	4	2,29,31,33
15	9	3,7,9,11,19,24,28,34,36
16	14	1,3,5,8,10,12,16,20,23,25,27,30,32,35
17	7	2,4,8,18,21,31,33
18	2	9,19
19	2	15,17
20	4	6,13,22,26
21	2	11,24

The single-neuron input term in eq. (12) is responsible for generating chaotic dynamics:

$$ky_{jk}(t) - z(t)(x_{jk}(t) - I_0) \quad . \tag{15}$$

The constraint input term in eq. (12) enforces the demand contraint:

$$\alpha\{-W_1 \sum_{q \neq k}^{M} x_{jq} + W_1 D_j\} \quad . \tag{16}$$

The interference input term in eq. (12) minimizes interference:

$$\alpha\{-W_2 \sum_{p=1, p \neq j}^{N} \sum_{q=1, q \neq k}^{M} P_{jp(|k-q|+1)} x_{pq}\} \quad . \tag{17}$$

Parameters (Table 4) are chosen so that the constraint energy term in eq. (13) is comparable in magnitude to the interference energy term in eq. (14). Similarly, the three neuronal input terms in eqs. (15) - (17) need also to be comparable in magnitude, so that each term can efficiently play its role.

Table 6. Channel assignment for HEX2 problem with interference 18.

Base Station No.	# Channels	Assigned channels
1	2	9,72
2	6	18,49,57,62,70,81
3	2	33,55
4	2	76,86
5	2	5,66
6	4	17,41,74,86
7	4	13,39,63,76
8	13	5,20,23,27,34,37,42,47,54,59,67,83,87
9	19	2,3,8,11,14,22,26,29,31,35,40,43,51,65,68,74,77,85,89
10	7	16,19,46,60,63,72,80
11	4	1,12,23,54
12	4	10,41,48,68
13	7	12,23,32,45,62,72,80
14	4	49,57,60,78
15	9	7,10,15,25,30,52,66,69,73
16	14	1,12,17,28,32,45,50,56,61,64,71,75,79,91
17	7	6,24,38,48,53,58,82
18	2	78,88
19	2	21,88
20	4	4,9,41,86
21	2	13,34

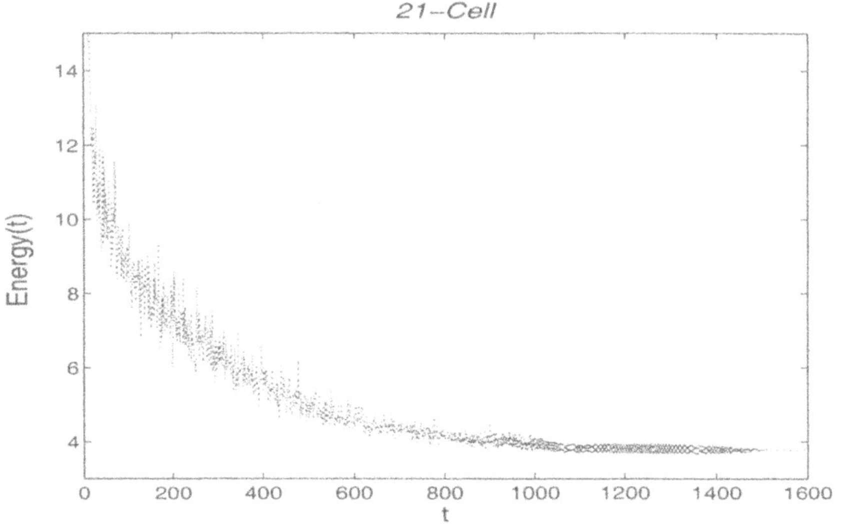

Fig. 2. The Energy (eq. (11)) as a function of time in HEX2.

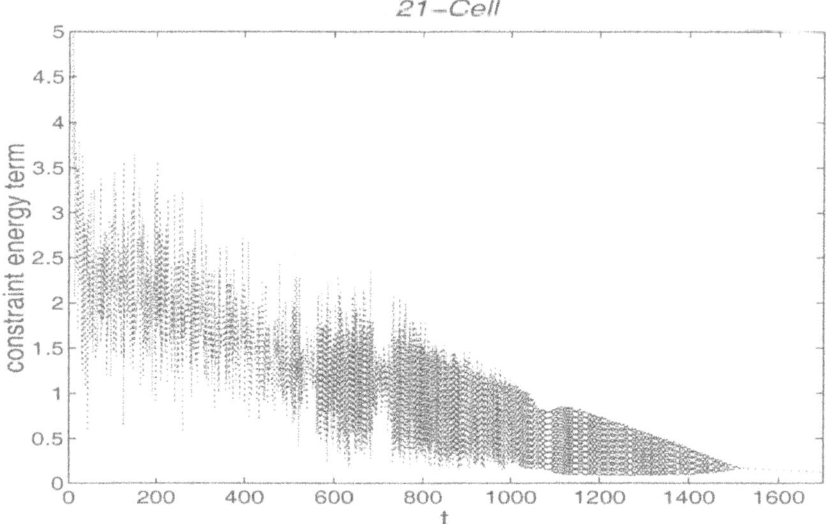

Fig. 3. The constraint energy term (eq. (13)) as a function of time in HEX2.

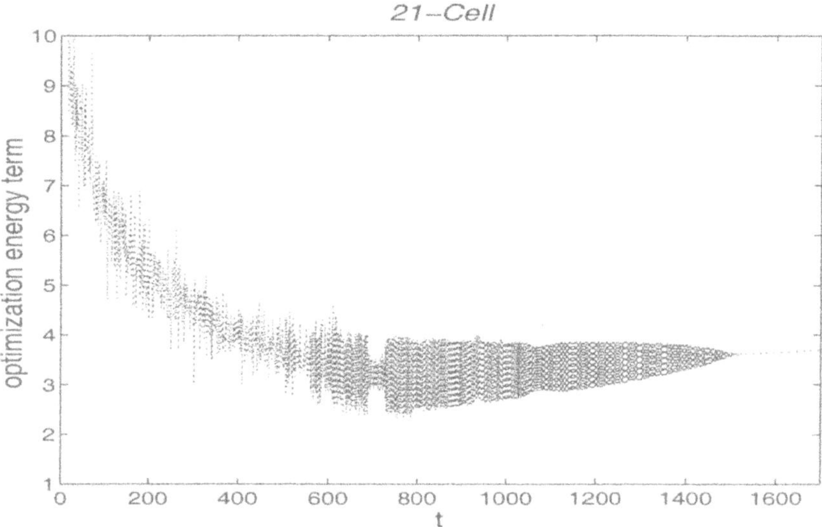

Fig. 4. The optimization (interference) energy term (eq. (14)) as a function of time in HEX2.

6 Conclusions

In this chapter, we have considered the CAP2 and demonstrated that chaotic simulated annealing (CSA) is able to further improve on results obtained by other algorithms in several benchmarking CAP2s.

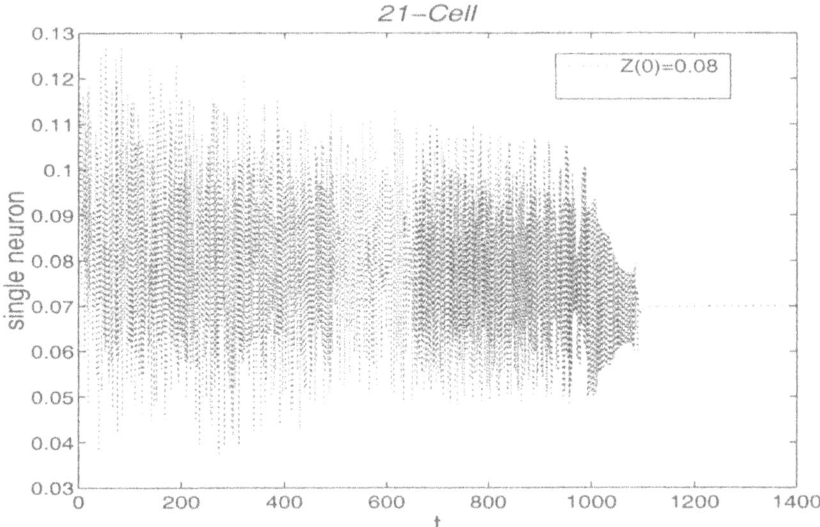

Fig. 5. The single-neuron input term (eq. (15)) as a function of time in HEX2.

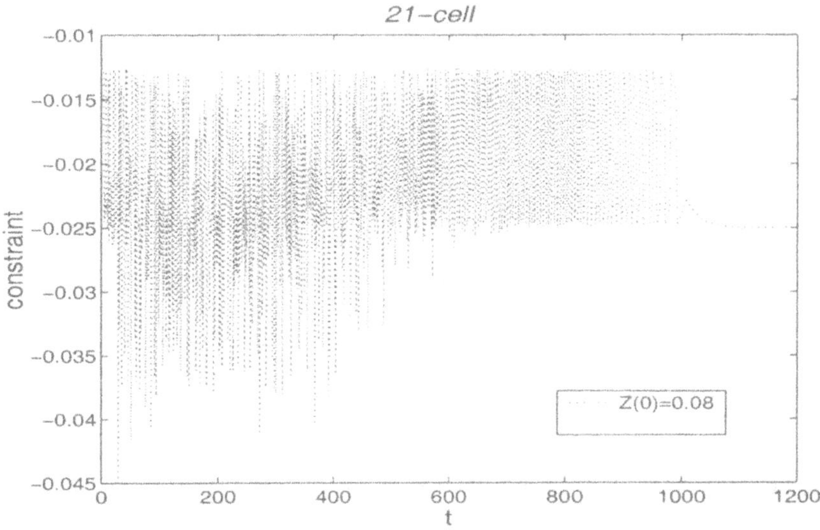

Fig. 6. The constraint input term (eq. (16)) as a function of time in HEX2.

Our work shows the potential of CSA in CAP2s, but it is concerned with only static CAPs. In a dynamic CAP, the demand becomes a function of time. Furthermore, CSA is deterministic and is not guaranteed to settle down at a global minimum. Therefore overcoming those deficiencies and implementation

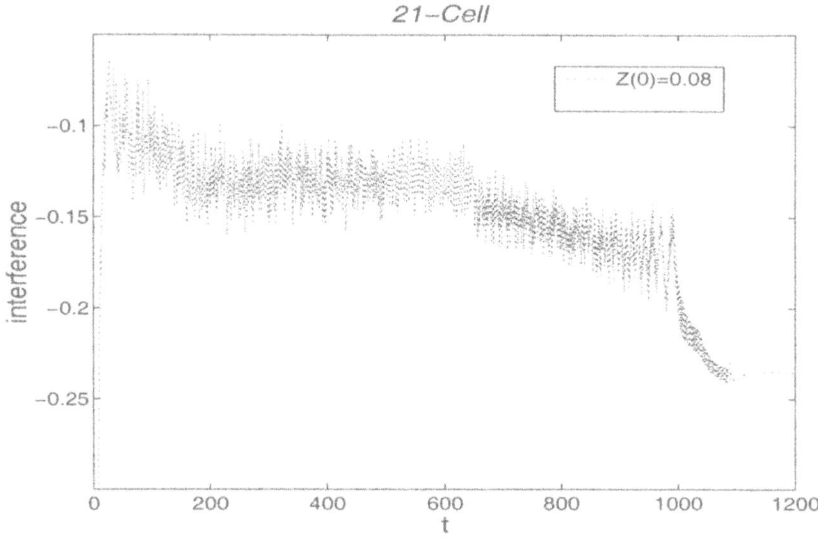

Fig. 7. The interference input term (eq. (17)) as a function of time in HEX2.

of CSA to solve other practical optimization problems, such as the dynamic CAP, will be studied in future work.

References

1. Gamst, A., Rave, W. (1982) On frequency assignment in mobile automatic telephone systems. Proc. GLOBECOM'82, 309-315
2. Kunz, D. (1991) Channel assignment for cellular radio using neural networks. IEEE Trans. Veh. Technol., **40**, 188-193
3. Sivarajan, K. N., McEliece, R. J., Ketchum, J.W. (1989) Channel assignment in cellular radio. Proc. 39th IEEE Veh. Technol. Soc. Conf., 846-850
4. Gamst, A. (1986) Some lower bounds for a class of frequency assignment problems. IEEE Trans. Veh. Tech., **VT-35**, 8-14
5. Hale, W. K. (1980) Frequency assignment: theory and application. Pro. IEEE, **68**, 1497-1514
6. Duque-Anton, M., Kunz, D., Ruber, B. (1993) Channel assignment for cellular radio using simulated annealing. IEEE Trans. Veh. Technol., **42**, 14 -21
7. Aihara, K., Takabe, T., Toyoda, M. (1990) Chaotic neural networks. Physics Letters A, **144**, 333-340
8. Smith, K., Palaniswami, M. (1997) Static and dynamic channel assignment using neural network. IEEE Journal on Selected Areas in Communications, **15**, 238-249
9. Funabiki, N., Takefuji, Y. (1992) A neural network parallel algorithm for channel assignment problems in cellular radio networks. IEEE Trans. Veh. Technol., **41**, 430-437

10. Chan, P., Palaniswami, M., Everitt, D. (1994) Neural network-based dynamic channel assignment for cellular mobile communication systems. IEEE Trans. Veh. Technol., **43**, 279 -288
11. Kim, J., Park, S. H., Dowdy, P. W., Nasrabadi, N. M. (1997) Cellular radio channel assignment using a modified Hopfield network. IEEE Trans. Veh. Technol., **46**, 957 -967
12. Chen L., Aihara, K. (1995) Chaotic simulated annealing by a neural network model with transient chaos. Neural Networks, **8**, 915-930
13. Chen L., Aihara, K. (1994) Transient chaotic neural networks and chaotic simulated annealing. in M. Yamguti(ed.), Towards the Harnessing of Chaos. Amsterdam, Elsevier Science Publishers B.V., 347-352
14. Chen L., Aihara, K. (1999) Global searching ability of chaotic neural networks. IEEE Trans. Circuits and Systems-I: Fundamental Theory and Applications, **46**, 974-993
15. Nozawa, H. (1992) A neural network model as a globally coupled map and applications based on chaos. Chaos, **2**, 377-386
16. Wang, L., Tian, F. (2000) Noisy chaotic neural networks for solving combinatorial optimization problems. Proc. International Joint Conference on Neural Networks (IJCNN 2000, Como, Italy, July 24-27, 2000), **4**, 37 -40
17. Wang, L., Tian, F., Fu, X. (2000) Solving channel assignment problems for cellular radio networks using transiently chaotic neural networks. Proc. International Conference on Automation, Robotics, and Computer Vision. (ICARCV 2000, Singapore)
18. Wang, L. (1996) Oscillatory and chaotic dynamics in neural networks under varying operating conditions. IEEE Transactions on Neural Networks, **7**, 1382-1388
19. Wang, L. K. Smith (1998) On chaotic simulated annealing. IEEE Transactions on Neural Networks, **9**, 716-718
20. Wang, L., Smith, K. (1998) Chaos in the discretized analog Hopfield neural network and potential applications to optimization. Proc. International Joint Conference on Neural Networks, **2**, 1679-1684
21. Kwok, T., Smith, K., Wang, L. (1998) Solving combinatorial optimization problems by chaotic neural networks. C. Dagli et al.(eds), Intelligent Engineering Systems through Artificial Neural Networks, **8**, 317-322
22. Kwok, T., Smith, K., Wang, L. (1998) Incorporating chaos into the Hopfield neural network for combinatorial optimization. Proc. 1998 World Multiconference on Systemics, Cybernetics and Informatics, N. Callaos, O. Omolayole, and L. Wang, (eds.) **1**, 646-651
23. Kohonen, T. (1982) Self-organized formation of topologically correct feature maps. Biol. Cybern., **43**, 59-69
24. Hopfiled, J.J. (1984) Neurons with graded response have collective computational properties like those of two-state neurons. Proc. Natl. Acad. Sci. USA, **81**, 3088-3092

Orthogonal Genetic Algorithms with Application to Multimedia Multicast Routing

Yiu-Wing Leung

Department of Computer Science, Hong Kong Baptist University
Kowloon Tong, Hong Kong.

Abstract. Some major steps of a genetic algorithm can be considered to be "experiments". For example, a crossover operator samples the genes from the parents to produce some potential offspring, and this operation can be considered to be a sampling experiment. Based on this observation, we recently propose to integrate *experimental design methods* into the genetic algorithms [1–4], so that the resulting algorithms can be more powerful and statistically sound. In this chapter, we summarize our recent results on this research direction [1,2]. In particular, we introduce the basic concept of experimental design methods and describe a well-known experimental design method called *orthogonal design*. Then we describe how to apply the orthogonal design to design a new type of genetic algorithms called *orthogonal genetic algorithms*. In additionally, we describe an orthogonal genetic algorithm for an optimization problem in communication networks called *multimedia multicast routing*. We present numerical results to demonstrate the effectiveness of orthogonal genetic algorithms.

Keywords: Evolutionary computation, experimental design methods, orthogonal design, orthogonal array, multimedia multicast routing.

1 Introduction

Optimization problems arise in almost every field of science, engineering, and business. Many of these problems cannot be solved analytically and consequently they have to be solved by numerical algorithms. Evolutionary algorithms are a promising methodology for these problems [5–7]. Unlike the conventional optimization algorithms, an evolutionary algorithm maintains a population of potential solutions instead of a single solution, and it evolves and improves the population iteratively. As a result, it can explore the search space effectively and reduce the chance of being trapped in the local minima.

In our recent studies [1–4], we observe that some major steps of an evolutionary algorithm can be considered to be "experiments". The following are two examples.

- *Example 1*: Usually, the first step of evolutionary algorithms is to generate an initial population. This initialization can be considered to be a sampling experiment, in which some initial points are sampled from the feasible solution space.

- *Example 2*: A crossover operator samples the genes from the parents to produce some potential offspring. This operation can also be considered to be a sampling experiment.

For the above reason, we can treat an evolutionary algorithm as an iterative experimental optimization method.

Experimental design method is a sophisticated branch of statistics [8,9]. It can be applied to reveal the behaviour of a system whose cost depends on controllable factors (each at several levels) and uncontrollable factors (such as noise). It samples a small portion of experimental points according to some pre-determined rules, and then analyzes the results of this experiment (e.g., identify the most influential factors on the cost, find the best combination of factor levels that minimizes the cost, etc). Experimental design method is extensively applied in various fields such as quality engineering.

We recently propose to integrate the experimental design methods into the evolutionary algorithms [1–4], so that we can apply this statistical methodology to handle the "experiments" in evolutionary computation. The resulting evolutionary algorithms can potentially be more powerful and statistically-sound.

In this chapter, we summarize our recent results on integrating experimental desgin methods into the evolutionary algorithms [1,2]. In particular, we describe how to integrate an experimental design method into the genetic algorithms, and apply this approach to design a genetic algorithm for an optimization problem in communication networks called *multimedia multicast routing*. This chapter is organized as follows.

- In section 2, we introduce the basic concept of experimental design methods and describe a well-known experimental design method called *orthogonal design*.
- In section 3, we describe how to apply the orthogonal design to design a new type of genetic algorithms called *orthogonal genetic algorithms*.
- In section 4, we consider the problem of multimedia multicast routing in communication networks. This problem arises in many multimedia communication applications such as multiparty multimedia teleconferences, multimedia-on-demand, and remote video lecture. We describe an orthogonal genetic algorithm for multimedia multicast routing, and present numerical results to demonstrate its effectiveness.

2 Experimental Design Methods

2.1 Basic Concept

In this section, we introduce the basic concept of experimental design methods. For more details, the readers can refer to [8,9]. To aid explanation, we consider the following example. The yield of a vegetable depends on: (1) the temperature, (2) the amount of fertilizer, and (3) the pH value of the soil. These three quantities are called *factors* of this experiment. Each factor has three possible values shown in Table 1, and we say that each factor has three *levels*.

Table 1. An experimental design problem with three factors and three levels per factor.

	Factors		
	Temperature	Amount of fertiliers	pH value
Levels	$20°C$ $25°C$ $30°C$	100 g/m^2 150 g/m^2 200 g/m^2	6 7 8

To find the best combination of levels for a maximum yield, we can do one experiment for each combination and then select the best one. In the above example, there are 3x3x3 = 27 combinations and hence there are 27 experiments. In general, when there are N factors and Q levels per factor, there are Q^N combinations. When N and Q are large, it may not be possible to do all Q^N experiments. Therefore, it is desirable to sample a small but representative set of combinations for experimentation.

The *orthogonal design* was developed for this purpose [8,9]. It provides a series of *orthogonal arrays* for different N and Q. We let $L_M(Q^N)$ be an orthogonal array for N factors and Q levels, where "L" denotes a Latin square and M is the number of combinations of levels. It has M rows where every row represents one particular combination of levels. For convenience, we denote $L_M(Q^N) = [a_{i,j}]_{M \times N}$ where the jth factor in the ith combination has level $a_{i,j}$ and $a_{i,j} \in \{1, 2, \ldots, Q\}$. The following is an example of orthogonal arrays:

$$L_9(3^4) = \begin{bmatrix} 1 & 1 & 1 & 1 \\ 1 & 2 & 2 & 2 \\ 1 & 3 & 3 & 3 \\ 2 & 1 & 2 & 3 \\ 2 & 2 & 3 & 1 \\ 2 & 3 & 1 & 2 \\ 3 & 1 & 3 & 2 \\ 3 & 2 & 1 & 3 \\ 3 & 3 & 2 & 1 \end{bmatrix} \qquad (1)$$

In $L_9(3^4)$, there are four factors, three levels per factor, and nine combinations of levels. In the first combination, the four factors have respective levels 1, 1, 1, 1; in the second combination, the four factors have respective levels 1, 2, 2, 2; etc.

In the above example on vegetable yield, there are 27 combinations to be tested. We apply the orthogonal array $L_9(3^4)$ to select nine representative combinations to be tested, and these nine combinations are shown in Table 2.

Table 2. Based on the orthogonal array $L_9(3^4)$, nine representative combinations are selected for experimentation.

Combination	Factors		
	Temperature	Amount of fertilizer	pH value
1st	$20°C$	$100\ g/m^2$	6
2nd	$20°C$	$150\ g/m^2$	7
3rd	$20°C$	$200\ g/m^2$	8
4th	$25°C$	$100\ g/m^2$	7
5th	$25°C$	$150\ g/m^2$	8
6th	$25°C$	$200\ g/m^2$	6
7th	$30°C$	$100\ g/m^2$	8
8th	$30°C$	$150\ g/m^2$	6
9th	$30°C$	$200\ g/m^2$	7

In general, the orthogonal array $L_M(Q^N)$ has the following properties:

(1) For the factor in any column, every level occurs $\frac{M}{Q}$ times.
(2) For the two factors in any two columns, every combination of two levels occurs $\frac{M}{Q^2}$ times.
(3) For the two factors in any two columns, the M combinations contain the following combinations of levels: $(1,1),(1,2),\ldots,(1,Q),(2,1),(2,2),\ldots,(2,Q),\ldots,(Q,1),(Q,2),\ldots,(Q,Q)$.
(4) If any two columns of an orthogonal array are swapped, the resulting array is still an orthogonal array.
(5) If some columns are taken away from an orthogonal array, the resulting array is still an orthogonal array with a smaller number of factors.

Consequently, the selected combinations are scattered uniformly over the space of all possible combinations. Fig. 1 shows an example. Orthogonal design has been proven optimal for additive and quadratic models, and the selected combinations are good representatives for all the possible combinations [10].

2.2 Construction of Orthogonal Array

As we shall explain, when we apply the orthogonal design to design genetic algorithms, we may need different orthogonal arrays for different problems. Although many orthogonal arrays have been tabulated in the literature (e.g., see [11]), it is impossible to store all of them. We will only need a special

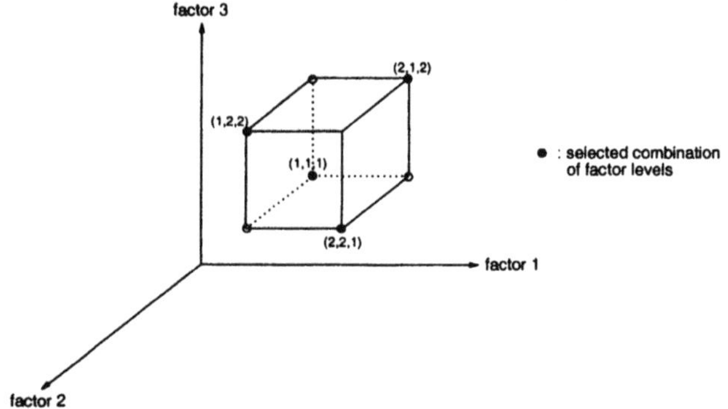

Fig. 1. Orthogonality of the orthogonal array $L_4(2^3)$.

class of orthogonal arrays $L_M(Q^N)$ where Q is odd and $M = Q^J$ where J is a positive integer fulfilling

$$N = \frac{Q^J - 1}{Q - 1} \tag{2}$$

In this subsection, we design a simple permutation method to construct orthogonal arrays of this class.

We denote the jth column of the orthogonal array $[a_{i,j}]_{M \times N}$ by $\boldsymbol{a}_j$. Columns $\boldsymbol{a}_j$ for $j = 1, 2, \frac{Q^2-1}{Q-1}+1, \frac{Q^3-1}{Q-1}+1, \ldots, \frac{Q^{J-1}-1}{Q-1}+1$ are called the *basic columns*, and the others are called the *nonbasic columns*. We first construct the basic columns and then construct the nonbasic columns. The details are as follows:

Algorithm 1: Construction of Orthogonal Array

Step 1: To construct the basic columns, compute the following quantities for all $1 \le k \le J$:

- $j = \frac{Q^{k-1}-1}{Q-1} + 1$
- For all $1 \le i \le Q^J$, compute $a_{i,j} = \lfloor \frac{i-1}{Q^{J-k}} \rfloor \ mod \ Q$

Step 2: To construct the nonbasic columns, compute the following quantities for all $2 \le k \le J$:

- $j = \frac{Q^{k-1}-1}{Q-1} + 1$
- For all $1 \le s \le j-1$ and $1 \le t \le Q-1$, compute $\boldsymbol{a}_{j+(s-1)(Q-1)+t} = (\boldsymbol{a}_s \mathrm{x} t + \boldsymbol{a}_j) \ mod \ Q$

Step 3: Increment $a_{i,j}$ by one for all $1 \le i \le M$ and $1 \le j \le N$.

3 Design of Orthogonal Genetic Algorithms

In this section, we describe an approach of designing genetic algorithms by applying the orthogonal design [1,2]. For different problems, this approach is the same but the details may be different. To illustrate this approach, we consider the following global optimization problem in this section:

$$\begin{aligned} &\text{Minimize} && f(\boldsymbol{x}) \\ &\text{Subject to} && \boldsymbol{l} \le \boldsymbol{x} \le \boldsymbol{u} \end{aligned} \tag{3}$$

where $\boldsymbol{x} = (x_1, x_2, \ldots, x_N)$ is a variable vector in $\Re_N$, $f(\boldsymbol{x})$ is the objective function, and $\boldsymbol{l} = (l_1, l_2, \ldots, l_N)$ and $\boldsymbol{u} = (u_1, u_2, \ldots, u_N)$ define the feasible solution space. We denote the *domain* of x_i by $[l_i, u_i]$, and the feasible solution space by $[\boldsymbol{l}, \boldsymbol{u}]$. We define $\boldsymbol{x} = (x_1, x_2, \ldots, x_N)$ to be a chromosome with cost $f(\boldsymbol{x})$. The optimization problem is equivalent to finding a chromosome of minimal cost.

3.1 Generation of Initial Population

Before an optimization problem is solved, we may have no information about the location of the global minimum. It is desirable that the chromosomes of the initial population be scattered uniformly over the feasible solution space, so that the algorithm can explore the whole solution space evenly. We observe that an orthogonal array specifies a small number of combinations that are scattered uniformly over the space of all the possible combinations. Therefore, orthogonal design is a promising method for generating a good initial population.

We define x_i to be the ith factor, and so each chromosome has N factors. These factors are continuous, but the orthogonal design is applicable to discrete factors only. To overcome this issue, we quantize each factor into a finite number of values. In particular, we quantize the domain $[l_i, u_i]$ of x_i into Q_1 levels $\alpha_{i,1}, \alpha_{i,2}, \alpha_{i,3}, \ldots, \alpha_{i,Q_1}$ where the design parameter Q_1 is odd and $\alpha_{i,j}$ is given by:

$$\alpha_{i,j} = \begin{cases} l_i & j = 1 \\ l_i + (j-1)\left(\frac{u_i - l_i}{Q_1 - 1}\right) & 2 \le j \le Q_1 - 1 \\ u_i & j = Q_1 \end{cases} \tag{4}$$

In other words, the difference between any two successive levels is the same. For convenience, we call $\alpha_{i,j}$ the jth level of the ith factor, and we denote $\alpha_i = (\alpha_{i,1}, \alpha_{i,2}, \ldots, \alpha_{i,Q_1})$. Fig. 2 shows an example of quantization.

After quantization, x_i has Q_1 possible values $\alpha_{i,1}, \alpha_{i,2}, \ldots, \alpha_{i,Q_1}$, and hence the feasible solution space contains Q_1^N points. We apply orthogonal design to select a small sample of points that are scattered uniformly over the feasible solution space.

We first construct a suitable orthogonal array. Recall that *Algorithm 1* can only construct $L_{M_1}(Q_1^N)$ where $M_1 = Q_1^{J_1}$ and J_1 is a positive integer

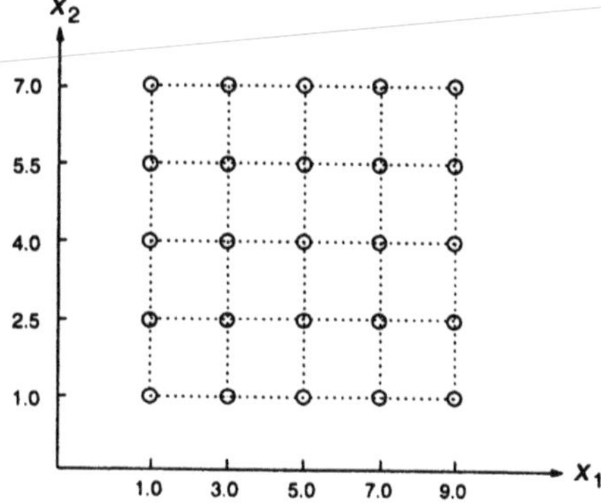

Fig. 2. Quantization for a two-dimensional optimization problem; the domains of x_1 and x_2 are $[1.0, 9.0]$ and $[1.0, 7.0]$ respectively. Each factor is quantized into five levels. After quantization, the feasible solution space contains 25 chromosomes.

fulfilling $\frac{Q_1^{J_1}-1}{Q_1-1} = N$(see equation (2)). Since the problem dimension N is given, there may not exist Q_1 and J_1 fulfilling this condition. We bypass this restriction as follows. We choose the smallest J_1 such that

$$\frac{Q_1^{J_1}-1}{Q_1-1} \geq N \tag{5}$$

We execute *Algorithm 1* to construct an orthogonal array with $N' = \frac{Q_1^{J_1}-1}{Q_1-1}$ factors, and then delete the last $N' - N$ columns to get an orthogonal array with N factors. The details are given in the following algorithm.

Algorithm 2: Construction of $L_{M_1}(Q_1^N)$

Step 1: Select the smallest J_1 fulfilling $\frac{Q_1^{J_1}-1}{Q_1-1} \geq N$.

Step 2: If $\frac{Q_1^{J_1}-1}{Q_1-1} = N$, then $N' = N$ else $N' = \frac{Q_1^{J_1}-1}{Q_1-1}$.

Step 3: Execute *Algorithm 1* to construct the orthogonal array $L_{Q_1^{J_1}}(Q_1^{N'})$.

Step 4: Delete the last $N' - N$ columns of $L_{Q_1^{J_1}}(Q_1^{N'})$ to get $L_{M_1}(Q_1^N)$ where $M_1 = Q_1^{J_1}$.

After constructing $L_{M_1}(Q_1^N) = [a_{i,j}]_{M_1 \times N}$, we get a sample of M_1 combinations out of Q_1^N combinations. We apply these M_1 combinations to generate the following M_1 chromosomes:

$$\begin{cases} (\alpha_{1,a_{1,1}}, \alpha_{2,a_{1,2}}, \ldots, \alpha_{N,a_{1,N}}) \\ (\alpha_{1,a_{2,1}}, \alpha_{2,a_{2,2}}, \ldots, \alpha_{N,a_{2,N}}) \\ \ldots \\ (\alpha_{1,a_{M_1,1}}, \alpha_{2,a_{M_1,2}}, \ldots, \alpha_{N,a_{M_1,N}}) \end{cases} \tag{6}$$

Among these M_1 potential chromosomes, we select G chromosomes having the smallest cost as the initial population, where G is the population size. In this manner, we have evenly scanned the feasible solution space once to locate potentially good points for further exploration in subsequent iterations.

When the feasible solution space is large, it may be desirable to generate more potential chromosomes for a better coverage. However, the value of M_1 depends on N and Q_1 and it cannot be increased arbitrarily. To overcome this issue, we divide the feasible solution space into S subspaces, where S is a design parameter. In particular, we choose the sth dimension such that

$$u_s - l_s = \max_{1 \leq i \leq N}\{u_i - l_i\} \tag{7}$$

and divide $[\boldsymbol{l}, \boldsymbol{u}]$ along the sth dimension into the following S subspaces $[\boldsymbol{l}(1), \boldsymbol{u}(1)], [\boldsymbol{l}(2), \boldsymbol{u}(2)], \ldots, [\boldsymbol{l}(S), \boldsymbol{u}(S)]$ where

$$\begin{cases} \boldsymbol{l}(i) = \boldsymbol{l} + (i-1)(\frac{u_s - l_s}{S})\mathbf{1}_s \\ \\ \boldsymbol{u}(i) = \boldsymbol{u} - (S-i)(\frac{u_s - l_s}{S})\mathbf{1}_s \end{cases} \quad i = 1, 2, \ldots, S \tag{8}$$

and $\mathbf{1}_s$ is an N-dimensional vector such that its sth element is one and all the other elements are zero. We apply $L_{M_1}(Q_1^N)$ to each subspace to generate M_1 chromosomes, so that we get a total of $M_1 S$ potential chromosomes for the initial population. We select G chromosomes having the smallest cost as the initial population.

The details for generating an initial population are given as follows.

Algorithm 3: Generation of Initial Population

Step 1: Divide the feasible solution space $[\boldsymbol{l}, \boldsymbol{u}]$ into S subspaces $[\boldsymbol{l}(1), \boldsymbol{u}(1)], [\boldsymbol{l}(2), \boldsymbol{u}(2)], \ldots, [\boldsymbol{l}(S), \boldsymbol{u}(S)]$ based on equations (7)-(8).

Step 2: Quantize each subspace based on equation (4), and then apply $L_{M_1}(Q_1^N)$ to select M_1 chromosomes based on equation (6).

Step 3: Among the $M_1 S$ chromosomes, select G chromosomes having the smallest cost as the initial population.

3.2 Orthogonal Crossover

We apply the orthogonal design to design a new crossover operator called *orthogonal crossover*. Our main idea is to apply the orthogonal design to sample the genes from the parents to form the potential offspring. In the following, we describe an orthogonal crossover operator that acts on two parents. Nevertheless, this operator can easily be generalized to act on multiple parents.

We quantize the solution space defined by its two parents into a finite number of points, and then apply the orthogonal design to select a small but representative sample of points as the potential offspring. Specifically, consider any two parents $\boldsymbol{p}_1 = (p_{1,1}, p_{1,2}, \ldots, p_{1,N})$ and $\boldsymbol{p}_2 = (p_{2,1}, p_{2,2}, \ldots, p_{2,N})$. These parents define the solution space $[\boldsymbol{l}_{parent}, \boldsymbol{u}_{parent}]$ where

$$\begin{cases} \boldsymbol{l}_{parent} = [\min(p_{1,1}, p_{2,1}), \min(p_{1,2}, p_{2,2}), \ldots, \min(p_{1,N}, p_{2,N})] \\ \boldsymbol{u}_{parent} = [\max(p_{1,1}, p_{2,1}), \max(p_{1,2}, p_{2,2}), \ldots, \max(p_{1,N}, p_{2,N})] \end{cases} \tag{9}$$

We quantize each domain of $[\boldsymbol{l}_{parent}, \boldsymbol{u}_{parent}]$ into Q_2 levels such that the difference between any two successive levels is the same. Specifically, we quantize the domain of the ith dimension into $\beta_{i,1}, \beta_{i,2}, \ldots, \beta_{i,Q_2}$ where

$$\beta_{i,j} = \begin{cases} \min(p_{1,i}, p_{2,i}) & j = 1 \\ \min(p_{1,i}, p_{2,i}) + (j-1)\left(\frac{|p_{1,i} - p_{2,i}|}{Q_2 - 1}\right) & 2 \leq j \leq Q_2 - 1 \\ \max(p_{1,i}, p_{2,i}) & j = Q_2 \end{cases} \tag{10}$$

We denote $\beta_i = (\beta_{i,1}, \beta_{i,2}, \ldots, \beta_{i,Q_2})$. As the population is being evolved and improved, the population members are getting closer to each other, so that the solution space defined by two parents is becoming smaller. Since Q_2 is fixed, the quantized points are getting closer and hence we can get increasingly precise results.

After quantizing $[\boldsymbol{l}_{parent}, \boldsymbol{u}_{parent}]$, we apply orthogonal design to select a small but representative sample of points as the potential offspring, and then select the ones with the smallest cost to be the offspring. Each pair of parents should not produce too many potential offspring in order to avoid a large number of function evaluations during selection. For this purpose, we divide the variables $x_1, x_2, \ldots, x_N$ into F groups where F is a small design parameter, and each group is treated as one factor. Consequently, the corresponding orthogonal array has a small number of combinations and hence a small number of potential offspring are generated. Specifically, we randomly generate $F-1$ integers $k_1, k_2, \ldots, k_{F-1}$ such that $1 < k_1 < k_2 < \ldots < k_{F-1} < N$, and then create the following F factors for any chromosome $\boldsymbol{x} = (x_1, x_2, \ldots, x_N)$:

$$\begin{cases} \boldsymbol{f}_1 = (x_1, \ldots, x_{k_1}) \\ \boldsymbol{f}_2 = (x_{k_1+1}, \ldots, x_{k_2}) \\ \ldots \\ \boldsymbol{f}_F = (x_{k_{F-1}+1}, \ldots, x_N) \end{cases} \tag{11}$$

Since $x_1, x_2, \ldots, x_N$ have been quantized, we define the following Q_2 levels for the ith factor $\boldsymbol{f}_i$:

$$
\begin{cases}
\boldsymbol{f}_i(1) = (\beta_{k_{i-1}+1,1}, \beta_{k_{i-1}+2,1}, \cdots, \beta_{k_i,1}) \\
\boldsymbol{f}_i(2) = (\beta_{k_{i-1}+1,2}, \beta_{k_{i-1}+2,2}, \cdots, \beta_{k_i,2}) \\
\cdots \\
\boldsymbol{f}_i(Q_2) = (\beta_{k_{i-1}+1,Q_2}, \beta_{k_{i-1}+2,Q_2}, \cdots, \beta_{k_i,Q_2})
\end{cases}
\tag{12}
$$

There are a total of Q_2^F combinations of levels. We apply the orthogonal array $L_{M_2}(Q_2^F)$ to select a sample of M_2 chromosomes as the potential offspring. We remind that *Algorithm 1* can only construct $L_{M_2}(Q_2^F)$ where Q_2 is odd and $M_2 = Q_2^{J_2}$ where J_2 is a positive integer fulfilling $\frac{Q_2^{J_2}-1}{Q_2-1} = F$ (see equation (2)). We bypass this restriction in a manner similar to that in *Algorithm 2*. Specifically, we select the smallest J_2 such that $\frac{Q_2^{J_2}-1}{Q_2-1} \geq F$, then execute *Algorithm 1* to construct an orthogonal array with $F' = \frac{Q_2^{J_2}-1}{Q_2-1}$ factors, and then delete the last $F' - F$ columns of this array. The resulting array has F factors only, and it is denoted by $L_{M_2}(Q_2^F) = [b_{i,j}]_{M_2 \times F}$. The details are given as follows:

Algorithm 4: Construction of $L_{M_2}(Q_2^F)$

Step 1: Select the smallest J_2 fulfilling $\frac{Q_2^{J_2}-1}{Q_2-1} \geq F$.

Step 2: If $\frac{Q_2^{J_2}-1}{Q_2-1} = F$, then $F' = F$ else $F' = \frac{Q_2^{J_2}-1}{Q_2-1}$.

Step 3: Execute *Algorithm 1* to construct the orthogonal array $L_{Q_2^{J_2}}(Q_2^{F'})$.

Step 4: Delete the last $F' - F$ columns of $L_{Q_2^{J_2}}(Q_2^{F'})$ to get $L_{M_2}(Q_2^F)$.

We apply $L_{M_2}(Q_2^F)$ to generate the following M_2 chromosomes out of Q_2^F possible chromosomes:

$$
\begin{cases}
(\boldsymbol{f}_1(b_{1,1}), \boldsymbol{f}_2(b_{1,2}), \dots, \boldsymbol{f}_F(b_{1,F})) \\
(\boldsymbol{f}_1(b_{2,1}), \boldsymbol{f}_2(b_{2,2}), \dots, \boldsymbol{f}_F(b_{2,F})) \\
\cdots \\
(\boldsymbol{f}_1(b_{M_2,1}), \boldsymbol{f}_2(b_{M_2,2}), \dots, \boldsymbol{f}_F(b_{M_2,F}))
\end{cases}
\tag{13}
$$

The details of orthogonal crossover are given as follows:

Algorithm 5: Orthogonal Crossover

Step 1: Quantize $[l_{parent}, u_{parent}]$ based on equation (10).

Step 2: Randomly generate $F-1$ integers $k_1, k_2, \ldots, k_{F-1}$ such that $1 < k_1 < k_2 < \ldots < k_{F-1} < N$. Create F factors based on equation (11).

Step 3: Apply $L_{M_2}(Q_2^F)$ to generate M_2 potential offspring based on equation (13).

3.3 Orthogonal Genetic Algorithm

We generate a good initial population with G chromosomes, and then evolve and improve the population iteratively. In each iteration, we apply orthogonal crossover and mutation to generate a set of potential offspring. Among these potential offspring and the parents, we select the G chromosomes with the smallest cost to form the next generation. We let $\boldsymbol{P}_{gen}$ be a population of chromosomes in the ith generation. The details of the overall algorithm are as follows:

Orthogonal Genetic Algorithm

Step 1: Initialization

Step 1.1: Execute *Algorithm 2* to construct $L_{M_1}(Q_1^N)$.

Step 1.2: Execute *Algorithm 3* to generate an initial population $\boldsymbol{P}_0$. Initialize the generation number *gen* to 0.

Step 1.3: Execute *Algorithm 4* to construct $L_{M_2}(Q_2^F)$.

Step 2: Population Evolution

WHILE (stopping condition is not met) DO
BEGIN

Step 2.1: **Orthogonal Crossover**
Each chromosome in $\boldsymbol{P}_{gen}$ is selected for crossover with probability p_c. When the number of selected chromosomes is odd, select one additional chromosome from $\boldsymbol{P}_{gen}$ randomly. Let the number of selected chromosomes be $2i$. Form i random pairs of chromosomes. Execute *Algorithm 5* to perform orthogonal crossover on each pair.

Step 2.2: **Mutation**
Each chromosome in $\boldsymbol{P}_{gen}$ undergoes mutation with probability p_m. To perform mutation on a chromosome, randomly generate an integer $j \in [1, N]$ and a real number $z \in [l_j, u_j]$ and then replace the jth component of the chosen chromosome by z to get a new chromosome.

Step 2.3: **Selection**
Among the chromosomes in $\boldsymbol{P}_{gen}$ and those generated by crossover and mutation, select the G chromosomes with the least cost to form the next generation $\boldsymbol{P}_{gen+1}$.

Step 2.4: Increment the generation number *gen* by 1.

END

In step 2, the population is evolved and improved iteratively until a stopping condition is met. Similar to the other genetic algorithms, there can be many possible stopping conditions. For example, one possible stopping condition is to stop when the best chromosome cannot be further improved in a certain number of generations.

4 Application to Multimedia Multicast Routing

In the last section, we describe an approach of designing genetic algorithms using the orthogonal design. This approach is general but its details are different for different problems. In this section, we describe how to apply this approach to design a genetic algorithm for multimedia multicast routing in communication networks.

4.1 Multimedia Multicast Routing

Many multimedia communication applications require a source to send multimedia data to multiple destinations through a communication network. The following are three examples:

1. In a multiparty multimedia teleconference, the conferees located at different sites can see and communicate with each other in real time. For this purpose, the video and voice captured at each conference site are sent to all the other conference sites [12,13].
2. In a remote video lecture for distant education, the video and voice of the instructor are sent to all the students through a communication network [14].
3. In video-on-demand systems, the major bottleneck is to retrieve video from the disks [15]. To relieve this bottleneck, we can batch a number of customer requests for the same video object and then use one I/O stream to serve multiple customers [15]. Using this method, multimedia information is sent from the video server to multiple customers.

To support these applications, it is necessary to determine a *multicast tree* of minimal cost [16,17] for every communication session. Through this multicast tree, the source node can send multimedia data to all the destination nodes. Since multimedia data cannot be excessively delayed [18], the total delay from the source node to any destination node must be smaller than a given requirement.

Example

Fig. 3(a) shows a communication network with 10 nodes where node i is located in city i. The cost and the delay of communication between every pair of adjacent nodes is 1.0. Suppose a user in city 1 wants to send multimedia information to five users in cities 2, 4, 8, 9 and 10 respectively. For this purpose, node 1 (the source node) sends this multimedia information to nodes 2, 4, 8, 9 and 10 (the destination nodes). Suppose the communication delay between the source node and any destination node cannot be larger than 5.0. Fig. 3(b) shows a multicast tree for this communication purpose. The source node sends the multimedia information to the five destination nodes through this multicast tree as follows:

- Node 1 sends the multimedia information to node 2.
- Node 2 sends the information to: (i) the user in city 2 and (ii) nodes 4 and 5.
- Node 4 sends the information to: (i) the user in city 4 and (ii) node 8.
- Node 5 sends the information to node 9.
- Node 8 sends the information to the user in city 8.
- Node 9 sends the information to: (i) the user in city 9 and (ii) node 10.
- Node 10 sends the information to the user in city 10.

In this manner, the users in cities 2, 4, 8, 9 and 10 can receive the multimedia information sent from the user in city 1.

The problem of determining multicast trees is known as *multimedia multicast routing* and has been proved to be NP-complete [16]. Therefore, good heuristic algorithms are of practical interest. To formulate the multimedia multicast routing problem, we model the communication network as a graph $G = (V, E)$, where V is a set of nodes and E is a set of edges. The cost and delay of edge e are denoted by $C(e)$ and $D(e)$ respectively, where $C(e)$ and $D(e)$ can be any positive real numbers. The cost $C(e)$ may measure the monetary utilization cost or the degree of congestion in edge e. For example, when most of the channels in an edge are being occupied, we can choose to assign a larger cost to this edge. This can encourage the new sessions to adopt the other edges, thereby balancing the traffic over the whole network. The delay $D(e)$ measures the time required to transmit a piece of information through edge e. We let the source node be node s and the set of destination nodes be S, where $\{s\} \cup S \subseteq V$. The delay requirement specifies that the total delay from the source node to *any* destination node must be smaller than or equal to Δ, where Δ can be any positive real number.

The multimedia multicast routing problem is to determine a multicast tree connecting the source node to every destination node such that the cost

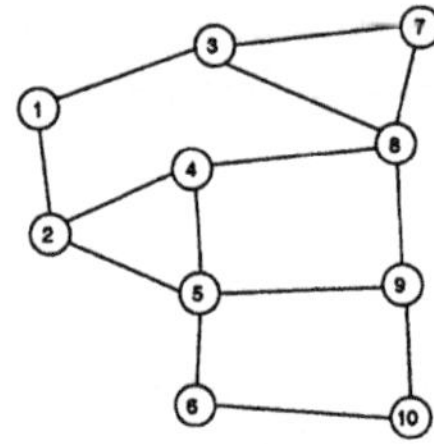

(a) A communication network in which the node i is located in city i .

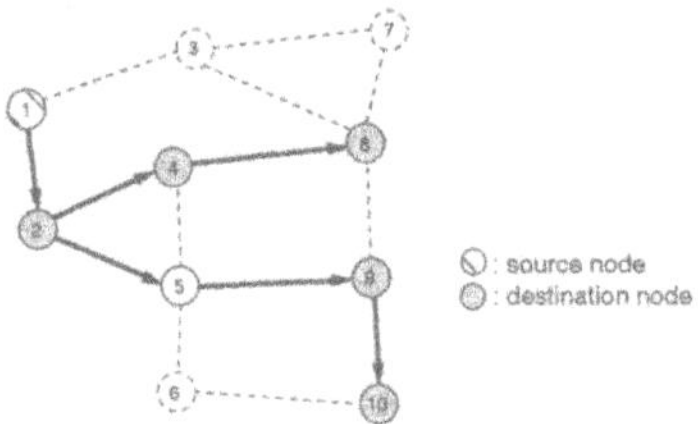

(b) A multicast tree through which multimedia information is sent from node 1 to nodes 2, 4, 8, 9 and 10.

Fig. 3. Example of a multicast tree.

of this tree is minimal while the total delay from the source node to *any* destination node is smaller than Δ. Mathematically, the problem is to find a tree $T = (V_T, E_T)$ (where $V_T \subseteq V$ and $E_T \subseteq E$) such that the total cost of this tree $\sum_{e \in E_T} C(e)$ is minimized subject to the following two constraints:

(1) $\{s\} \cup S \subseteq V_T$ and
(2) $\sum_{e \in P(s,v)} D(e) \leq \Delta$ for every $v \in S$, where $P(s,v)$ is a set of edges constituting the path from source node s to destination node v in the tree T.

4.2 Binary String Representation and Fitness Vector

We number the edges in the graph G from 1 to l. We can then represent any multicast tree T of the graph G as a binary string $(e_1, e_2, \ldots, e_l)$ where

$$e_i = \begin{cases} 1 & \text{if edge } i \text{ is included in the multicast tree } T \\ 0 & \text{otherwise} \end{cases} \tag{14}$$

We use the *fitness vector* $\boldsymbol{f} = (f_1, f_2)$ to measure the quality of a multicast tree where

$$\begin{cases} f_1 = \sum_{e \in E_T} C(e) \\ f_2 = \sum_{v \in S} \max\{0, \sum_{e \in P(s,v)} D(e) - \Delta\} \end{cases} \tag{15}$$

f_1 measures the cost of the multicast tree and f_2 measures how tight the delay constraints are fulfilled. A multicast tree is good if f_1 is small (i.e., a small cost) and f_2 is zero (i.e., the delay constraints are fulfilled). Given two multicast trees T_1 and T_2 with respective fitness vectors (f_1', f_2') and (f_1'', f_2''), we say that T_1 is better than T_2 if and only if

(1) $f_2' < f_2''$ or
(2) $f_2' = f_2''$ and $f_1' < f_1''$.

In this manner, we can minimize the cost of the multicast tree and enforce fulfilling the delay constraints.

4.3 Check and Repair Operation

In genetic algorithms, crossover and mutation are used to search the solution space. After performing these operations, the resulting binary string induces a subgraph which may not be a tree connecting the source node to all the destination nodes. To remedy this, we design a *check-and-repair operation.* This operation applies a simple method called *graph search method* [19] to traverse the subgraph. In particular, it applies this method to find a spanning tree from the subgraph and then connect the unconnected destination nodes to the tree. The details are as follows:

Algorithm 6: Check-and-Repair Operation

Step 1: If the subgraph induced by the input binary string is a tree connecting the source node to all the destination nodes, then stop.

Step 2: Apply the graph search method to find a spanning tree from the subgraph.

Step 3: For the destination nodes that are not included in the tree, apply the graph search method to connect them to the tree one after the other. The output binary string represents this tree.

The subgraph is only a part of the entire graph. The check-and-repair operation operates on this subgraph using the simple graph search method, and hence it can be executed quickly.

4.4 Orthogonal Crossover

In this subsection, we apply the orthogonal design to design a crossover operation for multimedia multicast routing. For illustration, we use the following orthogonal array $L_4(2^3)$:

$$L_4(2^3) = \begin{bmatrix} 1 & 1 & 1 \\ 1 & 2 & 2 \\ 2 & 1 & 2 \\ 2 & 2 & 1 \end{bmatrix} \tag{16}$$

Fig. 4 illustrates how to use $L_4(2^3)$ to sample the genes from two parents for crossover. Since $L_4(2^3)$ has three factors, we divide each parent string into three parts. We sample these parts from the two parents based on the four combinations of factor levels in $L_4(2^3)$, thereby producing four binary strings. Among these four binary strings, we select two of them to be the offspring. In general, for $L_m(n^k)$ with k factors and m combinations of factor levels, we divide each parent string into k parts, sample these parts from n parents based on the m combinations in $L_m(n^k)$ to produce m binary strings, and then select j of them to be the offspring. The details are given as follows:

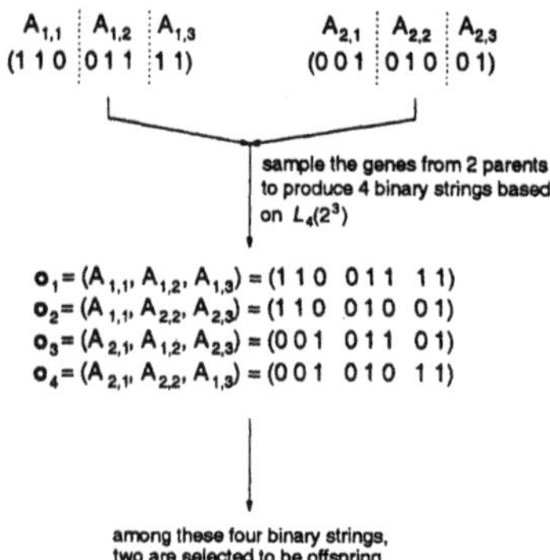

Fig. 4. The orthogonal array $L_4(2^3)$ is used to sample the genes from two parents for crossover.

Algorithm 7: n-to-j Orthogonal Crossover and Mutation

Step 1: Randomly and independently generate $p(i) \in \{1, 2, \ldots k\}$ for all $1 \le i \le l$.

Step 2: Based on the ith combination of factor levels $(a_1(i), a_2(i), \ldots, a_k(i))$ in the orthogonal array $L_m(n^k)$, produce the binary string $o_i = (x_{a_{p(1)}(i),1}, x_{a_{p(2)}(i),2}, \ldots, x_{a_{p(l)}(i),l})$ for all $1 \le i \le m$.

Step 3: Each $\boldsymbol{o}_i$ undergoes mutation with a small probability p_m. To perform mutation on a binary string, flip every bit in this string with a small probability p_f.

Step 4: Execute the check-and-repair operation on $\boldsymbol{o}_i$ for all $1 \leq i \leq m$.

Step 5: Evaluate the fitness vectors of $\boldsymbol{o}_1, \boldsymbol{o}_2, \ldots$ and $\boldsymbol{o}_m$, and then select j of them to be the offspring (these offspring are denoted by $\bar{\boldsymbol{o}}_1, \bar{\boldsymbol{o}}_2, \ldots$ and $\bar{\boldsymbol{o}}_j$).

In Step 1, we generate the indexes $p(i)(1 \leq i \leq l)$ to partition each parent string into k parts. In Step 2, we sample these parts from n parents based on $L_m(n^k)$ to produce m binary strings. In Step 3, we perform mutation with a small probability. In Step 4, we execute the check-and-repair operation on every string. In Step 5, we evaluate the fitness vector of every string and then select j of them to be the offspring.

In orthogonal crossover, one of the design choices is the orthogonal array. In general, a larger orthogonal array provides more combinations of factor levels and hence it can perform a better sampling for crossover. The resulting computational complexity, however, is also larger. As we shall demonstrate in section 4.6, $L_9(3^4)$ is already a good choice for multimedia multicast routing with practical problem sizes. Another design choice is the selection scheme because there are many possible ways to select j offspring among the m binary strings produced in Step 2. For example, to ensure a monotonic crossover operation (i.e., the best offspring is not poorer than the best parent), we can simply select the best parent to be an offspring and the best one among $\boldsymbol{o}_1, \boldsymbol{o}_2, \ldots$ and $\boldsymbol{o}_m$ to be another offspring. As another example, we can also introduce randomness into the selection scheme (say, $\boldsymbol{o}_i$ with a lower fitness can be chosen to be an offspring with a small probability). In our numerical experiments (see section 4.6), the best j offspring among $\boldsymbol{o}_1, \boldsymbol{o}_2, \ldots, \boldsymbol{o}_m$ are selected.

4.5 Orthogonal Genetic Algorithm

We apply the n-to-j orthogonal crossover and mutation operation to evolve and improve a population iteratively. Since n can be different from j, we perform this operation $\frac{N}{j}$ times in each generation in order to maintain a constant population size. The details are as follows:

Orthogonal Genetic Algorithm

Step 1: **Initialization**
Randomly create an initial generation of N binary strings $\boldsymbol{P}_0 = \{\boldsymbol{x}_1, \boldsymbol{x}_2, \ldots, \boldsymbol{x}_N\}$, and initialize the generation number *gen* to 0.

Step 2: **Population Evolution**
WHILE (stopping condition is not met) DO
BEGIN

Step 2.1: **n-to-j Orthogonal Crossover and Mutation**

DO $\frac{N}{j}$ times

Randomly select n parent strings from $\boldsymbol{P}_{gen}$, and perform n-to-j orthogonal crossover and mutation on them to generate j offspring $\bar{\boldsymbol{o}}_1, \bar{\boldsymbol{o}}_2, \ldots \bar{\boldsymbol{o}}_j$ for the next generation $\boldsymbol{P}_{gen+1}$.

END

Step 2.2 Increment the generation number gen by 1.

END

In Step 2, the population is evolved and improved iteratively until a stopping condition is met. One possible stopping criterion is to halt when the number of generations gen is equal to a given maximum value.

In many traditional genetic algorithms, global selection of parents is needed and this is a serious bottleneck in parallel implementation. In the above algorithm, selection is performed only within the orthogonal crossover operation but global selection is not needed. In each generation, all the $\frac{N}{j}$ orthogonal crossover and mutation operations can be executed in parallel. Therefore, the above algorithm is well suited for parallel implementation and execution.

To create a new generation, orthogonal crossover is executed $\frac{N}{j}$ times and mutation is executed $p_m N$ times on average. In every execution of orthogonal crossover and mutation, m and 1 new binary strings are produced respectively. Therefore, the algorithm has to handle an average of $(\frac{mN}{j} + p_m N)$ new binary strings in every generation. For the special cases $L_4(2^3)$ and $L_9(3^4)$, the factor levels of the first combination are all equal to 1 and hence the first binary string $\boldsymbol{o}_1$ is equal to the first parent string $\boldsymbol{x}_1$. In these special cases, three and eight new binary strings are respectively produced in every execution of orthogonal crossover.

4.6 Numerical Results

4.6.1 Test Problems : A test problem generator was proposed in [20] to generate test problems for multicast routing *without any delay constraint.* This generator needs three inputs (namely, the number of nodes, the number of links, and the number of source/destination nodes), and it will generate a test problem (which specifies a network topology, the link costs, a set of source and destination nodes, and an optimal multicast tree Γ with cost C^*). We now enhance this generator, so that it can generate test problems for multimedia multicast routing with delay constraints. The details are as follows:

Step 1: Execute the problem generator proposed in [20] to generate a test problem. Record the network, the set of source and destination nodes, the optimal multicast tree Γ and its cost C^*.

Step 2: Randomly assign a delay value to every link in this network.

Step 3: Compute the delay Δ_v from the source node to destination node v in the multicast tree Γ for all $v \in S$. The delay requirement Δ is chosen to be $\max_{v \in S}\{\Delta_v\}$.

In Step 1, we generate a test problem without any delay constraint. In Step 2, we randomly assign a delay value to every link in the network. In Step 3, we choose the delay requirement Δ such that the delay constraint can always be fulfilled. In this manner, the optimal multicast tree is still Γ with cost C^*. By executing the above three steps once, we get a test problem for multimedia multicast routing.

A real-world multimedia wide area network usually has several tens of nodes with a moderate connectivity. For example, the Arpanet has 20 nodes [21] and the Arpa2 has 21 nodes [22]. We generate two sets of test problems with two respective network sizes: (1) 50 nodes with 100 edges and (2) 80 nodes with 200 edges. In these networks, the number of possible multicast trees is already very large. Every set is further divided into eight groups of test problems, where every group consists of 100 different problem instances and different groups adopt different number of destination nodes. Therefore, there are a total of 1600 different test problems. Fig. 5 shows the test problems.

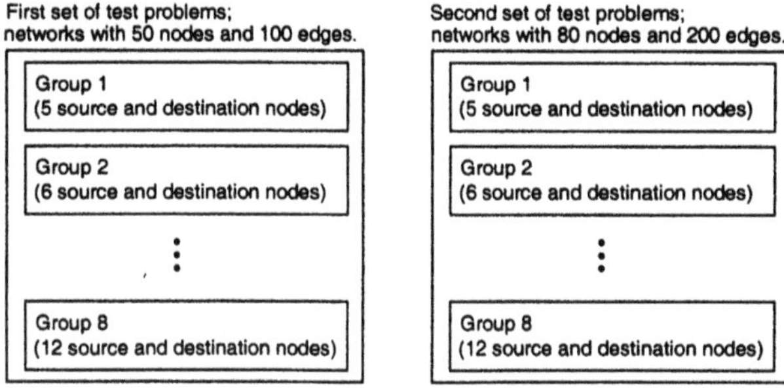

Fig. 5. Two sets of test problems. Every group consists of 100 test problems. There are a total of 1600 test problems.

4.6.2 Numerical Experiments and Performance Measures: We test two versions of the orthogonal genetic algorithm. The first version adopts 2-to-2 orthogonal crossover based on $L_4(2^3)$, and the second version adopts 3-to-2 orthogonal crossover based on $L_9(3^4)$. These two versions are referred to as OGA_4 and OGA_9 respectively. For comparison, we also consider the following traditional genetic algorithm (TGA), where TGA is basically the same as OGA_4 and OGA_9 except that TGA does not use the orthogonal design.

Table 3. Traditional genetic algorithm; first set of test problems.

1st Set of Test Problems	Percentage of runs in which the costs are within x% from optimal				
	$x = 20$	$x = 10$	$x = 5$	$x = 2$	$x = 0$
Group 1	0%	0%	0%	0%	0%
Group 2	0%	0%	0%	0%	0%
Group 3	1%	0%	0%	0%	0%
Group 4	0%	0%	0%	0%	0%
Group 5	0%	0%	0%	0%	0%
Group 6	0%	0%	0%	0%	0%
Group 7	0%	0%	0%	0%	0%
Group 8	0%	0%	0%	0%	0%

(a) Solution quality.

1st Set of Test Problems	Mean number of generations to find a solution whose cost is within x% from optimal				
	$x = 20$	$x = 10$	$x = 5$	$x = 2$	$x = 0$
Group 1	(NA)	(NA)	(NA)	(NA)	(NA)
Group 2	(NA)	(NA)	(NA)	(NA)	(NA)
Group 3	6.00	(NA)	(NA)	(NA)	(NA)
Group 4	(NA)	(NA)	(NA)	(NA)	(NA)
Group 5	(NA)	(NA)	(NA)	(NA)	(NA)
Group 6	(NA)	(NA)	(NA)	(NA)	(NA)
Group 7	(NA)	(NA)	(NA)	(NA)	(NA)
Group 8	(NA)	(NA)	(NA)	(NA)	(NA)

(b) Computational complexity: mean number of generations required.

1st Set of Test Problems	Mean execution time required (seconds) to find a solution whose cost is within x% from optimal				
	$x = 20$	$x = 10$	$x = 5$	$x = 2$	$x = 0$
Group 1	(NA)	(NA)	(NA)	(NA)	(NA)
Group 2	(NA)	(NA)	(NA)	(NA)	(NA)
Group 3	0.25	(NA)	(NA)	(NA)	(NA)
Group 4	(NA)	(NA)	(NA)	(NA)	(NA)
Group 5	(NA)	(NA)	(NA)	(NA)	(NA)
Group 6	(NA)	(NA)	(NA)	(NA)	(NA)
Group 7	(NA)	(NA)	(NA)	(NA)	(NA)
Group 8	(NA)	(NA)	(NA)	(NA)	(NA)

(c) Computational complexity: mean execution time required.

Traditional Genetic Algorithm

Step 1: Initialization
Randomly create an initial generation of N binary strings and initialize the generation number *gen* to 0.

Step 2: Population Evolution
WHILE (stopping condition is not met) DO
BEGIN

Table 4. Orthogonal genetic algorithm using $L_4(2^3)$; first set of test problems.

1st Set of Test Problems	Percentage of runs in which the costs are within x% from optimal				
	$x = 20$	$x = 10$	$x = 5$	$x = 2$	$x = 0$
Group 1	33%	16%	10%	6%	6%
Group 2	56%	35%	30%	20%	20%
Group 3	10%	4%	0%	0%	0%
Group 4	8%	1%	1%	1%	1%
Group 5	8%	1%	1%	1%	1%
Group 6	3%	1%	0%	0%	0%
Group 7	3%	0%	0%	0%	0%
Group 8	0%	0%	0%	0%	0%

(a) Solution quality.

1st Set of Test Problems	Mean number of generations to find a solution whose cost is within x% from optimal				
	$x = 20$	$x = 10$	$x = 5$	$x = 2$	$x = 0$
Group 1	92.60	72.87	78.00	86.83	86.83
Group 2	97.50	114.91	122.06	124.15	124.15
Group 3	99.50	107.50	(NA)	(NA)	(NA)
Group 4	113.75	46.00	46.00	46.00	46.00
Group 5	125.00	139.00	139.00	139.00	139.00
Group 6	142.33	34.00	(NA)	(NA)	(NA)
Group 7	139.66	(NA)	(NA)	(NA)	(NA)
Group 8	(NA)	(NA)	(NA)	(NA)	(NA)

(b) Computational complexity: mean number of generations required.

1st Set of Test Problems	Mean execution time required (seconds) to find a solution whose cost is within x% from optimal				
	$x = 20$	$x = 10$	$x = 5$	$x = 2$	$x = 0$
Group 1	11.14	8.58	9.08	10.13	10.13
Group 2	15.21	18.05	19.22	20.62	20.63
Group 3	12.11	13.49	(NA)	(NA)	(NA)
Group 4	16.08	7.81	7.81	7.81	7.81
Group 5	20.55	26.67	26.67	26.67	26.67
Group 6	20.12	4.82	(NA)	(NA)	(NA)
Group 7	20.49	(NA)	(NA)	(NA)	(NA)
Group 8	(NA)	(NA)	(NA)	(NA)	(NA)

(c) Computational complexity: mean execution time required.

Step 2.1: **Crossover**

DO $\frac{N}{2}$ times

Randomly select four parent strings from the current generation. Perform single-point crossover on the two best strings to generate two offspring for the next generation.

END

Table 5. Orthogonal genetic algorithm using $L_9(3^4)$; first set of test problems.

1st Set of Test Problems	Percentage of runs in which the costs are within x% from optimal				
	$x = 20$	$x = 10$	$x = 5$	$x = 2$	$x = 0$
Group 1	100%	100%	100%	95%	95%
Group 2	100%	100%	99%	99%	99%
Group 3	100%	99%	97%	82%	82%
Group 4	100%	95%	92%	89%	89%
Group 5	100%	98%	97%	95%	81%
Group 6	100%	100%	97%	91%	91%
Group 7	100%	100%	97%	90%	83%
Group 8	100%	100%	96%	78%	65%

(a) Solution quality.

1st Set of Test Problems	Mean number of generations to find a solution whose cost is within x% from optimal				
	$x = 20$	$x = 10$	$x = 5$	$x = 2$	$x = 0$
Group 1	14.56	21.15	22.28	24.76	24.76
Group 2	11.06	14.43	17.73	21.24	21.24
Group 3	15.81	20.86	27.60	39.01	39.01
Group 4	19.08	25.14	33.01	31.35	31.35
Group 5	17.43	23.66	25.76	26.89	33.85
Group 6	15.67	22.20	32.07	34.93	36.35
Group 7	15.15	21.62	30.78	33.64	35.45
Group 8	15.12	23.99	33.58	39.24	41.32

(b) Computational complexity: mean number of generations required.

1st Set of Test Problems	Mean execution time required (seconds) to find a solution whose cost is within x% from optimal				
	$x = 20$	$x = 10$	$x = 5$	$x = 2$	$x = 0$
Group 1	3.18	4.56	4.82	5.36	5.39
Group 2	2.46	3.18	3.88	4.62	4.62
Group 3	3.52	4.60	6.03	8.41	8.41
Group 4	4.33	5.65	7.35	7.02	7.02
Group 5	3.96	5.33	5.80	6.06	7.56
Group 6	3.80	5.33	7.60	8.27	8.61
Group 7	3.67	5.17	7.27	7.94	8.36
Group 8	3.70	5.78	7.99	9.32	9.81

(c) Computational complexity: mean execution time required.

Step 2.2: **Mutation**
Each string in the new generation undergoes mutation with a small probability p_m. To perform mutation on a binary string, flip every bit in this string with a small probability p_f.

Step 2.3: Execute the check-and-repair operation for the strings in the new generation.

Step 2.4: Increment the generation number *gen* by 1.

END

We adopt the following parameters or schemes for the experiments:

- *Population size*: We adopt the same population size for the TGA, OGA_4 and OGA_9, so that they have the same space complexity. In particular, the population sizes are experimentally chosen to be 30 and 80 for the first set and the second set of test problems respectively.
- *Stopping condition*: We note that routing must be done within a specified interval, so that the users can start a multimedia session without waiting indefinitely. Therefore, it is desirable to find a feasible and good multicast tree within a specified number of generations. To study the effectiveness of an algorithm, we evaluate the solution quality after it has been executed for a given number of generations. Since the ratios of the computational complexity per generation for the TGA, OGA_4 and OGA_9 are $2:3:8$, we choose the maximum number of generations for the TGA, OGA_4 and OGA_9 to be 400, 300 and 100 respectively. Since this stopping condition is based on the computational complexity, it implicitly takes the number of evaluations of the cost into account.
- *Selection*: In the orthogonal crossover operation, among the binary strings produced, we select the best two to be the offspring.
- *Mutation probability*: We choose $p_m = p_f = 0.1$, such that every bit undergoes mutation with probability 0.01.

To measure the solution quality, we record the percentage of runs in which the costs of the feasible solutions are within $x\%$ from the optimal one. To measure the computational complexity, we record two quantities for finding a feasible solution whose cost is within $x\%$ from the optimal one: (1) the mean number of generations required and (2) the mean execution time on a Sun Sparc 4 workstation.

4.6.3 Results: The results for the first set of test problems are shown in Tables 3-5. In particular, Tables 3(a), 4(a) and 5(a) show the solution quality when the algorithms have been executed for the given numbers of generations. We see that the TGA can only find one feasible solution out of 800 test problems, OGA_4 can find feasible solutions for many test problems, and OGA_9 can find feasible solutions for all the test problems. In addition, OGA_9 can give the closest-to-optimal solutions. For example, 95.63% of the solutions are within 5% of the optimum, and 89.75% of the solutions are within 2% of the optimum. This indicates the effectiveness of experimental design methods: if we use a larger orthogonal array for a more representative sampling of genes for crossover, we can get better solutions. Nevertheless, Tables 5(b)-5(c) show that OGA_9 only has a moderate computational complexity.

Table 6. Traditional genetic algorithm; second set of test problems.

2nd Set of Test Problems	Percentage of runs in which the costs are within x% from optimal				
	$x = 20$	$x = 10$	$x = 5$	$x = 2$	$x = 0$
Group 1	3%	0%	0%	0%	0%
Group 2	6%	4%	0%	0%	0%
Group 3	0%	0%	0%	0%	0%
Group 4	0%	0%	0%	0%	0%
Group 5	2%	2%	0%	0%	0%
Group 6	8%	3%	0%	0%	0%
Group 7	18%	2%	2%	0%	0%
Group 8	0%	0%	0%	0%	0%

(a) Solution quality.

2nd Set of Test Problems	Mean number of generations to find a solution whose cost is within x% from optimal				
	$x = 20$	$x = 10$	$x = 5$	$x = 2$	$x = 0$
Group 1	9.30	(NA)	(NA)	(NA)	(NA)
Group 2	9.00	9.25	(NA)	(NA)	(NA)
Group 3	(NA)	(NA)	(NA)	(NA)	(NA)
Group 4	(NA)	(NA)	(NA)	(NA)	(NA)
Group 5	8.00	8.00	(NA)	(NA)	(NA)
Group 6	10.00	9.00	(NA)	(NA)	(NA)
Group 7	1.027	10.00	10.00	(NA)	(NA)
Group 8	(NA)	(NA)	(NA)	(NA)	(NA)

(b) Computational complexity: mean number of generations required.

2nd Set of Test Problems	Mean execution time required (seconds) to find a solution whose cost is within x% from optimal				
	$x = 20$	$x = 10$	$x = 5$	$x = 2$	$x = 0$
Group 1	1.29	(NA)	(NA)	(NA)	(NA)
Group 2	1.28	1.30	(NA)	(NA)	(NA)
Group 3	(NA)	(NA)	(NA)	(NA)	(NA)
Group 4	(NA)	(NA)	(NA)	(NA)	(NA)
Group 5	1.29	1.29	(NA)	(NA)	(NA)
Group 6	2.00	1.58	(NA)	(NA)	(NA)
Group 7	1.89	1.77	1.77	(NA)	(NA)
Group 8	(NA)	(NA)	(NA)	(NA)	(NA)

(c) Computational complexity: mean execution time required.

For group 1 test problems, the mean number of generations required is 24.76 (equivalently, the mean execution time on a Sun Sparc 4 workstation is 4.16 seconds).

The results for the second set of test problems are shown in Tables 6-8. The network size is now larger. We observe that OGA_9 is still better than OGA_4, which is in turn better than TGA. In addition, the OGA_9 can find close-to-optimal solutions within moderate numbers of generations.

Table 7. Orthogonal genetic algorithm using $L_4(2^3)$; second set of test problems.

2nd Set of Test Problems	Percentage of runs in which the costs are within x% from optimal				
	$x = 20$	$x = 10$	$x = 5$	$x = 2$	$x = 0$
Group 1	18%	3%	2%	2%	2%
Group 2	15%	9%	9%	1%	1%
Group 3	1%	1%	1%	0%	0%
Group 4	0%	0%	0%	0%	0%
Group 5	1%	0%	0%	0%	0%
Group 6	0%	0%	0%	0%	0%
Group 7	0%	0%	0%	0%	0%
Group 8	0%	0%	0%	0%	0%

(a) Solution quality.

2nd Set of Test Problems	Mean number of generations to find a solution whose cost is within x% from optimal				
	$x = 20$	$x = 10$	$x = 5$	$x = 2$	$x = 0$
Group 1	116.66	106.00	94.50	94.50	94.50
Group 2	108.00	117.11	117.11	134.00	134.00
Group 3	75.00	75.00	75.00	(NA)	(NA)
Group 4	(NA)	(NA)	(NA)	(NA)	(NA)
Group 5	214.00	(NA)	(NA)	(NA)	(NA)
Group 6	(NA)	(NA)	(NA)	(NA)	(NA)
Group 7	(NA)	(NA)	(NA)	(NA)	(NA)
Group 8	(NA)	(NA)	(NA)	(NA)	(NA)

(b) Computational complexity: mean number of generations required.

2nd Set of Test Problems	Mean execution time required (seconds) to find a solution whose cost is within x% from optimal				
	$x = 20$	$x = 10$	$x = 5$	$x = 2$	$x = 0$
Group 1	69.73	59.16	53.83	53.84	53.87
Group 2	58.51	63.11	63.12	68.31	68.33
Group 3	63.20	63.20	63.20	(NA)	(NA)
Group 4	(NA)	(NA)	(NA)	(NA)	(NA)
Group 5	138.89	(NA)	(NA)	(NA)	(NA)
Group 6	(NA)	(NA)	(NA)	(NA)	(NA)
Group 7	(NA)	(NA)	(NA)	(NA)	(NA)
Group 8	(NA)	(NA)	(NA)	(NA)	(NA)

(c) Computational complexity: mean execution time required.

From the above results for these test problems, we can conclude two points:

- OGA_9 is better than OGA_4, which is better than TGA. This indicates that experimental design methods can effectively improve the genetic algorithms.

Table 8. Orthogonal genetic algorithm using $L_9(3^4)$; second set of test problems.

2nd Set of Test Problems	Percentage of runs in which the costs are within x% from optimal				
	$x = 20$	$x = 10$	$x = 5$	$x = 2$	$x = 0$
Group 1	100%	99%	99%	99%	99%
Group 2	100%	100%	100%	92%	92%
Group 3	100%	99%	93%	92%	92%
Group 4	100%	100%	100%	95%	95%
Group 5	100%	100%	99%	88%	88%
Group 6	100%	100%	100%	99%	46%
Group 7	100%	100%	100%	93%	67%
Group 8	100%	100%	86%	77%	65%

(a) Solution quality.

2nd Set of Test Problems	Mean number of generations to find a solution whose cost is within x% from optimal				
	$x = 20$	$x = 10$	$x = 5$	$x = 2$	$x = 0$
Group 1	13.97	20.03	20.81	20.81	20.81
Group 2	15.56	18.87	19.95	28.08	28.08
Group 3	19.27	22.40	25.74	27.80	27.80
Group 4	19.60	22.83	23.74	27.32	27.32
Group 5	21.08	24.49	26.46	28.85	28.85
Group 6	14.97	19.34	22.80	24.43	40.65
Group 7	16.16	20.93	25.27	32.02	38.56
Group 8	21.80	26.76	31.73	34.06	35.36

(b) Computational complexity: mean number of generations required.

2nd Set of Test Problems	Mean execution time required (seconds) to find a solution whose cost is within x% from optimal				
	$x = 20$	$x = 10$	$x = 5$	$x = 2$	$x = 0$
Group 1	13.26	18.53	19.27	19.27	19.27
Group 2	14.50	17.36	18.31	25.24	25.24
Group 3	18.24	21.01	23.95	25.76	25.76
Group 4	18.81	21.70	22.52	25.71	25.71
Group 5	20.33	23.43	25.22	27.40	27.40
Group 6	14.69	18.68	21.82	23.30	37.86
Group 7	16.39	20.88	24.94	31.16	37.31
Group 8	22.05	26.77	31.43	33.70	34.96

(c) Computational complexity: mean execution time required.

- OGA_9 can find close-to-optimal solutions within moderate numbers of generations. This indicates that the orthogonal array $L_9(3^4)$ is a good choice for multimedia multicast routing with practical problem sizes.

5 Summary

We recently proposed to apply experimental design methods to design evolutionary algorithms [1,2], so that the resulting algorithms can be more powerful and statistically-sound. In this chapter, we summarized our recent results on this research direction [1,2]. In particular, we described how to apply the orthogonal design to design a new class of genetic algorithms called orthogonal genetic algorithms, and described an orthogonal genetic algorithm for multimedia multicast routing. The numerical results show that the orthogonal genetic algorithm has a significantly better performance than the traditional genetic algorithm. This demonstrates that the orthogonal design can effectively enhance the genetic algorithms.

References

1. Zhang Q, Leung YW (1999) An orthogonal genetic algorithm for multimedia multicast routing. IEEE Trans. Evol. Comput., vol. 3, no. 1, pp. 53-62.
2. Leung YW, Wang Y (2001) An orthogonal genetic algorithm with quantization for global numerical optimization. IEEE Trans. Evol. Comput., vol. 5, no. 1, pp. 41-53.
3. Leung YW (1999) Interactive multiobjective programming using uniform design. FRG Research Grant Proposal, Hong Kong Baptist University.
4. Leung YW, Wang Y (2000) Multiobjective porgramming using uniform desgin and genetic algorithm. IEEE Trans. Syst. Man Cyber. Part C, vol. 30, no. 3, pp. 293-304.
5. Goldberg DE (1989) Genetic Algorithm in Search, Optimization and Machine Learning, Addison Wesley, Reading, MA.
6. Back T, Fogel DB, Michalewicz Z (ed.) (1997) Handbook of Evolutionary Computation, Institute of Physics Publishing.
7. Gen M, Cheng R (1999) Genetic Algorithms and Engineering Optimization, Wiley.
8. Montgomery DC (1991) Design and Analysis of Experiments, 3rd ed., John Wiley, NY.
9. Hicks CR (1993) Fundamental Concepts in the Design of Experiments, 4th ed., Saunders College Publishing, Texas.
10. Wu Q (1978) On the optimality of orthogonal experimental design. Acta Mathematical Applacatae Sinica, vol. 1, no. 4, pp. 283-299.
11. Math. Stat. Research Group in Chinese Academy of Science (1975) Orthogonal Design (in Chinese), People Education Pub., Beijing.
12. Leung YW, Yum TS (1996) Connection optimization for two types of videoconferences. IEEE Proc. Commun., vol. 143, no. 3, pp. 133-140.
13. Yum TS, Chen MS, Leung YW (1995) Video bandwidth allocation for multimedia teleconferences. IEEE Trans. Commun., vol. 43, no. 2, pp. 457-465.
14. IEICE Trans. Info. Syst. (1997) Special Issue on Educational System Using Multimedia and Communication Technology, vol. E80-D, no. 2.
15. Fonseca NLSD, Facanha RA (2002) The look-ahead-maximize-batch batching policy. IEEE Trans. Multimedia, vol. 4, no. 1, pp. 114-120.
16. Kompella VP, Pasquale JC, Polyzos GC (1993) Multicast routing for multimedia communication. IEEE/ACM Trans. Networking, vol. 1, no. 3, pp. 286-292.
17. Zhu Q, Parsa M, Aceves JJGL (1995) A source-based algorithm for delay-constrained minimum-cost multicasting. Proc. IEEE INFOCOM, pp. 377-385.

18. Halsall F (2001) Multimedia Communications, Addison Wesley.
19. Papadimitriou CH, Steiglitz K (1982) Combinatorial Optimization: Algorithms and Complexity, Prentice Hall.
20. Khoury BN, Pardalos PM, Du DZ (1993) A test problem generator for the Steiner problem in graphs. ACM Trans. Math. Soft., vol. 19, no. 4, pp. 509-522.
21. Ramaswami R, Sivarajan KN (1995) Routing and wavelength assignment in all-optical networks. IEEE/ACM Trans. Network., vol. 3, no. 5, pp. 489-500.
22. Lee KC, Li VOK (1993) A wavelength-convertible optical networks. IEEE J. Light. Tech., vol. 11, no. 5, pp. 962-970.

Solving the QoS Multicast Routing Problems Using Genetic Algorithms

Qijun Gu and Chao-Hsien Chu

School of Information Sciences and Technology, The Pennsylvania State University, 001 Thomas Building, University Park, PA 16802, USA

Abstract. Multicasting has been viewed as a viable means with high potential of improving service efficiency and quality in distributing multimedia information through networks. Because of its strategic importance, many interesting research topics have been articulated. One of such emerging topics is the multicast routing problem. This study explores the use of genetic algorithms (GAs) for solving the multicast routing problems when multiple quality of services (QoS) requirements are presented; compares for a small network the performance of two GAs with different representation schemes and constraints handling methods; and examines for a small- and a medium- size networks the effect of population sizes and mutation probabilities on solution quality. Our simulation results indicate that it is critical to select a suitable representation method and a set of appropriate parameters in order to obtain good performance.

Keywords. Genetic Algorithm, Quality of Services, Multicast, Multicast Routing

1 Introduction

The use of communication networks has increased significantly in the last decade due to the dramatic advance in information and network technologies. As the society transforms itself to an information society, the networks have evolved as the primary source for information creation, storage, distribution, and retrieval. The development of a reliable network infrastructure that meets a variety of service criteria becomes a critical issue. Especially, customers demand real time and other QoS (e.g., no delay, low loss rate, little delay variation, and low cost, etc.) in the network increases as Internet expands. To meet these emerging needs, many multimedia communication applications require a source to send multimedia information to multiple destinations through a communication network, such as video conferencing, distribution of multimedia documents to selected people, and requesting multimedia information from a distributed database. For example, in a video-on-demand system, the video at the source should be sent to all the users who made request. However, with a classic *unicast routing*, i.e. establishing a route to each user, it consumes tremendous bandwidth resources of the network when the number of users is large. It is also not proper to *broadcast* the information in the network, because only a group of selected users in the network need the information. Therefore, it is possible that we bunch a number of users requesting the same video object and use one stream to serve multiple users. Through such a multicast routing mechanism, the source node can send multimedia information to all of the destination nodes. The problem of determining an optimal multicast route is known as a NP-complete problem [13]. Therefore, good heuristic algorithms are of practical interest.

Genetic algorithm (GA) has emerged as a powerful tool from AI community to a wide variety of applications including engineering, economics, manufacturing, telecommunications and so on [9, 11, 17]. In recent years GA has also generated an increasing interest among scholars and engineers in the fields of network design, routing and management [2, 5, 6, 18, 20, 25, 26, 27]. However, the attempts of applying GA to solve the multicast routing problems have been lagged and limited in scope [3, 30, 33, 34, 35]; many technical issues yet remain to be explored. Here, we propose using GA for solving the multicast routing problems, which subject to meet four QoS requirements. The major objectives of this study are:

- To develop a GA solution to the QoS multicast routing problems.
- To evaluate and suggest a better representation scheme for our GA implementation.
- To explore the possible performance impact under different population sizes and mutation probabilities.

2 The Multicast Routing Problem

The multicast routing problem refers to a problem of searching a routing tree in a distributed network, where messages or information are to be sent from the source node to all destination nodes, while all requirements are met and the total cost is minimized. To define the multicast tree, it is necessary to introduce some basic terms used in graph problems.

Given a *connected graph* $G=(V,E)$ and a subset $V' \subseteq V$, the graph $G'=(V',E')$ is a subgraph of G, such that:

1. $E' \subseteq E$,
2. $(v_i,v_j) \in E' \Rightarrow v_i,v_j \in V'$, where (v_i,v_j) is the edge from node *i* to *j*,
3. $v_i,v_j \in V' \cap (v_i,v_j) \in E \Rightarrow (v_i,v_j) \in E'$,

Given a positive edge cost function $c: E \rightarrow R_+$, and a subset $U \subseteq V$, any acyclic connected subgraph $G'=(V',E')$ of G such that $U \subseteq V'$ and $c(G')$ is minimal is called a *Minimal Steiner Tree* (MStT).

The multicast routing problem is a special case of the MStT problem. In a network, $V=\{0, 1, 2, \cdots, N-1\}$, let the set of source nodes be S and the set of destination nodes be D and $e_{i,j}$ be the edge from node *i* to *j* in the network. $e_{i,j}=1$, if it was selected in the route; otherwise $e_{i,j}=0$. Therefore, $E'=\{all\ e_{i,j} : e_{i,j}=1\}$. The cost of a multicast route is $C = \sum_{e_{i,j} \in E'} c(e_{i,j})$. The multicast routing problem is

cost of a multicast route is $C = \sum_{e_{i,j} \in E'} c(e_{i,j})$. The multicast routing problem is to find a route, which is a MStT, $G' = (V', E')$, so that the cost C is minimal.

Fig. 1(a) shows an example of a bi-directional network, $V = \{0, 1, 2, 3, 4, 5\}$. Suppose the source node is 0, i.e., $S = \{0\}$, and the destinations nodes are 3 and 5, i.e., $D = \{3, 5\}$, then $U = \{0, 3, 5\}$. Fig. 1(b) depicts a legal multicast route, where, the multicast tree is in solid line from the source to all destinations; consequently, $V' = \{0, 2, 3, 4, 5\}$ and $E' = \{e_{0,2}, e_{2,5}, e_{2,4}, e_{4,3}\}$. The edges in dash line and node 1 are deleted from the tree.

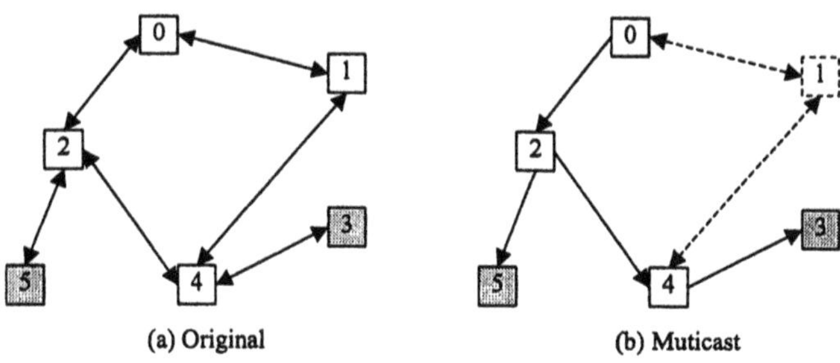

Fig. 1. Illustration of a multicast network

Although multicast routing is strongly related to the problem of finding the Minimum Spanning Tree (MST), it is actually more difficult to solve. This is because in multicast the number of participants can be less than N, which means the tree does not have to connect all nodes in the network and does not know which node, other than the source and destinations, should be selected in the tree.

Let N_U be the total number of source and destination nodes. There are two special (and easier to solve) cases for the multicast problem:

1. When $N_U = 2$, the multicast routing problem is reduced to *shortest path routing problem* between two terminals. The typical algorithms used to solve this type of problems are Dijkstra and Distance Vector [14].
2. When $N_U = N$, the number of participants equals the number of nodes in the network. The instance of the multicast problem is precisely the *minimum spanning tree problem*.

2.1 Quality of Services (QoS)

As Internet expands, a growing number of metrics have been added to gauge and assure service quality. For multicasting, common concerns include low bandwidth, delay, delay variation, and high packet loss rate, etc. [29].

2.1.1 Bandwidth

Let $EB_{i,j}$ be the bandwidth of the edge from node i to j, and CT_{BW}^{d} be the minimum bandwidth required by node d. If edge $e_{i,j}$ is selected in the route, $EB_{i,j} \geq CT_{BW}^{d}$ must be satisfied so that information requiring a certain bandwidth can be transmitted to the destination node d through this edge.

2.1.2 Delay

The delay has two components: the delay of edges and the delay of nodes in the route. Let $ED_{i,j}$ be the delay of edge from i to j, and ND_k be the delay of node k. DL_d is the delay of the sub-route from source 0 to destination d, which includes both the delay of intermediate edges and nodes in the sub-route. Therefore,

$$DL_d = \sum_{e_{i,j}^{d} \in E} e_{i,j}^{d} ED_{i,j} + \sum_{k^d \in D} ND_{k^d} \tag{1}$$

Where $e_{i,j}^{d}$ is the edge from node i to j and k^d is node k from source 0 to destination d in the sub-route. For destination d, it is required that $DL_d \leq CT_{DL}^{d}$, where CT_{DL}^{d} is the maximum delay node d can tolerate.

2.1.3 Jitter

Jitter is the delay variation. Assume that the delay at every node is independent and let NJ_k be the jitter of node k. The jitter at destination d is

$$JI_d = \sum_{k^d \in D} NJ_{k^d} \tag{2}$$

The jitter should satisfy that $JI_d \leq CT_{JI}^{d}$ so that the information is transmitted with a near constant rate. CT_{JI}^{d} is the maximum tolerable jitter at destination d.

2.1.4 Loss rate

Let $EL_{i,j}$ be the loss rate of the edge from node i to j, and LR_d be the total loss rate of the sub-route from source 0 to destination d.

$$LR_d = 1 - \prod_{e_{i,j}^d \in E} (1 - e_{i,j}^d EL_{i,j}) \quad (3)$$

The loss rate should satisfy that $LR_d \le CT_{LR}^d$ so that the data received by destination d is undamaged. CT_{LR}^d is the maximum loss rate tolerable by node d.

2.2 Solution Algorithms

Over the past decades, many unconstrained or simple constrained algorithms have been developed to solve the multicast routing problems. Typical approaches include (1) applying Dijkstra algorithm to find the shortest path, (2) seeking the minimum network cost using Steiner tree routing algorithm, and (3) finding multicast trees that the paths between source node and the destination nodes are connected and their cost is minimized. A state of the art review and analysis for simple multicast routing can be found, for example, in [23, 24, 32]. Recently, with the surge demand of fast and better quality of services, a number of efficient heuristic [10, 29] or nature-based algorithms [7, 8, 15, 16, 30, 31, 34, 36] have been proposed to solve the QoS multicast routing problem. An extensive review of QMR problem can be found in [32]. Among those traditional approaches, two typical algorithms -- DVMRP and MOSPF -- have been adopted as protocols in RFC documents.

Distance Vector Multicast Routing Protocol, known as DVMRP, is defined in RFC-1075. It is derived from the unicast routing protocol RIP (Routing Information Protocol). RIP calculates the next-hop toward a destination, however DVMRP computes the previous hop back toward a source. In addition, DVMRP performs the computation based on the unicast routing tables constructed by RIP. Therefore, it should be bound with RIP.

Multicast Open Shortest Path First Protocol, known as MOSPF, is defined in RFC-1584. It depends on OSPF to construct the unicast routing table. OSPF can use different types of a single link state metric to express the cost of a path. MOSPF complements OSPF's routing database with a new type of link state records: the group memberships. In this way, MOSPF can perform computations of the reverse path forwarding check, join and prune locally.

3 GA in Multicast Routing

GA refers to a class of adaptive search procedures based upon principles derived from natural evolution [11]. Over the last decade, GA has emerged as a new branch of heuristic to solve the MStT problems [3, 7, 18, 25, 26, 30, 33, 34, 35]. Because of its computational simplicity and parallel search capability, well-designed GAs exhibit good performance [18]. Esbensen [7] compared a GA with three other heuristic algorithms for Steiner problems, one of which is based on shortest path (SPH-I) and the other two are derived from the brand-and-cut method. His comparison shows that: (1) GA can find the global optimum in more than 77% of all runs and their solution is within 1% from optimum in more than 92% of all runs, and (2) the solution quality obtained by GA is always at least as good as that by SPH-I and the error ratio is an order of magnitude better.

Although the use of GA to any domain follows the same procedure – problem representation, population initialization, fitness evaluation and selection, reproduction (crossover and mutation), and replacement [9], the success of GA implementation lies in a number of key decisions [5]:

1. Selecting a suitable representation scheme to map the solution space to chromosomes.
2. Determining a suitable population size and a method of generating initial population.
3. Designing or applying effective reproduction operations.
4. Determining suitable crossover or mutation probabilities.
5. Finding a suitable method to repair or deal with illegal or infeasible chromosomes.
6. Determining when to terminate the evolution.

3.1 Representation

In a network design problem, the strings used for representing chromosome should include information on the link identification and network structure. The encoding should be able to uniquely represent all of possible solutions to the problem. Three types of schemes can be used to encode network related problems: edge-based [33, 35], node-based [1] and hybrid [18]. One can also borrow the coding scheme from the Spanning tree problem [12,19], which is node-based. Each scheme has its own good features. However, it is important to note that there is no "perfect" representation for a Steiner tree yet.

3.1.1 Edge code

Assume that there are m edges in a given bi-directional graph. Given a sequence number to every edge. The edge code EC is a string of m integers, where the j-th integer, 1 or 0, represents whether the j-th edge is selected or not in the route. Considering the original network in Fig. 1(a), an example of edge sequence is defined in Table 1.

Table 1. An example of edge coding

j	1	2	3	4	5	6
edge	0→1	1→0	0→2	2→0	1→4	4→1
j	7	8	9	10	11	12
edge	2→4	4→2	2→5	5→2	3→4	4→3

Suppose the edge code of the chromosome *EC* is "100010011001", we can obtain the tree as in Fig. 2. The edge code is the simplest one among all of the coding schemes. It is easy to calculate the fitness directly from this code. However, it may result in a tree with serious illegality, thus making the tree difficult to be repaired and the algorithm may take longer to converge.

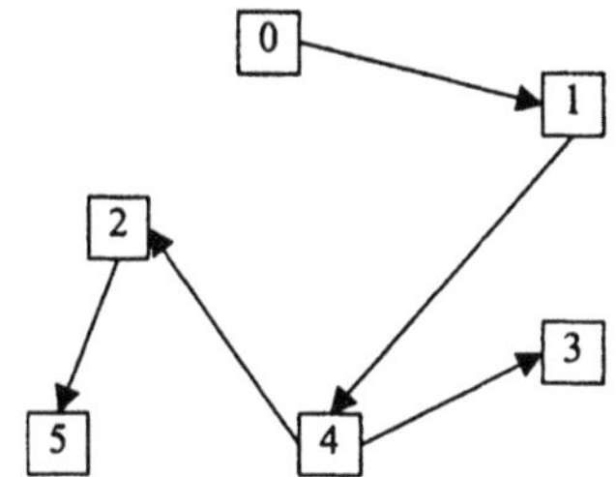

Fig. 2. The tree decoded from the edge code

3.1.2 Determinant code

This coding method is first proposed in [1]. To define a determinant code, we first define the edge-range vector *R* of a digraph $G(V,E)$ as $R(j)=\{i|D_{i,j}=-1\}$ for j=0 to n-1, where *n* is the number of nodes, and $D=\{D_{i,j}\}$ is the in-degree matrix.

$$D_{i,j}=\begin{cases} d_i & \text{if } i=j \\ -e & \text{if } i\neq j \end{cases} \tag{4}$$

Where d_i is the number of edges coming into node *i*, and *e* is the number of edges from *i* to *j*. The determinant code *DC* is a string of n-1 integers, in which the (j-1)th position in the string is selected from $R(j)\ for\ j=1\ to\ n-1$. Considering the network in Fig. 1(a), the in-degree matrix is:

$$D=\begin{array}{c} \\ 0\\1\\2\\3\\4\\5 \end{array}\begin{array}{c} \begin{array}{cccccc} 0 & 1 & 2 & 3 & 4 & 5 \end{array} \\ \begin{bmatrix} 2 & -1 & -1 & 0 & 0 & 0 \\ -1 & 2 & 0 & 0 & -1 & 0 \\ -1 & 0 & 3 & 0 & -1 & -1 \\ 0 & 0 & 0 & 1 & -1 & 0 \\ 0 & -1 & -1 & -1 & 3 & 0 \\ 0 & 0 & -1 & 0 & 0 & 1 \end{bmatrix} \end{array}$$

Therefore, the edge-range vector *R* can be obtained as in Table 2.

Table 2. The edge-range vector

j	1	2	3	4	5
$R(j)$	{0,4}	{0,4,5}	{4}	{1,2,3}	{2}

Suppose the determinant code *DC* is "04412", we can obtain the trees as shown in Fig. 3. The code is node-based. Although it is a little complex as compared with the edge code, it is not difficult to obtain the fitness from this code, because the code contains the edge information in an easy way.

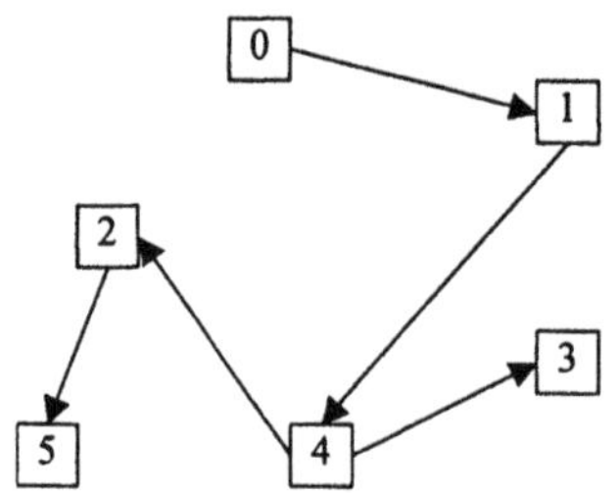

Fig. 3. The tree decoded from the determinant code

3.1.3 Prüfer code

Prüfer code [21] is originally used for the Spanning tree problem. However, since the Steiner tree is a sub-tree of a Spanning tree, it is possible to use it to encode a Steiner tree. A Prüfer code is a string of *n-2* integers, denoted as $v_1 v_2 ... v_{n-2}$. After it was decoded, we can get a Spanning tree. The degree of each node in the corresponding Spanning tree is one more than the number of times its label appears in the Prüfer code. Suppose a Prüfer code $v_1 v_2 ... v_{n-2}$ is "3442", and the degree of each node can be grouped in order as a vector $d = \{1,1,2,2,3,1\}$. Each entry of the vector is the degree of the corresponding node, i.e., node 0 has one degree, and node 4 has three degrees.

Assume that the source node is $i = 0$, given a Prüfer code $v_1 v_2 ... v_{n-2}$, the Prüfer decode algorithm works as follows:

Let d_i = the degree of node i, $0 \le i \le n-1$.

for $i = 1$ *to* $n-2$

$u = \min(j : d_j = 1, 0 \le j \le n-1)$

recordedge$(u \to v_i)$

$d_u = d_u - 1$

$d_{v_i} = d_{v_i} - 1$

endfor

$u_1 = \min(j : d_j = 1, 0 \leq j \leq n-1)$

$u_2 = \max(j : d_j = 1, 0 \leq j \leq n-1)$

$recordedge(u_1 \rightarrow u_2)$

Therefore, if $v_1 v_2 ... v_{n-2}$ = "3442", we can decode it to a tree following the steps outlined in Table 3. The final edges are $0 \rightarrow 3$, $1 \rightarrow 4$, $3 \rightarrow 4$, $4 \rightarrow 2$, $2 \rightarrow 5$ and the tree is as shown in Fig. 4.

Table 3. Decoding procedure of Prüfer code

i	u	v_i	d	edge
0			{1,1,2,2,3,1}	
1	0	3	{0,1,2,1,3,1}	$0 \rightarrow 3$
2	1	4	{0,0,2,1,2,1}	$0 \rightarrow 3, 1 \rightarrow 4$
3	3	4	{0,0,2,0,1,1}	$0 \rightarrow 3, 1 \rightarrow 4, 3 \rightarrow 4$
4	4	2	{0,0,1,0,0,1}	$0 \rightarrow 3, 1 \rightarrow 4, 3 \rightarrow 4, 4 \rightarrow 2$

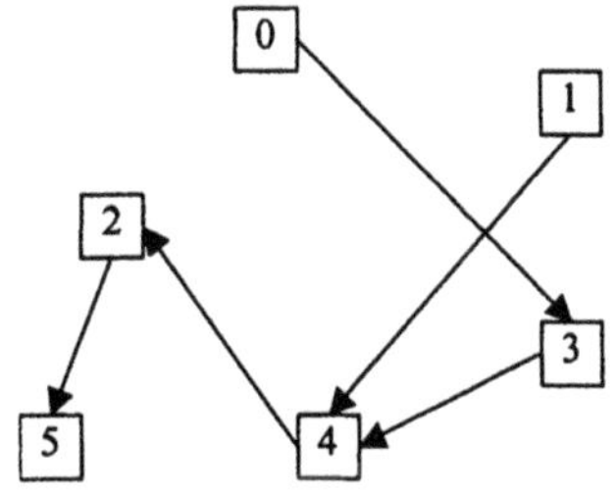

Fig. 4. The tree decoded from the Prüfer code

In [12, 19], some other coding schemes similar to the Prüfer code are proposed. These schemes can generate any spanning tree. However, the algorithm to decode a chromosome is complex and the result may produce a tree with some edges not existing in the network.

3.2 Initial Population

Generating initial population for evolutions is an important stage. There are three issues involved in initializing the population. First is to determine the size of population. There is no exact rule to determine the population size. Generally speaking, the size should increase as the network expands. [4] proved that a graph with n nodes can generate n^{n-2} distinct spanning trees. However, even if we know the number of nodes and edges in the network, no one knows how many Steiner trees the network can have. What we are sure is that the total number of Steiner trees is less than n^{n-2}. The second issue concerns the method used to generate the initial population. The simplest approach to use is random generation, in which every allele in the chromosome is randomly generated. Another common approach is to deploy some heuristic algorithms. The fact is that the randomly generated chromosomes may not necessarily be legal in the network, while the heuristic population is guaranteed to be legal and may exhibit lower cost. On the other hand, the random population may cover more diverse solutions. The last issue is concerned with how many chromosomes in the initial population are indeed the same. For example, some trees generated from decoding may have different illegal parts, but have the same legal parts, as shown in Fig. 5. In this case, the chromosomes, which generate these trees, are the same; consequently, the whole population may in fact only contain a small number of different chromosomes. In addition, whether the chromosomes are similar or not depends on the representation schemes and the reproduction operators used.

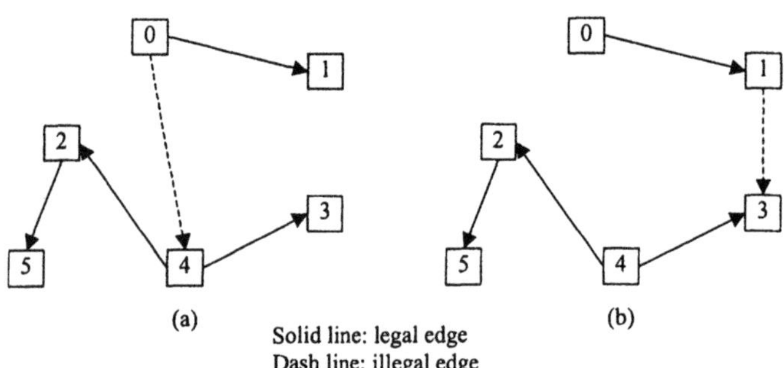

Fig. 5. Different trees decoded from the same chromosomes

3.3 Fitness

The design of fitness function should reflect the objective and constrains of the problem. Therefore, for the multiple QoS problems, the fitness F is a summation of several parts:

$$F = w_C F_C + w_N F_N + w_D F_D + w_B F_B + w_L F_L + w_J F_J \tag{5}$$

Where F_C is the cost of the multicast route, F_N is the fitness for connectivity, F_D is for the delay requirements, F_B is for the bandwidth requirements, F_L is for the loss rate requirements, and F_J is for the jitter requirements.w_i, $i \in \{C, N, D, B, L, J\}$, are their respective weights.

F_C is the primal fitness function as the main objective of multicast is to minimize cost. F_N is designed to punish a tree if it is illegal. A tree is illegal if some nodes in it are not connected or it has cycles. The other fitness/penalty functions are designed for meeting the QoS requirements of a multicast network. The alternative method is to repair the tree so that it is legal and satisfies the QoS requirements. The final selection between penalty and repair depends on the complexity of the network.

Weights serve two purposes. First, they are used to normalize all parts of the fitness and turn them into the same scale. In this study, a linear scaling is used. More methods to scale the fitness can be found in [9]. In addition, these weights can be used to balance the cost and penalties so that when F is minimized F_C is minimized as well. Generally speaking, the weights used for normalizing penalties are much larger than those of cost so that the best solution has less fitness even if other infeasible solutions have less cost.

3.3.1 Cost

Let $EC_{i,j}$ be the cost of edge from node i to j. The fitness of cost can be defined as:

$$F_C = \sum_{e_{i,j} \in E} e_{i,j} EC_{i,j} \tag{6}$$

It is obvious that if the edge $e_{i,j}$ is selected in the route, the cost of this edge will be counted into F_C.

3.3.2 Connectivity

Connectivity refers to the feature that the route starts from the source and goes through or finishes at all destinations. The route is disconnected if it is broken into several segments. Because $e_{i,j}$ is an edge from node i to j, node i has an out-degree introduced by $e_{i,j}$ while node j has an in-degree.

Given a multicast route (maybe broken), we can choose a sub-route from source 0 to destination d as a unicast route. Let $i^d \in D$ be the node i in this sub-route and $e_{i,j}^d = e_{i^d,j^d}$ be the edge from node i to node j in this sub-route. The fitness of connectivity for this sub-route can be expressed as:

$$NE_d = (\sum_{j^d \in D} e_{0,j^d} - 1)^2 + \sum_{i^d \in D} (\sum_{j^d \in D} e_{i,j}^d - \sum_{j^d \in D} e_{j,i}^d)^2 + (\sum_{j^d \in D} e_{j^d,d} - 1)^2 \tag{7}$$

Where $\sum_{j^d \in D} e_{0,j^d}$ is the total out-degree of source 0, $\sum_{j^d \in D} e_{i,j}^d$ and $\sum_{j^d \in D} e_{j,i}^d$ are the total out-degree and in-degree of intermediate node respectively in this sub-route, and $\sum_{j^d \in D} e_{j^d,d}$ is the total in-degree of destination d. If the sub-route is a connected unicast route from 0 to d, the out-degree of source 0 and the in-degree of destination d are 1, and the in-degree and out-degree of intermediate node are equivalent. The fitness of connectivity for the multicast route is $F_N = \sum_{d \in D} NE_d$.

3.3.3 Bandwidth

The fitness of bandwidth can be defined as:

$$F_B = \sum_{d \in D} \sum_{e_{i,j}^d \in E} \max(0, CT_{BW}^d - e_{i,j}^d EB_{i,j}) \tag{8}$$

Where $EB_{i,j}$ is the bandwidth of the edge from node i to j, and CT_{BW}^d is the minimum bandwidth required by node d. If the bandwidth of edge $e_{i,j}^d$ in the route to destination d is larger than CT_{BW}^d , the bandwidth fitness is 0.

3.3.4 Delay

The fitness of delay is defined as $F_D = \sum_{d \in D} \max(0, DL_d - CT_{DL}^d)$, where DL_d is the delay of the sub-route from source 0 to destination d and CT_{DL}^d is the maximum delay node d can tolerate. If the delay requirements are satisfied, i.e., $DL_d \le C_{DL}^d$, then the delay fitness for destination d is 0, otherwise it contributes $DL_d - CT_{DL}^d$ to F_D .

3.3.5 Jitter

The fitness of jitter is defined as $F_J = \sum_{d \in D} \max(0, JI_d - CT_{JI}^d)$, where JI_d is the jitter at node d and CT_{JI}^d is the maximum jitter. If the jitter at d is less than CT_{JI}^d, the jitter requirement is met.

3.3.6 Loss rate

The fitness of loss rate is $F_L = \sum_{d \in D} \max(0, LR_d - CT_{LR}^d)$, where LR_d is the total loss rate and CT_{LR}^d is the maximum loss rate tolerable by d. If the loss rate of destination d is less than CT_{LR}^d, the loss rate requirement is met and the fitness is 0.

3.4 Crossover

3.4.1 Conventional methods

Crossover methods, such as one-point, two-point, and uniform crossover, have been extensively used in GA applications for network design [6, 20]. In one-point crossover, a random position is generated for a pair of chromosomes and the alleles of the first chromosome from this fixed position to the end are exchanged with the second chromosome in the same range. In this process, the alleles of the second chromosome are transferred to the respective alleles in the first chromosome. Similarly, the two-point crossover generates two random positions, head and tail. The alleles of the first chromosome from the head position to the tail are exchanged with the second chromosome in the same range. Uniform crossover is a dynamic and nondeterministic method where a set of positions is selected for each of the chromosomes and their alleles are exchanged with each other based on the generated positions. Besides these conventional crossovers, some heuristic methods as discussed below have been proposed in [22, 33]. They were reported to be performed better than the traditional methods.

3.4.2 Logic-based crossover

This method is applicable only to edge code. The method is based on "and" and "or" logic operations. The main idea behind the method is as follows. First, select two parents *P1* and *P2* according to the elitist model. Then let *C1* = *P1* & *P2* bit by bit, and *C1* = *P1* | *P2* bit by bit, where "&" denotes "and" operation and "|" denotes "or" operation. Finally put one-point crossover operation on *C1* and *C2*. Obviously, this method makes the children inherit the same genes of parents: The "and" operation is the method where 0 is dominant, while "or" is where 1 is dominant.

3.4.3 Tree-based crossover

In this method, crossover is directly applied on the parent trees. Given two MStTs, T_M and T_F, the crossover algorithm constructs a child MStT T_O by identifying the links that are common to both T_M and T_F and retaining them in T_O. The above method of forming a child MStT is based on the heuristic that since both the selected parents are likely to have high values of fitness measure, the link that are common to them represent 'good traits' and must be passed on to the child. At this stage, T_O may be a disconnected graph and edges may have to be added to transform it into a tree. This is done by identifying the connected components in T_O and merging two of the connected components at a time until forming a tree. It may be possible that some of the leaf nodes of T_O are not destination nodes. Such nodes are then deleted to make it a MStT.

3.5 Mutation

Similarly, various mutation methods have been used to solve network design problem, including inversion, insertion, displacement, reciprocal exchange, and heuristic mutation [6, 20]. These methods process only on a chromosome. If the mutation procedure is directly put on a tree, it is to select a subset of nodes randomly and removes all the links that are incident on these nodes. If the resulted tree is broken, it can be reconstructed by repairing the disconnected graph as discussed in the next section.

3.6 Defects and Repair

The crossover and mutation can turn the tree to an illegal one. After decoding, a tree may result in five possible defect types. First, the tree may be separated into several unconnected sub-trees. Second, the tree may have cycles. Third, the tree may not connect all of the destination codes. Fourth, the tree may have extra nodes and edges, which should be deleted. Finally, the tree may not satisfy the QoS requirement any more.

3.6.1 Broken tree

The edge and determinant coding schemes may cause a tree broken, because when using these algorithms to encode a tree, we don't consider the relationship among nodes. Considering the determination code described above, DC = " 04152", the tree has broken into two parts as shown in Fig. 6.

Although Prüfer code can often generate a connected tree, some edges may not exist in the given graph. Considering the tree of Fig. 4, the edge from source 0 to node 3 doesn't exit. Thus it needs to be deleted, and the tree is broken into two parts as shown in Fig. 7.

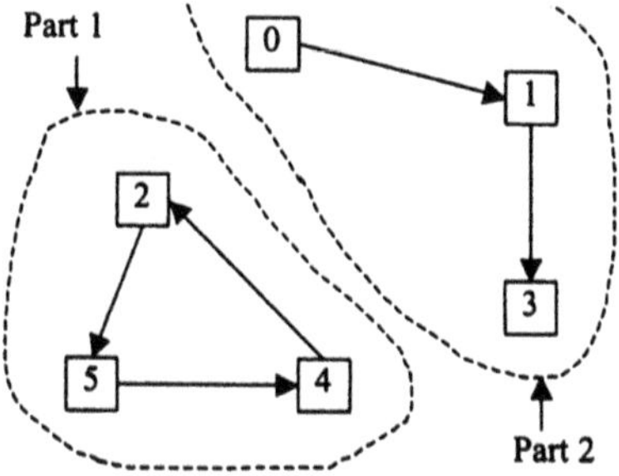

Fig. 6. Example of broken tree from determination coding

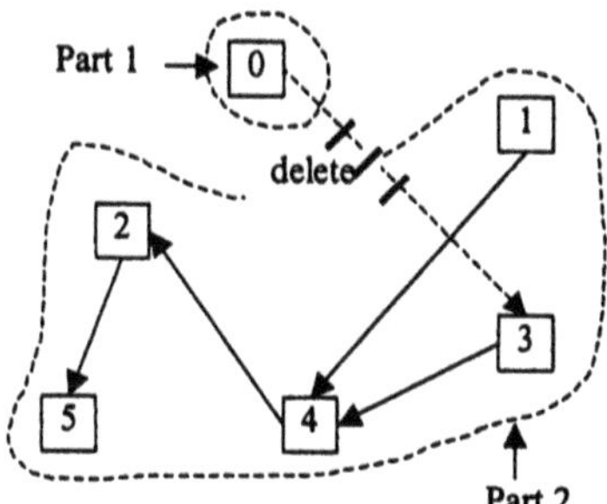

Fig. 7. Example of broken tree from Prüfer coding

3.6.2 Cycle

The tree generated by an edge code or a determinant code may result in cycles. Considering an edge code EC = "100110011001", the tree has a cycle as shown in Fig. 8. Please note that the Prüfer code immunes from such problem.

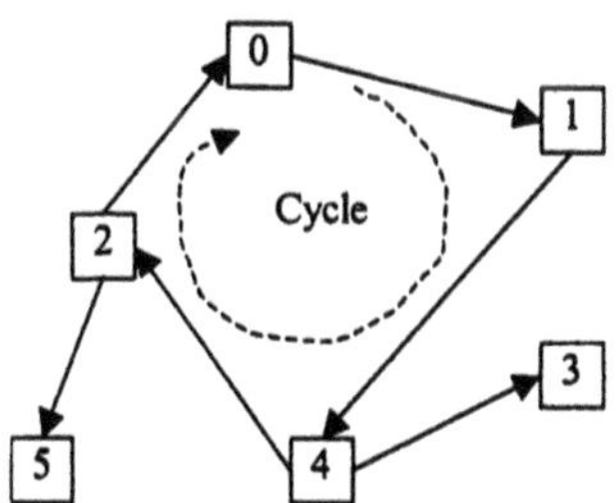

Fig. 8. Example of cycle from Prüfer coding

3.6.3 Disconnected destinations

Although the edge code can generate a tree, it may result in some disconnected destination nodes. For example, considering an edge code EC = "100010011000", node 3 became a destination that is not connected to the tree (see Fig. 9). The determinant and Prüfer codes won't cause this problem, because they are node-based codes.

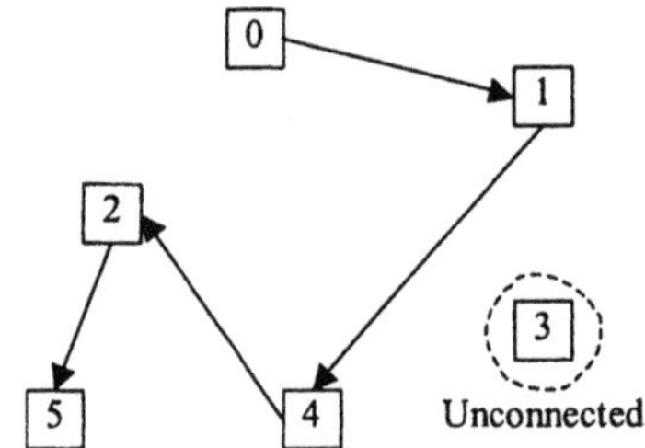

Fig. 9. Example of disconnected destinations

3.6.4 Extra nodes and edges

Even if a chromosome can generate a complete and acyclic tree, the tree may contains some extra nodes and edges, which should be deleted in order to form a Steiner tree. For example, the Prüfer code can derive a Spanning tree, which connect all of the nodes in the network. However, the set of nodes V' in the Steiner tree is only a subset of the whole set of nodes V, i.e., $V' \subseteq V$. Assume node 3, 4 and 5 are the destination nodes. Node 1 and the edge from node 3 to 1 is useless in the Steiner tree and should be deleted as shown in Fig. 10.

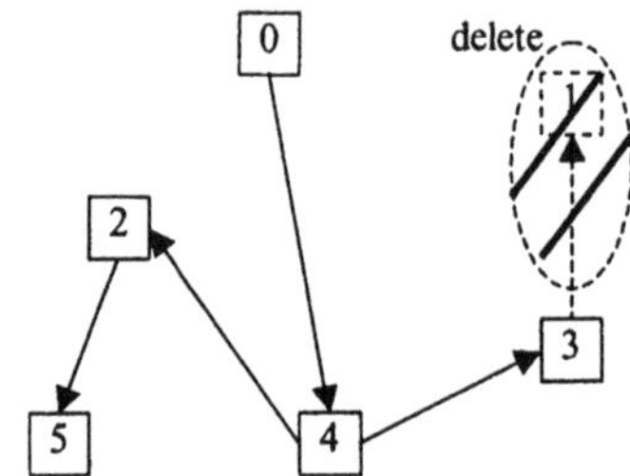

Fig. 10. Example of redundant nodes and edges

3.6.5 Repair

There are two methods to remedy the defects of a tree – penalize or repair. One viable approach is to use the fitness function of connectivity as a penalty. However, this method is passive, because the fitness function won't make the tree complete. Meanwhile, we have to wait until the crossover and mutation operations to generate a connected tree. In a large-scale network, it is not so practical, because generally there are more disconnected trees than connected ones [18]. The procedure of repairing the tree is adopted to accelerate the convergence of the GA. However, it is much more complex to repair a tree with QoS. Most of studies on QoS constraints use penalty methods. Therefore, a repair mainly focuses on making a tree legal. In this study, we propose to use the following procedure to repair a tree after decoding a chromosome.

S1) Delete the nonexistent edges in the graph.
- a) Given a graph $G_1 = (V_1, E_1)$
- b) Delete any edge $e_{i,j} \in E_1$, if it does not exist in the original network.
- c) The new graph $G_2 = (V_2, E_2)$ is formed.

S2) Change in-degree of every node to 1, except the source node to 0.
- a) Check all nodes in the network.
- b) If the in-degree of node $v \in V_2$ is more than 1, delete all except one flow-in edge of this node.
- c) If the in-degree of node $v \in V_2$ is 0, select a flow-in edge to this node.
- d) Delete all flow-in edges of source node 0.
- e) The new graph $G_3 = (V_3, E_3)$ is formed.

S3) Remove cycles in the graph.
- a) Choose a node $v \in V_3$
- b) Follow the flow-in edge to the upper node and record the path.
- c) If the upper node is already in the recorded path, the path has a cycle.
- d) Break the cycle at $e_{i,j}$, where node *j* has another possible flow-in edge $e_{k,j}$ that is from an upper node *k* not in the path.
- e) Delete $e_{i,j}$ and select $e_{k,j}$ as the flow-in edge of node *j*.
- f) Repeat a) to e) until no cycle exists.
- g) The new graph $G_4 = (V_4, E_4)$ is formed.

S4) Remove all redundant nodes and edges in the tree.
- a) Choose a leaf node $v \in V_4$.
- b) If it is not a destination node, delete it and its flow-in edge.
- c) Repeat a) to b) until all leaf nodes are destination nodes.

Fig. 11 shows an example of repairing a tree. Graph (a) is the original network, which is an undirected graph, and has node 0 as the source and nodes 1, 2, 3 as the destination nodes. Graph (b) is the graph after decoding a chromosome. Graph (f) is the final tree after repair.

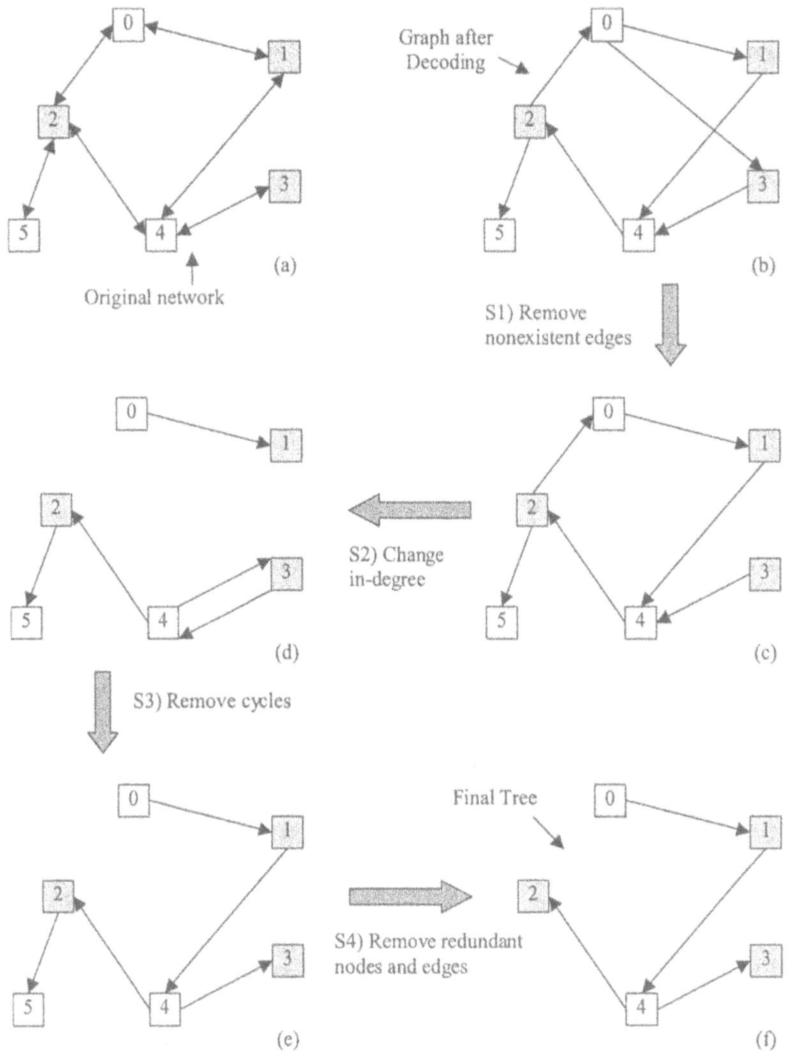

Fig. 11. Illustration of repairing procedure

3.7 Selection

There are many methods to select the chromosomes for reproduction. Each has its advantages and disadvantages [17]. In this study, two typical strategies are considered: $(\mu+\lambda)$ and (μ,λ). In $(\mu+\lambda)$ strategy, μ parents and λ offsprings compete for survival and the μ best solutions are selected for the next generation. In (μ,λ) strategy, we select the μ best solutions out of λ offspring solutions

$(\lambda > \mu)$. Using the roulette wheel method, the probability that a parent i is selected is given by:

$$p_i = \frac{fitness^{-1}(i)}{\sum_{j \in \{population\}} fitness^{-1}(j)} \tag{9}$$

4 Experiment and Computational Analyses

We consider two experiments for assessing the effectiveness of our proposed GA implementation. The first experiment involves the comparison of two different representation schemes and methods of dealing with QoS constraints. The other experiment concerns the selection of proper settings for parameters. Determining proper parameter setting is a critical decision for any GA implementation. In general, four system parameters need to be determined: population size, crossover rate, mutation rate and number of generations to terminate. In this study, we set crossover rate as "1", because we use $(\mu + \lambda)$ selection strategy [6]. We also keep the program running sufficiently long before termination.

Data collected from two networks are used for exploration. The first data is a small-size network (see Fig. 12), which has 6 nodes and 8 edges. The cost of each edge is the number besides the edge. The source is node 0 and the destinations are node 4 and 5. The bold line shows one of the optimal multicast routes, which have the minimum cost of 6.

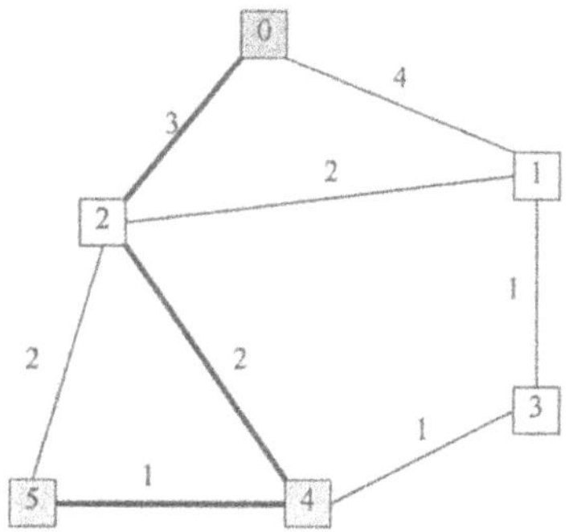

Fig. 12. Small-size network

The other data set is a medium-size network as shown in Fig. 13, which has 50 nodes and 63 edges. The data can be downloaded from [28]. The source is node 0 and the destinations are nodes 48, 21, 34, 26, 11, 36, 33 and 24. The bold line shows one of the optimal multicast routes, which have the minimum cost of 82.

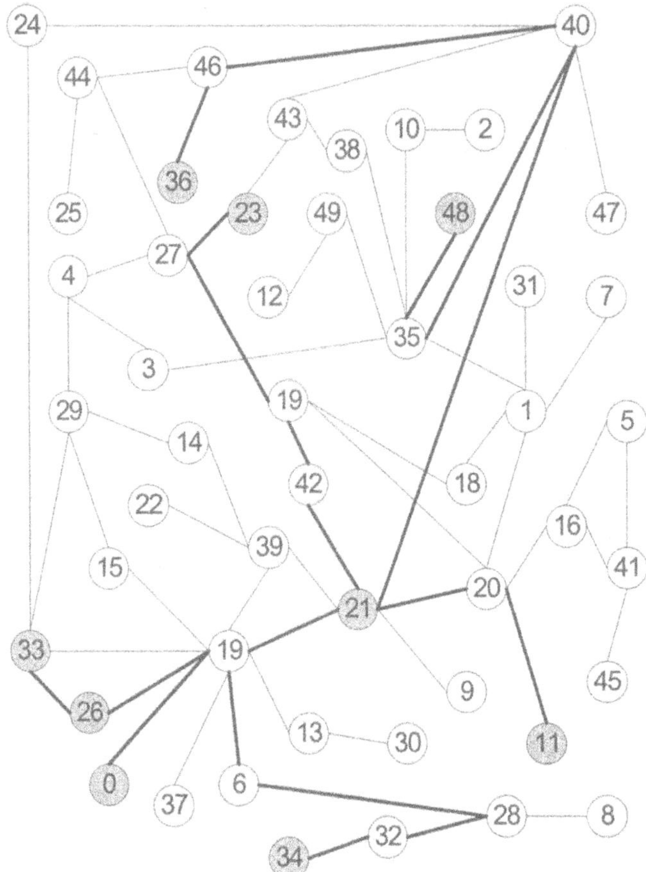

Fig. 13. Medium-size network

4.1 Comparison of Representation Schemes

Two configurations of GA, programmed in Matlab, are selected for comparison. The first algorithm uses edge coding for chromosome representation and applies penalty function to punish an illegal tree. The second algorithm applies determinant coding for representation and uses repair method to repair an illegal tree. All other GA components are kept the same. They are: population size is 8, crossover is one-point crossover with crossover rate 1, mutation probability is 0.02, and selection strategy is $(\mu + \lambda)$. These settings have been carefully examined to ensure fairness for both algorithms.

We use the small-size network for our comparison. The computational results are depicted in Fig. 14, which show that the second algorithm performed better than the first one. As shown, the second algorithm reaches the minimum cost around the fifth iteration, while the first one still stays at a high cost level. Moreover, the first algorithm seems converge more slowly than the second one does. One possible explanation for the slow convergence is that the edge code is longer than the determinant code. Assume that a network has n nodes and its average degree is d, the estimation of search space size for an edge code is 2nd, but it is n^{d} for a determinant code. Therefore, the first algorithm requires more number of iterations to converge to the optimal solution. This experiment shows that the GA with determinant coding scheme and repair function outperformed the GA with edge code and penalty function in terms of convergence speed. Therefore, it is critical to choose the right representation method for your GA implementation.

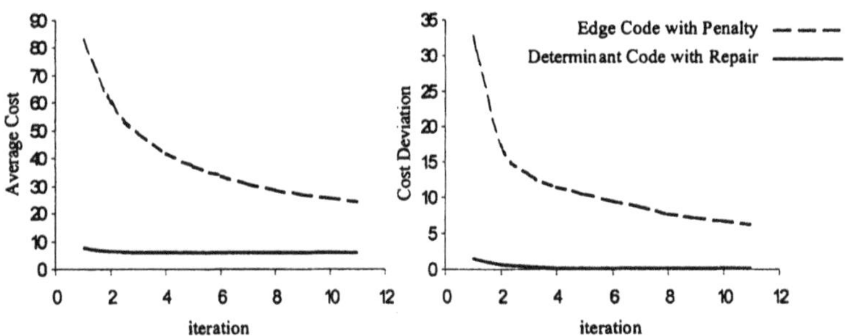

Fig. 14. Performance comparison of GAs

4.2 Effects of Population Size

The population size determines the number of solutions available for exploration in each iteration. Generally speaking, a larger population size is needed for a larger network in order to fully explore the possible solution space. Fig. 15 and Fig. 16 illustrate the effect of population size while mutation probability is set at 0.02. As shown, a larger population size can drive GA converge to the optimal solution more quickly. However, as the size increases to some level, the convergence speed does not increase much. In both sizes of the networks, the differences between population size of 8 and 16 are much less than that of 2 and 4. Therefore, we recommended to use the value between 8 and 16 as the population size for both sizes of networks.

4.3 Effects of Mutation Probability

Most literature suggests to use a small mutation probability such as 0.001 for most applications. However, based on our experiment, if the value was set to too small, the current solution may be stuck at a local optimal. On the contrary, if its value was set to too large, the population may become unstable that no optimal solution can be found before the current genes are mutated. Fig. 17 and Fig. 18 illustrate the effects of mutation probability, while population size is set at 8. As shown in Fig. 17, for the small network, we experience little difference among four different mutation probabilities. In contrast, for the medium-size network, the probability from 0.01 to 0.1 seems work better than those of too small or too large do (see Fig. 18). Therefore, we recommend selecting a mutation probability in the range of 0.01 and 0.1 for both sizes of networks.

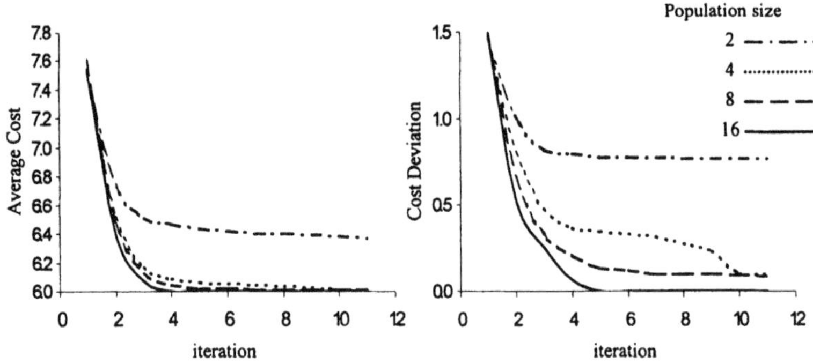

Fig. 15. Effect of population size in a small-size network

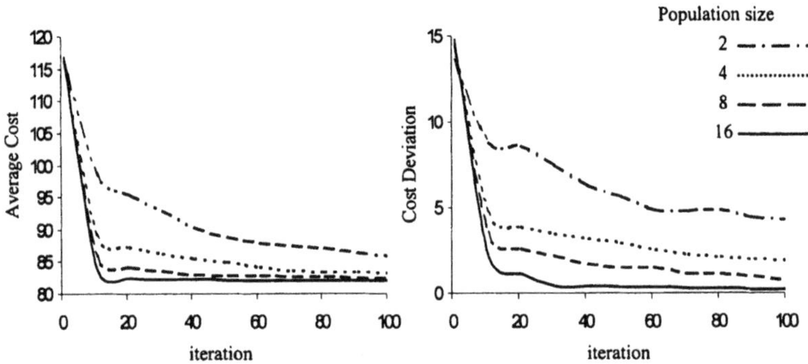

Fig. 16. Effect of population size in a medium-size network

5 Conclusion

The essence of multicast is that it can be used to efficiently send the same message to a group of selected users simultaneously with one stream. In this study, we have applied GA to solve the multicast routing problems with four QoS requirements. The quality metrics considered are: bandwidth, delay, delay jitter, and loss rate. Although the multicast routing problem is a special case of a Steiner tree problem, the QoS requirements make it more complex to solve.

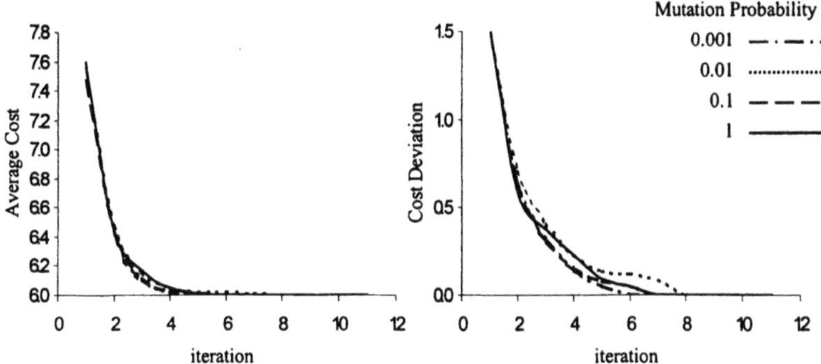

Fig. 17. Effect of mutation probability in a small-size network

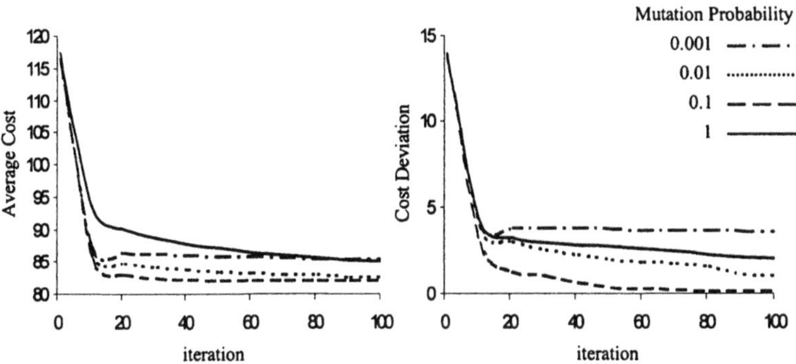

Fig. 18. Effect of mutation probability in a medium-size network

To test the relative performance of our GA implementation, we selected two representing schemes with corresponding constraint handling method – edge encoding with penalty function and determinant coding with repair method – for comparison. Our computational results show that the GA with determinant coding and repair method outperformed the GA with edge coding and penalty function in terms of convergence speed. Therefore, it is essential to examine the possible coding schemes before implementing the GA applications.

In our effort of identifying proper system parameters, we found that: (1) With ($\mu + \lambda$) selection strategy, we can set the crossover probability at value of "1" in order to maintain solution quality and efficiency. (2) For a small network, the mutation probability has little impact on the solution quality; however, for a medium-size network, the best performance can often be obtained when we set the mutation probability in the range of 0.01 and 0.1. (3) While a larger population size can drive GA converge to the optimal solution more quickly, the convergence speed does not increase much as the size increase. For our implementation, a population size between 8 and 16 chromosomes can efficiently obtain optimal solutions. (4) With determinant coding scheme and repair function, the GA can always converge to optimal solution in less than 20 iterations (generations). In summary, with proper selection of mutation probability and population size, our proposed GA implementation can solve the multicast routing problems with multiple QoS requirements efficiently and effectively.

Although in this study we did not compare the performance of the GA with other heuristics, there are sufficient evidences in literature to support that well-designed GAs exhibit good performance [6, 7, 18]. For instance, Esbensen [7] compared a GA with three other heuristic algorithms. His comparison shows that (1) GA can find the global optimum in more than 77% of all runs and their solution is within 1% from optimum in more than 92% of all runs and (2) the solution quality obtained by GA is always at least as good as heuristic and the error ratio is an order of magnitude better.

References

1. Abuali FN, Wainwright RL, Schoenefeld DA (1995) Determinant factorization: a new encoding scheme for spanning trees applied to the probabilistic minimum spanning tree problem. Proceeding of 6th International Conference on Genetic Algorithms 470-477
2. Arabas J, Kozdrowski S. (2001) Applying an evolutionary algorithm to telecommunication network design. IEEE Transactions on Evolutionary Computation **5**: 309-322
3. Banerjee N, Das SK (2001) Fast determination of QoS-based multicast routes in wireless networks using genetic algorithm. IEEE International Conference on Communications 2588-2592
4. Cayley A (1889) A theorem on trees. Quarterly Journal of Mathematics 376-378
5. Chou H, Premkumar G, Chu CH (2001) Genetic algorithms for communications network design—an empirical study of the factors that influence performance. IEEE Transaction on Evolutionary Computation **5**: 236-249
6. Chu CH, Premkuma G, Chou H (2000) Digital data networks design using genetic algorithms. European Journal of Operational Research **127**: 140-158
7. Esbensen H (1995) Computing near-optimal solutions to the Steiner problem in a graph using a genetic algorithm. Networks **26**: 173-185
8. Gelenbe E, Ghanwani A, Srinivasan V (1997) Improved neural heuristics for multicast routing. IEEE Journal on Selected Areas in Communications **15**: 147-155

9. Goldberg D (1989) Genetic algorithms in search, optimization & machine learning. Addison-Wesley, New York
10. Guo L, Matta I (1999) QDMR: an efficient QoS dependent multicast routing algorithm. Proceedings of the Fifth IEEE Real-Time Technology and Applications Symposium 213-222
11. Holland J (1995) Adaptation in natural and artificial systems. MIT Press, Massachusetts
12. Julstrom B (2001) The blob code: a better string coding of spanning trees for evolutionary search. 2001 Genetic and Evolutionary Computation Conference Workshop Program
13. Kompella VP, Pasquale JC, Polyzos GC (1993) Multicast routing for multimedia communication. IEEE/ACM Trans. Networking **1**:286–292
14. Kurose J, Ross K (1999) Computer networking: a top-down approach featuring the Internet. Addison Wesley Longman, Massachusetts
15. Lu G., Liu Z, Zhou Z (2000) Multicast routing based on ant algorithm for delay-bounded and load-balancing traffic. 25th Annual IEEE Conference on Local Computer Networks 362-368
16. Lu G., Liu Z (2000) Multicast routing based on ant algorithm with delay and delay variation constraints. The 2000 IEEE Asia-Pacific Conference on Circuits and Systems 243-246
17. Mitchell M (1997) An introduction to genetic algorithms. MIT Press, Massachusetts
18. Palmer CC, Kershenbaum A (1995) An approach to a problem in network design using genetic algorithms. Networks **26**: 151-163
19. Picciotto S (1999) How to encode a tree. PhD thesis University of California, San Diego
20. Premkuma G, Chu CH, Chou H (2000) Telecommunications network design comparison of alternative approaches. Decision Sciences **31**: 483-506
21. Prüfer H (1918) Neuer beweis eines satzes uber permutationen, arch. Math. Phys **27**: 742-744
22. Ravikumar CP, Bajpai R (1998) Source-based delay-bounded multicasting in multimedia networks. Computer Communications **21**: 126-132
23. Sahasrabuddhe LH, Mukherjee B (2000)Multicast routing algorithms and protocols: a tutorial. IEEE Network **January/February**: 90-102
24. Salama HF, Reeves DS, Viniotis Y (1997) Evaluation of multicast routing algorithms for real-time communication on high-speed networks. IEEE Journal on Selected Areas in Communications **15**: 332-345
25. Saltouros M, Theologou M, Angelopoulos M, Ricudis CS (1999) An efficient evolutionary algorithm for (near-) optimal Steiner tree calculation: an approach to routing of multipoint connections. Proceedings. Third International Conference on Computational Intelligence and Multimedia Applications 448-453
26. Shan D, Ishii N (2000) An online genetic algorithm for dynamic Steiner tree problem. 26th Annual Conference of the IEEE **2**: 812-817
27. Sinclair M (1999) Evolutionary telecommunications: a summary. Proceeding of 1999 Genetic and Evolutionary Computation Conference

28. SteinLib Testdata Library, http://elib.zib.de/steinlib/steinlib.php
29. Wang Z, Crowcroft J (1996) Quality of service for supporting multimedia applications. IEEE Journal on Selected Areas in Communications **14**: 1228-1234
30. Wang Z, Shi B (2001) Solution to QoS multicast routing problem based on heuristic genetic algorithm. Journal of Computer **24**: 55-61 (in Chinese)
31. Wang Y, Xie J (2000) Ant colony optimization for multicast routing. The 2000 IEEE Asia-Pacific Conference on Circuits and Systems 54-57
32. Wang B, Hou JC (2000) Multicast routing and its QoS extension: problems, algorithms, and protocols. IEEE Network **January/February**: 22-36
33. Xiang F, Junzhou L, Jicyi W, Guanqun G (1999) QoS routing based on genetic algorithm. Computer Communications **22**: 1392-1399
34. Zhou X, Chen C, Zhu G (2000) A genetic algorithm for multicasting routing problem. Proceedings of Communication Technology **2**: 1248-1253
35. Zhang Q, Leung YW (1999) An orthogonal genetic algorithm for multimedia multicast routing. IEEE Transactions on Evolutionary Computation **3**: 53-62
36. Zhang S, Liu Z (2000) A QoS routing algorithm based on ant algorithm. 25th Annual IEEE Conference on Local Computer Networks 574-578

Base Station Location Optimization in Cellular Wireless Networks using Heuristic Search Algorithms

Bhaskar Krishnamachari[1] and Stephen Wicker[2]

[1] Department of Electrical Engineering-Systems, University of Southern California, Los Angeles, CA 90089, USA
[2] Department of Electrical and Computer Engineering, Cornell University, Ithaca, NY 14853, USA

Abstract. Heuristic Search techniques such as Genetic Algorithms, Simulated Annealing, Tabu Search and Random Walk Algorithms have been proposed as useful approaches for difficult global optimization problems. We focus on one such problem that arises in the design of cellular mobile communication systems: the optimal location of base stations. We investigate important parameter settings for each algorithm and then compare them with each other using a common neighborhood definition to ensure fairness. The results suggest that Tabu Search and Genetic Algorithms perform well on this problem, with Tabu Search providing high-quality solutions consistently.

Keywords. Heuristic Search, Genetic Algorithms, Tabu Search, Simulated Annealing, Random Walk, Base Station Location, Optimization, Cellular Wireless Networks

1 Introduction

Mobile personal communications networks consist essentially of a number of cells, each corresponding to the radio footprint of a distinct base station. A number of difficult optimization problems arise in the design of such networks, many of which have been tackled in recent years through the use of heuristic search techniques such as genetic algorithms and simulated annealing [1,8].

We study here one challenging problem that arises when setting up a new cellular network -- the optimal location of base stations in the service area. There are two conflicting objectives in locating base stations -- the first is to maximize the coverage area by using more base stations, and the second is to minimize the equipment cost by restricting the number of base stations.

We can assume that a list of potential sites has been generated *a priori*. This is a reasonable assumption since the number of possible sites where a cellular operator may locate base stations is often restricted by a number of consideratgions such as aesthetics, geographical layout, etc. We also assume that the radio propagation characteristics of the service area are known, specifically the area in which transmitters placed at each of these potential sites result in acceptable signal strength. Such information can be obtained either through field measurements or through the use of sophisticated ray tracing software. We will use a cost function formulation in which the conflicting objectives of maximizing coverage area while minimizing the equipment cost are combined through the use of a weighing parameter. This problem has previously been shown to be NP-hard by relating it to the minimum dominating set problem in [2], where a genetic algorithm approach is used. A similar problem is also studied using simulated annealing in [3].

We aim to study and compare the performance of various heuristic search algorithms including simulated annealing, tabu search and genetic algorithms on this problem. We begin by briefly reviewing these search algorithms in the next section.

2 Heuristic Search Algorithms

Heuristic search algorithms are iterative mechanisms for optimization. At each step, they proceed from the current point x to a new point x' that is in the set $N(x)$, the neighborhood of x. These search algorithms are primarily differentiated from each other by the way in which the considered points are accepted. The acceptance criteria in these algorithms are often based on the cost function values $f(x)$ and $f(x')$ [9].

2.1 Random Walk

The Random Walk algorithms are perhaps the simplest of all local search techniques [4]. At each iteration a neighbor x' that is generated is accepted unconditionally if it has a lower or same cost function as the current point (i.e. if $f(x')$ $f(x)$, and conditionally with a probability p if it has a higher cost function (i.e. if it is uphill). Thus these algorithms range from a purely greedy search ($p = 0$) to a completely random search ($p = 1$). For the greedy case, often a restart strategy is used to start the search afresh when a local minimum is reached.

2.2 Simulated Annealing

Simulated Annealing is similar to Random Walk in permitting uphill moves stochastically, except that the uphill move probabilities depend on the difference in cost functions and a temperature parameter which is decreased during the search

[5]. The most commonly used criterion for uphill moves, the Metropolis criterion, is as follows: for x' $N(x)$, the probability that x' is selected is

$$P_{x \quad x'} = min(1, exp((f(x')-f(x))/T)) \qquad (1)$$

When T is high initially, there is a greater probability of making uphill moves, which allows the search greater freedom to explore the space. As it is decreased, the search becomes more focused and rapidly descends to a minimum.

2.3 Tabu Search

Tabu Search is based on the premise that intelligent problem solving requires incorporation of adaptive memory [6]. In TS, a finite list of forbidden moves called the tabu list T is maintained. The tabu list may also consist of attributes instead of specific solutions. At any given iteration, if the current solution is x, its neighborhood $N(x)$ is searched aggressively to yield the point x' which is the best neighbor such that it is not on the tabu list. Very often though, to reduce complexity, instead of searching all the points in $N(x)$, a subset of these points called the *candidate list* is considered at each step. The use of this candidate list allows the search to test multiple possibilities at each iteration and choose the best of these. As each new solution x' is generated, it is added to the tabu list and the oldest member of the tabu list is removed. The tabu list prevents cycling by disallowing repetition of moves within a finite number of steps (determined by the size of the list). This, along with the acceptance of higher cost moves, prevents entrapment in local minima.

2.4 Genetic Algorithms

These algorithms derive their inspiration from the natural process of biological evolution [7]. Solutions are encoded (often in binary) into strings or chromosomes. The algorithm operates on a population of these chromosomes, which evolve to the required solution through successive operations of fitness-based selection, reproduction with crossover and mutation. The crossover and mutation operations are essentially a form of neighborhood generation. Genetic algorithms permit the parallel exploration of different parts of the search because they act on entire populations of points at a time.

3 Problem Description and Parameters

We model the service area as a square that is broken up into a grid of 100x100 points spaced a unit distance apart. From this set of 100,000 possible points, 51 locations where base transmitting stations could be potentially be located were generated *randomly* in order to test the algorithms. These 51 possible sites can be seen in Figure 1. As mentioned before, for real world problems, this initial list of potential base station locations would depend upon geographical considerations, physical constraints, and the ease with which radio transmitters can be installed in those locations.

For each potential location a service coverage area is obtained using a log-normal shadow fading model for radio propagation. In this model, the power loss in dB at a distance *d* from the base station is given by the equation

$$P_{loss}(d) = A + Blog(d) + N \quad (2)$$

, where *N* is a zero-mean Gaussian random variable with a variance s^2. The parameters are chosen to be $A = 50$, $B = 40$, $s^2 = 10$. For each of the possible base station locations, the radio path loss, $P_{loss}(d)$, is computed for all 100,000 points. Then, a cutoff value $P^* = 100\ dB$ is chosen for the path loss, such that if $P_{loss}(d)$ exceeds P^* for any point it is deemed that the radio coverage is not sufficient. Figure 2 shows the radio coverage for one such transmitter located at one of the 51 possible locations. If all possible locations were selected, the net coverage would be 100\% but there would be a significant overlap between different base stations. The overlap factor *OF*, which represents the average number of base stations providing coverage to each point in the area, was found to be approximately 6.59.

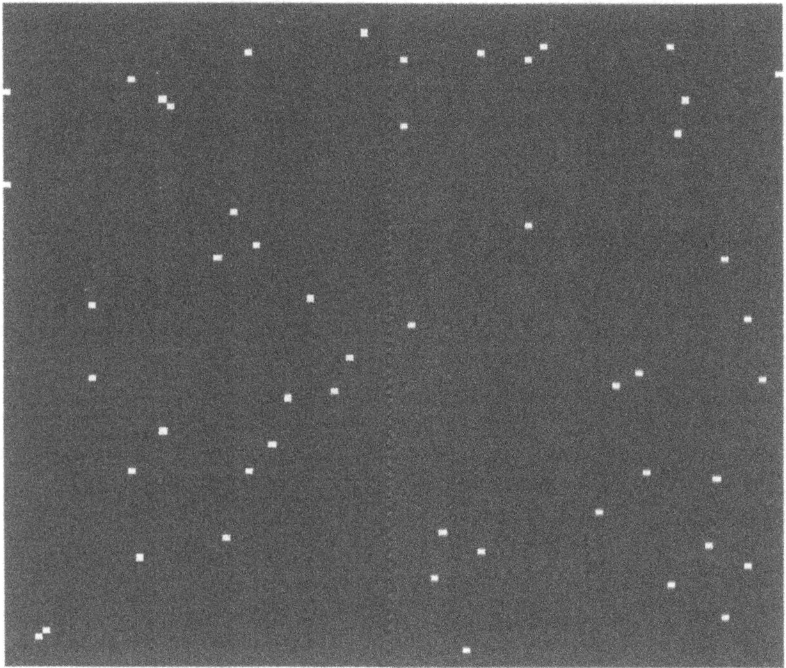

Fig. 1. Potential base station locations

Thus clearly that many of the possible locations provide redundant coverage and that far fewer than 51 base stations would be required to provide good coverage in the area.

For this optimization each point in the search space would represent which subset of the 51 potential locations were actually chosen. This can be represented by a 51 bit binary string with each bit corresponding to one of the potential locations. Each bit is one if a base station is placed at the corresponding location and zero otherwise. The size of the search space is therefore 2^{51}~2.25 10^{15}. The cost function chosen is similar to that used in [2]:

$$f = k\, N_{BTS}\, R^{\cdot} \qquad (3)$$

..., where N_{BTS} is the number of base transmitting stations selected,R is the radio coverage (percentage of locations in service area which are are covered by at least one base station) provided by these selected stations, reflects the weight attached to maximizing coverage as opposed to minimizing the number of base stations, and k is a scaling factor. For this problem the values of the parameters were chosen to be $= 3$, $k = 10^4$.

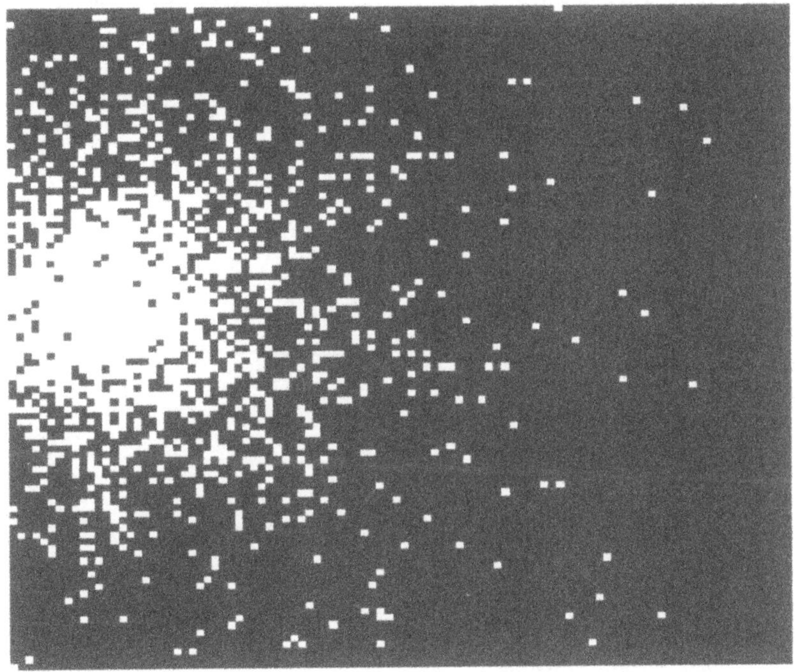

Fig. 2. Sample radio coverage for one potential location

Using the cost function described in equation (3), a histogram of costs was generated from 10^5 random samples and is seen in Figure 3. The histogram shows a long tail, this is explained by the fact that there are a number of possible choices of locations closely spaced such that if they were selected there would be a high number of N_{BTS} that corresponds to low radio coverage. A significant number of search points have cost values between 0.2 and 0.5. The peak around 0.23 corresponds to choices like (19 base stations, 95% coverage) and (16 base stations, 90% coverage). The most optimal points are those with low cost and the lowest values seen in the histogram are about 0.18 which corresponds to solutions like (11 base stations, 85% coverage). It is expected that the search techniques would yield solutions with costs at least good as 0.18, if not better. Only then could their performance be said to be better than a random sampling of the search space.

4 Neighborhood Definition

Recall that a solution for this problem consists of a bit string indicating which base station locations are selected. We considered 3 neighborhood definitions:

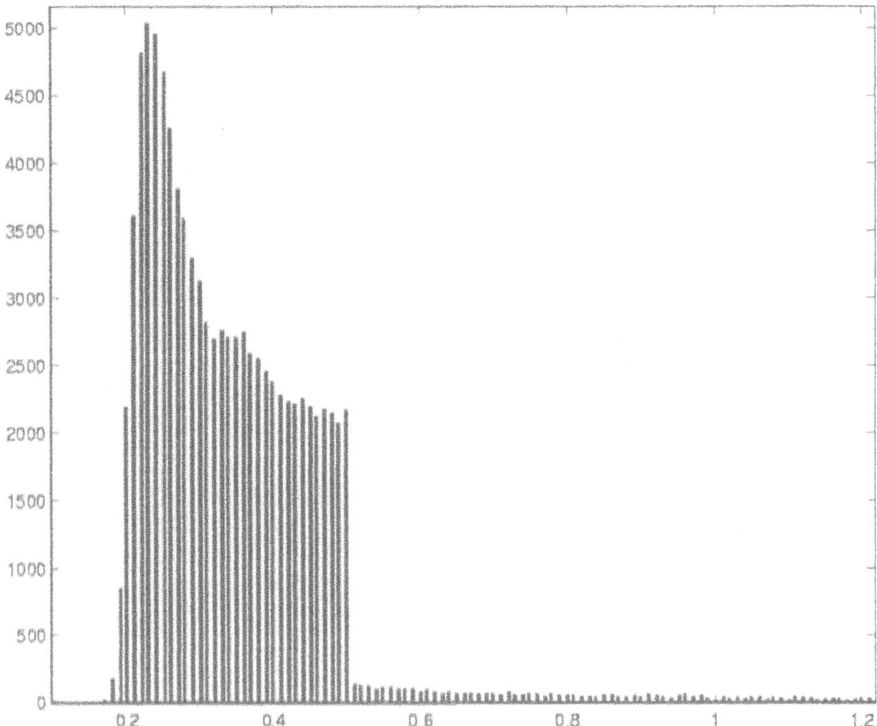

Fig. 3. Histogram of costs for base station location optimization (x-axis indicates costs , y axis indicates frequency)

Neighborhood definition A: flip any one bit of the current solution string at random.

Neighborhood definition B: with a low "mutation" probability P_m (typically about 0.01), flip each bit in the binary string representing the current solution.

Neighborhood definition C: randomly pick a new point from the entire search space. This is the completely random neighborhood where $N(x) = S$.

Since definition A was utilized for most of the experiments performed in this study, it is assumed in the descriptions below that the neighborhood is defined as per definition A, unless specified otherwise. Also for the experiments described below the initial starting point for the various algorithms was chosen randomly in such a way that the number of initial locations was 26 (approximately half the total number of possible locations).

We now begin describing our experiments with the various search algorithms.

5 Performance of Random Walk

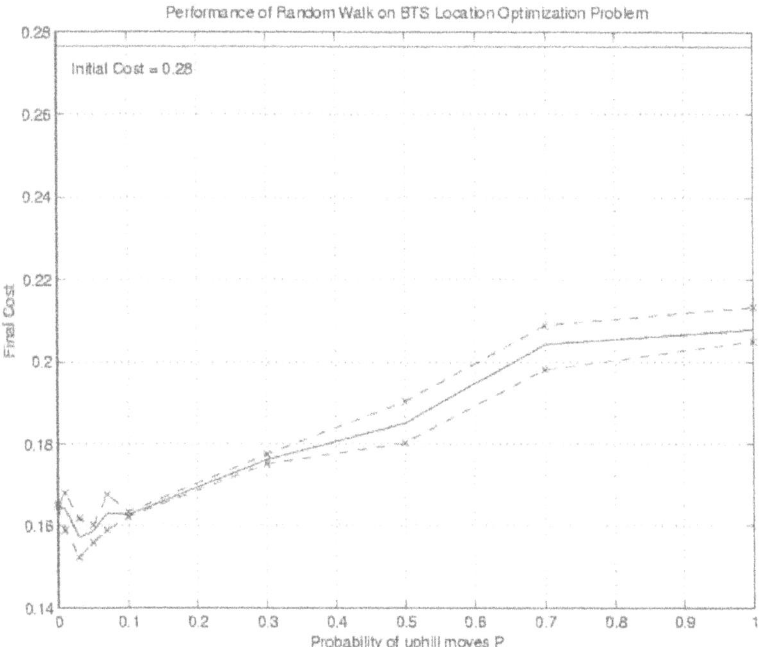

Fig. 4. Performance of Random Walk algorithms for base station location optimization

Figure 4 shows the average (solid), minimum and maximum (dotted) final costs obtained. An acceptance probability level $p = 0.03$ is found to be optimal, which implies the existence of local minima that may be escaped by allowing a small percentage of uphill moves. If p is too high, on the other hand, the search becomes unfocused and the algorithm is less likely to locate minimum cost solutions efficiently.

An experiment was conducted comparing the random walks for different values of p. For each p, the Random Walk algorithm was run 10 times for 1000 iterations. Figure 4 shows the average final cost obtained in these runs for values of p ranging from 0 to 1.

It may be noticed that the final solutions provided by all Random Walk algorithms are all quite good, and certainly much better than what could be obtained via random sampling of the space. While in general the quality of solutions deteriorates as p increases, there is a small dip at low values of p. It is seen that for low non-zero values of p, the algorithms provide better final solutions than the greedy search ($p = 0$).

6 Performance of Simulated Annealing

An SA algorithm with a geometric cooling schedule (T_{new} = ${}_cT_{old}$) and the Metropolis acceptance criterion was used. As suggested in [5], the initial temperature was selected to be T_0= 10 , where is the standard deviation of costs obtained by a purely random search.

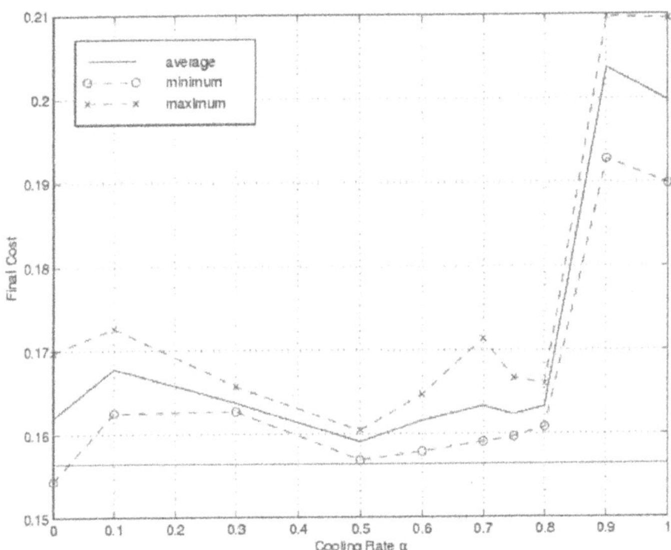

Fig. 5. Performance of Simulated Annealing for base station location optimization

The initial temperature T_0 needs to be high enough. It was determined by first obtaining L cost values via a purely random search (RW with $p = 1$) and finding the standard deviation of these L costs. If the cooling rate $_c$ is too low, then the Simulated Annealing process would go down to zero temperature quickly and perform a greedy search. If, on the other hand, $_c$ is too high, the algorithm would not cool quickly enough and its behaviour would be more like a Random Walk algorithm with p=1. To identify the optimum value of $_c$ the algorithm was compared for values ranging from 0 to 1 (10 runs for each value).

If I is total number of iterations to run the algorithm and n the number of discrete temperature levels in each run, the length L of the homogeneous Markov chains was set to be

$$L = In^{-1} \tag{4}$$

The number of iterations I for this comparison was chosen to be 1000, and the lengths of the isothermal Markov chains L = 50. Hence the number of discrete temperatures seen in each run was 20.

All runs were compared for the same initial point in the search space. The results are shown in figure 5 which shows the average, minimum and maximum final costs for each value of $_c$. For comparison, the average performance of RW with p = 0.03 (the solid flat line with cost = 0.157) is also shown. As expected, for low values of $_c$, SA performs about as well as (but no worse than) greedy search. For high values of $_c$ (greater than 0.8), the performance deteriorates significantly. The optimum value for which SA performs well is found to be $_c \sim 0.5$. However, it does not appear that SA performs any better than RW with p = 0.03.

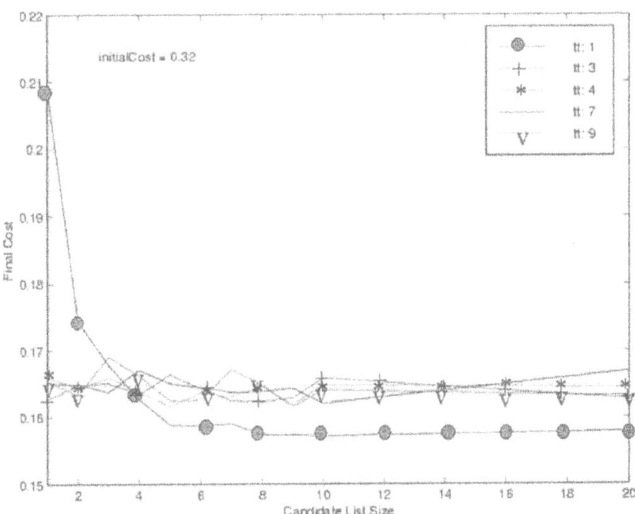

Fig. 6. Performance of Tabu Search for base station location optimization

7 Performance of Tabu Search

A simple TS with the following tabu definition is tested: any base station that is selected within the last *tt* iterations is considered tabu and may not be selected at the current iteration. The value of the tabu tenure *tt* is chosen to be 1. An important parameter that can be varied is the the size of the candidate list *v*, which determines how many neighbors are considered at each step in the search. One might expect that there would some improvement as the size of the candidate list increases since one can better sample the neighborhood of the current point in order to determine a good point to move to.

To determine the effect of the values of *tt* and *v* on the performance of Tabu Search, the algorithms were compared for different parameter values. This was done for the same initial starting point for 10 runs each consisting of 1000 function evaluations (the number of iterations $I = 1000v^{-1}$. The algorithms were compared for *tt* = [1, 3, 4, 7, 9] and *v* = [1, 2, 3, 4, 5, 6, 7, 8, 9, 10, 20]. One would expect that if the Tabu Tenure *tt* is high, given the above definition of tabu moves, the performance would deteriorate. This is because if the tenure is high, then if the new base station located at some iteration in the search is a good one, it won't be considered for the next *tt* generated solutions (which may not be as good). In terms of the candidate list size *v*, one might expect that there would some improvement as the size of the candidate list increases since one can better sample the neighborhood of the current point in order to determine a good point to move to.

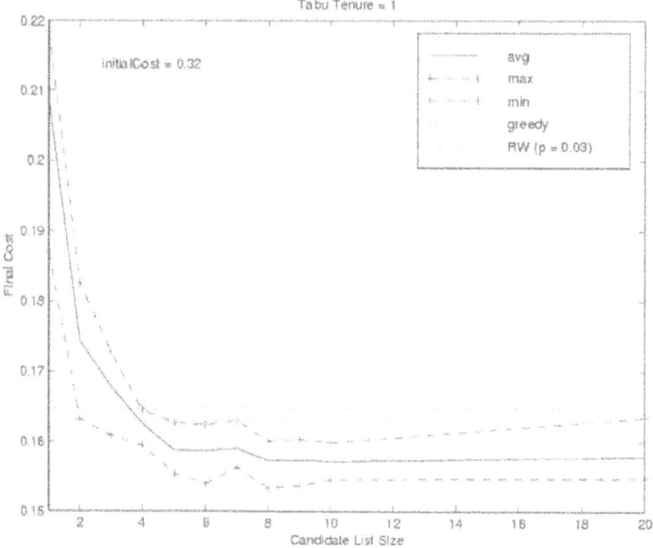

Fig. 7. Performance of Tabu Search for base station location optimization (*tt*=1)

The results of the experiments are shown in Figure 6. It is observed that indeed Tabu Search performs best for the lowest possible value of the tabu tenure ($tt = 1$). The performance with higher values of tt is somewhere between the performance of greedy search and RW with p=0.03 (the two dashed lines) which appear to be stuck in relatively poorer regions in the search space. Figure 7 shows the performance for this tenure with additional details on the minimum and maximum final cost values observed in the 10 runs. Further, for the case where $tt = 1$, Tabu Search is seen to perform progressively better as v is increased, reaching the best performance for a candidate list size of 10.

8 Performance of Genetic Algorithms

The Genetic Algorithms considered here do not implement crossover and utilize a mutation operator that is equivalent to the neighborhood definition described before.Given the neighborhood definition A (described in section 4, the parameters for Genetic Algorithms that can affect performance are the type of selection used and the size of the populations in each generation. Three types of selection methods were compared with each other:

Proportional Selection: At each generation the probability of selecting an individual chromosome in the population for reproduction is proportional to its fitness value. One problem with this selection technique is that it is not invariant with respect to the scaling of the cost-functions, which means that the fitness function must be a scaled inverse of the cost function.
Exponential scaling was utilized with an exponent of 100.

Tournament Selection: A tournament is conducted by picking t chromosomes in the population at a time randomly and selecting the best chromosome among those for reproduction. This tournament is repeated until the next generation can be created. The tournament size t was fixed to be 2 (*binary Tournament Selection*). This selection scheme is scaling invariant.

Rank-based Selection: In this all the chromosomes in a population are ranked according to their fitness and the probability of selecting a chromosome is made proportional to its rank. One parameter for Rank-based selection is $_{rank}$ which is defined to be the expected number of offspring of the best chromosome in the population. For the experiments here, $_{rank}$ was assigned the value 1.5 which is a typical value. One potential problem with Rank-based selection is that in some sense the exact value of the cost functions for each solution is ignored. While this is useful for objective functions where the exact cost values are not easy to obtain, it could result in sub-optimal performance when the exact cost values are known.

Another factor that can influence the performance of Genetic Algorithms is the size of the population. Figure 8 shows a comparison of the average performance Genetic Algorithms for the three different selection techniques and for population size values of PS = [10, 30, 50]. All runs were conducted for 1000 cost function evaluations. It is observed that Rank-based Selection performs quite poorly (and performs the same for all three population sizes tested), while Tournament Selection performs the best of the three selection techniques. With both Tournament and Proportional Selection the performance seems to deteriorate linearly with increasing population size with the best performance at a value of 10. This may be because of the smaller number of generations G which depends inversely upon the population size: $G = 1000/PS$.

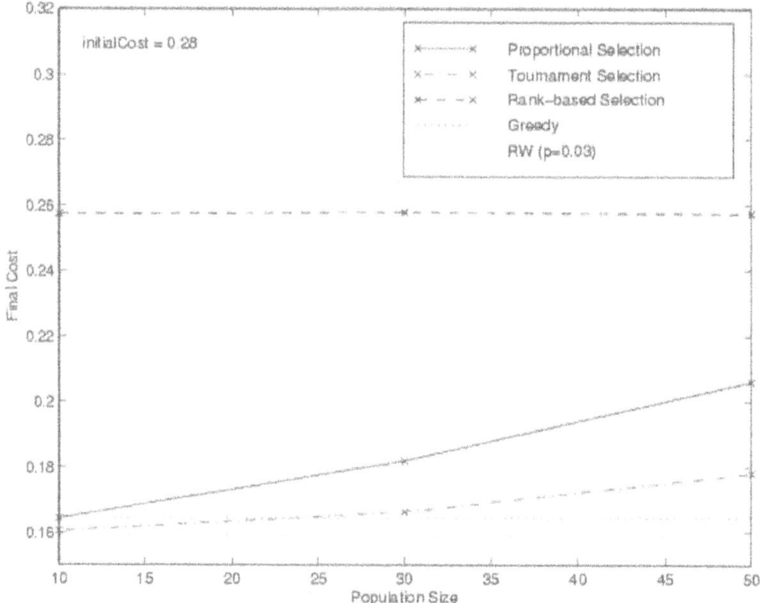

Fig. 8. Performance of Genetic Algorithms for base station location optimization

9 Comparison of Algorithms

The different search techniques were first compared with each other for their performance on the problem using the best parameters from the previous experiments. Figure 9 is a bar chart comparing the performance of Random Walk with $p = 0$ (labeled '1'), Random Walk with $p = 0.03$ ('2'), Simulated Annealing ('3'), Tabu Search ('4'), and Genetic Algorithms ('5') respectively. These algorithms are compared for Neighborhood A, each was run 10 times for 10,000 evaluations. The figure shows the maximum, average and minimum final cost values obtained by each search technique. We observe first that all techniques provide roughly the same performance though the results of Greedy Search show a high variance indicating that it has a tendency to get trapped in local minima. Also the minimum final values obtained by all of them in 10 runs are nearly identical. The performance of Simulated Annealing both in terms of variance and mean is better than Greedy

Search but slightly worse than Random Walk with p=0.03, which in turn appears slightly worse Genetic Algorithm. The performance of Tabu Search is observed to be the best of them all, both in terms of average final cost and variance.

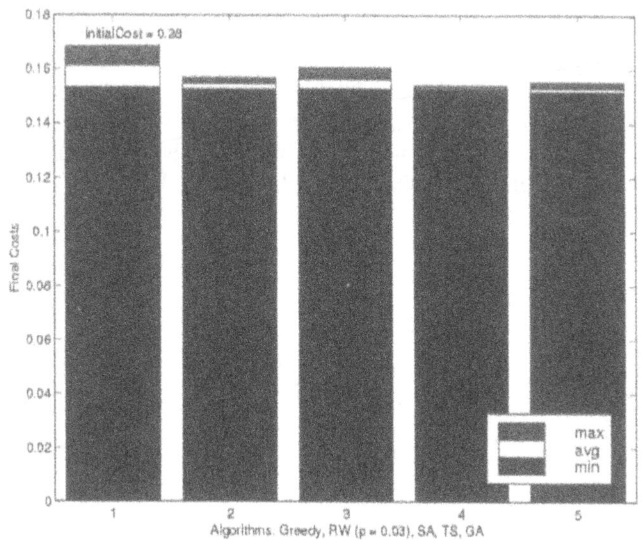

Fig. 9. Comparison of search algorithms for base station location optimization

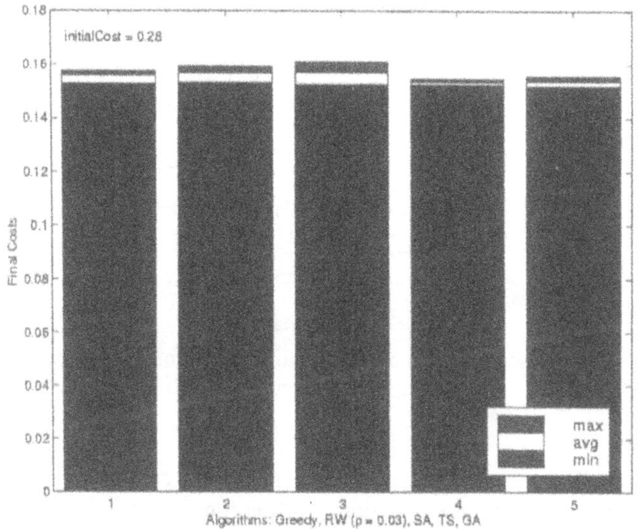

Fig. 10. Comparison of search algorithms for base station location optimization (with multi-start Random Walk)

It was observed that the Random Walk Algorithms (both for Greedy Search and for p=0.03) tended to settle down to solutions quite early in the search and not move from there. Since this is wasteful of function evaluation calls, one approach is to re-start this algorithm from a randomly generated point in the search space. This was done for RW, modifying them so that they re-start if no change in the current point is seen for more than 50 iterations. The multi-start RW algorithms were then compared with SA, TS and GA (again for 10 runs). This is seen in Figure 10. This clearly improves the performance of Greedy Search (even to the point of a small improvement over RW with p = 0.03) which is now comparable to the performance with GA and better than SA. Tabu Search, however, still offers the best performance.

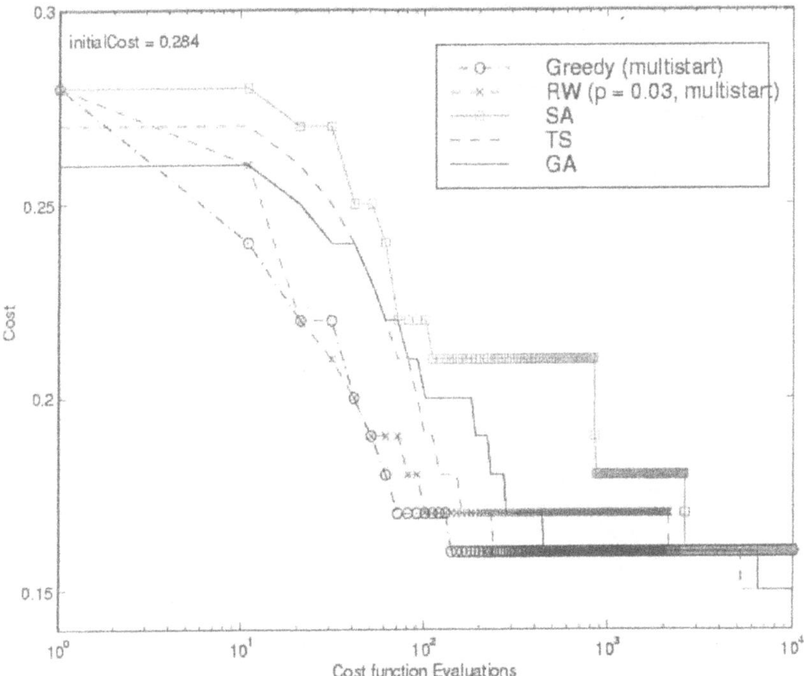

Fig. 11. Comparison of search algorithms for base station location optimization (1 sample run)

Figure 11 shows one sample run of each algorithm comparing the best cost to date vs. number of function evaluations. In this run, TS and GA provide the best final cost while all the other converge to about the same final cost, which is still quite close to that obtained by TS.

10 Effect of Neighborhood Definition

The definition of neighborhood used in the search procedure can be the major factor affecting performance. To investigate this, experiments were performed to compare the performance of each search technique for other neighborhood definitions. Figures 12-14 show the results of utilizing neighborhood definition B (see section 4) with varying values of P_m, the probability of "mutating" or flipping each bit. The three figures are for P_m= 0.001, 0.01 and 0.1 respectively. Increasing the probability of mutation makes the neighborhood more random. When P_m = 0.1, for example, there is a one in a ten probability of changing each bit in one step which means that neighboring points will on average differ by about 5 bits. For each P_m the algorithms were run 10 times for 10,000 function evaluations.

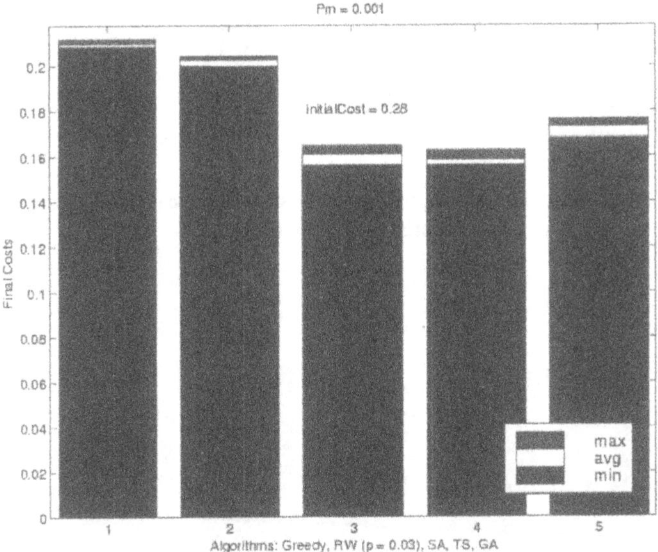

Fig. 12. Comparison of search algorithms for base station location optimization (neighborhood B, $P_m = 0.001$)

The results are interesting. For P_m = 0.001, as seen in figure 12, we find that while Tabu Search still provides the best overall results, the performance of Simulated has been improved considerably, making it perform even better than GA. The Random Walk algorithms are seen to perform rather poorly. This is perhaps due to the restrictive nature of the neighborhood definition - for this low value of P_m, only one in two neighbors will differ by a bit. This means there is a greater chance for the Random Walk algorithms to be stuck in a small portion of the search space and provide locally optimal solutions.

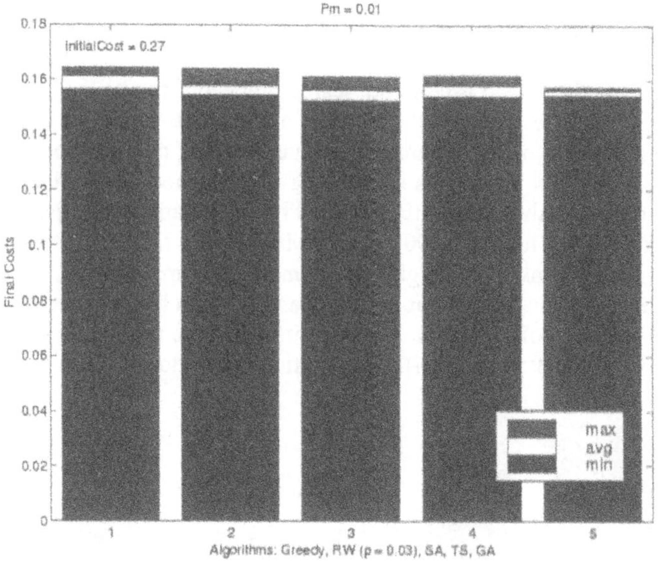

Fig. 13. Comparison of search algorithms for base station location optimization (neighborhood B, $P_m = 0.01$)

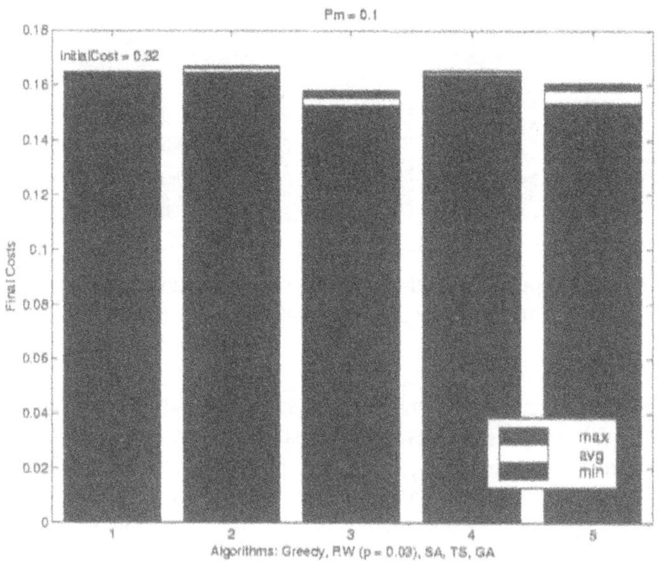

Fig. 14. Comparison of search algorithms for base station location optimization (neighborhood B, $P_m = 0.1$)

Tabu search is still able to perform well because the tabu list directs it out of poor regions. The Random Walk algorithms perform better as P_m is increased, and as seen in figure 14, their performance is comparable to the other algorithms. Relative to the other algorithms, Simulated Annealing performs much better at $P_m = 0.1$, providing the lowest average final cost. Another result observed is that Tabu Search provides a slightly worse final cost for this value of P_m but in the ten runs observed, the final cost obtained shows little or no variance. These results demonstrate clearly the dependence of the performance of search algorithms on the definition of the neighborhood for a given problem.

It can be argued that when the neighborhood is completely random i.e. $N(x) = S$, x S, all algorithms provide the same performance. This effect was verified by utilizing neighborhood C (described in section 4). Figure 15 shows the result of comparing the search algorithms for 10,000 function evaluations for 10 runs each. As predicted the performance is nearly identical for all algorithms for this neighborhood. It is also worth noting that the final cost in all runs was about 0.27, significantly worse than the performance for more structured, localized, neighborhood schemes. This shows that search algorithms with reasonable neighborhood schemes do provide much better results than randomly sampling the problem space.

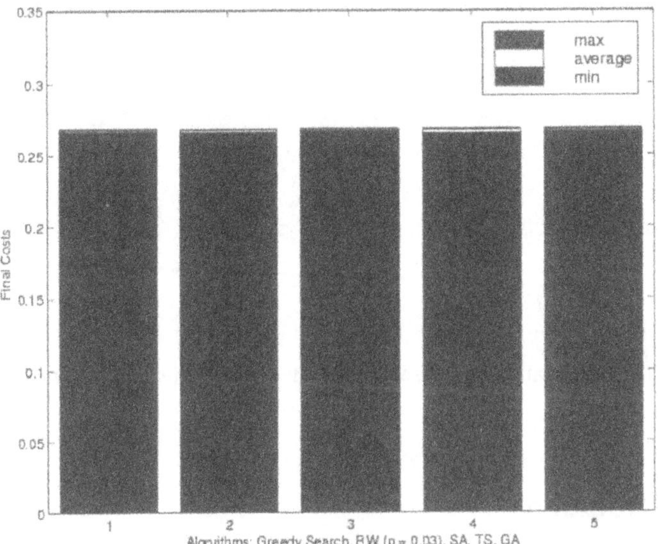

Fig. 15. Comparison of search algorithms for base station location optimization (neighborhood C)

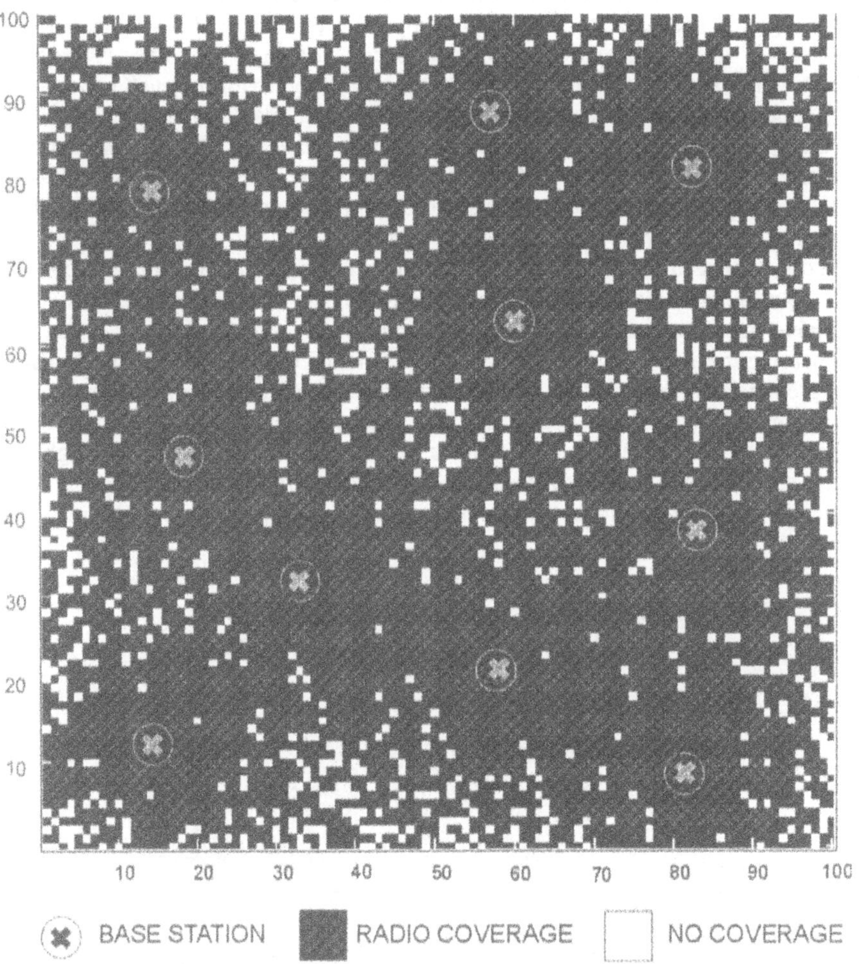

Fig. 16. Best final solution discovered for the base station location problem

11 Final Solution

The best overall solution in these experiments was found using Tabu Search (with tabu tenure 1 the size of candidate list 10) and Neighborhood type A. The final cost of this solution was 0.153 and it corresponds to 10 base stations offering 86.82% coverage of the area. This near-optimal solution is shown in figure 16. The crosses indicate the positions of the final base stations. Points that are not covered are indicated by white area, while the covered areas are indicated in black.

12 Conclusions

We have presented an experimental analysis of the performance of four search algorithms for the problem of designing the optimal location of base stations in a mobile communication system - Random Walk, Simulated Annealing, Tabu Search and Genetic Algorithms. We investigated the effect of varying the value of important parameters for each algorithm to determine suitable values. We then compared the performance of these algorithms using a common neighborhood definition, the same number of cost function evaluations, and the same initial conditions. It is found that Genetic Algorithms and Tabu Search both perform well, with Tabu Search providing the most consistent final cost value on multiple runs.

References

1. Krishnamachari B, Wicker SB (2000) Global search techniques for problems in mobile communications. In: Corne D *et al.* (eds) Telecommunications Optimisation: Adaptive and Heuristic Methods, John Wiley and Sons
2. Calegari P *et al.* (1997) Genetic approach to radio network optimization for mobile systems. IEEE 47th Vehicular Technology Conference 2:755-759
3. Anderson HR, McGeehan JP (1994) Optimizing Microcell Base Station Locations Using Simulated Annealing Techniques. IEEE 44th Vehicular Technology Conference 2:858-862
4. Selman B, Kautz HA, Cohen B (1994) NoiseStrategies for Improving Local Search. Proceedings of Twelfth National Conference on Artificial Intelligence 1:337-343
5. Aarts EHL, Korst JHM (1989) Simulated Annealing and Boltzmann Machines, Wiley
6. Glover F, Laguna M (1997) Tabu Search. Kluwer Academic Publishers
7. Bäck T, Fogel DB, Michalewicz Z (1997) Handbook of Evolutionary Computation. IOP Publishing and Oxford University Press
8. Gibson D (1999) The Mobile Communication Handbook, CRC Press, 2nd edition
9. Aarts E, Lenstra JK (1997) Local Search in Combinatorial Optimization, John Wiley and Sons

The Synthesis and Design of Communication Antennas Using Genetic Algorithms

Douglas H. Werner [1,2], Matthew G. Bray [1], Rene J. Allard [2], and Pingjuan L. Werner [1]

[1] Department of Electrical Engineering, Pennsylvania State University
University Park, PA 16802, USA
[2] The Applied Research Laboratory, Pennsylvania State University
State College, PA 16804, USA

Abstract. This chapter discusses the design and optimization of communication antenna arrays through genetic algorithms. A genetic algorithm approach to communication antenna design is presented by considering two different applications of array synthesis. The first application will investigate a method of creating thinned aperiodic linear phased arrays that will have suppressed grating lobes with increased scan angles. In addition, the genetic algorithm will place restrictions on the driving point impedance of each array element so that they are well behaved during scanning. The second application involves using a domain-decomposition / reciprocity procedure in conjunction with genetic algorithms to design microstrip patch antenna arrays mounted on an arbitrarily-shaped three-dimensional metallic platform, such as a base station tower. In this application a genetic algorithm synthesis procedure is introduced that is capable of determining the optimal set of element excitation phases required to yield a desired or specified far-field radiation pattern.

Keywords. Communication antenna arrays, genetic algorithms, thinned arrays, conformal arrays, aperiodic arrays, phased arrays, base station antennas

1 Introduction

Antenna array technology is becoming an increasingly important component in commercial communication systems. This is primarily due to the continuing demand for additional functionality and improved performance from wireless communication services. Smart and adaptive antenna arrays have received a considerable amount of recent attention for their ability to enhance the coverage and capacity of wireless systems at the base station [15,17,20,26]. The design of such arrays typically depends on a large number of parameters and therefore following a trial-and-error approach will almost certainly lead to antenna systems with non-optimal performance characteristics. In this chapter a powerful tool for the design optimization of advanced communication antenna arrays will be introduced that is based on Genetic Algorithms (GAs). Two different applications of the GA to communication antenna array design will be considered here. The first application involves using the GA to design thinned arrays with optimal radiation pattern and driving point impedance performance verses scan angle. One of the main features of this technique is that it is very versatile and can be applied to the design of communication antennas for operation in either the transmit mode or the receive mode. The second important application of the GA that will be considered involves the design of conformal microstrip antenna arrays for optimal performance in the presence of a complex three-dimensional mounting platform, such as a base station tower.

2 Background

Genetic algorithms were developed in the 1970's by John Holland in an attempt to model the natural processes found in Darwinian evolution [5]. Natural evolution is a form of optimization which operates on an individual's chromosomes through the process of mating and genetic mutation to make that individual better suited to their environment. Holland mimicked this process by creating algorithms which mapped strings of binary digits representing an optimization problem onto a set of artificial chromosomes. A simulated evolution on a population of these chromosomes was subsequently carried out. The result was an extremely versatile and robust algorithm capable of solving complex optimization problems, many of which could not be solved using any of the more traditional optimization techniques. In the years since their inception, GAs have been applied to a wide range of problems from many different scientific and engineering disciplines.

The intent of this chapter is not to discuss the detailed theory behind how genetic algorithms operate, but rather to demonstrate their importance in the design and synthesis of advanced communication antenna arrays. The interested reader is referred to [5,7,10,16] for an in-depth guide to the theory and applications of genetic algorithms toward a general set of optimization problems. The review articles [9, 11, 27] provide an excellent overview of the use of GAs in solving a variety of problems in electromagnetics. A collection of chapters is contained in [18] that deal with the application of GAs to a number of important electromagnetic optimization problems, including antenna array and conformal antenna design. In addition, chapter 8 in [30] is devoted to the subject of beamforming in antenna arrays through the use of GAs and neural networks. Finally, an early application of GAs to the design of thinned arrays can be found in [8].

3 Grating Lobes in Thinned Periodic Arrays

The traditional way of creating linear thinned arrays has been to take a fully populated half wavelength array and remove a certain percentage of the elements. The first thinned arrays were created in this manner by removing elements randomly or by trial and error [25,26]. The problem with this technique was the high maximum sidelobe levels that resulted from non-optimal placement of the radiating elements. More recent work on thinned arrays makes use of different optimization algorithms to remove elements in such a way as to have the lowest possible sidelobe levels. Simulated annealing and genetic algorithms have been shown to be fairly successful at accomplishing this task [12,15,17,23].

Most work in thinning arrays has only considered optimizing an array to have low sidelobe levels at broadside. As these arrays are scanned a small angle away from broadside, grating lobes will begin to appear. To create an array that will have reduced grating lobes during scanning, one must optimize the array when it is steered to the maximum desired scan angle [3,8]. Now as the array is steered between broadside and the maximum scan angle, the sidelobes will be bounded by the maximum sidelobe level at the furthest scan angle. The tradeoff for increasing the scan range is that the maximum sidelobe level increases as the array is optimized to operate in a range further from broadside. The technique introduced in [8] starts out with a periodic array and uses a genetic algorithm to determine which elements should be removed or "turned off" to yield the lowest maximum relative sidelobe level.

This chapter will address the problem of grating lobes during scanning by considering a different representation of the thinned array. Instead of removing elements from a fully populated half wavelength array, the thinned array will be based on a periodic linear phased array with an interelement spacing greater than a half wavelength. The optimization parameter is a perturbation added to each interelement spacing in such a way as to create an aperiodic array that maintains low sidelobes during scanning. In addition to increasing the scan angle by reducing grating lobes, a driving point impedance restriction will be placed on each element in the array. Considering driving point impedance makes matching element impedances easier and can prevent, for example, the occurrence of blind angles during scanning. A similar approach can also be used to optimize the performance of adaptive receiving arrays for communication systems. In this case, the array element spacings would be optimized to provide the desired scan range while, at the same time, minimizing the influence of mutual coupling.

A major problem in the development of thinned phased arrays has been the limited scan range due to the appearance of grating lobes. The maximum angle that a thinned linear phased array can be scanned from broadside may be derived from a well-known relation in [14]:

$$\theta = \sin^{-1}\left(\frac{\lambda}{d} - 1\right) \tag{1}$$

where θ is the maximum steerability from broadside, and d is the uniform element spacing. From this equation it can be seen that half wavelength spaced arrays will have a complete theoretical scan range of 90°. However as the uniform spacing of the array is increased beyond a half wavelength the scan range of the array begins to be severely reduced. For example, with a uniform spacing of just 0.8 wavelengths the maximum scan angle of the array without grating lobes has been reduced to only 14.5° from broadside. Figure 1 shows that scanning beyond this angle to 45° will result in a grating lobe that has the same magnitude as the main beam. Even though thinned arrays have a reduced scan range due to grating lobes, they are very important because they require fewer elements to fill a given aperture, compared to half wavelength spaced arrays. The ability to reduce the number of elements required for a given aperture is an important practical design consideration for many communication systems, since it can lead to less expensive and lighter weight antennas.

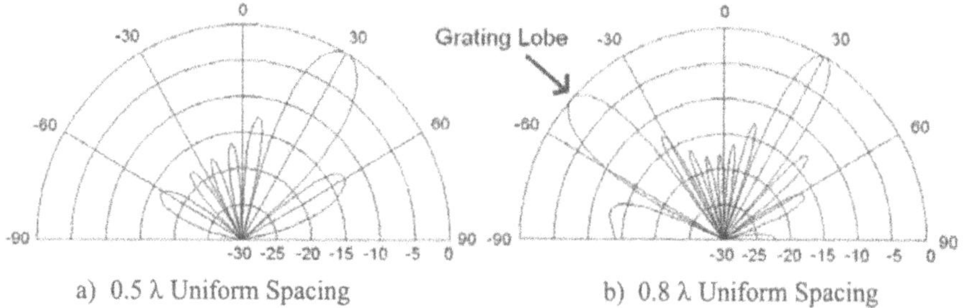

Fig. 1. A large grating lobe appears for an 8 element array with a uniform spacing of 0.8λ when it is steered past 14.5° to 45°.

4 Increased Steerability Using Thinned Aperiodic Phased Arrays

The limiting factor on the maximum scan angle in thinned periodic phased arrays, as seen in the previous section, is the appearance of grating lobes. These grating lobes are the result of the superposition of the element patterns when the array is steered to an angle larger than that predicted by (1). This grating lobe may be reduced at larger scan angles; however, by perturbing the spacings of the periodic array in such a way so that the patterns from each element are only completely in phase in the direction of the main beam. Figure 2 shows a thinned periodic array (solid elements) with a uniform spacing *d*. In order to reduce the grating lobes in the periodic array, it is transformed into an aperiodic array (hollow elements) by adding a perturbation δd_n to the position of each element in the array.

If the elements in the array are taken to be isotropic sources, the pattern of these two arrays can then be described by their array factors. The array factor for the conventional periodic array is given by

$$AF(\theta) = \sum_{n=1}^{N} I_n e^{ikd(n-1)(\sin\theta - \sin\theta_0)} \quad (2)$$

The array factor of the aperiodic array is created by adding in the appropriate element position perturbations, δd_n, such that [4]

$$AF(\theta) = \sum_{n=1}^{N} I_n e^{jkd_n(\sin\theta - \sin\theta_0)} \tag{3}$$

where

$$d_n = d(n-1) + \delta d_n \tag{4}$$

In these equations, θ_o is the angle that the array is steered from broadside and I_n is the excitation current on each element in the array. The optimization process developed in this chapter for generating arrays that have radiation patterns with reduced grating lobes will be based on (3).

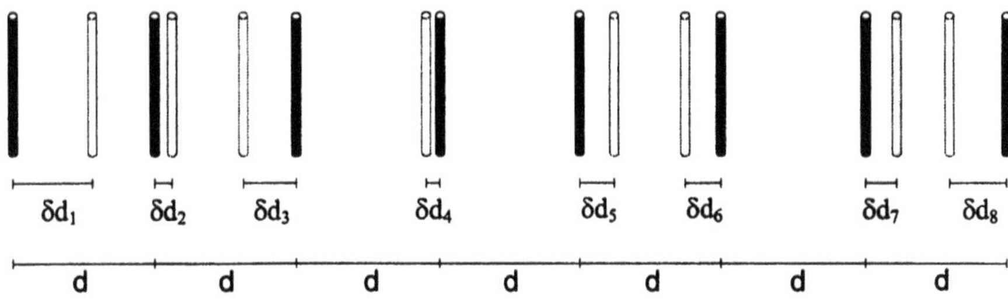

Fig. 2. Geometry of a thinned aperiodic array (hollow elements), which is created by adding perturbations to the position of each element in a thinned periodic array (solid elements).

5 Optimization of Element Positions Using a Genetic Algorithm

The geometry of Figure 2 shows how the positions of each element in the periodic array are perturbed to create an aperiodic array that will have reduced grating lobes. An N element array of this type will have N separate positional perturbations. The solution space of this problem is very large and its optimization lends itself to genetic algorithms, which are robust and can effectively search through large solution spaces.

The genetic algorithm used for this optimization represents the phased array as a list of element position perturbations. This list is called a chromosome and is what the GA will be optimizing. All other parameters of the array are kept constant. For a given optimization the number of elements N, uniform spacing d, and scan angle from broadside θ_o, are chosen. The current on each element of the array can be left constant, or can be included in the genetic algorithm for optimization. For this present discussion we shall consider arrays with only uniform current distributions.

The GA used has two operators, crossover and mutation [6,9,10,18,27]. The crossover technique used is a two-point crossover. This commonly used method chooses two of the best members of the population at random and picks two points on the chromosome. The data in-between these two points are then exchanged and two new chromosomes are created. If the fitness of these chromosomes is better than the average fitness they are included into the population. A constant population model is used here, so that every time a new member is placed into the population, a member of the population that has a low fitness is replaced. The mutation operator used enters new data by mutating a small portion of the genes in the system. A flowchart of the basic GA is shown in Figure 3.

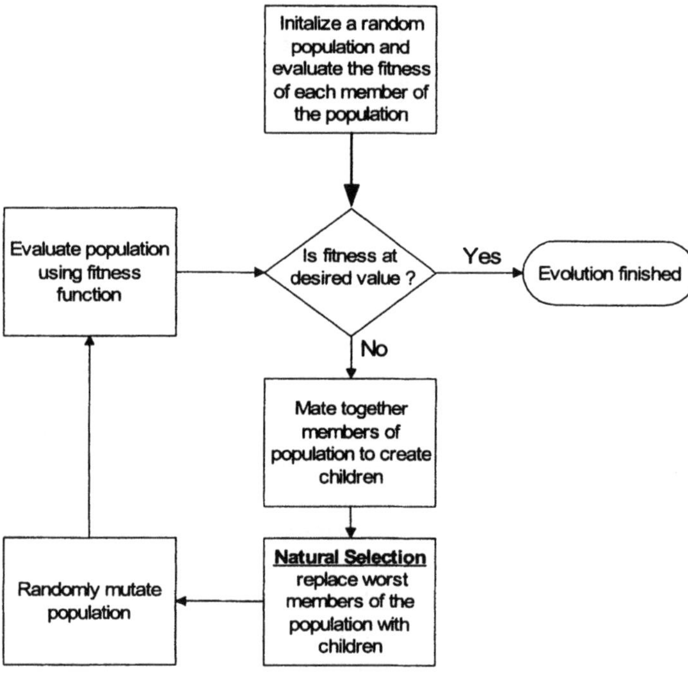

Fig. 3. Flowchart of a basic Genetic Algorithm.

The fitness of each member in the population is calculated by using the array factor given in (3). The maximum grating lobe is found in the normalized array factor, and one minus this value is assigned to the fitness, as illustrated in Figure 4. Therefore, as the grating lobe is reduced, the fitness of the array increases. This type of fitness function will evolve a flat sidelobe profile that is at a minimum in the optimized array.

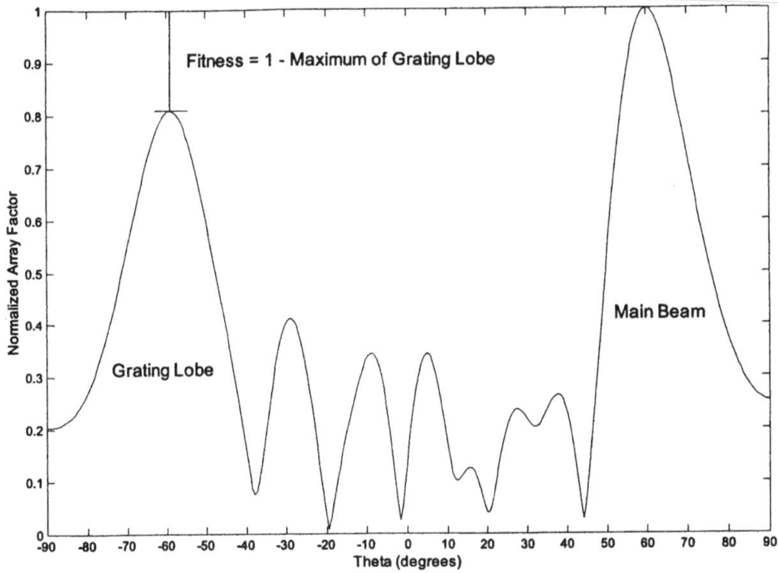

Fig. 4. The fitness of the array is calculated by finding the maximum grating lobe level and assigning one minus this value to the fitness.

6 Optimization Results

The optimization procedure described above was performed on an eight-element array with a uniform spacing of 0.8 wavelengths and a uniform current excitation. From (1) it can be seen that a periodic array with this spacing has a maximum scan angle of 14.5° from broadside without generating grating lobes. A scan angle of 60° was chosen for the optimization, well beyond the scan angle limitation of the periodic array. Figure 5 shows the results obtained from the GA in comparison to that of the periodic array. The GA clearly reduces the grating lobe, creating a pattern with a maximum sidelobe level of about -10 dB. The optimized array can now be scanned from 0° to 60° from broadside without the sidelobe level rising above about -10 dB, a vast improvement over the steerability of the thinned periodic array.

Even further sidelobe suppression results when the number of elements is increased. The solid line in the top pattern of Figure 5 shows a 16 element periodic array steered to 60° with a uniform spacing of 0.8 wavelengths. The bottom patterns show a GA optimized 16 element array with the 8 element aperiodic pattern superimposed upon it. The sidelobes of the 16 element array are about 3 dB lower than the maximum sidelobe level in the 8 element array. The greater sidelobe suppression in the larger array is due to more degrees of freedom in choosing element positions.

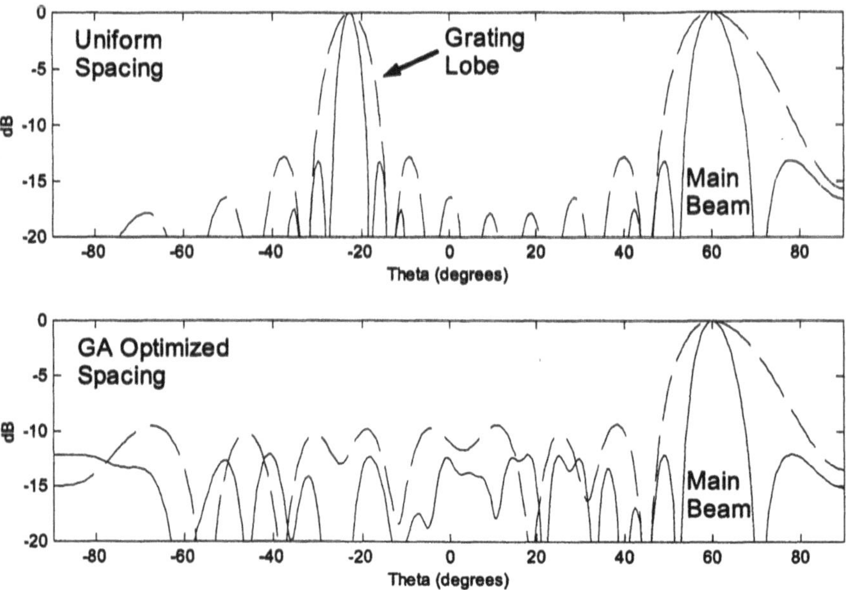

Fig. 5. The top pattern shows a clear grating lobe in a thinned 8 (dashed line) and a 16 (solid line) element periodic array with a uniform spacing of 0.8 wavelengths steered to 60°. The bottom pattern shows an 8 (dashed line) and 16 (solid line) GA optimized aperiodic array based on the same uniform spacing.

The maximum sidelobe level in the optimized array is mainly dependent upon the number of the elements in the array, and the angle chosen for optimization. It will remain constant, at the value it was optimized at, as it is scanned from broadside to the angle of optimization. For the above simulation, if the optimization angle is increased beyond 60° the maximum sidelobe level will increase sharply; however, if the optimized scan angle is decreased, the maximum sidelobe level will decrease, improving the performance of the array. Further suppression of the sidelobes can be achieved by increasing the number of elements in the array.

Figure 6 shows the maximum sidelobe suppression that can be achieved for three different sized arrays optimized by starting from a 0.8 wavelength periodic array. These curves were created by running many different optimizations at scan angles ranging from 0 to 90 degrees. Because the maximum sidelobe level remains constant as the array is scanned, each point on a curve represents the maximum sidelobe level over the entire scan range of the array (up to and including its optimization angle). These curves show that the optimized aperiodic thinned arrays have a much larger scan range than their periodic counterparts. The dotted line at 14.5 degrees represents the maximum possible scan angle of a periodic array without the appearance of grating lobes. From these curves it can be seen that the scan range of the aperiodic arrays are extended to about 60 degrees for the 8 element array and to at least 72 degrees for the 16 and 24 element arrays. The trend in the curves seems to indicate that by increasing the amount of elements in the array, the scan range is increased and the maximum sidelobe level drops.

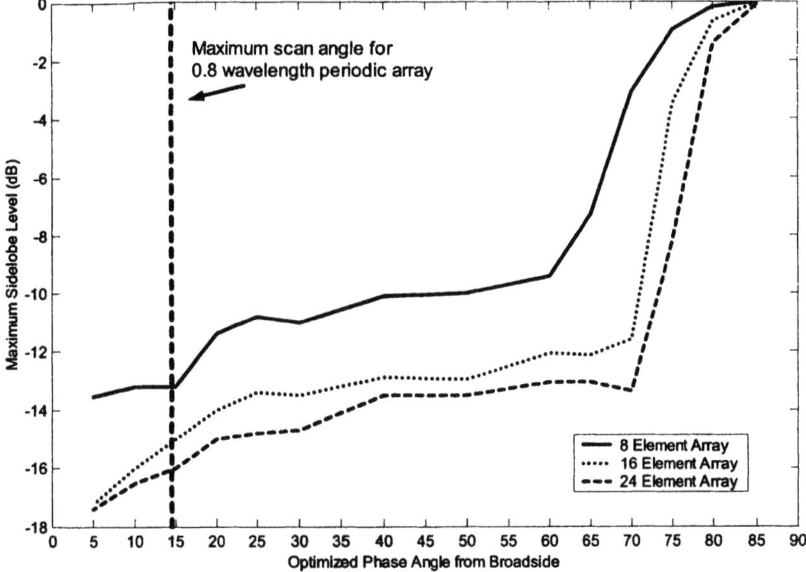

Fig. 6. Maximum sidelobe level that can be achieved over a range of different optimization angles.

Additional sidelobe suppression can be achieved with the inclusion of variable element current amplitudes into the GA, however the improvement is small. Preliminary results show that on the eight element array allowing the GA to pick currents within a ratio of 20 improves the sidelobe suppression by only 0.6 dB. Increasing the ratio between the maximum and minimum currents to 100 improves the sidelobe levels by about 1.5 dB.

7 MoM Pattern Verification of Aperiodic Arrays

The radiation patterns used in the optimization algorithm were calculated from the array factor expression given in (3) and assuming a half-wave dipole element pattern. This represents the most efficient approach since it neglects any secondary effects of mutual coupling on the radiation patterns. However, this assumption requires justification, especially for aperiodic arrays where the coupling environment seen by each element can be significantly different [13,22]. To verify the assumption that coupling effects are negligible, the directivity pattern of the array was calculated based on (3) and compared to that determined by using a numerically rigorous Method of Moments (MoM) technique [2]. In particular, the MoM simulations were carried out for an 8 element GA optimized array of half-wave dipoles, generated via perturbations of a 0.8 wavelength periodically spaced array. Figure 7 shows the directivity pattern based on (3) superimposed upon the pattern resulting from the MoM simulation. The resulting patterns are nearly identical, thereby providing justification for the use of aperiodic array factor theory in the genetic algorithm.

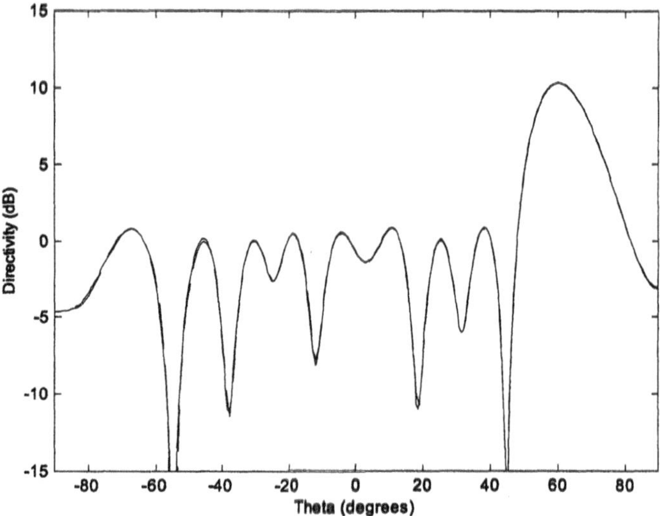

Fig. 7. The theoretical pattern calculated from the array factor (solid line) is compared with the pattern resulting from a MoM simulation (dashed line).

8 Driving Point Impedance Restrictions

Once it had been verified that the GA could reduce grating lobes during scanning, the practical issue of driving point impedance was considered. In conventional optimization procedures, only the pattern of the antenna is considered without regard to the resulting driving point impedance behavior. This may lead to designs that have very low and/or high input impedances that cannot be practically implemented. Very low or high values of input impedance may be difficult to compensate for in matching networks, potentially leading to "blind angles" during the scanning of phased array systems. To circumvent this problem, the real part of the driving point impedance for an optimized aperiodic array of dipoles was limited between 20 and 250 ohms over the entire desired scanning range of the array (in this case for all angles between 0° and 60°). These restrictions will result in reasonable component values for matching networks that do not tend toward the extremes of very large or very small.

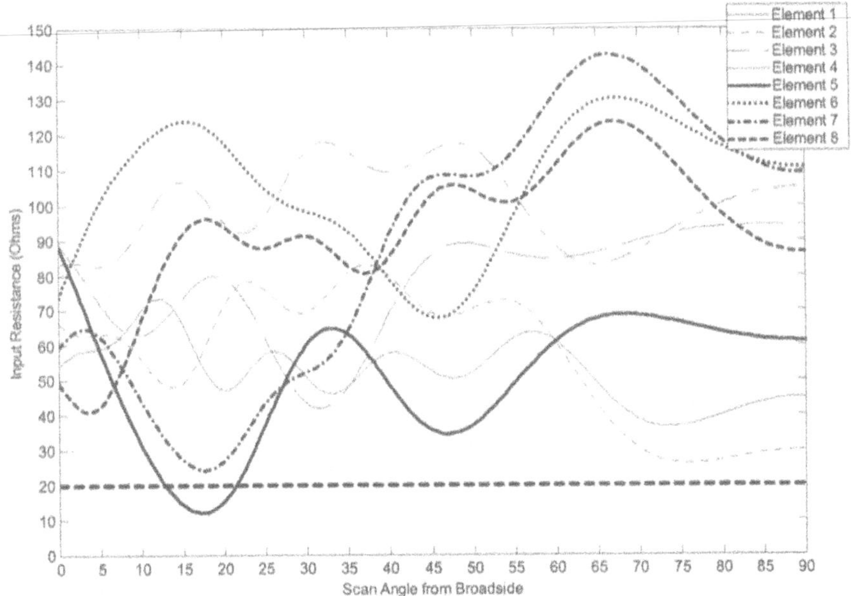

Fig. 8. Input resistance for each element of an 8 element 0.8 wavelength optimized aperiodic array plotted over scan angles ranging from 0 to 90 degrees.

The driving point impedance of each element was calculated using the induced EMF method outlined in [1] for an array of side-by-side half-wave dipoles. This method has the advantage of being more efficient than a numerically rigorous MoM technique while, at the same time, yielding results that are comparable in accuracy [1]. These properties of the induced EMF method make it attractive for use in conjunction with a GA. When the driving point impedance for the 8 element aperiodic array in Figure 5 was calculated for all angles between broadside and the optimization angle of 60° it was found that the input resistance of one of the elements dropped below the limit of 20 ohms. Figure 8 shows the driving point resistance plotted for each element of the array for scan angles ranging from 0 to 90 degrees. In this array, element 5 drops to about 12 ohms at about 17 degrees. To compensate for this, a driving point impedance restriction was placed within the GA.

After optimization, the input impedance now fell within the prescribed bounds; however, the sidelobe levels of the array increased, which is clearly an undesirable effect. Considering this problem from the matching network point of view, the driving point impedance of an element can be rendered in either a parallel or a series form. From circuit theory there exists a simple transform between these two representations. The GA was then modified to choose either a parallel or series representation for driving point impedance on each element of the array depending upon which view of the impedances remained best within the prescribed bounds. The resulting pattern was found to have the same scan range and sidelobe levels as the aperiodic array evolved without the input impedance restriction.

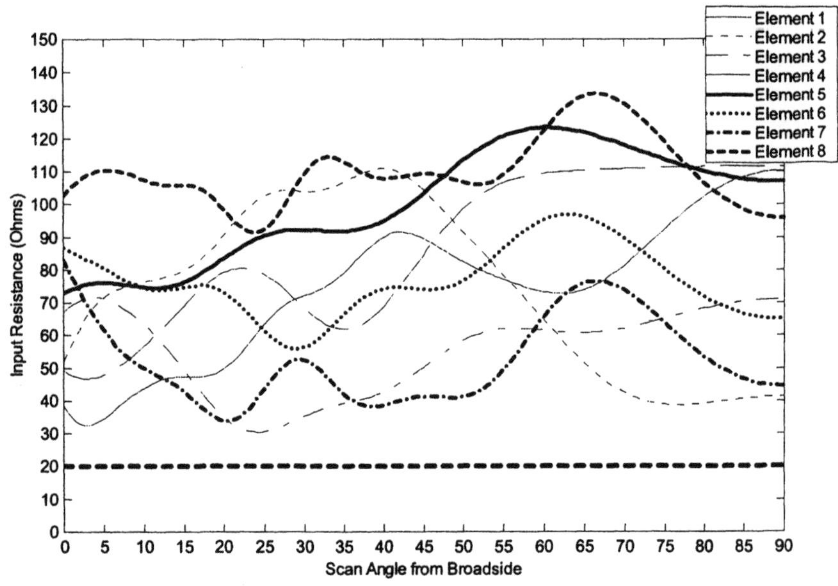

Fig. 9. Input resistance for each element of an 8 element 0.8 wavelength optimized aperiodic array plotted over scan angles from 0 to 90 degrees with impedance bounds placed within the GA.

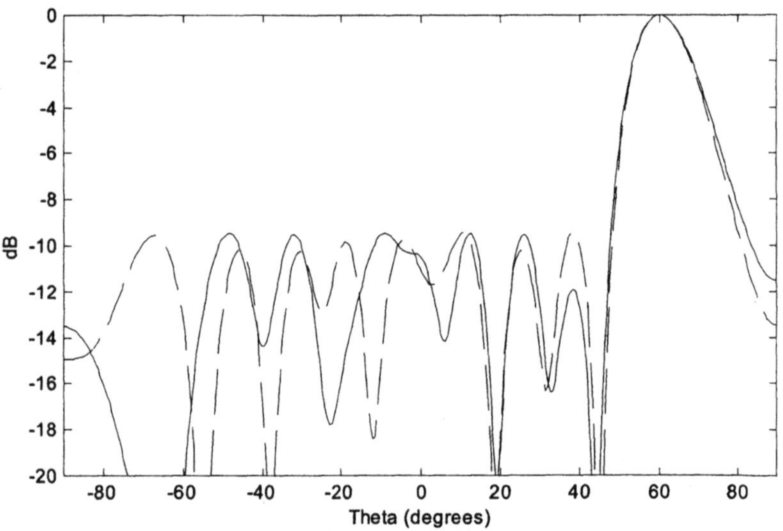

Fig. 10. Normalized array factor of an eight element array optimized at 60° with input impedance restriction (solid line) and without input impedance restriction (dashed line).

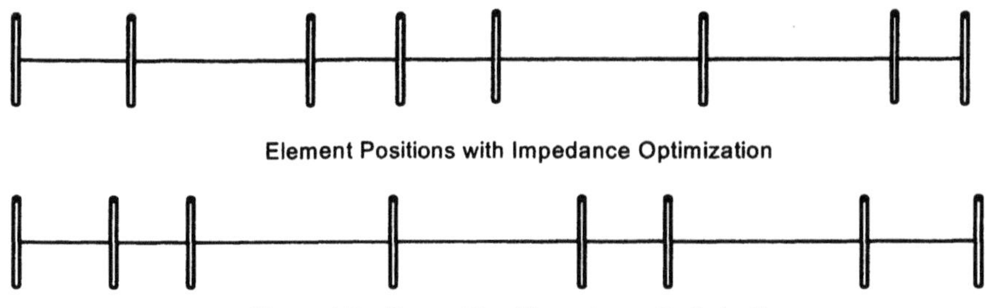

Fig. 11. The overall size of the aperiodic array remains nearly the same when the input impedance restriction is imposed.

Elements	1	2	3	4	5	6	7	8
With	0.0	0.598	1.532	1.992	2.490	3.561	4.537	4.910
Without	0.0	0.500	0.901	1.899	2.941	3.370	4.383	4.980

Table 1. Positions of the elements in wavelenghts for the arrays evolved with and without an impedance optimzation.

Figure 9 shows the real part of the driving point impedance plotted against the scan angle for an impedance restricted 8 element array optimized at 60 degrees. The real part of the impedance now falls within the given bounds with little change to the radiation pattern characteristics of the array. Figure 10 shows the pattern of the 8 element array optimized with the input impedance restriction superimposed upon an array optimized without the input impedance restriction.

From these patterns it can be seen that while the sidelobe positions move around, the maximum sidelobe level does not change. In addition, the total length of the array was found to remain essentially the same. Figure 11 shows the element positions of the array with and without the driving point impedance restriction. The two arrays have nearly the same length, with the element positions slightly shifted to compensate for the impedance restriction. The exact positions of these elements referenced to element 1 are shown in Table 1. These results show that reasonable input impedance restrictions can be placed on an array without any major change in the pattern characteristics or overall size of the array.

9 Evolving Matching Networks for Minimized VSWR Over Scan Angle

As stated previously, one consequence of placing restrictions on the input resistance of each of the array elements is that it makes the design of matching networks easier. By constraining the variations in the input resistance over scan range, it becomes possible to design a simple matching network for each element in the array so that the VSWR seen at the input to each port is below 2:1. An example will now be considered where a three element reactive network with a Π topology is adopted for the purpose of matching the antenna impedance to 50 ohms (see Figure 12). A balun is located between the antenna and the matching network that is assumed to act as an ideal 1:1 impedance transformer.

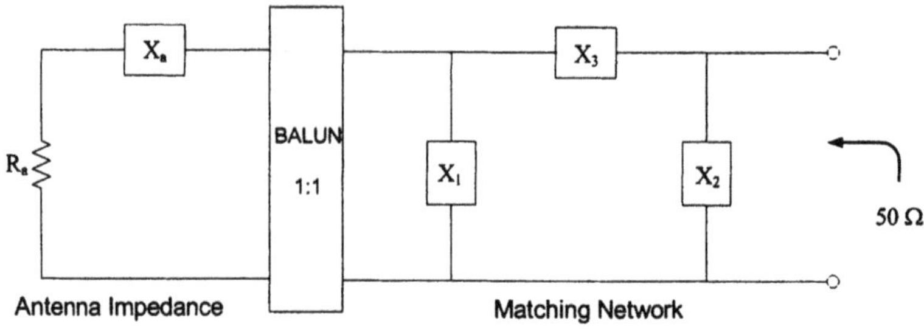

Fig. 12. A Π network is used to match the input impedance of each antenna element of the array to 50 ohms.

Each component of the matching network is assumed to be purely reactive and is chosen via a GA. The value of each component is bounded between –300 and 300 ohms, and must also lie outside the range of –4 to 4 ohms. These limits are chosen to illustrate how the GA may be constrained in such a way to prevent it from choosing component values that are difficult to realize in practice. The fitness function of the GA in this case is calculated as the inverse of the maximum VSWR over the entire scan range for each element. Therefore as the maximum VSWR lowers, the fitness correspondingly increases. Table 2 contains a listing of matching network component values obtained for the 8 element GA optimized aperiodic array with impedance bounds illustrated in Figure 11 (the input resistance versus scan angle for this array is shown in Figure 9). The VSWR verses scan angle for each element in the array, with respective matching network, is shown in Figure 13. This figure demonstrates that the VSWR essentially meets the performance criteria of below 2 for each element over the entire scan range, only slightly raising above 2 at a few angles for element 1.

Element	Maximum VSWR	Average VSWR	X_1	X_2	X_3
1	2.045	1.620	-155.37j	-142.58j	-30.52j
2	1.396	1.206	249.85j	249.63j	-84.99j
3	1.604	1.371	-151.79j	-175.93j	-17.76j
4	1.977	1.426	-144.17j	-143.68j	52.12j
5	1.696	1.335	-167.36j	-154.63j	52.12j
6	1.704	1.429	-284.30j	-182.16j	-9.46j
7	1.974	1.474	-116.36j	-112.12j	21.15j
8	1.360	1.152	-161.92j	-194.55j	72.94j

Table 2. Component values and VSWR for each matching network.

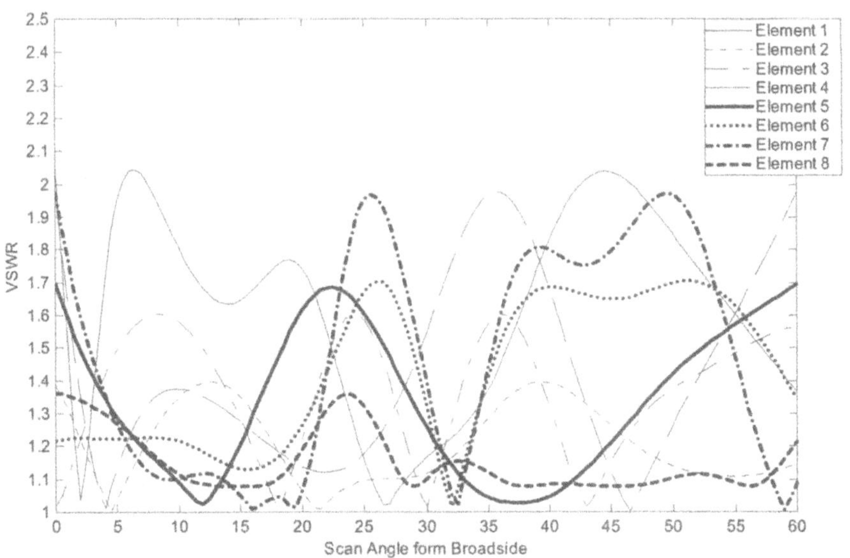

Fig. 13. VSWR verses scan angle for each element in the array.

10 Optimization of Conformal Antenna Arrays

Conformal antennas have become increasingly popular for use in communication systems because they are low profile and can be mounted directly onto flat or curved platforms [20]. One of the most widely used types of conformal elements are microstrip patch antennas. Arrays of conformal antennas are also being considered as a means for improving the performance and capacity of wireless communication systems [31]. A common problem encountered in the design of conformal antenna arrays is that their performance can be significantly degraded once they are mounted on a host platform, such as a base station tower. In this section a design optimization technique based on a genetic algorithm will be introduced that takes into account the presence of the mounting platform on the performance of an array of conformal microstrip patch antennas. Moreover, the potentially adverse effects of the platform on array performance can be compensated for as a result of this new GA design synthesis process.

In order to illustrate the advantages of the GA synthesis technique for conformal arrays, we will begin by considering a simple linear array of conformal antenna elements. Suppose the far-field radiation pattern produced by such an array with N elements may be expressed in the general form:

$$E(\theta,\phi)=\sum_{n=0}^{N-1} I_n EP_n(\theta,\phi)e^{j(nkd\cos\theta+\beta_n)} \tag{5}$$

where I_n are the element excitation current amplitudes, β_n are the excitation current phases, $EP_n(\theta,\phi)$ are the individual array element patterns, d is the separation distance between elements, and k is the free-space wavenumber. The normal procedure used to design arrays such as these is to assume that all of the array elements have identical element patterns, in which case the expression in (5) may be factored and rewritten as

$$E(\theta,\phi)=EP(\theta,\phi)AF(\theta) \tag{6}$$

where $EP(\theta,\phi)$ represents the element pattern and $AF(\theta)$ is the associated array factor, given by

$$AF(\theta)=\sum_{n=0}^{N-1} I_n e^{j(nkd\cos\theta+\beta_n)} \tag{7}$$

If this assumption is made, the phasing parameter β_n may then be selected to steer the main beam of the array radiation pattern by selecting its value according to [22]

$$\beta_n = n\alpha = -nkd\cos\theta_0 \tag{8}$$

where θ_0 is the desired steering angle.

For most practical platforms and applications, the assumption that all of the elements in an array have identical element patterns is invalid, and a more general expression must be used to compute the array far-field radiation pattern instead (e.g., equation (5)). This is a consequence of the fact that for three-dimensional objects, such as finite-length circular cylinders, the radiation patterns of the individual array elements are influenced by their location on the platform. This considerably complicates the radiation pattern synthesis procedure since, in general, each individual array element pattern may be different. Under such conditions, the uniform progressive phase shift as defined in (8) may not be used to perform beam steering since this equation depends on the element patterns being identical. Instead, a genetic algorithm pattern synthesis technique is introduced here for determining the optimal set of excitation phases required in order to compensate for platform effects on the individual array element patterns. The GA radiation pattern synthesis technique will be illustrated by considering a specific example involving a 9-element conformal array of microstrip patch antennas mounted axially along a finite-length, circular cylinder assumed to be a Perfect Electric Conductor (PEC).

In order to compute the radiation patterns for the individual conformal patch elements of the array mounted on the finite-length circularly-cylindrical platform, the computationally efficient domain-decomposition/reciprocity procedure presented in [28] is applied. The geometries of the platform and the array elements are illustrated in Figure 14, along with a labeling scheme for the patch antennas. The cylinder has a length l of 60 cm and a radius a of $0.5\lambda_0$ or 7.5 cm, and an operating frequency of 2 GHz is assumed. The microstrip antennas are probe-fed square patches that are $0.5\lambda_r$ on a side at a frequency of 2 GHz, with both a substrate and a superstrate made of duroid ($\varepsilon_r = 2.33$ and $\lambda_r = 9.83$ cm) that are each $0.05\lambda_r$ in thickness. Nine different patch positions are considered, beginning at the center element, which is placed equidistant along the cylinder axis from both endcaps and is denoted as Element 0. The remaining elements are positioned along the axis of the cylinder with center-to-center patch separation distances of $d = 0.6\lambda_r$.

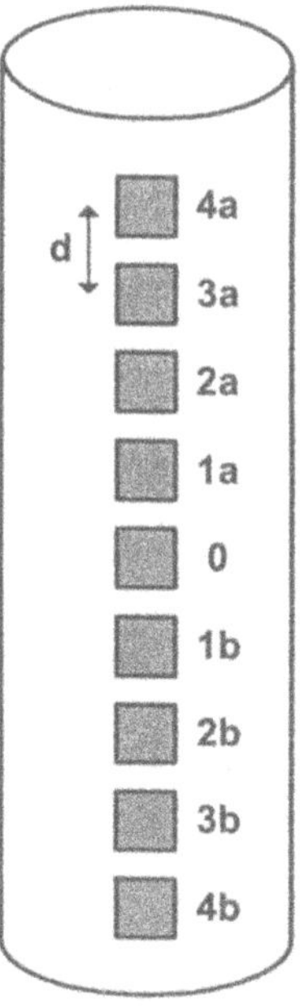

Fig. 14. Configuration of 9 element array placed on finite length PEC cylinder, with the center element denoted as Element 0, the patches above the center element having the naming suffix "a", and the patches below the center element having the naming suffix "b".

Figure 15 shows the θ-component of the axial radiation patterns for each of the nine elements in the $\phi = 0°$ plane, when no other elements are present. As expected, the center element pattern, shown in Figure 15(e), is symmetric due to its location on the cylinder, and the relatively large back lobe observed has been enhanced due to the effects of truncation [28]. The remaining radiation patterns of the elements are also plotted, with Figure 15(a)-15(d) being elements 1a through 4a, and Figure 15(f)-15(i) being elements 1b through 4b. This sequence of plots clearly illustrates that the radiation pattern of an individual element is strongly dependent upon the location at which it is placed on the platform. It is clear from these plots that the closer to the end of the cylinder an element is placed the more distorted its pattern becomes. Under these conditions, the assumption that all of the array elements have identical radiation patterns is invalid, and the simple expression based on pattern multiplication for the array radiation pattern given in (6) cannot be used. Instead, the pattern must be calculated using the more general form of the expression provided in (5).

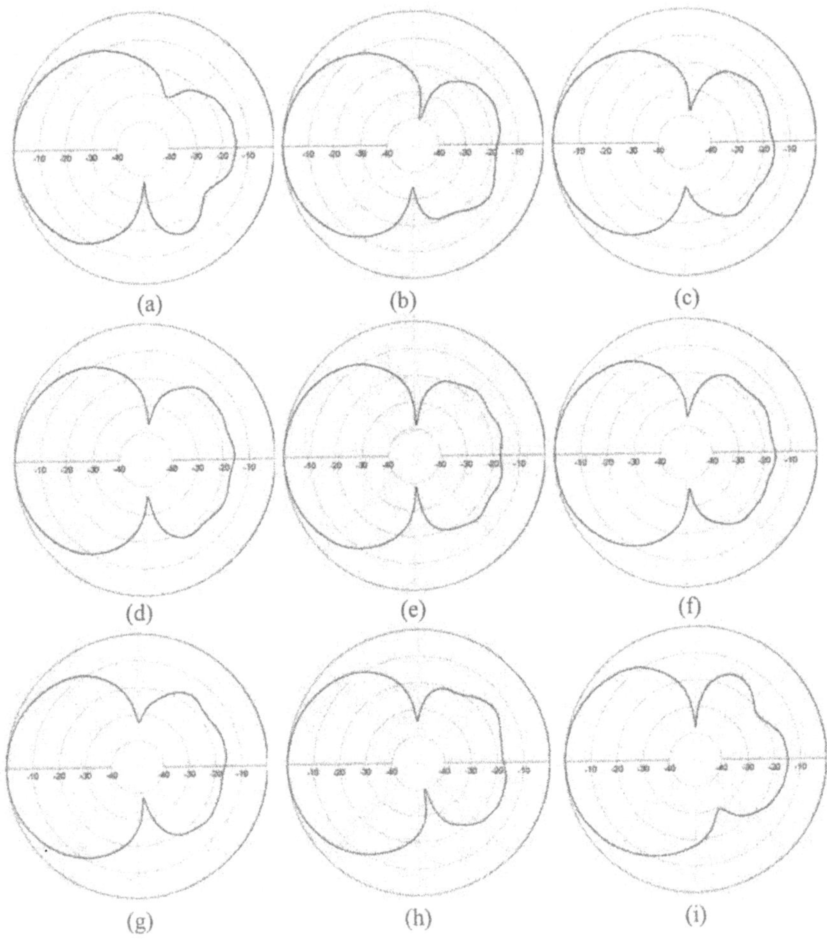

Fig. 15. Element patterns for each of the nine array elements mounted on the finite-length PEC cylinder illustrated in Figure 1. Shown are the patterns for Elements (a) 4a, (b) 3a, (c) 2a, (d) 1a, (e) 0, (f) 1b, (g) 2b, (h) 3b, and (i) 4b.

If the effects of the platform truncation on the element patterns are first ignored, and it is assumed that each element has the pattern of Element 0, the expression given in (6) along with the array factor for this array produces the radiation pattern as indicated by the solid curve in Figure 16. This pattern has a maximum sidelobe level of about –14 dB and a back lobe that is 17 dB down from the main beam. Next, the effects of cylinder truncation are included in the array pattern computation through the use of the actual element patterns, which means that the more general form of the array radiation pattern expression as given in (5) must be used. The resulting array radiation pattern is shown as the dashed curve in Figure 16 for the case where all elements are fed in phase at 0° (broadside). It is clear from these results that the effects of platform truncation include a broadening of the array main beam as well as an increase in the intensity level of the back lobe. Next the important practical case is considered where it is desired to steer the main beam of the array to a position other than broadside, which is accomplished by finding the appropriate set of excitation current phases (i.e., all values of β_n for n = 0,1,2,...,N-1). For conventional arrays in which all elements have identical radiation patterns, it is common practice to use a progressive phasing of the form given in (8). The solid curve shown in Figure 17 is the radiation pattern that would result from a progressively phased 9-element antenna array with the main beam steered to θ_0 = 60°, where each individual element is assumed to have the radiation pattern of Element 0. However, the dashed curve in Figure 17 shows what happens if the same progressive phasing is applied to the more realistic case of using the actual element patterns in the array computations. The resulting radiation pattern does not have its main beam steered to 60°, but instead there is nearly a null in the desired location of the main beam. This clearly demonstrates that a standard progressive phasing approach is not adequate in general for arrays of conformal antennas mounted on arbitrarily shaped three-dimensional platforms.

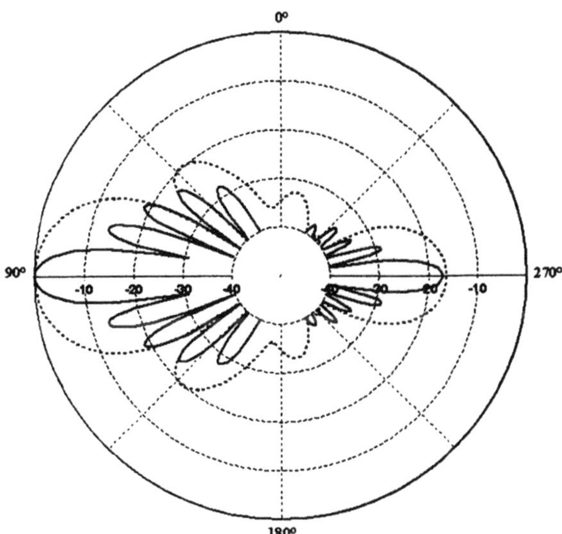

Fig. 16. Radiation pattern determined using the array factor expression (6) from a cylinder-mounted array of 9 identical microstrip patch elements (solid curve). The dashed curve shows the radiation pattern for an array of 9 microstrip patch antenna elements mounted on a finite-length PEC circular cylinder that was calculated using (5). This expression takes into account the actual patterns of the individual array elements.

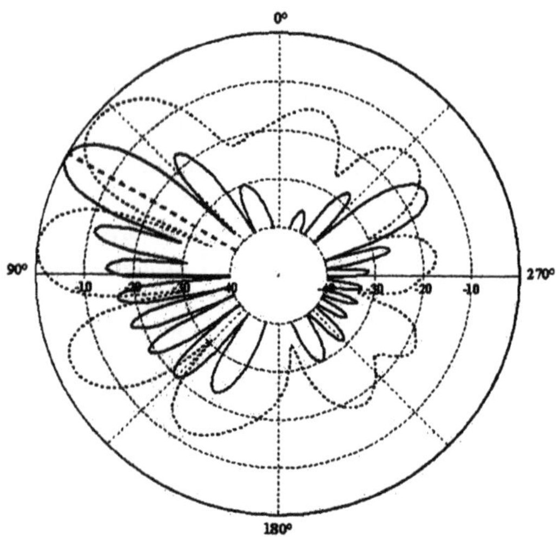

Fig. 17. Radiation pattern determined using the array factor expression from a linear array of 9 cylinder-mounted microstrip patch antennas with identical element patterns and uniform progressive phase shift chosen to point the main beam in the direction $\theta_0 = 60°$ (solid curve). The dashed curve shows the radiation pattern resulting when the actual element patterns are used along with progressive phase shifting to steer the main beam of the array to $\theta_0 = 60°$.

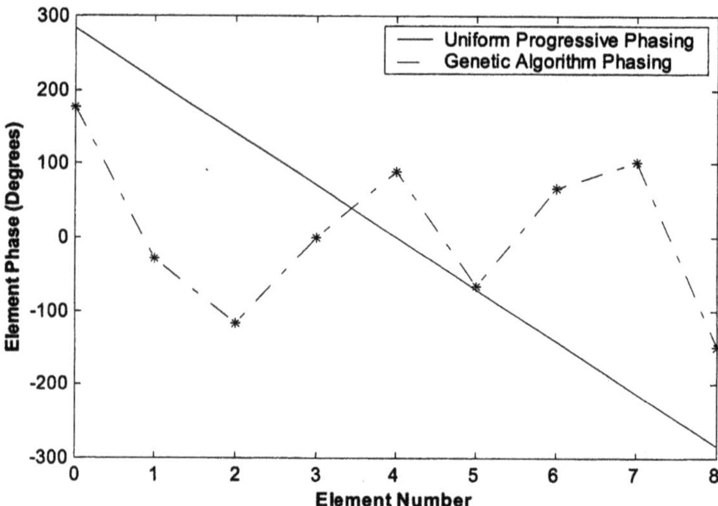

Fig. 18. Plot of element excitation phases as a function of element position used to steer the main beam of the 9-element conformal microstrip array assuming identical element patterns (solid curve) and with actual element patterns (dashed curve). Note that element 0 and element 8 correspond to element 4b and element 4a, respectively, as shown in Figure 14.

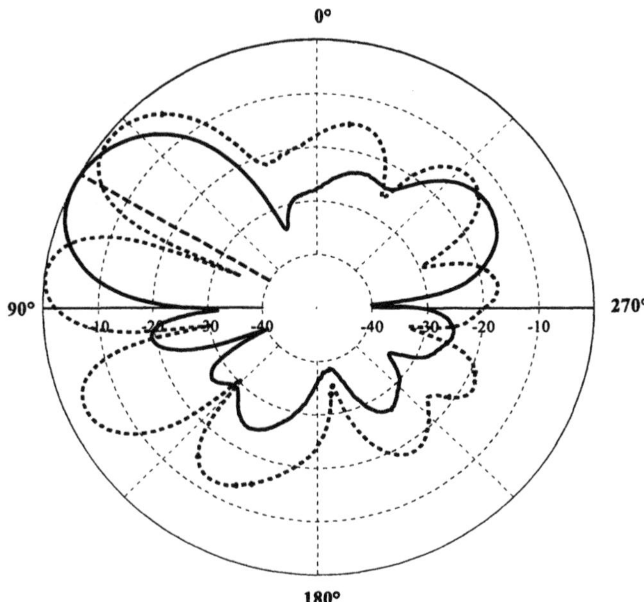

Fig. 19. Comparison between the GA optimized array radiation pattern (solid curve) and the radiation pattern resulting from the use of the uniform progressive phase shift (dashed curve) for an array of 9 elements mounted along the axis of a truncated PEC circular cylinder. For this optimization, the ideal cosine pattern objective was used with the parameters $\theta_0 = 60°$, $\delta = 25°$, $\theta_a = 35°$, and $\theta_b = 85°$. Note that each radiation pattern has been normalized by its respective maximum.

In order to compensate for the effects of platform truncation on the synthesis of conformal microstrip patch antenna array radiation patterns, a genetic algorithm approach is presented here to choose the optimal excitation phase parameter β_n for each of the array elements. This procedure is demonstrated by applying it to steer the main beam of the patch antenna array considered in Figure 14. One of the most important parts of any GA is the selection of a suitable objective function used to test the fitness of possible solutions generated during the optimization process. For this specific case, the pattern chosen as an objective is the ideal cosine pattern steered to the angle 60°. The general mathematical representation for the ideal cosine pattern is given by [29]

$$f(\theta) = \begin{cases} \cos\left[\dfrac{\pi}{2}\left(\dfrac{\theta - \theta_0}{\delta}\right)\right], & \text{for } \theta \in [\theta_a, \theta_b] \subseteq [-\pi/2, \pi/2] \\ 0, & \text{elsewhere} \end{cases} \tag{6}$$

where θ_0 is the desired steering angle, δ is the half-power beam width, and θ_a and θ_b are limits on the extent of the main beam. In this specific instance, these parameters have been chosen to be $\theta_0 = 60°$, $\delta = 25°$, $\theta_a = 35°$, and $\theta_b = 85°$. The

GA was encoded using a population size of approximately 150 to 250 members, a crossover probability of 50%, and a mutation probability of 2.5%. The resulting element excitation phases obtained from the GA for this case are shown by the dashed line in Figure 18, where they are compared to the uniform progressive phases represented by the solid line and computed using (8). In Figure 19, the optimized radiation pattern obtained using the phases selected by the GA is plotted as a solid curve, while the dashed curve again depicts the radiation pattern that results from using the conventional uniform progressive phase shift. It is evident from these results that the uniform progressive phase shift is not adequate to steer the beam of this array. However, when the GA-selected excitation phases are used, the main beam of the pattern now points in the desired direction and the resulting sidelobe levels are also within acceptable limits.

11 Conclusions

This chapter documents a method of developing thinned aperiodic communication antenna arrays with reduced grating lobes when steered over large scan angles. The aperiodic arrays were optimized via a genetic algorithm for a given scan angle and were shown to be steerable with constant sidelobe levels up to and including that angle. It was also shown that by increasing the size of the array that the maximum scan angle increased while at the same time the maximum sidelobe level was reduced. Practical considerations in the design of these aperiodic arrays were taken into account by placing restrictions on the input impedance of the array elements. These restrictions limit the driving point impedance of the array without adversely affecting the performance or overall size of the array. An example was presented illustrating how matching networks could be evolved via a GA for each individual array element.

A domain-decomposition/reciprocity procedure was presented in this chapter which provides the ability to accurately and efficiently compute the radiation patterns for microstrip patch antennas mounted on arbitrarily shaped metallic platforms, such as base station towers. The utility of the technique was demonstrated by considering an array of nine conformal microstrip patch antenna elements mounted axially on a finite-length PEC circular cylinder, where it was clearly shown that the position of an element along the cylinder axis greatly influences its resulting radiation pattern. It was demonstrated through this example that the assumption commonly made in the array design process, namely that all elements of an array have identical element patterns, is not valid in general for conformal microstrip antenna arrays mounted on arbitrary three-dimensional platforms. This fact was shown to considerably complicate the design of such conformal phased array antenna systems. In order to circumvent these problems, a genetic algorithm approach to radiation pattern synthesis was developed based on the computationally efficient domain-decomposition/reciprocity procedure. This synthesis technique yields the optimal set of excitation current phases required to compensate as much as possible for any mounting platform effects on the individual array element patterns.

References

1. Balanis CA (1997) Antenna Theory, Analysis and Design. 2nd edn. John Wiley and Sons, Inc. New York
2. Burke GJ, Poggio AJ (1981) Numerical Electromagnetics Code (NEC) – Method of Moments. Rep. UCID 18834, Lawrence Livermore Lab. Livermore, CA
3. Chang BK, Ma X, Sequeira HB (1994) Minimax-Maxmini: A New Approach to Optimization of the Thinned Antenna Arrays. 1994 IEEE Antennas and Propagation Society International Symposium. AP-S Digest. **1**, 514-517
4. Cohen N, Hohlfeld RG (2000) Array Sidelobe Reduction by Small Position Offsets of Fractal Elements. Proceedings of The Applied Computational Electromagnetic Society (ACES). Naval Postgraduate School, Monterey, CA, 822-828
5. Davis L (1996) Handbook of Genetic Algorithms. International Thomson Computer Press, London
6. Elias JG, Chang B (1992) A Genetic Algorithm for Training Networks with Artificial Dendritic Trees. Proceedings of the International Joint Conference on Neural Networks. Baltimore, Maryland, **1**, 652-657
7. Goldberg DE (1989) Genetic Algorithms in Search, Optimization, and Machine Learning. Addison-Wesley Publishing Company, Inc. Reading, Massachusetts
8. Haupt RL (1994) Thinned Arrays Using Genetic Algorithms. IEEE Trans. Antennas Propagat. **42**, 993-999
9. Haupt RL (1995) An Introduction to Genetic Algorithms for Electromagnetics. IEEE Antennas Propagat. Magazine. **37**, 7-15
10. Haupt RL, Haupt SE (1998) Practical Genetic Algorithms. John Wiley and Sons, Inc. New York
11. Johnson JM, Rahmat-Samii Y (1997) Genetic Algorithms in Engineering Electromagnetics. IEEE Antennas Propagat. Magazine. **39**, 7-25
12. Junker GP, Kuo SS, Chen CH (1998) Genetic Algorithm Optimization of Antenna Arrays with Variable Interelement Spacings. 1998 IEEE Antennas and Propagation Society International Symposium, AP-S Digest. **1**, 50-53
13. Kang YW, Pozar DM (1985) Correction of Error in Reduced Sidelobe Synthesis due to Mutual Coupling. IEEE Trans. Antennas Propagat. **33**, 1025-1028
14. Lo YT, Lee SW (1988) Antenna Handbook: Theory, Applications, and Design. Van Nostrand Reinhold Company. New York
15. Meijer CA (1998) Simulated Annealing in the Design of Thinned Arrays Having Low Sidelobe Levels. COMSIG'98, Proceedings of the 1998 South African Symposium on Communications and Signal Processing. 361-366
16. Mitchell M (1996) An Introduction to Genetic Algorithms. The MIT Press. Cambridge, Massachusetts
17. O'Neill DJ (1994) Element Placement in Thinned Arrays Using Genetic Algorithms. OCEANS '94, Oceans Engineering for Today's Technology and Tomorrows Preservation. **2**, 301-306
18. Rahmat-Samii Y, Michielssen E (ed) (1999) Electromagnetic Optimization by Genetic Algorithms. John Wiley and Sons, Inc. New York
19. Rappaport TS (ed) (1998) Smart Antennas: Adaptive Arrays, Algorithms, and Wireless Position Location. The Institude of Electrical and Electronics Engineers, Inc. Piscataway, New Jersey
20. Sainati RA (1996) CAD of Microstrip Antennas for Wireless Applications. Artech House. Boston, Massachusetts

21. Saunders SR (1999) Antenna and Propagation for Wireless Communication Systems. Wiley, New York, Ch. 17
22. Stutzman WL, Thiele GA (1998) Antenna Theory and Design. 2nd edn. John Wiley and Sons, New York
23. Trucco A, Murino V (1999) Stochastic Optimization of Linear Sparse Arrays. IEEE Journal of Oceanic Engineering. **24**, 291-299
24. Tsoulos GV (ed) (2001) Adaptive Antennas for Wireless Communications. IEEE Press. Piscataway, New Jersey
25. Unz H (1956) Linear Arrays with Arbitrarily Distributed Elements. Electronic Res. Lab. University of California, Berkeley, Report Serial No. 60, Issue No. 168
26. Unz H (1960) Linear Arrays with Arbitrarily Distributed Elements. IRE Trans. Antennas Propag. **8**, 222-223
27. Weile DS, Michielssen E (1997) Genetic Algorithm Optimization Applied to Electromagnetics: A Review. IEEE Antennas Propagat. Magazine. **45**, 343-353
28. Werner DH, Allard RJ, Martin RA, Mittra R (2003) A Reciprocity Approach for Calculating Radiation Patterns of Arbitrarily Shaped Microstrip Antennas Mounted on Circularly-Cylindrical Platforms. Accepted for publication in IEEE Trans. Antennas Propagat.
29. Werner DH, Ferraro AJ (1989) Cosine Pattern Synthesis for Single and Multiple Main Beam Uniformly Spaced Linear Array. IEEE Trans. Antennas Propagat. **37**, 1480-1484
30. Werner DH, Mittra R (2000) Frontiers in Electromagnetics. The Institute of Electrical and Electronics Engineers, Inc. New York
31. Winters JH, (1998) Smart Antennas for Wireless Systems. IEEE Personal Communications. **1**, 23-27

QoS Adaptation Based on Fuzzy Theory

Cristian Koliver[1]*, Jean-Marie Farines[2]**, and Klara Nahrstedt[3]***

[1] University of Caxias do Sul, Caxias do Sul RS Brazil
[2] Federal University of Santa Catarina, Florianópolis, SC Brazil
[3] University of Illinois at Urbana-Champaign, Urbana IL USA

Abstract. More and more distributed multimedia applications are becoming an integral part of our computing and communication environment. To achieve this goal, the multimedia applications must be delivered with high Quality of Service (QoS). This is a challenge as the distributed multimedia applications run on top of general purpose operating systems and networks that have been developed for best-effort data processing and transmission. Hence, they suffer high level of perturbations in resource allocation when handling continuous media. To solve this challenge, many approaches for QoS adaptation to the distributed multimedia system context have been proposed.

Any QoS adaptation approach needs to address the following problems: (1) QoS specification to capture users' QoS requirements and preferences; (2) QoS mapping to translate representations of QoS at different layers of a distributed multimedia system; and (3) QoS control to provide real-time traffic control and shaping of flows, based on available network bandwidth.

To address the first problem, QoS adaptation approaches must often deal with varying user's point of view of quality, taking in account the vagueness and subjectivity inherent to the user's evaluation of quality. The usual solutions for the second problem use measurement-based techniques, obtained from samples of applications or simulations of traffics patterns, which is a time-consuming process. Finally, to solve the third problem, QoS adaptation approaches must deal with the considerable uncertainty related to congestion determination, as the network load oscillates very suddenly and drastically.

In this context, soft computing techniques, such as fuzzy control, artificial neural networks and neuro-fuzzy control, seem a promising approach to be used for QoS adaptation, since these techniques deal with vagueness and uncertainty.

In this chapter, we discuss how soft computing techniques can be used for QoS adaptation in distributed multimedia systems, specially in assisting QoS control of multimedia end-to-end delivery activities through Fuzzy Logic.
Key words: quality of service, QoS, adaptation, multimedia, soft computing, fuzzy control.

* This work was funded by the FAPERGS funding agency - Brazil.

** This work has been partially supported by a grant from CNPq - Brazil.

*** This work was funded by the DARPA funding agency, under contract number F30602-97-2-0121, and by the NSF funding agency, under contract number CCR-9988199 grant, both USA grants.

1 Introduction

The entities composing a distributed multimedia system (DMS) architecture are characterized by performance attributes called Quality of Service (QoS) parameters. Some relevant QoS parameters are: video frame rate, audio sample rate, sample size, level of resolution, number of colors and response time (application layer); processor cycles, task periods and deadlines and memory requirements (system layer); display resolution, power processing and memory capacity (devices layer); and bit rate, network delay, jitter and packet loss rate (network layer).

QoS parameters can suffer uncontrolled variations in their values during delivery of continuous media data mainly due to network congestion. Such variations affect the perceived quality and, consequently, the user's satisfaction. As a consequence, distributed multimedia applications must be controlled by QoS adaptation mechanisms that adjust the bit rate to the network state (load), by changing one or more QoS parameters values.

In DMS whose network platform is best-effort type, approaches for QoS adaptation are usually based on closed loop [1] [2] [3] [4] [5]. The new bit rate of the distributed multimedia application is computed using linear functions. Each linear function is, in fact, a linear controller (LC) which acts inside a precise range bounded by two values of the feedback information (e.g., packet loss rate between 5 and 10%). The adjustment of the multimedia application bit rate often is done by using dropping algorithms, which discard some video frames, or by changing the quantization factor value of the encoder.

However, LC's do not deal with an important feature of a DMS: *uncertainty*. The network load oscillates very suddenly and drastically. This non-linear behavior may lead the controller to use outdated feedback information and, consequently, a very inaccurate available bandwidth estimate.

In addition, many approaches for QoS adaptation either (1) do not consider how the available bandwidth estimate should be used by the multimedia application (i.e., how bit rate is mapped into application QoS parameters values), or (2) deal with the quality as an one dimensional phenomenon, changing a single application QoS parameter in order to adjust the multimedia stream to the available bandwidth. However, under the user's perspective, quality is a multidimensional phenomenon, in which a quality dimension (or application QoS parameter) may influence the perception of the other ones. Furthermore, the main goal of QoS adaptation is to maximize the quality according to current DMS context. Therefore, approaches should use the vagueness and subjectivity properties inherent to quality evaluation, to adjust bit rate without changing the user's perception of quality.

In this context, techniques of soft computing (SC) [6] are a promising approach for a user-centered QoS adaptation, specially for QoS control, since SC-based controllers, such as fuzzy and neurofuzzy controllers, and artificial neural networks, have been used successfully to control non-linear plants with high levels of perturbations. These features match features of DMS's whose network platform is best-effort type.

This chapter discusses how SC techniques can be used for QoS adaptation, specially in QoS control through Fuzzy Logic. The chapter is outlined as follows: we provides an overview of QoS adaptation in DMS in Sect. 2. In Sect. 3, we describe briefly how the main SC techniques can be used for QoS adaptation and we exemplify describing some approaches that have been proposed. In Sect. 4, we describe our framework for QoS adaptation based on fuzzy control. We conclude in Sect. 5.

2 Distributed Multimedia Systems and QoS Adaptation

Despite the multitude of approaches, the overall goal of QoS adaptation in DMS context is always the same and consists in extending the range of conditions over which a distributed multimedia application performs acceptably [7]. Let us assume two domains, the performance domain (P) and the resource domain (R), as shown in Fig. 1. The first one contains n-tuples of application QoS parameters values; the second one contains m-tuples of QoS parameters values representing amount of a resource[1]. The sub-domain P' contains the tuples of application QoS parameters values that represent acceptable qualities of the multimedia stream. Without QoS adaptation, the mapping $M : P \leftrightarrow R$ maps P' ($P' \subseteq P$) onto R'' ($R'' \subseteq R'$). Introducing QoS adaptation changes the M mapping so that P' is mapped onto a larger sub-domain R'. This means that with QoS adaptation, there might be more values of a given resource that provide acceptable quality for the application (e.g., more values of network bandwidth that are still enough for providing a good quality).

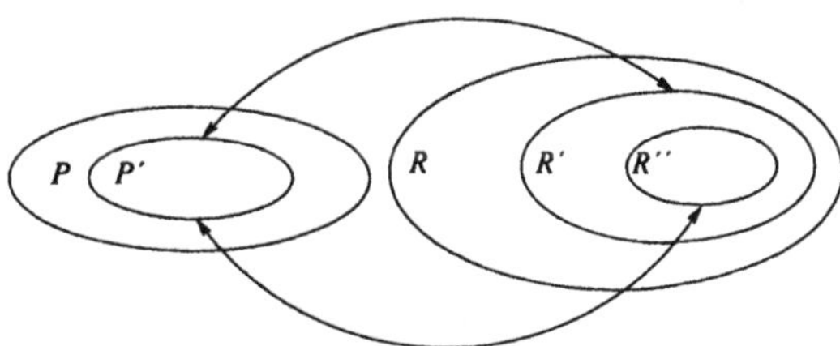

Fig. 1. Mapping between QoS application parameters and resources.

A common scenario in a DMS that can move the operating point outside P' is network congestion. If network is congested, then some packets might be

[1] Theoretically, both sets can be infinite but, in fact, they are finite sets due to technology restrictions.

dropped in buffers of intermediary nodes (routers, gateways, ATM switches etc.). In receivers' end-systems, lack of some packets impedes the assembly of some video frames or audio samples, reducing the video frame rate, sometimes "freezing" the image when video frame rate goes down close to 0 frames per second (fps), and causing sound lapses. In addition, some packets might wait a (relatively) long time in buffers, increasing the end-to-end delay. The increasing of end-to-end delay implies missing deadlines of decompression tasks and discarding of video frames and audio samples in the receivers' end-systems.

Dynamically, QoS adaptation in DMS works through a set of mechanisms whose goal is to keep the operating point within P'. A monitor observes QoS parameters (from P or R) to detect if the operating point is still inside P'. If not, a controller initiates some corrective action in order to bring the operating point back to P' [7].

2.1 Example: Videoconference Acceptable Quality

Let a very simple scenario: a videoconference application that requests video to 25 fps at least. Suppose that the image can also be scaled in three resolution modes (all of them acceptable to the application): SQCIF (128 pixels per line by 96 lines), QCIF (176 pixels per line by 144 lines), and CIF (352 pixels per line by 288 lines). Hence, P' is:

$$\begin{aligned} P' = \{ &< 25 \text{ fps, CIF} >, < 26 \text{ fps, CIF} >, ..., < 30 \text{ fps, CIF} >, \\ &< 25 \text{ fps, QCIF} >, < 26 \text{ fps, QCIF} >, ..., < 30 \text{ fps, QCIF} >, \\ &< 25 \text{ fps, SQCIF} >, < 26 \text{ fps, SQCIF} >, ..., < 30 \text{ fps, SQCIF} > \}. \end{aligned} \tag{1}$$

We note that the most of users does not perceive differences among rates above 30 fps.

Suppose that the main resource needed to reach acceptable qualities is network bandwidth, given by bits per second (bps) and the bit rates for the acceptable video frame rates (CIF resolution) are:

$$R'' = \{805\text{K bps}, 850\text{K bps}, ..., 1105\text{K bps}\}. \tag{2}$$

However, if the network bandwidth is scarce, then the application can be adapted by reducing resolution. Hence, the set of bit rates associated to acceptable qualities is extended to:

$$\begin{aligned} R' = \{ &215\text{K bps}, 272\text{K bps}, ..., 339\text{K bps}, ..., \\ &515\text{K bps}, 550\text{K bps}, ..., 739\text{K bps}, ..., \\ &805\text{K bps}, 850\text{K bps}, ..., 1105\text{K bps}\}. \end{aligned} \tag{3}$$

The domain $R' - R''$ contains bit rates for QCIF and SQCIF resolutions.

2.2 QoS Control

As discussed in Sect. 1, approaches for QoS adaptation are generally based on closed loop, in which a linear controller (LC) changes, directly or indirectly, some observed QoS parameters. The feedback information of controller, when QoS adaptation is a response to network congestion, is usually one or more communication QoS parameters, such as packet loss rate [2] [1] and end-to-end delay [8]. This information is used to estimate network load. According to the network load estimate, the controller decides to decrease, keep or increase application bit rate. In order to change the application bit rate, some application QoS parameters values are changed. Generally, adjusted application QoS parameters are quantization factor used by encoder [9], video frame rate [10], and level of resolution [11].

2.3 Example: Videoconference QoS Control

Let the videoconference of the example of Sect. 2.1. The rules used by a controller used for adjusting the application bit rate could be:
if *packet loss rate* $\leq$ 10% **then** *resolution* $\leftarrow$ *CIF*.
if 10% $<$ *packet loss rate* $\leq$ 20% **then** *resolution* $\leftarrow$ *QCIF*.
if 20% $<$ *packet loss rate* **then** *resolution* $\leftarrow$ *SQCIF*.

Therefore, if the packet loss rate observed across the network increases, the QoS controller will decrease the bit rate reducing resolution of the multimedia stream in according with R'. Then, the application bit rate is decreased and the loss rate lowers. On the other hand, if the packet loss rate decreases, the application bit rate is increased by improving resolution.

We want to stress again, that feedback information is imprecise, giving just a vague idea of the network load. Vagueness is a keyword for using SC techniques. In the above example, packet loss rate does not inform of *how much* bandwidth (in bps) is available for the videoconference application. Packet loss rate permits just that the controller infers if the application bit rate is excessive or not for current context of the SMD.

3 Soft Computing and QoS Adaptation

In this section, we describe briefly how certain SC techniques can be (and have been) used for QoS adaptation and we exemplify them by describing some approaches that have been proposed in the literature.

3.1 Introduction

Initially, the term soft computing was coined by L. Zadeh [6] to be an extension to fuzzy logic by merging it mainly with artificial neural networks and evolutionary computing. Nowadays, SC can be viewed as a collection of techniques - fuzzy logic, neuro-computing and probabilistic reasoning, with the latter subsuming genetic algorithms, chaotic systems, belief networks and parts of learning theory - capable of dealing with imprecise, uncertain or vague information. The essence of SC is that, unlike the traditional ("hard") computing, it is aimed at an compromise with the pervasive imprecision of the real world. Thus, the guiding principle of SC is to exploit the tolerance for imprecision, uncertainty and partial truth to achieve easiness of using, robustness, low solution cost and better rapport with reality.

3.2 SC for QoS Adaptation

As presented in Sect. 1, LC's does not deal well with uncertainty on determining network load because of its sudden and bursty oscillations. Hence, the resulting available bandwidth estimate might be very inaccurate. In addition, the overall goal of QoS adaptation is to extend the range of conditions over which a distributed multimedia application performs acceptably. In fact, that means to keep a ***good quality*** according to users' perspective. Good quality concept is subjective, varying from user to user.

The uncertainty on determining the network load, the difficulty of building analytical model of a DMS, the non-linearity in the relationship among many QoS parameters, and the vagueness and subjectivity of quality evaluation point out SC techniques as suitable ones for QoS adaptation. In the following, we examine the potential of some SC techniques (specifically, artificial neural networks, fuzzy control and neuro-fuzzy control) for QoS adaptation, detailing how they have been used to deal with the problems mentioned in Sect. 1.

Artificial Neural Networks Artificial neural networks (ANN's) represent an important class of numerical learning tools, inspired by the human learning capabilities.

ANN's imitate the brain parallelism using highly interconnected simple processing units (called neurons or nodes) to store and process information. Each aspect of a neuron is represented mathematically by real numbers. A neuron is characterized by input and output connections, an activation level, an output level, and a bias value. The output level is determined according to a function of the activation level, which is a weighted sum of the signal from the input connections. ANN's have many characteristics such as non-linear mapping, self-organization and learning. The learning is achieved through learning algorithms that optimize a number of network parameters. Basically, given a number samples, where a sample is an input and the desired output, the learning algorithm aims to find a set of weights to generate the output corresponding to the input.

The ANN's features make them highly useful for numerical modelling and control, particularly for systems like a DMS, where there is little knowledge about dynamic and operating environment.

ANN can be used for QoS control, like in [12]. In this work, the authors propose to utilize a backpropagation ANN to control the bit rate of a VoD (video on demand) application transmitted over an ATM network. The ANN computes the MPEG[2] quantization factor in a linear way according to the available network bandwidth estimate.

ANN's can also be used to aid to solve the problem of mapping between QoS parameters. In [14], an ANN is used to map different network configurations and loads into communication QoS parameters values (e.g., packet loss rate, network delay and jitter). The results permit to the QoS adaptation mechanism to determine acceptable load and configuration given bounds of QoS parameters values. As presented above, estimate of QoS parameters values through the simulation of traffic requires considerable computational resources. On the other hand, ANN's can produce estimate in order of milliseconds, as the results shown in [14].

In addition, ANN's can be used to capture the user's point of view in terms of quality. In [15], the authors propose to use ANN's for creating *users' QoS profilers*. A user's QoS profiler holds the user's preferences based on the user's feedback and experience. User's QoS profiler allows to personalize QoS management. Whenever some QoS controllers are activated in response to certain events, the user's QoS profiler sends a notification message to the

---

[2] MPEG [13] is a family of video encoding algorithms. It compresses the image frames by splitting them into 8×8 blocks of pixels. A block can be one of three types: luminance, red chrominance, or blue chrominance. The algorithm includes the following steps: (1) discrete cosine transform (DCT), (2) quantization, and (3) run-length encoding. Since image blocks and prediction-error blocks have high spatial redundancy, MPEG algorithm transforms 8×8 blocks of pixels or 8×8 blocks of error terms from the spatial domain to the frequency domain with DCT in order to reduce this redundancy. The quantization step consists of normalizing each of the DCT coefficients by dividing it with a predefined value (the quantization factor). Hence, in this step the data accuracy is reduced.

The combination of DCT and quantization results in many of the frequency coefficients being zero, specially high spatial frequencies coefficients. To take maximum advantage of this, the coefficients are organized in a zigzag order to produce long runs of zero. The coefficients are then converted to a series of run-amplitude pairs, each pair indicating a number of zero coefficients and the amplitude of a non-zero coefficient. These run-amplitude pairs are then coded with a variable-length code (Huffman Encoding), which uses shorter codes for commonly occurring pairs and longer codes for less common pairs.

Some blocks of pixels need to be coded more accurately than others. For example, blocks with smooth intensity gradients need accurate coding to avoid visible block boundaries. To deal with this inequality between blocks, the MPEG algorithm allows the amount of quantization to be modified for each macroblock of pixels (16 × 16 segment of pixels in a frame).

user and requests a feedback. The user may give the feedback in the form of a percentage, which represents the degree of his or her satisfaction about the QoS management services.

Fuzzy Controllers Fuzzy control is a control method based on Fuzzy Set Theory [16] and originated from the research of H. Mamdani [17]. Formally, a fuzzy set A of the domain X is defined by a function

$$\mu_A : X \mapsto [0,1].$$

Such function μ_A - called *membership function* - associates a compatibility degree with the concept expressed by A to each element $x \in X$:
if $\mu_A(x) = 1$, x is completely compatible with A;
if $\mu_A(x) = 0$, x is completely incompatible with A;
if $0 < \mu_A(x) < 1$, x is partially compatible with A, with a degree $\mu_A(x)$.
From this definition, Classical Set Theory can be seen as a subset of Fuzzy Set Theory. While in Fuzzy Set Theory the compatibility of an element in a domain X with a subset A can be expressed with membership degrees in $[0,1]$, in Classical Set Theory this compatibility can be expressed just by a value in $\{0,1\}$.

Each element $x \in X$ is a *linguistic variable*; X is the *universe of discourse*; the fuzzy set A is labeled by a *linguistic value* (such as $FULL$, $EMPTY$, $GOOD$, $FAIR$, BAD etc.).

According to [18], Fuzzy Logic can be described simply as computing with words rather than numbers and fuzzy control can be described simply as control with sentences rather than equations. The sentences are empirical rules in the format `if` < *conditions* > `then` < *conclusion* >. An example of a rule is

if *network* is $CONGESTED$ then *quantization factor* is LOW.

Fuzzy controllers (FC's) can be used to QoS control, since they are robust in the presence of perturbations, easy to design and implement, and efficient for systems that deal with continuous variables [19]. Note the complexity in building an analytical model is another aspect of a DMS. The feedback information (QoS parameters from different DMS architecture layers) plays the role of linguistic variables (FC inputs). Linguistic values represent a data granulation, where the values are overlapping fuzzy sets [6], resulting in gradual adaptation. (QoS adaptation approaches based on LC's generally use intervals, resulting in abrupt adaptation.)

An example of QoS adaptation based on fuzzy control is found in [20]. Two FC's provide the input and output rates of a shaper whose role is to smooth a MPEG video stream, transmitted over an ATM network, to reduce its burstiness. To avoid a long delay at the shaper, the first FC aims to tune the output rate of the shaper based on the occupancy of the shaper's buffer in the video frame time-scale. Based on the average occupancy of the shaper's buffer and its variance, the second FC tunes the input rate to the shaper over a much larger time-scale by applying a closed loop MPEG encoding scheme.

Another example of QoS adaptation based on fuzzy control is found in [21]. In this work, a distributed multimedia application is seen as a task pipeline where each task is characterized by input quality, output quality and utilized resources. Every application task is accompanied by an observation task, an adaptation task and a state which indicates the status of the application. The observation task monitors the requested rate (state) of the application task, and in the case of low resource(s) availability, the adaptation task throttles the requested rate of the application task, hence decreasing the request of resources. The work proposes to use Fuzzy Set Theory to translate throttled requested rate values into tuning actions.

Neuro-Fuzzy Systems According to [22], SC techniques can be combined since each technique can complement the other. The most widely research hybrid system in the area of SC is neuro-fuzzy systems [23], where the learning capabilities of ANN's are exploited within FC's, particularly in generating and tuning the membership functions. Note the performance of a FC is critically dependent on the membership functions parameters that are usually defined with the assistance of a human expert, which is, in general, a manual trial-and-error procedure. This procedure is time-consuming and is typically characterized by poor optimization performance, specially in ill defined and unpredictable systems like a DMS.

The above strategy for generating and tuning the membership functions is used in a QoS control approach described in [24]. The authors use an ANN to provide the learning ability to a FC used for estimating the required bandwidth in ATM networks. The membership functions are built by learning, in a fine tuning process.

4 Illustration of SC Application onto QoS Adaptation

The former section has provided an overview of how SC techniques are been used for QoS adaptation. In this section, we focus on QoS adaptation based on Fuzzy Theory through an approach that we have proposed in [25].

There are many approaches for QoS adaptation where QoS control is based on using SC techniques [24] [26] [20]. However, most of them are bounded to ATM networks not addressing the mapping between bit rate and quality, according to user's perspective. In contrast, one of the main features of our QoS adaptation approach is generality. Besides generality, two other features are user-awareness and simplicity.

The QoS adaptation approach is general because (a) it does not assume any specific platform, protocol, nor operating system, (b) it does not assume any specific compression algorithms or specific QoS parameters, and (c) it can be used in best-effort environments as well as in QoS-aware environments.

The QoS adaptation approach is user-aware because our model uses a function to map users' preferences, represented by various application QoS parameters, into a neutral, unified metric, called quality degree as defined in [25]. The quality degree then drives a controller that estimates and derives new QoS parameters values according to the current resources availability.

The QoS adaptation approach is simple because the control is performed by a FC, that is intrinsically simple and that provides a solution for the non-linear DMS behavior.

4.1 Assumptions and End-to-end Control Model

Our QoS adaptation model considers 1:N applications (one sender delivers a multimedia stream to N receivers) but it can be extended to N:M ones.

Each stream is characterized by a set of QoS application parameters ($\{\rho_1, \rho_2, ..., \rho_n\}$) which will be mapped, using the quality degree function, into one unified quality measure, called quality degree, in the range [0,1].

The quality degree function represents a quality mapping considering users' preferences, which are implicitly present in the set of application QoS parameters.

Hence, each multimedia stream is assumed to be sent out with a *quality degree of emission* ($Q\hat{o}S_e$) and it arrives at the kth receiver with a *quality degree of reception* ($Q\hat{o}S_{r_k}$, $k = 1, 2, ..., N$) so that $Q\hat{o}S_{r_k} \leq Q\hat{o}S_e$. In the kth receiver, after the decompression process, the multimedia stream is displayed with a quality degree of visualization ($Q\hat{o}S_{v_k}$), so that $Q\hat{o}S_{v_k} \leq Q\hat{o}S_{r_k}$. Fig. 2 provides an overview of our QoS adaptation model.

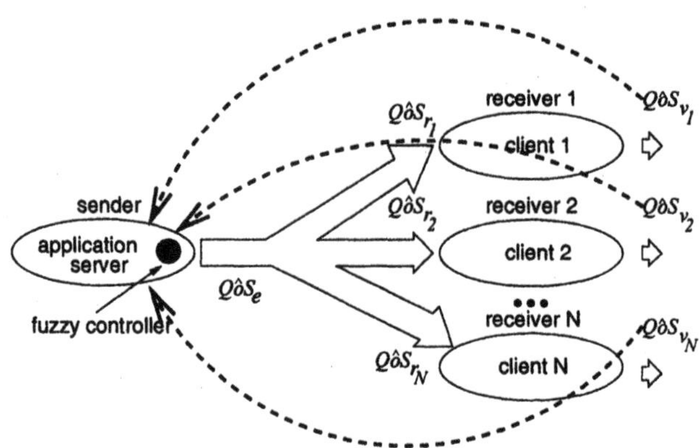

Fig. 2. QoS Adaptation model.

Since the quality degree along the end-to-end path from sender to receivers might change due to the variability of resources availability (e.g., decrease of bandwidth due to network congestion), we need quality adaptation to adjust the QoS parameters values (and consequently the quality degree) for each multimedia stream. Our approach for QoS adaptation assumes a comprehensive QoS adaptation control at the sender and monitors with feedback at the receivers. The adaptation, performed at the sender, integrates the quality degree function, a fuzzy controller (FC) and enforcement algorithms to execute the QoS adjustment.

In the following, we describe the *quality degree function* that is used in our approach to compute $Q\hat{o}S_e$ and $Q\hat{o}S_{v_k}$.

4.2 The Quality Degree Function

The quality degree function is a very significant component of our QoS adaptation approach. The goal of this function is to quantify the quality. Though the quality degree function, the QoS adaptation mechanism can select the (theoretically) best combination of application QoS parameters values for a given network bit rate. The definition of a choice criterion is important since different combinations of QoS parameters values can have similar requirements of bit rate representing completely distinct qualities.

The quality degree function will be described after initial definitions.

Definition 1. The *domain of values for a QoS parameter* ρ_i is a set

$$P_{\rho_i} = [\rho_{i_{min}}, \rho_{i_{max}}] \; (i = 1, 2, ..., n). \tag{4}$$

(All considered sets P_{ρ_i} are finite sets.)

Definition 2. A *QoS level* L is a multidimensional variable denoted by a n-tuple $< \rho_1, \rho_2, ..., \rho_n >$ representing a combination of values of the n QoS parameters. In our case, QoS levels refer just to the application layer QoS parameters. P is the domain of all QoS levels, i.e., it is the Cartesian product of the n domains P_{ρ_i}, i.e.,

$$P = P_{\rho_1} \times P_{\rho_2} \times ... \times P_{\rho_n}. \tag{5}$$

Definition 3. The *quality degree* is a quality metric used to quantify the quality of QoS levels in agreement with user's perception. It is represented by $Q\hat{o}S$ and is arbitrarily included in $[0, 1]$ domain. The value of $Q\hat{o}S$ is obtained using the quality degree function

$$\mathcal{Q}o\mathcal{S} : P \mapsto [0, 1] \tag{6}$$

which maps the set P of QoS levels into the [0,1] domain. For the QoS level $L_j =< \rho_{1_j}, \rho_{2_j}, ..., \rho_{n_j} >$, we use the notation $Q\hat{o}S_j$ to represent the application of the quality degree function in L_j, i.e., $Q\hat{o}S_j = \mathcal{Q}o\mathcal{S}(\rho_{1_j}, \rho_{2_j}, ..., \rho_{n_j})$.

Two important assumptions, related to the quality degree and used by our QoS adaptation approach, are:

$$\exists L_i, L_j | (L_i \neq L_j \Rightarrow Q\hat{o}S_i = Q\hat{o}S_j) \quad (7)$$

and

$$\exists L_i, L_j | (Q\hat{o}S_i > Q\hat{o}S_j \Rightarrow B_i < B_j) \quad (8)$$

where B_p is the bit rate required by the QoS level L_p ($p = 1, 2, ..., |P|$). Equation 7 means two or more different QoS levels can have the same quality degree. This assumption is reasonable, because in some cases users are not usually capable of perceiving differences among QoS values inside some ranges. For example, most users do not perceive differences in the frame rate between 25 and 30 fps. Equation 8 means the bandwidth requested by a QoS level L_i is not necessarily greater than the bandwidth requested by a QoS level L_j just because the quality degree of L_i is greater than the quality degree of L_j. This assumption comes from the fact that a QoS parameter can impact strongly on the quality without having high impact on the bit rate and vice versa. The properties described by Equations 7 and 8 allow for a smooth adaptation and, many times, for changes in the current bit rate without changing the quality.

In our framework, quality degree function $\mathcal{QoS}$ is the way to deal with the quality as a whole and to capture the vagueness inherent to quality evaluation. The function is used to quantify the end-to-end quality of the multimedia stream according to user's perspective. We use $Q\hat{o}S_e$, $Q\hat{o}S_{r_k}$ and $Q\hat{o}S_{v_k}$ ($k = 1, 2, ..., N$) to define instances of the quality degree respectively for emission, reception and visualization of the multimedia stream.

4.3 Framework

Our QoS adaptation approach is based on a FC and a table whose records represent QoS levels with quality degrees and bit rates associated like in [27]. Therefore, each entry of the table is a $(n + 2)$-tuple

$$< \rho_1, \rho_2, ..., \rho_n, Q\hat{o}S, B > . \quad (9)$$

The construction of the FC and the table is a composite process of several steps, shown in Fig. 3. The following describes each step.

First Step. Definition of the Target-application Since the QoS adaptation policy uses information closely related to application type, it is necessary to define the target-application (e.g., videoconference, VoD, remote sensoring, Internet TV, voice transmission, remote training, distance medicine, video email, collaborative film and video production, etc.). An important information obtained in this step is the set of n application QoS parameters ρ_i and their domains of values

$$P_{\rho_1}, P_{\rho_2}, ..., P_{\rho_n} . \quad (10)$$

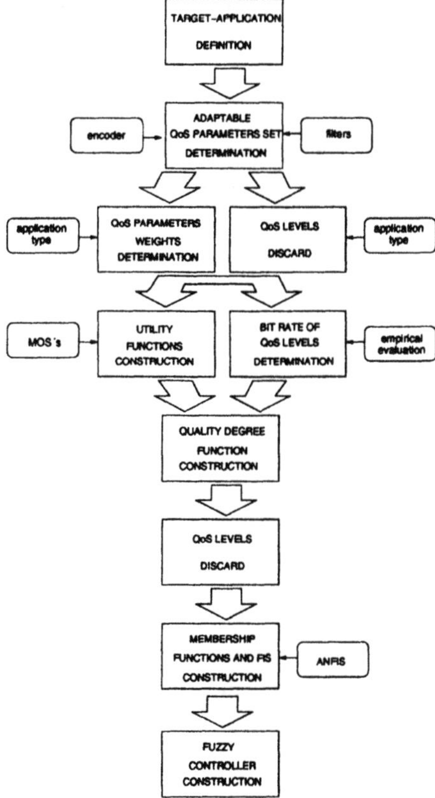

Fig. 3. Framework for QoS Adaptation.

The Cartesian product of the domains represents the initial set of QoS levels (P).

Second Step. Determination of Adaptable QoS Parameters Just a subset of $\{\rho_1, \rho_2, ..., \rho_n\}$ contains application QoS parameters able to have their values changed dynamically. Thus, it is necessary to determine the adaptable QoS parameters subset. For simplicity, this subset will be represented by $\{\rho_1, \rho_2, ..., \rho_m\}$, such that

$$\{\rho_1, \rho_2, ..., \rho_m\} \subseteq \{\rho_1, \rho_2, ..., \rho_n\}. \tag{11}$$

This subset is closely related to the compression algorithm and the compression algorithm implementation (encoder) as well as the available application QoS parameters filters[3].

[3] A filter is a mechanism that allows to change dynamically the value of a QoS parameter.

The Cartesian product of the m domains of values of QoS parameters ρ_i $(i = 1, 2, ..., m)$ represents a subset of QoS levels $P' \subseteq P$.

Example. Suppose that there are filters for two application QoS parameters: DCT coefficient (D) and quantization factor (Q). They are used by MPEG encoders and they are related to the image spatial resolution. Hence, the adaptable QoS parameters subset is:

$$\{D, Q\} \tag{12}$$

where

$$P_D = \{1, 2, 3, ..., 62, 63\}$$
$$P_Q = \{0, 1, 2, ..., 15, 16\}.$$

Third Step. Discard of Some QoS Levels and Determination of Weights Application type can determine several restrictions related to bounds of application QoS parameters values[4]. Thus, some application QoS parameters domains might be reduced as follows:

$$P_{\rho_i} = [\rho_{i_{min'}}, \rho_{i_{max'}}],\ \rho_{i_{min'}} \geq \rho_{i_{min}}, \rho_{i_{max'}} \leq \rho_{i_{max}}\ (i = 1, 2, ..., m). \tag{13}$$

This reduction of domains implies in discarding QoS levels containing application QoS parameters values outside the new domains. Therefore, the new subset of QoS levels will be $P'' \subseteq P'$.

Application type can also determines relative importance of each application QoS parameter in quality composition[5]. Hence, it is necessary to assign weights to the application QoS parameters. The assignment can be chosen arbitrarily or intuitively or determined by previous experiments related to a type of application. For each QoS parameter ρ_i, is assigned a weight ω_{ρ_i}, such that

$$\sum_{i=1}^{m} \omega_{\rho_i} = 1. \tag{14}$$

Fourth Step. Determination of Bit Rate and Construction of Utility Functions An empirical evaluation of bit rate of several clips with different QoS levels is needed to associate bandwidth to QoS levels. This association permits to the QoS adaptation mechanism to select the QoS level more compatible with the available network bandwidth when adapting.

For each QoS level in P'', it should be taken the worst case with regards to required bandwidth[6]. The worst case might be gotten by using clips formed

[4] For example, a VoD application requires a video frame rate above 25 fps.

[5] In an ordinary videoconference, for example, the QoS parameters related to sound are more important than those ones related to image.

[6] The measurements could be extended to other dimensions of cost (resources consumption), such as memory and processor cycles.

only by I-type frames (if the compression algorithm is from MPEG family), what reduces drastically the compression rate.

Besides associating bit rate to QoS levels, another activity of this step is to build ***utility functions*** [28] for each application QoS parameter. The utility functions are needed to build the quality degree function.

A utility function is a two dimensional function which associates a utility degree (in fact, a measure of quality), between 0 and 1, to a value of a QoS parameter based on users' preferences.

For an application QoS parameter ρ_i, the utility function is

$$v_{\rho_i} : P_{\rho_i} \mapsto [0,1] \; (i = 1, 2, ..., m). \tag{15}$$

Utility functions can be obtained though Mean Opinion Scores (MOS's) [29]. A group of users evaluates several clips with different QoS levels. The clips to be evaluated should be chosen in compliance with the target-application. For example, if the target-application is an ordinary videoconference, the clips might be "talking heads" type. The results are normalized into the range [0,1], since MOS-based evaluation quantifies quality using a five points scale.

Fifth Step. Construction of the Quality Degree Function In order to build $\mathcal{QoS}$, let $v_{\rho_1}, v_{\rho_2}, ..., v_{\rho_m}$ be the utility functions (constructed in the former step) and let $\omega_{\rho_1}, \omega_{\rho_2}, ..., \omega_{\rho_m}$ be the weights of the application QoS parameters $\rho_1, \rho_2, ..., \rho_m$ (determined in the third step). For a QoS level

$$L_j = < \rho_{1_j}, \rho_{2_j}, ..., \rho_{m_j} >,$$

$$\begin{gathered} \mathcal{QoS}(\rho_{1_j}, \rho_{2_j}, ..., \rho_{m_j}) = \\ min(v_{\rho_1}(\rho_{1_j}) \times \omega_{\rho_1}, v_{\rho_2}(\rho_{2_j}) \times \omega_{\rho_2}, ..., v_{\rho_m}(\rho_{m_j}) \times \omega_{\rho_m}). \end{gathered} \tag{16}$$

The above strategy for building the quality degree function never permits $Q\hat{o}S_j = 1$. The upper quality degree will be

$$Q\hat{o}S_j = \frac{1}{m} \tag{17}$$

if $v_{\rho_1}(\rho_{1_j}) = v_{\rho_2}(\rho_{2_j}) = ...v_{\rho_m}(\rho_{m_j}) = 1$ and $\omega_{\rho_1} = \omega_{\rho_2} = ... = \omega_{\rho_m} = \frac{1}{m}$. In order to keep the definition of the quality degree function given by Equation 6, the results obtained using the strategy of Equation 16 must be normalized into the range [0,1].

Fig. 4 shows the utility functions for the application QoS parameters of Equation 12 and the resulting quality degree function for a video frame rate $F = 25$ fps. For the quantization factor, $Q = 0$ represents the highest utility and $Q = 16$ the lowest one.

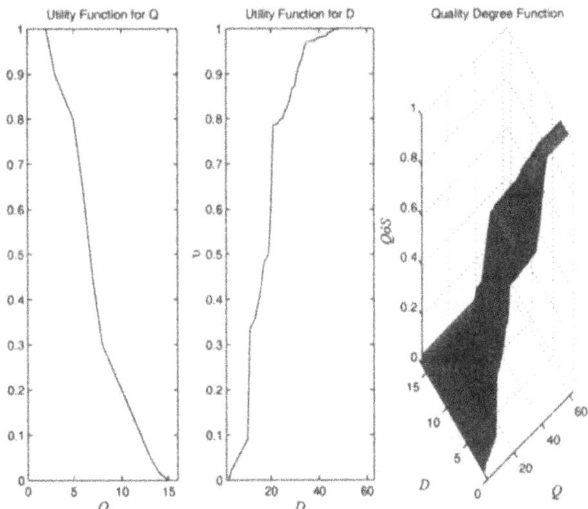

Fig. 4. Utility functions for quantization factor and DCT coefficient and the resulting quality degree function $Q\hat{o}S$ (video frame rate = 25 fps).

Sixth Step. Discard of Some QoS Levels The assumptions of Equations 7 and 8 are used to discard many QoS levels for reducing the table used by our framework. If there are two QoS levels L_i and L_j and $Q\hat{o}S_i \geq Q\hat{o}S_j$ but $B_i \leq B_j$, the table will held only L_i. That is, it should be kept only QoS levels with quality degrees as highest as possible and requiring less bandwidth.

Normally, some kind of applications are more tolerant to quality degradation than other ones. For example, an ordinary videoconference application accepts image and sound quality lower than a VoD application. For that reason, it should be imposed a minimum quality degree according to application type. All QoS levels with quality degree less than the lower bound should be discarded.

The above discards of QoS levels shrinks P''. The new domain of QoS levels will be $P''' \subseteq P''$.

The remaining QoS levels (in the set P'''), with associated quality degree and bit rate, can be represented by a table, where j^{th} record is

$$< \rho_{1_j}, \rho_{2_j}, ..., \rho_{m_j}, Q\hat{o}S_j, B_j > \quad (18)$$

Therefore, the memory required by our QoS adaptation approach (the number of entries of the table) is proportional to the cardinality of P''' ($|P'''|$).

Fig. 5 details the quality degree function of Fig. 4. The marked points represent the remaining QoS levels if $Q\hat{o}S_{min} = 0.5$. While the cardinality of P' is 63×17=1071, the cardinality of P''' is 315.

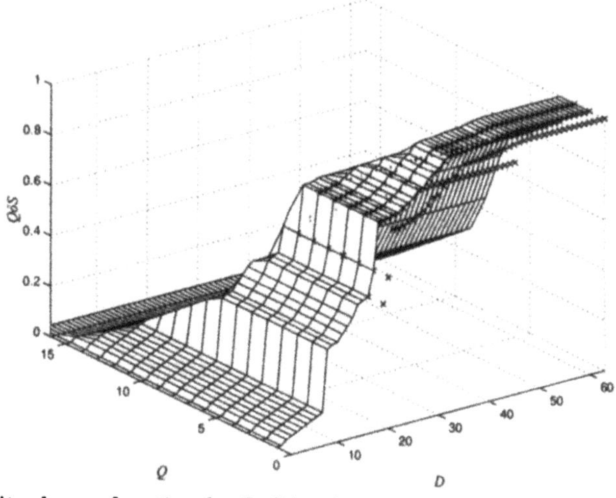

Fig. 5. Quality degree function for QoS levels composed by quantization factor and DCT coefficient. Marked points represent QoS levels with quality degree higher or equal to 0.5.

Seventh Step. Construction of the Fuzzy Controller The FC is the kernel of our approach. It is used, when adapting, to deal with uncertainty for network load determination. In our FC, linguist variables are quality degree and error. In order to keep an association between fuzzy sets and the five scales of subjective quality for multimedia applications, defined by the International Telecommunication Union [30], linguist values for quality degree could be $EXCELLENT$, $GOOD$, $FAIR$, $POOR$ and BAD.

For building the membership functions of the FC, fuzzy c-means clustering method [31] is used. Fuzzy c-means is a data clustering technique wherein each data point belongs to a cluster to some degree that is specified by a membership grade. The membership degree of an element in a fuzzy cluster can be viewed as the similarity degree of the cluster's center to the element.

4.4 FC Performance

In order to compare the FC performance with a LC performance, we have implemented a strategy for QoS adaptation based on the above framework to adjust the bit rate of a video distribution application (VDA) [32] to the available network bandwidth. The LC uses a common approach for QoS adaptation [2] [8] [1]: according to the losses (video frame losses, in this case), the LC increases or decreases the bit rate based on linear functions. There is a dead zone where the bit rate is not changed.

In the absence of actual data, the utility functions, used to construct QoS were built using polynomial interpolation. The breakpoints were assigned intuitively and we did not consider weights for the application QoS parameters.

Each QoS level $L = < D, Q, F >$ is associated with the quality degree ($Q\hat{o}S$) and the bit rate (B) needed to reach the level. The bit rates for the different QoS levels have been measured empirically for the worst case (video streams containing just I-frames[7]).

The FC has two inputs, the current quality degree of emission ($Q\hat{o}S_e$) and the error(ε), and an output, the new quality degree of emission ($Q\hat{o}S'_e$). The error is given by:

$$\varepsilon = Q\hat{o}S_e - \overline{Q\hat{o}S}_v \tag{19}$$

where $\overline{Q\hat{o}S}_v$ is an aggregation of quality of degrees of visualization obtained from N receivers ($Q\hat{o}S_{v_k}$, $k = 1, 2, ..., N$) computed according to a pre-defined policy (e.g., $\overline{Q\hat{o}S}_v$ can be the worst case, the result of the weighted mean and so on).

The domain of quality degree is divided in five fuzzy sets: $EXCELLENT$, $GOOD$, $FAIR$, $POOR$ and BAD, but the FC considers only the three first ones; the domain of error is divided in three fuzzy sets: $LARGE$, $MEDIUM$ and $SMALL$.

In the implementation, we have used the rule base shown in Table 1 and, for simplicity, triangular membership functions.

Table 1. Rule Base

Rule	$Q\hat{o}S_e$	ε	$Q\hat{o}S'_e$
0	*EXCELLENT*	*SMALL*	*EXCELLENT*
1	*EXCELLENT*	*MEDIUM*	*GOOD*
2	*EXCELLENT*	*LARGE*	*FAIR*
3	*GOOD*	*SMALL*	*EXCELLENT*
4	*GOOD*	*MEDIUM*	*FAIR*
5	*GOOD*	*LARGE*	*FAIR*
6	*FAIR*	*SMALL*	*GOOD*
7	*FAIR*	*MEDIUM*	*FAIR*
8	*FAIR*	*LARGE*	*FAIR*

The overall QoS adaptation control flow is shown in Fig. 6. According to this figure, the QoS adaptation control flow follows the direct control scheme, where the QoS adaptation mechanism is in the forward path at the

[7] A MPEG video stream [13] can have three types of frames: I, P and B. The first one is the most "important" because in a sequence of video frames, the I-frame decompression is completely independent of other frames; a P-frame decompression needs the former frame of the sequence; a B-frame is decompressed using the former and the next ones.

feedback control system. In time t of the VDA lifetime, the N receivers deliver to the sender their quality degrees of visualization ($Q\hat{o}S_{v_k}$, $k = 1, 2, ..., N$) computed using the quality degree function. The application QoS parameters used are the DCT coefficient (D), the quantization factor (Q) and the video frame rate (F). Therefore, a QoS level is $< D, Q, F >$. $Q\hat{o}S_{v_k}$ provides the QoS adaptation mechanism with an approximation of the network load. The mechanism performs control actions over the VDA to change the D and Q values in order to adjust de multimedia stream bit rate (that corresponds to the quality degree of emission) to the available bandwidth. The F value is not changed by the QoS adaptation mechanism in the sender but it can be lower in receivers because of the network load.

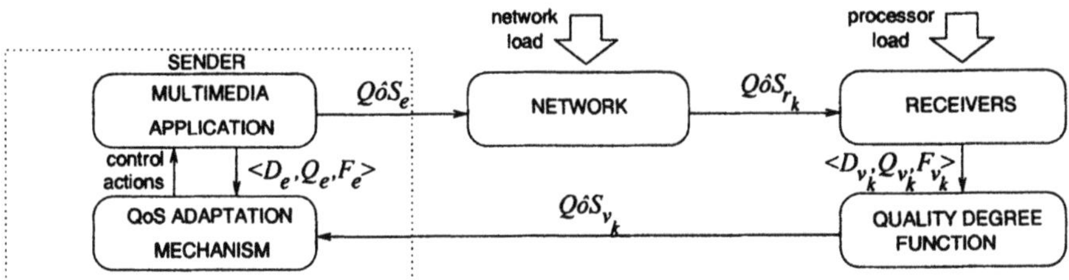

Fig. 6. QoS adaptation control flow.

Fig. 7 details the QoS adaptation mechanism. The values $Q\hat{o}S_{v_k}$ are used to compute the aggregated quality degree of visualization ($\overline{Q\hat{o}S}_v$). Then, $\overline{Q\hat{o}S}_v$ is compared with the reference ($Q\hat{o}S_e$), and depending on the error, the FC reacts according to the control strategy to generate $Q\hat{o}S'_e$. A mapping interface queries in the table the first record whose quality degree is less or equal to $Q\hat{o}S'_e$, assuming the table is sorted by quality degree in descendant order. This QoS level will be the new QoS level of emission ($L'_e =< D'_e, Q'_e, F_e >$; as said, F_e is constant). Finally, an actuator assists the distributed multimedia application in configuring itself according to the QoS parameters values of L'_e by invoking filtering operations.

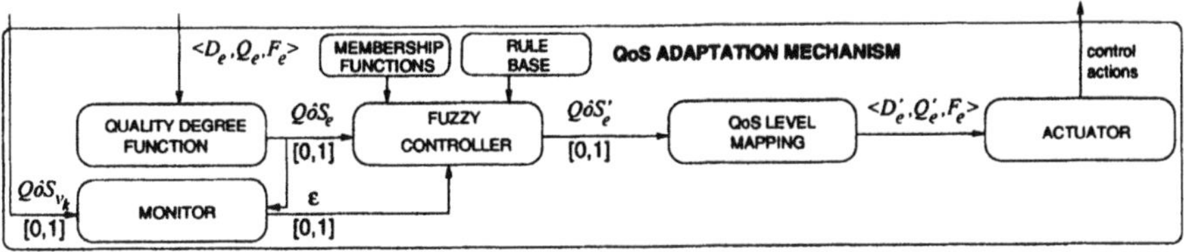

Fig. 7. QoS adaptation mechanism.

The experiments have considered two scenarios: (1) VDA with fuzzy control, and (2) VDA with linear control. In both scenarios, we have measured and compared the following metrics: video frame loss rate (L) and quality degree of visualization ($Q\hat{o}S_v$). Fig. 8 shows the comparison between the FC and the LC. According to graphics of Fig. 8 (a) and (b), during the overall VDA lifetime, the FC keeps a much lower video frame loss rate than the LC, avoiding bandwidth waste; with respect to the quality degree of visualization; graphic of Fig. 8 (c) shows that the FC keeps $Q\hat{o}S_v \geq 0.5$ during the entire VDA lifetime. Graphic of Fig. 8 (d), on the other hand, shows that the LC causes high oscillations of $Q\hat{o}S_v$ and, many times, permits that it reaches too low values.

The results have showed: (1) the FC allows a better identification of network congestion than the LC; (2) the FC allows to keep a higher and more stable quality than the LC; and (3) the use of more than a single QoS parameter for QoS adaptation, through the quality degree function, allows a better utilization of the network bandwidth.

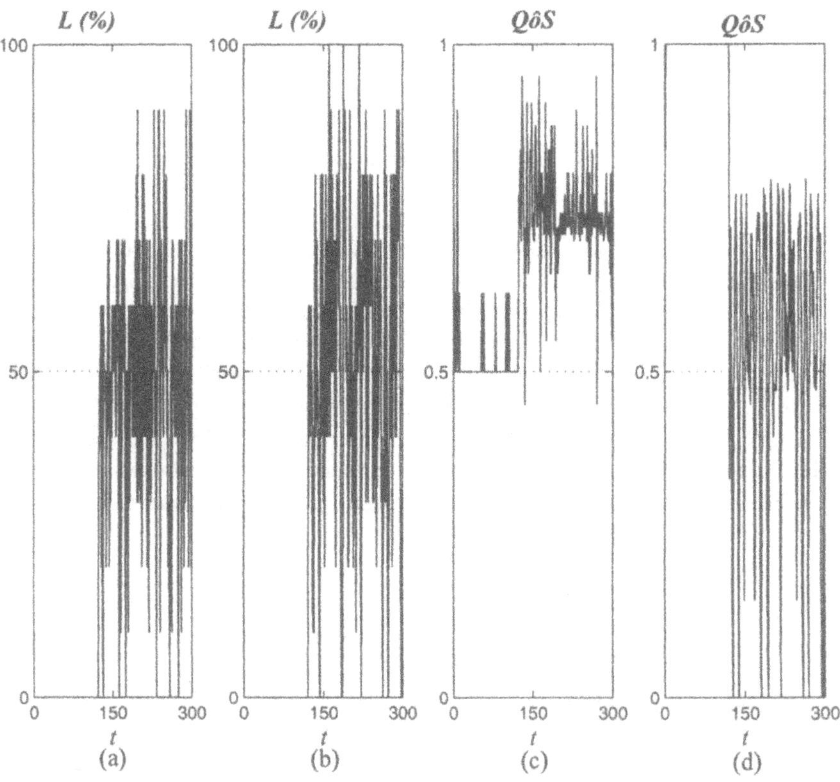

Fig. 8. Video frame loss rate with (**a**) fuzzy control and (**b**) linear control during the VDA lifetime; quality degree of visualization at the client with the VDA controlled by the (**c**) FC and the (**d**) LC. Network is overloaded by $t = 150$.

5 Conclusions

The research community has been perceiving the potential of SC techniques for QoS adaptation in best-effort DMS and many approaches for QoS adaptation based on SC have been developed. Some reasons that point to suitability of SC are: difficulty of building mathematical models for a DMS, non-linearity in the relationship among many QoS parameters and subjectivity in the analysis of the performance criterion of the system. The most efforts have been made to solve the QoS control problem, using mainly FC's and ANN's. Other SC techniques, such as genetic algorithms and evolutionary computing, have not yet been exploited in great depth.

Despite the suitability of using SC techniques for QoS adaptation, there is still a lack of comprehensive performance comparisons between QoS adaptation based on hard computing techniques and the ones involving SC techniques [33]. The comparisons performed usually have been undertaken in simplified networking scenarios. Testing on real hardware has not been performed extensively yet, except for some partial implementations, as we have done in [25]. As Sekercioglu et al. observe in [33], we also think important, for an acceptation of using SC techniques for QoS adaptation, a greater emphasis on a rigorous demonstration of the advantages of them over traditional techniques.

References

1. Busse, I., Deffner, B., Schulzrinne, H.: Dynamic QoS of Multimedia Applications Based on RTP. In: 1st International Workshop on High Speed Networks and Open Distributed Platforms, St. Petesburg, Russia (1995)
2. Sisalem, D.: Fairness of Adaptive Multimedia Applications. In: IEEE International Conference on Communications (ICC'98), Atlanta, USA (1998)
3. Lakshman, V., Misshra, P.P., Ramakrishnan, K.K.: Transporting Compressed Video over ATM Networks with Explicit Rate Feedback Control. In: IEEE INFOCOM'97, Kobe, Japan (1997) 38–47
4. Duffield, N., Ramakrishnan, K., Reibman, A.: SAVE: an Algorithm for Smoothed Adaptive Video over Explicit Rate Networks. IEEE Transactions on Networking **6** (1998) 717–728
5. Hull, D., Shankar, A., Nahrstedt, K., Liu, J.W.S.: An End-To-End QoS Model and Management Architecture. In: IEEE Workshop on Middleware for Distributed Real-Time Systems and Services, San Francisco, USA (1997) 82–89
6. Zadeh, L.A.: Fuzzy Logic, Neural Netwoks, and Soft Computing. Communication of the ACM **37** (1994) 77–Fuzzy Systems
7. Gecsei, J.: Adaptation in Distributed Multimedia Systems. IEEE Multimedia (1997) 58–95
8. Sisalem, D., Schulzrinne, H.: The Direct Adjustment Algorithm: a TCP-Friendly Adaptation Scheme. In: 1st International Workshop Quality of Future Internet Services (QofIS'2000), Berlin, Germany (2000)

9. Ortega, A., Ramchandran, K.: Forward-Adaptive Quantization with Optimal Overhead Cost for Image and Video Coding with Applications to MPEG Video Coders. In: ST/SPIE Digital Video Compression '95, San Jose, California (1995)
10. Walpole, J., Liu, L., Maier, D., Pu, C., Krasic, C.: Quality of Service Semantics for Multimedia Database Systems. Technical Report DS-8, Oregon Graduate Institute, Oregon, USA (1999)
11. Gonçalves, P.A.S., Rezende, J.F., Duarte, O.C.M.B.: An Active Service for Multicast Video Distribution. Journal of the Brazilian Computer Society **7** (2000) 43–51
12. Silveira, R.M., Ruggiero, W.V.: Using Neural Networks for Adjusting the Quality of a Video on Demand System. In: XIX Brazilian Symposium of Computer Networks (SBRC'2001), Florianópolis, Brazil (2001) 33–49 (in Portuguese).
13. Gall, D.L.: MPEG: A Video Compression Standard for Multimedia Applications. Communication of the ACM **34** (1991)
14. Levin, B., Holst, A., Lansner, A., Haraszti, Z.: Simulation Support and ATM Performance Prediction. In: International Conference on Artificial Neural Networks (ICANN 98), Skvde, Sweden (1998) 263–268
15. Gu, X., Nahrstedt, K.: An Event-Driven, User-Centric, QoS-aware Middleware Framework for Ubiquitous Multimedia Applications. In: 9th ACM Multimedia Middleware Workshop), Ottawa, Canada (2001)
16. Zadeh, L.A.: Fuzzy Sets. Information & Control (1965) 338–353
17. Mamdani, E.H., Baaklini, N.: Prescriptive Method for Deriving Control Policy in a Fuzzy Logic Controller. Eletronic Letters **11** (1975) 625–626
18. Jantzen, J.: Design of Fuzzy Controllers. Technical Report TR 98-E-864, Technical University of Denmark, Lyngby, Denmatk (1998)
19. Akbarzadeh-T, M.R., Tunstel, E., Kumbla, K., Jamshidi, M.: Soft Computing Paradigms for Hybrid Fuzzy Controllers: Experiments and Applications. In: IEEE World Congress on Computational Intelligence. Volume 2., Anchorage, Alaska, USA (1998) 1200–1205
20. Tsang, D.H.K., Bensaou, B., Lam, S.T.C.: Fuzzy-Based Rate Control for Real-Time MPEG Video. IEEE-FS **6** (1998) 504
21. Li, B., Nahrstedt, K.: A Control-based Middleware Framework for Quality of Service Adaptation. IEEE Journal on Selected Areas in Communications (JSAC) **17** (1999) 1632–1650
22. Azvine, B., Azarmi, N., Tsui, K.C.: Soft Computing: a Tool for Buiding Intelligent Systems. BT Technology Journal **14** (1996) 37–45
23. Jang, J.S.R.: ANFIS: Adaptive-Network-Based Fuzzy Inference Systems. IEEE Transactions on Systems, Man and Cybernetics **23** (1993) 665–685
24. Al-sharhan, S., Karray, F., Gueaieb, W.: Tools of Computational Intelligence as Applied to Bandwidth Allocation in ATM Networks. In: IEEE International Conference on Communication (ICC 2001), Helsinki, Finland (2001) 2907–2911
25. Koliver, C., Nahrstedt, K.O., Farines, J.M., Fraga, J.S., Sandri, S.: Specification, Mapping and Control for QoS Adaptation. The Journal of Real-Time Systems **23** (2002) 143–174
26. Bogatinovski, M., Trajkovski, G., Spasenovski, B.: Fuzzy Controller for Video Conference Traffic in B-ISDN. In: IEEE 6th International Workshop on Computer-Aided Modeling, Analysis and Design of Communication Links and Networks (CAMAD'98), São Paulo, Brazil (1998) 55–59

27. Vogel, A., Kerherv, B., von Bochmann, G., Gecsei, J.: Distributed Multimedia Applications and Quality of Service: a Survey. IEEE MultiMedia **2** (1995) 10–19
28. Krasic, C., Walpole, J.: QoS Scalability for Streamed Media Delivery. Technical Report CSE-99-011, Oregon Graduate Institute os Science and Technology, Oregon, USA (1999)
29. International Telecommunication Union: Interactive Test Methods for Audiovisual Communications. (2000) Recommendation ITU-T P.920.
30. ITU: Methods for Subjective Determination of Transmission Quality. International Telecommunication Union. (1996) Recommendation ITU-T P.800.
31. Bezdek, J.C.: Pattern Recognition with Fuzzy Objective Function Algorithms. Plenum Press (1981)
32. Yeadon, N., Garcia, F., Hutchinson, D., Mauthe, A.: Filters QoS Support Mechanisms for Multipeer Communications. IEEE Journal on Selected Areas in Communications (JSAC) **14** (1996) 1245–1262
33. Sekercioglu, Y.A., Pitsillides, A., Vasilakos, A.: Computational Intelligence in Management of ATM Networks: A Survey of Current State of Research. Soft Computing Journal **5** (2001) 257–263

Fuzzy Chaotic Synchronization and Communication — Signal Masking and Encryption

Kuang-Yow Lian, Peter Liu, and Chian-Song Chiu

Department of Electrical Engineering,
Chung-Yuan Christian University, Chung-Li, 32023, Taiwan

Abstract–In this chapter, we address synthesis approaches for signal synchronization and secure communications of chaotic systems by using LMI-based fuzzy system design methods. Following a general form of Takagi-Sugeno fuzzy chaotic models, the structure of the response system is firstly proposed. Synthesizing from the observer and controller points of view, two developed drive-response systems achieve asymptotic synchronization. For chaotic communications, the asymptotical recovering of messages is ensured by the same framework. As a further application, a chaotic cryptosystem is proposed. The synchronization design is applied to transmit and decrypt the ciphertext. The combination of both chaotic signal and cyptosystem characteristics achieves communication with a higher level of security. Finally, several well-known chaotic systems are used in numerical simulations and DSP-based experiments.

Keywords– Chaotic synchronization, chaotic communication, cryptosystem, T-S fuzzy model.

1 Introduction

Many fuzzy model-based design, in recent years, are carried out using Takagi-Sugeno (T-S) fuzzy models [1]-[4]. The main concept is that the fuzzy IF-THEN rules have consequent parts which represent local linear models. Then the overall output of the fuzzy system represents the input to output relationship of any general nonlinear system in an interested region. Using analogous discussions to that of linear control, many objectives are achievable in a unified manner. The stability of such fuzzy model-based systems relies on finding a common symmetric positive matrices resulting in a Linear Matrix Inequalities (LMIs) [5] problem. Powerful numerical toolboxes are then used to solve these problems. Since this control approach proposes the advantage straighforwardness and unity for nonlinear systems with ripe analysis tools of linear control, mass research has focused on this topic.

On the other hand, chaotic dynamics are deterministic but extremely sensitive to initial conditions. Even infinitesimal changes in initial condition will lead to an exponential divergence of orbits. The pioneering work of Carroll and Pecora [6, 7] has led to many works regarding synchronization of two chaotic systems [8, 9] where two chaotic systems with suitable coupling produce identical oscillations. Many theories [8]-[12] have been proposed to achieve the synchronized manner from master-slave configuration. This master-slave configuration consists of the original chaotic system as a *drive system* to provide a *driving signal* to drive another system called the *response system* to synchrony. In addition, chaotic signals are typically broadband, noise-like, and difficult to predict, therefore they can be used in various context for masking information-bearing waveforms. They can also be used as modulating waveforms in spread spectrum systems. The idea of chaotic masking [13, 14] is to directly add the message in a noise-like chaotic signal at the transmitter's end, while chaotic modulation [15]-[18] is by injecting the message into a chaotic system as a spread-spectrum transmission. Later, at the receiver, a coherent detector with some signal processing is employed to recover the message.

As an extension of works [19, 20], this chapter addresses two types of drive-response structures to accommodate the issues of chaotic synchronization and communication. First, we introduce how to present chaotic systems by T-S fuzzy models. The proposed method of building T-S fuzzy model is applicable to following chaotic systems: Chua's circuit, Rössler and transformed Rössler system [8], Lorenz system, Hénon map, and Lozi map, etc. Next, in light of the fact that synchronization issues are closely related to the observer/controller design, the generalized fuzzy response system are proposed in this chapter to solve synchronization and secure communication by using two types of driving signals from observer and controller points of view. Furthermore, the proposed driving signals would be employed as applications to different signal masking such that the chaotic communications are achieved based on the results of synchronization. Hence the existence of the solution for the LMI conditions theoretically implies that the message can be perfectly recovered.

To further extend this fuzzy model-based chaotic communications, a higher level security methodology is proposed as follows. A chaotic signal, which can be flexibly chosen to be an output or any state of the chaotic drive system, is used to generate a superincreasing sequence. The plaintext (message) is encrypted using the superincreasing sequence at the drive system side which results in the ciphertext. In light of the previous sections, the ciphertext is added to the output of the drive system. Then this ciphertext embedded scalar signal is sent to the response system end. Following the design of a response system, chaotic synchronization between the drive and response system is achieved by solving LMIs. Since synchronization is ensured, which means internal states of the drive and response system are same, we regenerate the same superincreasing sequence and recover the ciphertext at the response system end. Using the regenerated superincreasing sequence, the ciphertext is decrypted into the plaintext at the response system.

2 General T-S Fuzzy Modeling of Chaotic Systems

Here, we consider a general chaotic system represented by

$$sx = f(x) \tag{1}$$

where sx denotes $\dot{x}(t)$ and $x(t+1)$ for continuous and discrete-time systems, respectively; $x = [x_1\ x_2\ \ldots\ x_n]^T \in R^n$ is the state vector; $f(x)$ are smooth nonlinear functions. Note that chaotic systems, by nature, are not control systems. Hence, the output of chaotic systems can be chosen in various ways. This flexibility will affect the outcome of the T-S fuzzy modeling representation and have a direct impact on the analysis/synthesis of the synchronization methodology. Detailed explanation will be more intuitive after the T-S fuzzy modeling of chaotic systems is given. First, for the original chaotic system (1), we choose a general chaotic output as:

$$y = h(x) \tag{2}$$

where $y \in R$ will be taken as the coupling or drive signal in synchronization and communication.

2.1 T-S Fuzzy Model

The T-S fuzzy rules of (1) and (2) are given as:

$$\begin{aligned}
&\textit{Rule } i: \text{IF } z_1 \text{ is } F_{1i} \text{ and } \cdots \text{ and } z_g \text{ is } F_{gi} \text{ THEN}\\
&sx = A_i x + b_i\\
&y = C_i x,\ i = 1, 2, \ldots, r,
\end{aligned} \tag{3}$$

where $z_1 \sim z_g$ are the premise variables which may consist of the states of the system; F_{ji} for $j = 1, 2, \ldots, g$ are the fuzzy sets; A_i and C_i are system matrices of appropriate dimensions; b_i are bias terms arising from the exact modeling process; and r is the number of fuzzy rules. Using the singleton fuzzifier, product fuzzy inference, and weighted average defuzzifier, the inferred output of (3) is:

$$\begin{aligned}
sx &= \sum_{i=1}^{r} \mu_i(z)\{A_i x + b_i\}\\
y &= \sum_{i=1}^{r} \mu_i(z) C_i x
\end{aligned} \tag{4}$$

where $z = [z_1\ z_2\ \ldots\ z_g]^T$. From the fuzzification, fuzzy inference, and defuzzification procedure, we have the properties $\mu_i(z) = \bar{\mu}_i(z) / \sum_{i=1}^{r} \bar{\mu}_i(z)$ with $\bar{\mu}_i(z) = \prod_{j=1}^{g} F_{ji}(z)$; and $\sum_{i=1}^{r} \mu_i(z) = 1$ for all t, where $\mu_i(z) \geq 0$ are normalized weights.

2.2 Modeling Approach

Here we give brief guidelines to construct the T-S fuzzy model. To this end, we represent (1) in the form:

$$sx = A_0 x + a(x) \quad (5)$$

where $A_0 x$ and $a(x)$ are accordingly the linear and nonlinear terms of $f(x)$. The modeling for the output function $h(x)$ is analogous to the modeling of $a(x)$. Hence, for brevity, only discussion of modeling for $a(x)$ is carried out.

Modeling a Scalar Function—

Consider a scalar nonlinear function $a(x_k)$ dependent on one state variable x_k, where it is assumed that x_k varies among a bounded interval Ω. Assume that the nonlinear term can be represented in the form $\phi(x_k)x_k$. The function $\phi(x_k)$ is well defined if $\frac{a(x_k)}{x_k}$ exists (if not, the bias term rises). Then the nonlinear term is represented by

$$a(x_k) = \sum_{i=1}^{r} \mu_i(x_k) d_i x_k$$

where the weighting function satisfies $1 \geq \mu_i(x_k) \geq 0$ and $\sum_{i=1}^{r} \mu_i(x_k) = 1$. To represent $\phi(x_k) = \sum_{i=1}^{r} \mu_i(x_k) d_i$ exactly, we must suitably assign $\mu_i(x_k)$ and d_i. For clarity, we show an example for $r = 2$. From the assumptions on the weighting functions, we have

$$\begin{aligned} \mu_1 + \mu_2 &= 1 \\ \mu_1 d_1 + \mu_2 d_2 &= \phi(x_k). \end{aligned} \quad (6)$$

Rewriting (6), the following is obtained

$$\mu_1 = \frac{-d_2}{d_1 - d_2} + \frac{1}{d_1 - d_2}\phi(x_k),\ \mu_2 = 1 - \mu_1.$$

Care must be taken to determine the value of d_1 and d_2 such that $\mu_i(x_k) \in [0\ 1]$ for all $x \in \Omega$. For instance let $d_1 = -d_2 = d$ in which d is the upper bound of $\phi(x_k)$, i.e., $d = \sup_{x_k \in \Omega} |\phi(x_k)|$. This results in $\mu(x_k) = \frac{1}{2}(1 + \frac{1}{d}\phi(x_k))$ and $\mu_2(x_k) = \frac{1}{2}(1 - \frac{1}{d}\phi(x_k))$. Also, it is reasonable that when $\phi(x_k) \geq 0$ for all $x_k \in \Omega$, the weighting functions can be chosen as

$$\mu_1(x_k) = \frac{1}{d}\phi(x_k),\ \mu_2(x_k) = 1 - \frac{1}{d}\phi(x_k),$$

with $d_1 = d$ and $d_2 = 0$.

Modeling a Vector Function—

Consider a vector function $a(x)$ with a nonlinear term $\phi(z)x_k$ appearing at some $l - th$ entry. According to the method of modeling a scalar function, $f(x)$ may be expressed as

$$f(x) = \sum_{i=1}^{r} \mu_i(z)\bar{A}_i x + A_0 x$$

where $A_0 x$ denotes the linear part and $\sum_{i=1}^{r} \mu_i(z)\, \bar{A}_i x$ is to represent $a(x)$. This means that all entries of $\bar{A}_i$ are zeroes except for the $(l,\ k)$-entry is d_i. Then we have

$$f(x) = \sum_{i=1}^{r} \mu_i(z)\, A_i x$$

where $A_i = \bar{A}_i + A_0$. The modeling approach can be extended to deal with $f(x)$ involving many nonlinear terms.

Depending on the application and the chaotic system's basic characteristics, the system output is not a unique choice. However, the observability condition must be satisfied (in the linear subsystems of the T-S representation (3)). On the other hand, for output selection, we find it natural to use the chosen output as the premise variable for the T-S modeling [19]of chaotic systems. This concept is valid from investigation on a large class of continuous-time and discrete-time chaotic systems. The reason is that the nonlinear term(s) of the chaotic system usually is dependent on one common variable. This variable, once chosen as the output signal, will greatly lessen the complexity of the T-S fuzzy modeling procedure and in turn the observer design is more straightforward. T-S fuzzy models for several well-known continuous-time and discrete-time chaotic systems are given in Table I and the details can be found in [19].

Table I

Chaotic Systems	Dynamical Equations	Fuzzy Sets	System Matrices	Bias Terms
Rössle 's system	$\dot{u} = -v - w$ $\dot{v} = u + 0.2v$ $\dot{w} = 0.2 + uw - 5w$	$F_1(u) = \frac{1}{2}(1 + u/d)$ $F_2(u) = \frac{1}{2}(1 - u/d)$ $d = 10.5$	$A_1 = \begin{bmatrix} 0 & -1 & -1 \\ 1 & 0.2 & 0 \\ 0 & 0 & d-5 \end{bmatrix}$ $A_2 = \begin{bmatrix} 0 & -1 & -1 \\ 1 & 0.2 & 0 \\ 0 & 0 & -d-5 \end{bmatrix}$	$b_1 = b_2$ $= \begin{bmatrix} 0 \\ 0 \\ 0.2 \end{bmatrix}$
Trans-formed Rössle 's system	$\dot{x}_1 = -x_2 - f_1(x_3)$ $\dot{x}_2 = x_1 + 0.2x_2$ $\dot{x}_3 = x_1 + 0.2 f_2(x_3) - 5$ $f_1(x_3) = \exp(x_3)$; $f_2(x_3) = \exp(-x_3)$;	$F_1(x_3) = \frac{1}{4}(1 + \frac{f_1(x_3)}{d})(1 + \frac{f_2(x_3)}{d})$ $F_2(x_3) = \frac{1}{4}(1 + \frac{f_1(x_3)}{d})(1 - \frac{f_2(x_3)}{d})$ $F_3(x_3) = \frac{1}{4}(1 - \frac{f_1(x_3)}{d})(1 + \frac{f_2(x_3)}{d})$ $F_4(x_3) = \frac{1}{4}(1 - \frac{f_1(x_3)}{d})(1 - \frac{f_2(x_3)}{d})$ $d = 75$	$A_i = \begin{bmatrix} 0 & -1 & 0 \\ 1 & 0.2 & 0 \\ 1 & 0 & 0 \end{bmatrix}$ $i = 1,\ 2,\ 3,\ 4$	$b_1 = \begin{bmatrix} -d \\ 0 \\ -5+0.2d \end{bmatrix}$ $b_2 = \begin{bmatrix} -d \\ 0 \\ -5-0.2d \end{bmatrix}$ $b_3 = \begin{bmatrix} d \\ 0 \\ -5+0.2d \end{bmatrix}$ $b_4 = \begin{bmatrix} d \\ 0 \\ -5-0.2d \end{bmatrix}$
Lorenz's system	$\dot{x}_1 = -10x_1 + 10x_2$ $\dot{x}_2 = 28x_1 - x_2 - x_1 x_3$ $\dot{x}_3 = x_1 x_2 - \frac{8}{3}x_3$	$F_1(x_1) = \frac{1}{2}(1 + x_1/d)$ $F_2(x_1) = \frac{1}{2}(1 - x_1/d)$ $d = 30$	$A_1 = \begin{bmatrix} -10 & 10 & 0 \\ 28 & -1 & -d \\ 0 & d & -\frac{8}{3} \end{bmatrix}$ $A_2 = \begin{bmatrix} -10 & 10 & 0 \\ 28 & -1 & d \\ 0 & -d & -\frac{8}{3} \end{bmatrix}$	$b_1 = b_2$ $= 0$
Hénon map	$x_1(t+1) = 1.4 - x_1^2(t) + 0.3x_2(t)$ $x_2(t+1) = x_1(t)$	$F_1(x_1) = \frac{1}{2}(1 + x_1/d)$ $F_2(x_1) = \frac{1}{2}(1 - x_1/d)$ $d = 2$	$A_1 = \begin{bmatrix} -d & 0.3 \\ 1 & 0 \end{bmatrix}$; $A_2 = \begin{bmatrix} d & 0.3 \\ 1 & 0 \end{bmatrix}$	$b_1 = b_2$ $= \begin{bmatrix} 1.4 \\ 0 \end{bmatrix}$

3 Chaotic Synchronization

The synchronization problem is to design the response system and the output of the drive system to force the response system to same internal states as the drive system. Fuzzy synchronization for chaotic continuous-time fuzzy systems(CFS) and discrete-time fuzzy systems(DFS) are investigated in this section. Take (4) as the drive system, then the response system may be designed as the following fuzzy rules:

Response System Rule i : IF $\hat{z}_1$ is F_{1i} and $\cdots$ and $\hat{z}_g$ is F_{gi} THEN

$$s\hat{x} = A_i\hat{x} + b_i + L_i(y - \hat{y})$$
$$\hat{y} = C_i\hat{x} \tag{7}$$

where "$\wedge$" denotes the estimates of the corresponding variables and $L_i \in R^n$ are observer to be designed at a later stage. The inferred output of (7) is:

$$\begin{aligned} s\hat{x} &= \sum_{i=1}^{r} \mu_i(\hat{z}) \{A_i\hat{x} + b_i + L_i(y - \hat{y})\} \\ \hat{y} &= \sum_{i=1}^{r} \mu_i(\hat{z}) C_i\hat{x} \end{aligned} \tag{8}$$

where $\hat{z} = [\hat{z}_1 \ \hat{z}_2 \ \ldots \ \hat{z}_g]^T$. From the fuzzification, fuzzy inference, and defuzzification procedure, we have the properties $\mu_i(\hat{z}) = \bar{\mu}_i(\hat{z}) / \sum_{i=1}^{r} \bar{\mu}_i(\hat{z})$ with $\bar{\mu}_i(\hat{z}) = \prod_{j=1}^{g} F_{ji}(\hat{z})$; and $\sum_{i=1}^{r} \mu_i(\hat{z}) = 1$ for all t, where $\mu_i(\hat{z}) \geq 0$ are normalized weights. Define the synchronization error $e = x - \hat{x}$. Therefore, the error dynamics between the drive system (4) and response system (8) are:

$$\begin{aligned} se = & \sum_{i,j=1}^{r} \mu_i(z)\mu_j(z)\{A_i - L_iC_j\}x - \sum_{i,j=1}^{r} \mu_i(\hat{z})\mu_j(\hat{z})\{A_i - L_iC_j\}\hat{x} \\ & + \sum_{i=1}^{r} (\mu_i(z) - \mu_i(\hat{z}))\{b_i + L_iy\}. \end{aligned} \tag{9}$$

3.1 Synchronization with Fuzzy Driving Signals

Consider a class of chaotic systems which have common bias term in the fuzzy representation, that is, the drive system (4) has $b_i = b$. If the drive signal y is generated by the local linear combinational states, namely, the *fuzzy driving signal*, then vectors C_i, for $i = 1, 2, ..., r$, must be designed in the response system (8). To this end, let L_i equal a given vector L. Then the error dynamics is reduced as:

$$se = \sum_{i=1}^{r} \mu_i(z)\{A_i - LC_i\}x - \sum_{i=1}^{r} \mu_i(\hat{z})\{A_i - LC_i\}\hat{x} \tag{10}$$

Similar to the concept of the exact linearization (EL) in [3], with a given vector L, there exist C_i such that

$$\{(A_1 - LC_1) - (A_i - LC_i)\}^{\mathrm{T}} \{(A_1 - LC_1) - (A_i - LC_i)\} = 0,\ 2 \leq i \leq r. \tag{11}$$

Then, the overall error dynamics is linearized as $se = Ge$, where $G = A_1 - LC_1 = A_i - LC_i$, for $2 \leq i \leq r$. The following results for CFS and DFS are stated to ensure the stability of the overall system.

Theorem 1-CFS *Consider the CFS synchronization error system described by (10). The system is asymptotically stable if there exist a common positive definite matrix P and C_i for $i = 1, 2, ..., r$, by solving the following eigenvalue problem (EVP):*

$$\min_{M_i,\ X} \ \varepsilon$$

subject to $X > 0,\ \varepsilon > 0$

$$-A_i X - X A_i^{\mathrm{T}} + M_i^{\mathrm{T}} L^{\mathrm{T}} + L M_i > 0, \text{ for all } i, \tag{12}$$

$$\begin{bmatrix} \varepsilon I & R_i^{\mathrm{T}} \\ R_i & I \end{bmatrix} > 0,\ 2 \leq i \leq r, \tag{13}$$

where $R_i \equiv \{A_1 X - L M_1 - (A_i X - L M_i)\}$ and $M_i \equiv C_i P^{-1}$, for $i = 1, 2, ..., r$, and $X \equiv P^{-1}$.

Proof: For the EL conditions (11), there exist a positive definite matrix X and a small constant $\varepsilon > 0$ such that

$$\varepsilon I - R_i^{\mathrm{T}} R_i > 0,\ 2 \leq i \leq r.$$

This means if all elements in εX^{-2} are near zero in above inequalities for a proper choice of $\varepsilon > 0$, $X > 0$, i.e., $\varepsilon X^{-2} \approx 0$, then the EL conditions (11) are achieved. This implies that the error dynamics (10) can be expressed as $se = Ge$ once the inequalities (13) are held. Therefore the error system with C_i should be designed to guarantee stability of the linearized error system (10). To this end, define a Lyapunov function candidate as $V(e) = e^{\mathrm{T}} P e$ with $P > 0$, and take the time derivative of $V(e)$ along the overall error dynamics. This yields

$$\dot{V}(e) = e^{\mathrm{T}} \left(G^{\mathrm{T}} P + P G \right) e. \tag{14}$$

Thus if the inequalities (12) are satisfied, then $\dot{V}(e) < 0$, which implies that error e asymptotically converges to zero as $t \to \infty$. □

Theorem 2-DFS *Consider the DFS synchronization error system described by (10). The system is asymptotically stable if there exist a common positive definite matrix P and C_i, for $i = 1, 2, ..., r$, by solving the following eigenvalue problem:*

$$\min_{M_i,\ X} \ \varepsilon$$

subject to $X > 0,\ \varepsilon > 0$

$$\begin{bmatrix} X & (A_iX - LM_i)^{\mathrm{T}} \\ A_iX - LM_i & X \end{bmatrix} > 0 \text{ for all } i \tag{15}$$

$$\begin{bmatrix} \varepsilon I & R_i^{\mathrm{T}} \\ R_i & I \end{bmatrix} > 0, \; 2 \leq i \leq r, \tag{16}$$

where $M_i \equiv C_i P^{-1}$, *for* $i = 1, 2, ..., r$, *and* $X \equiv P^{-1}$.

Proof: The proof is similar to Thm. 1. If the solutions of the LMI design problem stated in (15) and (16) are feasible, the EL technique is realized by minimizing $\|\varepsilon X^{-2}\|$ near to zero. Moreover, there exists a Lyapunov function candidate for DFS as $V(\widetilde{x}) = \widetilde{e}^{\mathrm{T}} P \widetilde{e} > 0$ with difference

$$\Delta V(e) = e^{\mathrm{T}}\{G^{\mathrm{T}} PG - P\}e \tag{17}$$

where $P > 0$. The conditions (15) ensure that $G^{\mathrm{T}} PG - P < 0$, for $i = 1, 2, ..., r$, i.e., $\Delta V(\widetilde{e}) < 0$. Then, the error system (10) has $\lim_{t \to \infty} e = 0$. □

When the feasible solutions of above LMIs are obtained, the vectors C_i, for $i = 1, 2, ..., r$, are obtained by $C_i = M_i P$ from the solutions of X and M_i.

3.2 Synchronization with Crisp Driving Signals

In order to relax EL conditions, the typical driving signal is considered. Since the typical driving mechanism is designed by properly selecting a crisp output, the signal is called the *crisp driving signal.* Here, the crisp driving signal is chosen to be same as the premise variable of the fuzzy model for the corresponding chaotic systems, i.e., $y = Cx = z$, and a known vector $C = C_i$ for all i. Moreover, this implies that the premise variable of the fuzzy response system can be set same as the driving signal, that is $z = \widehat{z} = y$. Therefore the error dynamics of (9) is:

$$se = \sum_{i=1}^{r} \mu_i(z) \{A_i - L_i C\} e. \tag{18}$$

The stability conditions for (18) are addressed here.

Theorem 3-CFS: *The error system of chaotic synchronization is described by (18) for CFS and is asymptotically stable if there exist a common positive definite matrix* P *and gains* L_i, *for* $i = 1, 2, ..., r$, *such that the following LMIs, with* $N_i \equiv PL_i$,

$$-A_i^{\mathrm{T}} P - PA_i + C^{\mathrm{T}} N_i^{\mathrm{T}} + N_i C > 0, \text{ for all } i, \tag{19}$$

have feasible solutions.

Proof: Define the Lyapunov function candidate $V(e) = e^{\mathrm{T}} Pe$ with $P > 0$, then the time derivative of $V(e)$ along the error dynamics (18) is

$$\dot{V}(e) = \sum_{i=1}^{r} \mu_i(z) e^{\mathrm{T}} \{(A_i - L_i C)^{\mathrm{T}} P + P(A_i - L_i C)\} e. \tag{20}$$

Combining the conditions of (19) and $\mu_i(z) \geq 0$, there exists a minimum positive definite matrix on the left hand side of (19) denoted by Q. It follows that

$$\begin{aligned} \dot{V}(e) &\leq -\sum_{i=1}^{r} \mu_i(z) e^{\mathrm{T}} Q e \\ &= -e^{\mathrm{T}} Q e < 0. \end{aligned}$$

Hence $e = 0$ is asymptotically stable. □

Theorem 4-DFS *The error system of the fuzzy chaotic synchronization described by (18) for DFS is asymptotically stable if there exist a common positive definite matrix P and gains L_i, for $i = 1, 2, ..., r$, such that the following LMIs, with $N_i \equiv PL_i$,*

$$\begin{bmatrix} P & (PA_i - N_i C)^{\mathrm{T}} \\ PA_i - N_i C & P \end{bmatrix} > 0, \text{ for all } i, \tag{21}$$

have feasible solutions.

Proof: The proof is similar as Theorem 3. Given a Lyapunov function candidate for DFS as $V(e) = e^T P e > 0$, we have

$$\begin{aligned} \Delta V(e) &= \sum_{i=1}^{r} \mu_i^2(z) e^{\mathrm{T}} [\bar{A}_i^{\mathrm{T}} P \bar{A}_i - P] e \\ &\quad + \sum_{i<j}^{r} \mu_i(z) \mu_j(z) e^{\mathrm{T}}(t) [\bar{A}_i^{\mathrm{T}} P \bar{A}_j + \bar{A}_j^{\mathrm{T}} P \bar{A}_i - 2P] e, \end{aligned} \tag{22}$$

where $P > 0$, and $\bar{A}_i = A_i - L_i C$. Notice that if $\bar{A}_i^{\mathrm{T}} P \bar{A}_i - P < 0$, then $\bar{A}_i^{\mathrm{T}} P \bar{A}_j + \bar{A}_j^{\mathrm{T}} P \bar{A}_i - 2P < 0$. This means if there are P and L_i such that the conditions (21) are held, then $\bar{A}_i^{\mathrm{T}} P \bar{A}_i - P < 0$. Hence $\Delta V(e) < 0$ and the synchronization error e asymptotically converges to zero as $t \to \infty$. □

Note that the gains L_i, for $i = 1, 2, \cdots, r$, are obtained by $L_i = P^{-1} N_i$ from the solutions of P and N_i.

4 Chaotic Communications with Signal Masking

In chaotic communications, the message is masked by driving signals, and the modulation process is carried out by injecting the masking signal into the fuzzy chaotic transmitter. Then the masked signal is sent to the fuzzy chaotic receiver, where the message is extracted according to the masking methods. Inspired by previous works of modulated chaotic communication [15]-[18], the fuzzy modulated chaotic transmitter and the signal masking are designed for the general fuzzy modeling form of chaotic systems (3). The fuzzy chaotic transmitter with message m embedded is

$$\begin{aligned} \textit{Transmitter Rule } i: \quad & \text{IF } z_1 \text{ is } F_{1i} \text{ and } \cdots \text{ and } z_g \text{ is } F_{gi} \text{ THEN} \\ & sx = A_i x + b_i + L_i m, \;\; i = 1, 2, ..., r, \end{aligned}$$

with the masking mechanism:

$$\begin{aligned} \textit{Masking Rule } i: \quad & \text{IF } z_1 \text{ is } F_{1i} \text{ and } \cdots \text{ and } z_g \text{ is } F_{gi} \text{ THEN} \\ & \bar{y} = C_i x + m, \;\; i = 1, 2, ..., r, \end{aligned}$$

where $\bar{y}$ is the masked signal, and vectors L_i and C_i are to be designed later. The fuzzy inferred transmitter can be expressed in the form:

$$sx = \sum_{i=1}^{r} \mu_i(z) \{A_i x + b_i + L_i m\} \tag{23}$$

$$\bar{y} = \sum_{i=1}^{r} \mu_i(z) \{C_i x + m\}. \tag{24}$$

The overall transmitter consists of message embedded chaotic system (23) and signal masking system (24). The modulation form (23) and (24) can be regarded as an extension of modulated chaotic communications. To recover the message, the fuzzy receiver is designed as (7) with $\bar{y}$ instead of y, which yields the error dynamics:

$$\begin{aligned} se &= \sum_{i,j=1}^{r} \mu_i(z)\mu_j(z)(A_i - L_i C_j)x - \sum_{i,j=1}^{r} \mu_i(\widehat{z})\mu_j(\widehat{z})(A_i - L_i C_j)\widehat{x} \\ &\quad + \sum_{i=1}^{r} (\mu_i(z) - \mu_i(\widehat{z})) \{b_i + L_i \bar{y}\} \end{aligned} \tag{25}$$

$$\widetilde{y} = \sum_{i=1}^{r} \mu_i(z) C_i e + \sum_{i=1}^{r} (\mu_i(z) - \mu_i(\widehat{z})) C_i \widehat{x} + m, \tag{26}$$

where $\widetilde{y} \equiv \bar{y} - \widehat{y}$.

When different types of driving signals are applied in the masking mechanism, the conditions for recovering the message to be recovered are given in the following theorems. First, we consider the message embedded in the fuzzy driving signal, namely, *fuzzy signal masking*. Accordingly $b_i = b$, and L_i equals a given L in this case. The solution to design C_i and stability of the error dynamics are addressed as follows:

Theorem 5 *For the fuzzy signal masking structure, the error system described by (25) for CFS or DFS is asymptotically stabilized if the corresponding conditions in Thms.* 1 *or* 2 *are satisfied. Therefore,* $e \to 0$ *and* $\widetilde{y} \to m$ *as* $t \to \infty$.

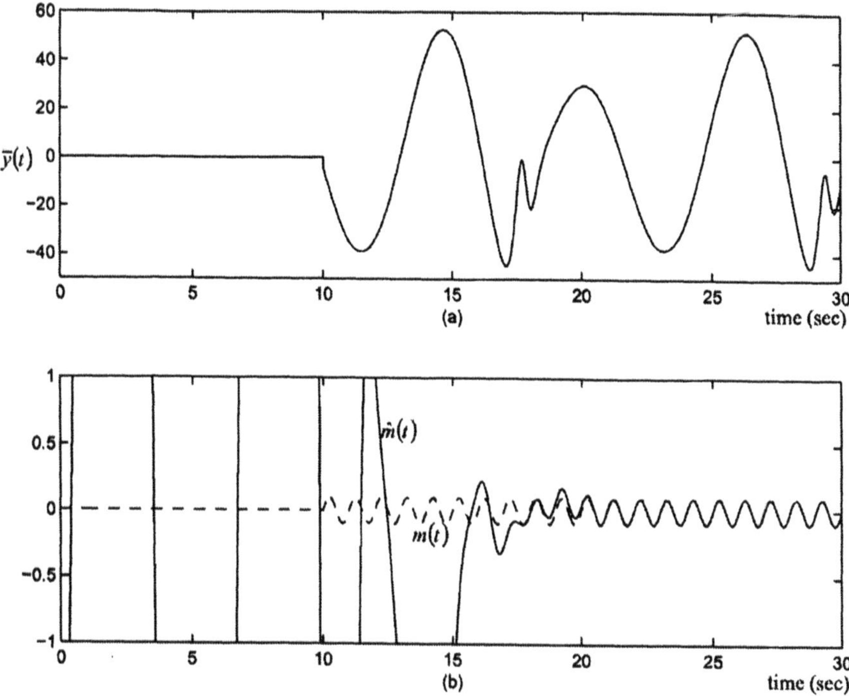

Figure 1: (a) Coupling signal and (b) original message and using Rossler's system with fuzzy signal masking activated at $t \geq 10$.

The proofs for CFS and DFS are the same as Thms. 1 and 2, respectively. It is noted that when e converges to zero as $t \to \infty$, then $\lim_{t\to\infty}(z - \widehat{z}) = 0$ in (26), and $\lim_{t\to\infty} \widetilde{y} = m$.

To enhance the convergence rate of recovering the message, the decay rate of errors is carefully considered. The following LMI design problems for CFS and DFS are performed according to Thm. 5.

Fuzzy Signal Masking Communication with Decay Rate—CFS:

$$
\begin{aligned}
&\min_{M_i,\, X} \ \varepsilon \\
&\max_{M_i,\, X} \ \alpha \\
&\text{subject to } X > 0,\ \varepsilon > 0,\ \alpha > 0 \\
&-A_i X - X A_i^{\mathrm{T}} + M_i^{\mathrm{T}} L^{\mathrm{T}} + L M_i - 2\alpha X > 0, \text{ for all } i, \\
&\qquad\qquad \begin{bmatrix} \varepsilon I & R_i^{\mathrm{T}} \\ R_i & I \end{bmatrix} > 0, \text{ for } 2 \leq i \leq r.
\end{aligned}
$$

Under this setting, (14) becomes $\dot{V}(e(t)) \leq -2\alpha V(e)$ which further implies that e exponentially converges to zero with decay rate α.

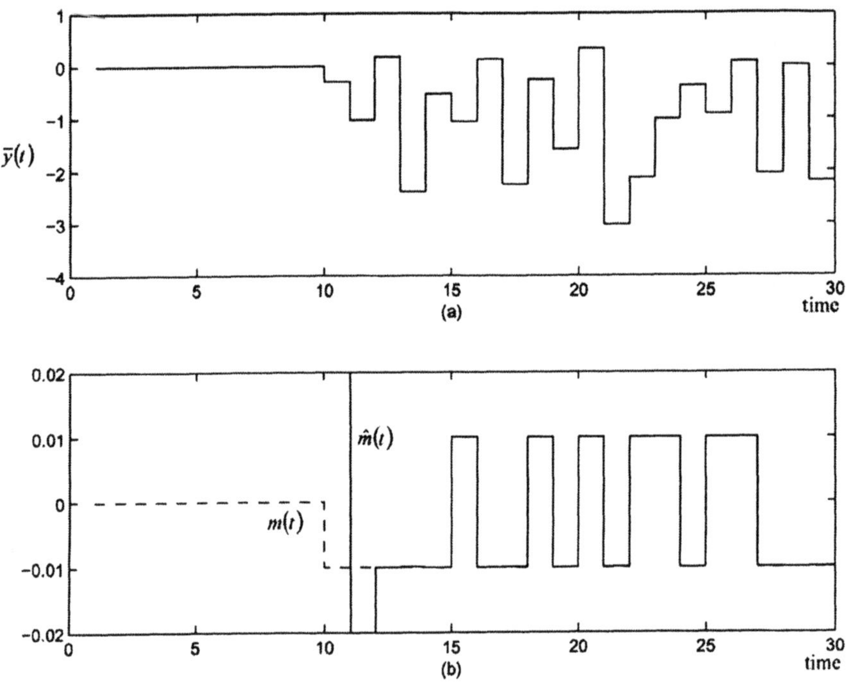

Figure 2: (a) Coupling signal and (b) original message and using Henon map with fuzzy signal masking activated at $t \geq 10$.

Fuzzy Signal Masking Communication with Decay Rate—DFS:

$$
\begin{aligned}
&\min_{M_i,\, X} \ \varepsilon \\
&\max_{M_i,\, X} \ \beta \\
&\text{subject to } X > 0,\ \varepsilon > 0,\ 0 < \beta < 1 \\
&\begin{bmatrix} \beta X & (A_i X - L M_i)^{\mathrm{T}} \\ A_i X - L M_i & X \end{bmatrix} > 0, \text{ for all } i, \\
&\begin{bmatrix} \varepsilon I & R_i^{\mathrm{T}} \\ R_i & I \end{bmatrix} > 0, \text{ for } 2 \leq i \leq r.
\end{aligned}
$$

Then Eq. (17) becomes $\Delta V(e) \leq -(1-\beta)V(e)$ with parameter β tuning the decay rate.

Example 1: Using the fuzzy signal masking structure, the secure communications employing Rössler's system and Hénon map are considered (system matrices and fuzzy sets are given in Table I). The message m is a sine wave and is considered to be low powered. The initial values of $x(0)$ are set different from those of $\widehat{x}(0)$ and the fuzzy response system is activated at $t \geq 10$

(second). The masked signal $\bar{y}$, original message m and recovered message $\hat{m}$ of corresponding systems are shown in Figs. 1~2.

Now we consider the message embedded in the crisp driving signal, namely, *crisp signal masking.* Accordingly, a common C is known in this case. The solution to design L_i and the stability of the error dynamics are given as follows:

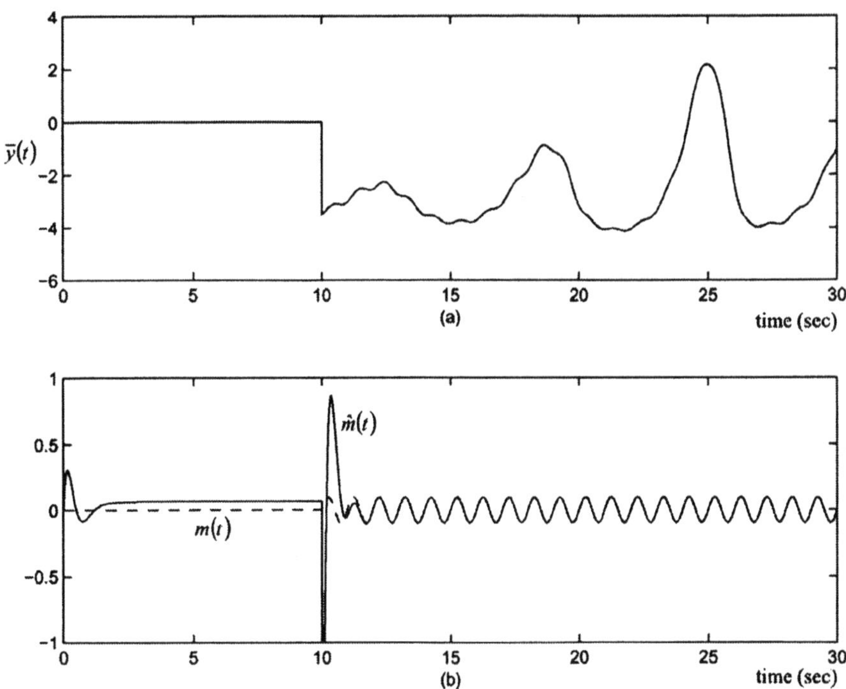

Figure 3: (a) Coupling signal and (b) original message and using transformed Rossler's system with crisp signal masking activated at $t \geq 10$.

Theorem 6 *Consider crisp signal masking structure, the extracted message of the communication system represented by* $\widetilde{y} = C\widetilde{x} + m$ *for CFS or DFS asymptotically converges to* m *as* $t \to \infty$ *if there exist the gains* L_i *such that the corresponding Thms. 3 or 4 are satisfied.*

As for the decay rate design for e in the previous fuzzy signal masking structure, it is possible to use an analogous approach for this crisp signal masking case. Note here that it is sufficient to emphasize the decay rate on only Ce, which in turn will lead to relaxed conditions as shown below.

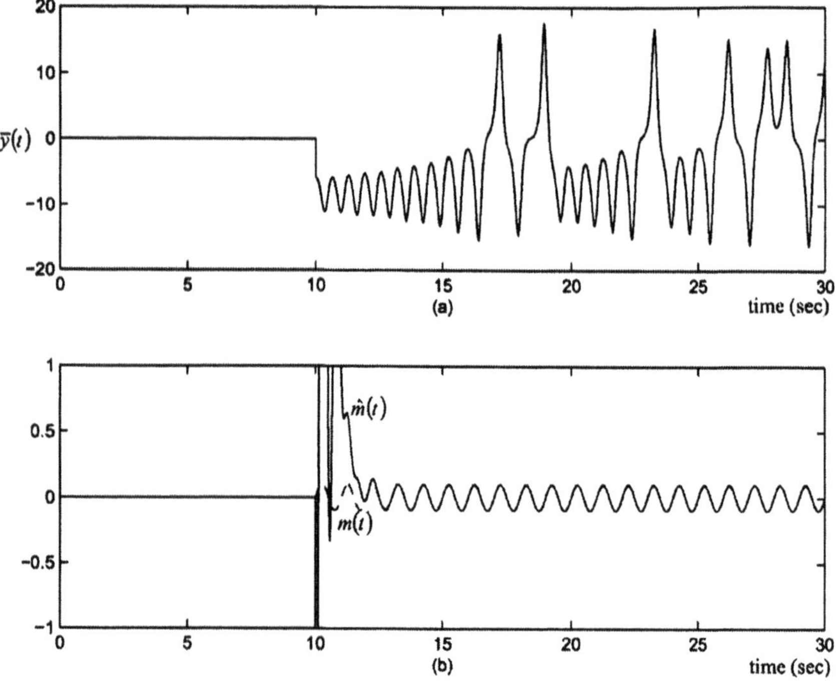

Figure 4: (a) Coupling signal and (b) original message and using Henon map with fuzzy signal masking activated at $t \geq 10$.

Crisp Signal Masking Communication with Decay Rate—CFS:

$$\begin{array}{l} \max\limits_{M_i,\ X} \ \alpha \\ \text{subject to } P > 0,\ \alpha > 0 \\ \qquad -A_i^{\mathrm{T}} P - P A_i + C^{\mathrm{T}} N_i^{\mathrm{T}} + N_i C - 2\alpha C^{\mathrm{T}} C > 0, \text{ for all } i. \end{array}$$

Consequently, (20) becomes $\dot{V}(e) \leq -2\alpha e^{\mathrm{T}} C^{\mathrm{T}} C e$ with parameter α tuning its performance.

Crisp Signal Masking Communication with Decay Rate—DFS:

$$\begin{array}{l} \max\limits_{M_i,\ X} \ \beta \\ \text{subject to } P > 0,\ 0 < \beta < 1 \\ \qquad \begin{bmatrix} P - \beta C^{\mathrm{T}} C & (P A_i - N_i C)^{\mathrm{T}} \\ P A_i - N_i C & P \end{bmatrix} > 0, \text{ for all } i. \end{array}$$

Hence, (22) becomes $\Delta V(e) \leq -\beta e^{\mathrm{T}} C^{\mathrm{T}} C e$ with parameter β tuning its performance.

Example 2: Using the crisp signal masking structure, the secure communications employing transformed Rössler's system with $y = x_3$ and Lorenz's equation with $y = x_1$ are considered (system matrices and fuzzy sets are given in Table I). The simulation conditions are set same as Example 1. The masked signal $\bar{y}$, original message m and recovered message $\hat{m}$ of corresponding systems are shown in Figs. 3~4.

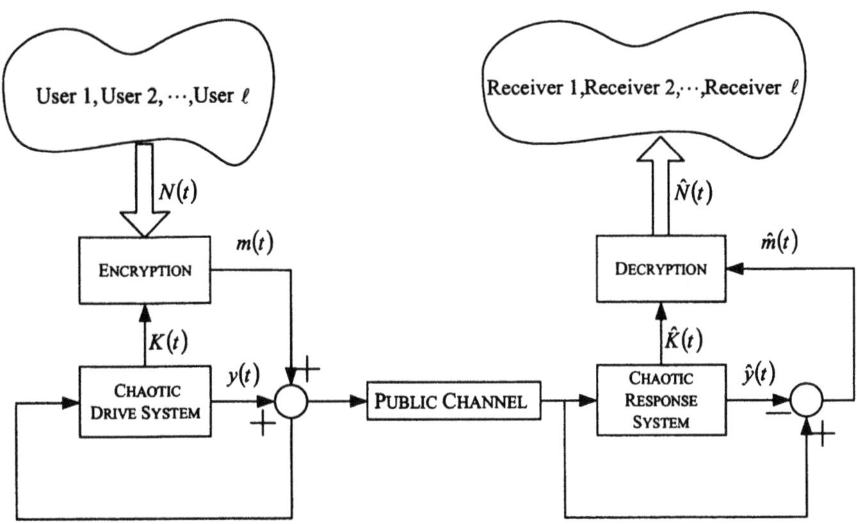

Figure 5: Block diagram of chaotic encryption methodology.

5 Chaotic Communication with Encryption

In traditional chaotic communications, the synchronization of drive-response system was exploited. This direct signal masking based approach often proposes setbacks in security, as stated in [21]. Therefore applying cryptosystem theory, the message encoded using a chaotic signal provides a higher level of security and multiuser capabilities. Now, we introduce some basic cryptosystem terminology.

The message N to be transmitted is the *plaintext*, which is encoded by using the *superincreasing sequence* S_i. The encoded plaintext results in the *ciphertext* $E(\cdot)$. The process of recovering the plaintext from the ciphertext is the decryption function $\widehat{E}(\cdot)$. Encryption and decryption process use the keys K and $\widehat{K}$, respectively. Here, the definition of a superincreasing sequence is given.

Definition — A real sequence $\{S_i\}_{i=1}^{\ell}$ is called a superincreasing sequence if the following is satisfied $S_j > \sum_{i=1}^{j-1} S_i$, $\ell \geq j > 1$ and all $S_i > 0$.

Note that in traditional superincreasing problems [22, 23], the sequence is a set of positive integers. The superincreasing sequence used here is modified where a set of positive real numbers is considered.

The proposed secure communication framework consists of three main components:

a) Chaotic model component: Consists of a T-S fuzzy model representing discrete-time chaotic systems.

b) Encrypting component: Consists of a superincreasing sequence $\{S_i\}$, $i = 1, \cdots, \ell$, where ℓ is the number of messages; and a plaintext $N = [n_1\ n_2 \cdots n_\ell]$ with $n_i \in \{0,\ 1\}$ for all i. The superincreasing sequence along with plaintext combines into a ciphertext as follows:

$$E(N(t), K(t), K(t-1), \ldots, K(t-\ell+1)) = S(t)\, N(t)^T \equiv E(t)$$

where $E(\cdot)$ is an encryption function that makes use of the superincreasing sequence $S(t) = [\ S_1(t) \quad S_2(t) \quad \cdots \quad S_\ell(t)\ ]$ formed by the key signal $K(t-i)$, $i = 0, \ldots, \ell - 1$. The sequence of the key, i.e. $K(t), K(t-1), \ldots, K(t-\ell+1)$, may be (i) output of the T-S fuzzy chaotic drive system; or (ii) any state in which the synchronization error approaches zero.

c) Decrypting component: Consists of a T-S fuzzy model discrete-time chaotic response system generating the same key as the encryption component. Achieving $\widehat{E}(\cdot) \to E(\cdot)$ and $\widehat{K}(\cdot) \to K(\cdot)$, the plaintext is obtained from decrypting the ciphertext $\widehat{E}(t)$ as follows:

$$\widehat{N}(t) = D(\widehat{K}(t), \widehat{E}(N(t), \widehat{K}(t), \widehat{K}(t-1), \ldots, \widehat{K}(t-\ell+1)))$$

where $D(\cdot)$ is an decryption function that makes use of the recovered key $\hat{K}(t-i)$, $i = 0, \ldots, \ell - 1$.

The structure above is depicted in a block diagram in Fig. 5. In details, we explain the algorithm for the chaotic cryptosystem.

Drive system—

1. Generate a superincreasing sequence $\{S_i(t)\}$, where $S_1(t) = |K(t)| + \tau$, $S_j(t) = \sum_{i=1}^{j-1} S_i(t) + |K(t-j+1)| + \tau\ \forall\, j = 2, \ldots, \ell$, and $\tau > 0$;
2. Form the encryption function $E(t) = S(t)\, N(t)^T$;
3. Modify encryption function $E(t)$ into $m(t) = (E - H(t)/2)/(\gamma H(t)/2)$ where $H = \sum_{i=1}^{\ell} S_i$ and γ is a scalar such that $m(t) \in (-0.01\ 0.01)$ is sufficiently small as to not destroy the chaotic characteristics of the masking signal;
4. Add $m(t)$ to the masking signal and send a scalar coupling signal to response system.

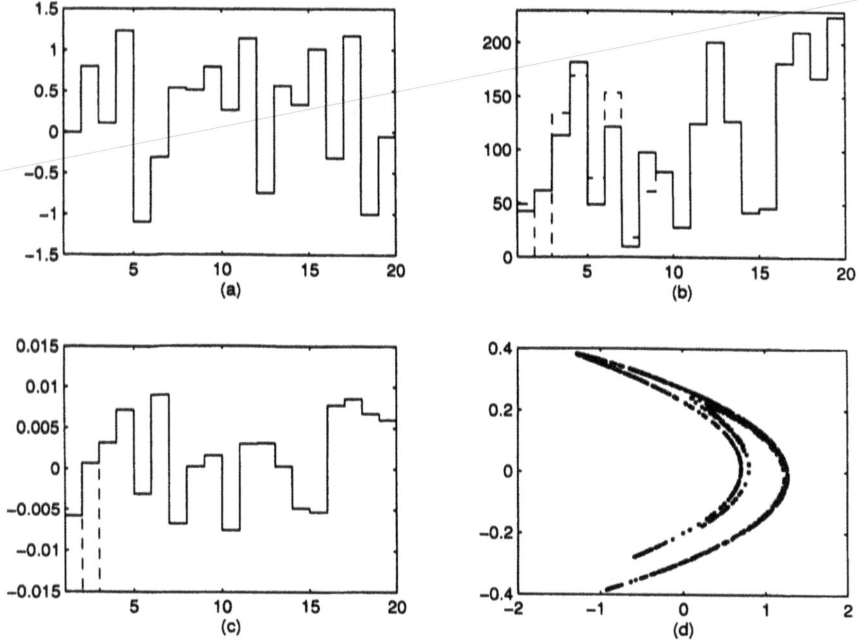

Figure 6: (a) Chaotic coupling signal; (b) encryption function $E(\cdot)$, $\hat{E}(\cdot)$(dotted line); (c) scaled ciphertext $m(t)$, $\hat{m}(t)$; and (d) phase portrait.

Response system—

1. Design a T-S fuzzy response system to recover the signal $\hat{m}(t)$ from synchronization;
2. Generate a superincreasing sequence $\{\widehat{S}_i(t)\}$, $i = 1, \cdots, \ell$ from the key signal obtained from synchronization, i.e., $\widehat{S}_1(t) = |\widehat{K}(t)| + \tau$, $\widehat{S}_j(t) = \sum_{i=1}^{j-1} \widehat{S}_i(t) + |\widehat{K}(t - j + 1)| + \tau$, $\forall\, j = 2, \ldots, \ell$, and $\tau > 0$;
3. Demodify $\hat{m}(t)$ to obtain $\widehat{E}(t) = \hat{m}(t)(\gamma \widehat{H}(t)/2) + \widehat{H}(t)/2$.
4. Decrypt the message using the superincreasing sequence from Step 3 of the response system design and the following algorithm:

$\widehat{V} = \widehat{E}$
For $i = \ell$ down to 1
Begin
If $\widehat{V} - \widehat{S}_i > -\varepsilon$
$\hat{n}_i = 1$
$\widehat{V} = \widehat{V} - \widehat{S}_i$
Else $\hat{n}_i = 0$
END

where $0 < \varepsilon < \tau$.

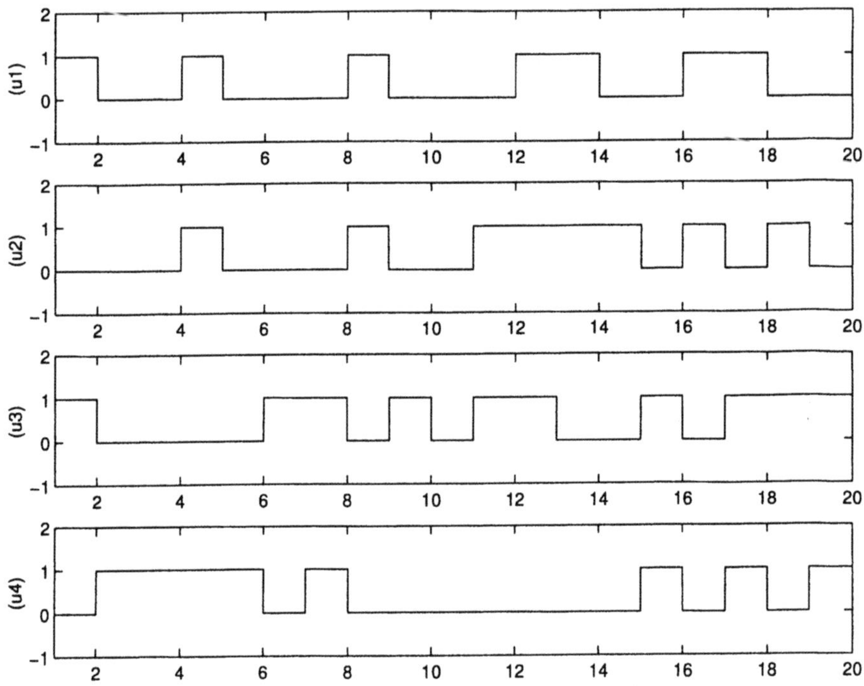

Figure 7: Plaintext transmitted my users 1 thru 4 (total 8 users).

Here notice that if $n_\ell(t) = 1$ (or $n_\ell(t) = 0$), then $E(t) \geq S_\ell(t)$ or $(E(t) \leq S_\ell(t) - \tau)$. After the transient time, the response system is synchronized to the drive system. This means that $\widehat{E}(t) = E(t)$ and therefore $\widehat{S}_\ell(t) = S_\ell(t)$. Finally $\hat{n}_\ell(t) = 1$ (or $\hat{n}_\ell(t) = 0$) from the above algorithm and the other plaintext $\hat{n}_{\ell-1}(t) \cdots \hat{n}_1(t)$ are recovered by the iteration loop.

Here we use the Henon map to illustrate the above chaotic cryptosystem design. According to the crisp signal masking structure (in this case $y = x_1$), the equivalent fuzzy dynamic model with ciphertext added in output can be constructed as:

$$\begin{aligned}
&Transmitter\ Rule\ i : \text{IF } y(t) \text{ is } F_i \text{ THEN} \\
&x(t+1) = A_i x(t) + b_i + L_i m(t) \\
&\bar{y}(t) = Cx(t) + m(t),\ i = 1, 2,
\end{aligned}$$

where $x(t) = [\ x_1(t) \quad x_2(t)\ ]^T$; the fuzzy sets and system matrices can be found in Table I; and $C = [\ 1 \quad 0\ ]$. The response system is designed as follows:

$$\begin{aligned}
&Response\ Rule\ i : \text{IF } \hat{y}(t) \text{ is } F_i \text{ THEN} \\
&\hat{x}(t+1) = A_i \hat{x}(t) + b_i + L_i (\bar{y}(t) - \hat{y}(t))
\end{aligned}$$

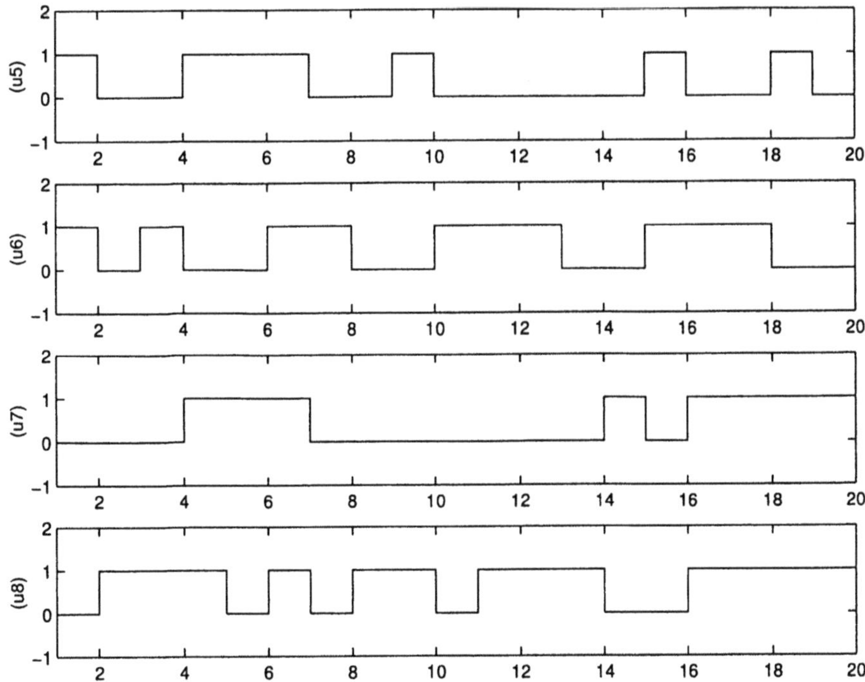

Figure 8: error between original message and recovered message of user 1 thru 4.

$$\hat{y}(t) = C\hat{x}(t)$$

where $\hat{x}(t)$ is state estimate; L_i are response gains; and $\hat{y}(t)$ is the output estimate. Define the output error $\tilde{y}(t) = \bar{y}(t) - \hat{y}(t)$, then we have $\tilde{y}(t) = Ce(t) + m(t)$. If synchronization is achieved, the estimate of $m(t)$ is thus $\hat{m}(t) \equiv \tilde{y}(t)$. Then $\hat{m}(t) \to m(t)$ and $\widehat{K}(t) \to K(t)$ after the transient time. In application of the ciphertext masked by output, we have a variety of choices for the secure key $K(\cdot)$ other from the coupling signal $\bar{y}(t)$. For example, one of the internal states $x_1(t), \ldots, x_n(t)$ can be used. Since the internal states are not made public, this in turn achieves higher security.

To verify the theoretical results, we carry out numerical simulations on the chaotic cryptosystem. The Henon map discrete-time chaotic system is used whereas the state $x_1(t)$ is the coupling channel which generates the secure key $K(t)$. The cyphertext is added at the output of the drive system $y(t)$. The response system structure is designed according to the T-S fuzzy model response system in the previous section whereas the decoding algorithm mentioned in the beginning of this section is implemented. Parameters of the T-S fuzzy decay rate and cryptosystem are set as $\beta = 0.1$ and $\gamma = 100$, $\tau = 0.01$, respectively. The response gains solved by LMIs are

$L_1 = [-2.1384\ 5.6608]^{\mathrm{T}}$; $L_2 = [2.1384\ 5.6608]^{\mathrm{T}}$. The number of multiusers are set to 8 and the plaintexts are randomly binary [0 1]. Therefore in Fig. 6a-6d, we illustrate the chaotic coupling signal; encrypting function $E(\cdot)$, $\widehat{E}(\cdot)$ (dotted line); scaled ciphertext $m(\cdot)$, $\hat{m}(\cdot)$ (dotted line); and chaotic phase portrait, respectively. In Figs. 7 and 8, $u1 \sim u4$ are the first four user plaintext transmitted (total of eight users) and error between the recovered plaintext and original plaintext of users $u1 \sim u4$.

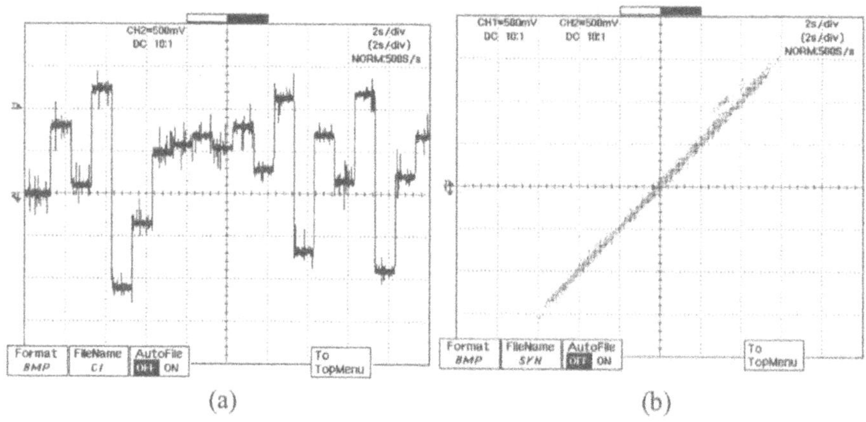

Figure 9: Oscilloscope images of (a) chaotic coupling and (b) synchronization in $x_1 - \hat{x}_1$ plane.

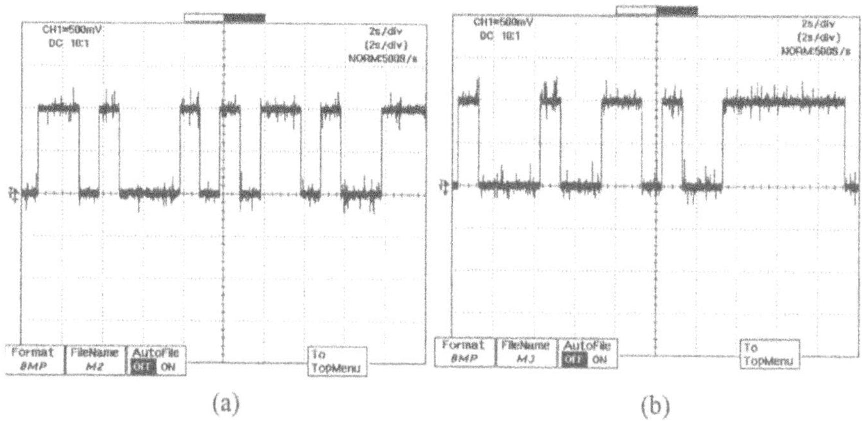

Figure 10: Oscilloscope images of (a)user 2 plaintext and (b) user 3 plaintext.

Now we carry out DSP based experiments using the same parameters as numerical simulations. Here, the hardware used is the DSpace DS1102 single board system which is based on Texas Instruments TMS320C31 DSP.

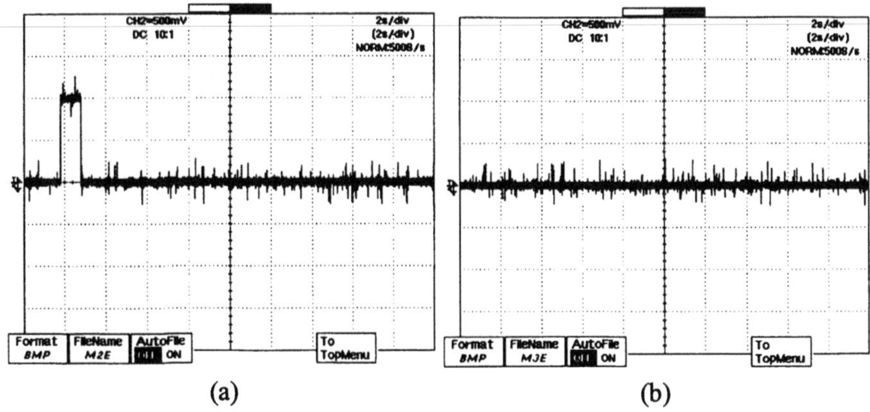

(a) (b)

Figure 11: Oscilloscope images of (a) error between user 2 original plaintext and recovered plaintext and (b) error between user 3 original plaintext and recovered plaintext.

Note that a minimum bit rate of the DSP is necessary to avoid quantization errors. In other words, we must have at least a bit rate of 12 to ensure that the signal generated by the DSP maintains chaotic characteristics. This is an important feature in order to achieve synchronization and generate the superincreasing sequence. To increase the performance, the step size of t may be set to a small value. But for simplicity, the step size of t is set as 1sec. The coupling signal is externally connected from the output of the drive system through D/A channel to the A/D channel leading to the response system. The chaotic system structure and parameters are same as the numerical simulations above. In Fig. 9a and 9b, we illustrate the chaotic coupling signal. and the synchronization of $x_1(t)$ and $\hat{x}_1(t)$. For simplicity and the maximum channels allowed on this card, we only show in Fig. 10a, 10b and Fig. 11a, 11b the messages and error of recovered message for the second and third users, respectively. From the oscilloscope images after the transient responses, the plaintext is recovered exactly. Note here that the recovering of ciphertext and decoding into plaintext is implemented in a real-time sense. Therefore comparing the numerical and experimental results, we are able to conclude the consistency of the theoretical derivation.

6 Conclusions

We have presented a general framework of fuzzy model-based chaotic synchronization and communications. Using fuzzy or crisp coupling scalar signals, synchronization between the transmitter and response system is achieved once sufficient stability conditions are satisfied. Extending the properties of synchronization, chaotic communications have been proposed in the sense of signal masking and encryption. The multiobjectives mentioned above has been approached in a systematic manner whereas the final results are expressed by solving LMIs. In addition, both numerical and DSP-based experiments on several well-known continuous-time and discrete-time chaotic systems, have shown that zero synchronization error is achieved and message recovered in the chaotic communication application.

Acknowledgment

This work was supported by the National Science Council, R.O.C., under Grant NSC-89-2213-E033-006.

References

1. Takagi T, Sugeno M (1985) Fuzzy identification of systems and its applications to modeling and control. IEEE Trans Syst Man Cybern 15: 116-132
2. Wang HO, Tanaka K, GriffinMF (1996) An approach to fuzzy control of nonlinear systems: stability and design issues.IEEE Trans Fuzzy Systems 4: 14-23
3. Tanaka K, Ikeda T, Wang HO (1998) A unified approach to controlling chaos via an LMI-based fuzzy control system design.IEEE Trans Circuits Syst 45: 1021-1040
4. Tanaka K, Ikeda T, Wang HO (1998) Fuzzy regulators and fuzzy observers: relaxed stability conditions and LMI-based designs. IEEE Trans Fuzzy Systems 6: 250–265
5. Boyd S, Ghaoui LE, Feron E, Balakrishnan V (1994) Linear matrix inequalities in system and control theory. SIAM, Philadelphia
6. Pecora LM, Carroll TL (1990) Synchronization in chaotic systems. Phys Rev Lett 38: 821-824
7. Carroll TL, Pecora LM (1991) Synchronizing chaotic circuits. IEEE Trans Circuits Syst 38: 453-456
8. Chen C, Dong X (1998) From chaos to order methodologies, perspectives and applications. World Scientific, Singapore
9. Lakshmanan M, Murali K (1996) Chaos in nonlinear oscillators: controlling and synchronization. World Scientific, Singapore
10. Morgül O, Solak E (1996) Observer based synchronization of chaotic systems. Phys Rev E 54: 4803-4811
11. Huijberts HJC, Nijmeijer H, Pogromsky AYu(2000) An observer point of view on synchronization of discrete-time systems. Proc ISCAS 2000: 491-494.
12. Grassi G, Mascolo S (1999) Synchronizing hyperchaotic systems by observer design. IEEE Trans Circuits Syst II 46: 478-483

13. Cuomo KM, Oppenheim AV, Strogatz SH (1993) Synchronization of Lorenz-based chaotic circuits with applicatons to communications. IEEE Trans Circuits Syst II 40: 626-633
14. Kocarev LJ, Halle KD, Eckert K, Chua LO, Parlitz U (1992) Experimental demonstration of secure communications via chaotic synchronization. Int J Bifurcation Chaos 2: 709-713
15. Liao TL, Huang NS (1999) An observer-based approach for chaotic synchronization with applications to secure communications. IEEE Trans Circuits Syst I 46: 1144-1150
16. Wu CW, Chua LO (1993) A simple way to synchronize chaotic systems with applications to secure communication systems. Int J Bifurcation Chaos 3: 1619-1627
17. Halle KS, Wu CW, Itoh M, Chua LO (1993) Spread spectrum communication through modulation of chaos. Int J Bifurcation Chaos 3: 469-477
18. Lian KY, Chiang TS, Liu P (2000) Discrete-time chaotic systems: applications in secure communications. Int J Bifurcation Chaos 10: 2193-2206
19. Lian KY, Chiang TS, Liu P, Chiu CS (2001) LMI-based fuzzy chaotic synchronization and communication. IEEE Trans Fuzzy Systems 9: 539-553
20. Lian KY, Chiang TS, Chiu CS, Liu P (2001) Synthesis of fuzzy model-based design to synchronization and secure communication for chaotic systems. IEEE Trans Syst Man Cybern Part B 31: 66-83
21. Short K (1994) Steps toward unmasking secure communications. Int J Bifurcation Chaos 4: 959-977
22. Garey MR, Johnson DS (1976) Computers and intractability: a guide to the theory of NP-completeness. WH Freeman & Co, San Francisco
23. Denning DER (1982) Cryptography and data security. Addison Wesley, New York

Intelligent Channel Assignment Schemes for Hierarchical Cellular Systems

Kuen-Rong Lo[1,2], Chung-Ju Chang[2], and Yih-Shen Chen[2]

[1] Wireless Communication Technology Department, Telecommunication Laboratories Chunghwa Telecom Co., Ltd., Taoyuan, Taiwan
[2] Department of Communication Engineering, National Chiao Tung University, Hsinchu, Taiwan

Abstract. In this chapter, two intelligent channel assignment schemes using soft computing techniques: *fuzzy channel allocation controller* (FCAC) and *neural fuzzy resource manager* (NFRM), are studied. The FCAC employs the concept of fuzzy logic control system, and it mainly contains a fuzzy channel allocation processor (FCAP) which is designed to be in a two-layer architecture: a fuzzy admission threshold estimator in the first layer and a fuzzy channel allocator in the second layer. In the fuzzy admission threshold estimator, *Sugeno's position-gradient type reasoning method* is applied to adjust the admission threshold adaptively. In the fuzzy channel allocator, the *max-min* inference method is used to balance the utilization between macrocells and microcells to achieve a higher overall system utilization. On the other hand, The NFRM mainly contains a *neural fuzzy channel allocation processor* (NFCAP) which is in a two-layer architecture: a *fuzzy cell selector* (FCS) in the first layer, and a *neural fuzzy call-admission and rate controller* (NFCRC) in the second layer. The NFCRC adopts a structure of five-layer neural fuzzy controller. Based on the knowledge obtained from CCA and FCAC, the structure of the NFCRC is constructed. In order to initialize the parameters of the membership functions, Kohonen's feature-maps algorithm is applied for statistical clustering. Also, a reinforcement learning algorithm is adopted for the NFCRC and the handoff failure probability is used as a reinforcement signal. The performance comparison between conventional schemes and the two intelligent schemes are performed, from the aspects of the system utilization, the new-call blocking probability, the handoff failure probability, the forced termination probability, and the handoff rate. Simulation results show that the intelligent schemes, FCAC and NFRM, outperform the conventional schemes such as overflow channel allocation (OCA) and combined channel allocation (CCA).

Keywords: channel assignment, fuzzy logic system, neural fuzzy network, and hierarchical cellular system.

1 Introduction

The wireless communication services have been gained great success in global telecommunication markets. To accommodate the fast growing population, it is essential to re-configure the existing cellular system into a hierarchical structure for enhancing system capacity and improving coverage. The hierarchical cellular system consists of microcells and macrocells. The overlaid microcells is for high-teletraffic area while overlaying macrocells is for low-teletraffic region [1], [2].

Generally speaking, the channel assignment scheme for cellular networks is to appropriately assign channels to mobile terminals such that a tradeoff between new call blocking probability and handoff call blocking probability can be well achieved. For the hierarchical cellular systems, the problem is more complicated because the available channels channel pools are in either macrocells or microcells. The channel assignment schemes normally assign the microcell channels to high mobility terminals and the macrocell channels to low mobility terminals. Once no channel is available in the macrocells (microcells), the channel assignment schemes could turn to the channels in the microcells (macrocells) instead. However, to efficiently allocate the system resource, a sophisticated channel assignment scheme for hierarchical cellular systems should also consider the mobility behavior and service request of terminals.

Rappaport and Hu proposed an *overflow channel allocation* (OCA) scheme that allows a new or handoff call which has no channel available in the overlaid microcell to overflow to use free channel in the overlaying macrocell [3] [4]. We also proposed a *combined channel allocation* (CCA) mechanism for hierarchical cellular systems in [5]. It combines overflow, underflow, and reversible schemes, where new or handoff calls having no idle channel to use in the overlaid microcell can *overflow* to use free channels in the overlaying macrocell, handoff calls from neighboring macrocell can *underflow* to use free channels in the overlaid microcell, and handoff attempts from macrocell-only region to a microcell in the same macrocell can be *reversed* to use free channels in the microcell. Both OCA and CCA schemes can reduce the blocking probability of new calls and the forced termination probability of handoff calls.

To guarantee QoS requirement for handoffs, these conventional channel assignment schemes employ the *guard channel* and *handoff queue* methods [1], [6]. Unfortunately, these protection methods are hard to deal with multimedia wireless services. High traffic burstiness and high mobility are the two main traffic characteristics of the wireless multimedia services in future cellular systems [7], [8]. With such protection mechanism, these conventional schemes can guarantee the QoS requirement of the handoff failure probability, but the channel utilization would degrade. Due to the fundamental limitations by unpredictable statistical fluctuations within new and handoff call arrivals, these channel assignment schemes are ineffective. Also, the channel assignment for mobile multimedia services is difficult, if not impossible, to drive an accurate mathematical model to obtain the solution.

On the other hand, soft computing techniques have been applied successfully to many large-scale telecommunication control problem [9]-[14]. The soft computing techniques are aimed at exploiting the tolerance for imprecision, uncertainty, and partial truth to achieve tractability, robustness, and low solution cost [16]. Though there are many techniques in this field, *fuzzy logic control* and *neural network* techniques are employed in this chapter.

In the past decade, fuzzy logic control systems have replaced conventional technologies in many scientific applications and engineering systems, especially in control systems and pattern recognition. They can provide decision-support and expert systems with powerful reasoning capabilities bound by a minimum of rules. The major feature of the fuzzy logic is its ability to express the amount of ambiguity in human thinking and subjectivity in a comparatively undistorted manner. When a mathematical model of the process does not exist, it is appropriate to use fuzzy logic [15]. On the other hand, neural networks are a new generation of information processing systems that are constructed to utilize some of the organizational principles which characterize the human brain. They are able to learn arbitrary nonlinear input/output mapping directly from training data; they can sensibly interpolate input patterns that are new to the network; and they can automatically adjust their connection weights to optimize their behaviors as controllers, predictors, pattern recognizers, decision makers, etc. Neural networks are good at tasks such as pattern matching and classification, function approximation, optimization, vector quantization, and data clustering [15]. Furthermore, neural fuzzy network, combining fuzzy system and neural network, possesses the capability of linguistic modelling from fuzzy system and self-tuning from neural network. Both fuzzy systems and neural networks are numerical model-free estimators and dynamical systems; they are the intelligent techniques that can improve systems working in uncertain and nonstationary environments.

We have also proposed two intelligent channel assignment schemes with soft computing techniques: *fuzzy channel allocation controller* (FCAC) in [12] and *neural fuzzy resource manager* (NFRM) in [14]. The FCAC employs the concept of fuzzy logic control system. The NFRM improves furtherly the system performance with the synergy of fuzzy logic system and neural network, based on the domain knowledge obtained from FCAC.

The FCAC consists of a fuzzy channel allocation processor (FCAP), a *resource estimator*, a *performance evaluator*, and a *base-station interface modules*. It estimates available resources in macrocell and microcells, evaluates system performance, and determines whether and how to allocate resources to a call, based upon the call's QoS requirement, resource availability, and mobility. The FCAP is a two-layer fuzzy logic controller that consists of the *fuzzy admission threshold estimator* in the first layer and the *fuzzy channel allocator* in the second layer. The fuzzy admission threshold estimator employs *Sugeno's position-gradient type reasoning method* to adjust adaptively the admission threshold for the fuzzy channel allocator. The fuzzy channel allocator aims to balance the utilization between macrocell and microcells for a higher overall system utilization. The domain knowledge is based upon CCA mechanism proposed in [5].

The NFRM utilizes the learning capability of the neural network to reduce the decision error of these conventional channel assignment schemes resulted from modelling, approximation, and unpredictable statistical fluctuations of the system. It also employs the control rule structure of fuzzy logic, which absorbs benefits of those conventional channel assignment schemes, to provide robust operation and to prevent operating errors due to the learning of incorrect training data. The *neural fuzzy channel allocation processor* (NFCAP) is the core of the NFRM. NFCAP is a two-layer neural fuzzy logic controller that consists of a *fuzzy cell selector* (FCS) in the first layer and a *neural fuzzy call-admission and rate controller* (NFCRC) in the second layer. The FCS takes handoff failure probability, resource availability, and user mobility as input linguistic variables, and applies the *max-min* interference method to determine which cell, macrocell or microcell, to serve the call request; FCS intends to enhance the channel utilization by balancing utilization between macrocell and microcells. The NFCRC takes the handoff failure probability and the resource availability of the selected cell as input variables; NFCRC intends to guarantee the QoS and provides an appropriate rate for users.

The rest of this chapter is organized as follows. In section 2, the functional blocks of FCAC and the design of FCAP are described. Section 3 presents the design of NFRM. Also, the architecture of the neural fuzzy controller and its reinforcement learning algorithm are presented. Section 4 shows simulation results and discussions of the intelligent channel assignment schemes for hierarchical cellular systems. Finally concluding remarks are given in Section 5.

2 Fuzzy Channel Allocation Controller (FCAC)

2.1 System Architecture

Fig. 1 shows the functional block diagram of our proposed fuzzy channel allocation controller (FCAC) for hierarchical cellular systems [12]. The hierarchical cellular system consists of a large geographical region tessellated by cells, referred to as macrocells, and each of which overlays several microcells. The overlaying macrocell is denoted by cell 0 and its overlaid microcells are denoted by cell $1, \cdots, N$. For cell i, a number of channels C_i is allocated, $0 \leq i \leq N$. FCAC contains functional blocks such as *base-station interface module* (BIM), *performance evaluator*, *resource estimator*, and *fuzzy channel allocation processor* (FCAP). It is installed in either base station controller (BSC) or mobile switching center (MSC). Note that for simplicity, FCAC is drawn to do the channel allocation for one macrocell only.

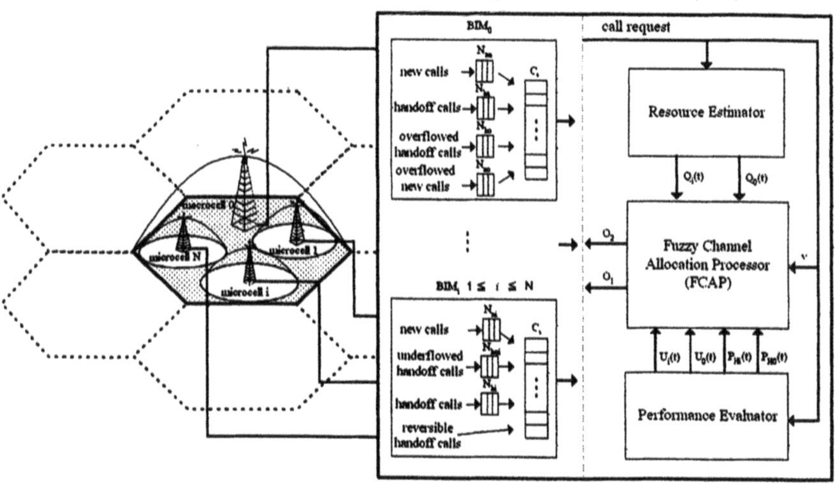

Fig. 1. The fuzzy channel allocation controller for hierarchical cellular systems

BIM is to interface with the base station of macrocell or microcell. It provides complete-partitioning buffers for queueing new and handoff calls which are originated in the corresponding cell and temporarily have no free channel to use. Whenever BIM_i receives a call request, $0 \leq i \leq N$, it sends the necessary service profile to the resource estimator, the performance evaluator, and the FCAP.

The resource estimator calculates the available resources in macrocell 0 and microcell i when it receives service profile of the call from BIM_i at the time instant t, which are denoted by $Q_0(t)$ and $Q_i(t)$, respectively.

Performance evaluator is to calculate the channel utilization and the handoff failure probability. The channel utilization of macrocell (microcell i) at time t, is denoted by $U_0(t)$ ($U_i(t)$), while the handoff failure probability in macrocell (microcells) at time t, is denoted by $P_{H0}(t)$ ($P_{Hi}(t)$). These four performance measures are fed to the FCAP.

FCAP performs the channel allocation using fuzzy logic control to attain high channel utilization and keep the QoS requirement guaranteed. Here, a threshold is designed for channel allocation, and it can be adaptively adjusted to cope with the input traffic fluctuation.

2.2 The Design of FCAP

As Fig. 2 shows, FCAP mainly consists of a fuzzy admission threshold estimator and a fuzzy channel allocator.

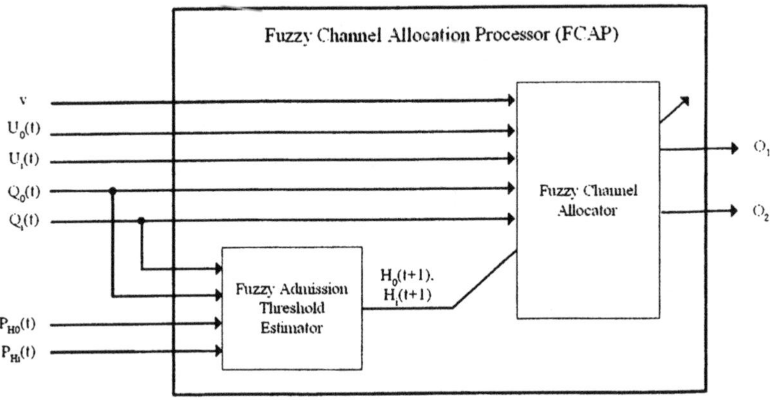

Fig. 2. The block diagram of fuzzy channel allocation processor

Fuzzy Admission Threshold Estimator The fuzzy admission threshold estimator is to determine adaptively the decision threshold for the fuzzy channel allocator so that the QoS requirement can be guaranteed and a high channel utilization can be achieved. Here the system QoS requirement is defined as the maximum handoff failure probability, P_H^*. The admission thresholds for macrocell 0 and microcell i during time period $(t, t+1]$, denoted by $H_0(t+1)$ and $H_i(t+1)$, are sent to the fuzzy channel allocator if a new call occurs in macrocell 0 or microcell i at time t. Note that "t+1" denotes the time instant the next new call arrives.

For the fuzzy admission threshold estimator, $P_{H0}(t)$ ($P_{Hi}(t)$) and $Q_0(t)$ ($Q_i(t)$) are chosen as input linguistic variables to determine $H_0(t+1)$ ($H_i(t+1)$) for macrocell 0 (microcell i). Since the fuzzy rule set will be commonly used by macrocell 0 and microcell i, hereafter in this subsection we simply use $H(t+1)$, $P_H(t)$ and $Q(t)$, instead of $H_0(t+1)$ or $H_i(t+1)$, $P_{H0}(t)$ or $P_{Hi}(t)$, and $Q_0(t)$ or $Q_i(t)$.

Term sets for input variables are designed as $T(P_H(t))$={*Low*, *High*}= {L, H}, and $T(Q(t))$={*Enough*, *Not Enough*}={E, NE}. And a trapezoidal function $g(x; x_0, x_1, a_0, a_1)$ is chosen to be their membership function, which is given by

$$g(x; x_0, x_1, a_0, a_1) = \begin{cases} \frac{x-x_0}{a_0} + 1, & \text{for } x_0 - a_0 < x \leq x_0; \\ 1, & \text{for } x_0 < x \leq x_1; \\ \frac{x_1-x}{a_1} + 1, & \text{for } x_1 < x \leq x_1 + a_1; \\ 0, & \text{otherwise}, \end{cases} \quad (1)$$

where $x_0(x_1)$ is the left (right) edge of the trapezoidal function; a_0 (a_1) is the left (right) width of the trapezoidal function.

Denote $\mu_L(P_H(t))$ and $\mu_H(P_H(t))$ to be the membership functions for L and H in $T(P_H(t))$, respectively, and define $\mu_L(P_H(t))$ and $\mu_H(P_H(t))$ as

$$\mu_L(P_H(t)) = g(P_H(t); 0, L_e, 0, L_w), \tag{2}$$

$$\mu_H(P_H(t)) = g(P_H(t); H_e, 1, H_w, 0). \tag{3}$$

H_e would be set to be P_H^*, L_w and H_w could be the safety margin provided to tolerate the dynamic behavior of handoff failure probability and guarantee the QoS requirement, and the edge L_e is $L_e = H_e - L_w$.

Similarly, denote $\mu_E(Q(t))$ and $\mu_{NE}(Q(t))$ to be the membership functions for E and NE in $T(Q(t))$, respectively, and define $\mu_E(Q(t))$ and $\mu_{NE}(Q(t))$ as

$$\mu_E(Q(t)) = g(Q(t); E_e, R_e, E_w, 0), \tag{4}$$

$$\mu_{NE}(Q(t)) = g(Q(t); 0, R_n, 0, NE_w). \tag{5}$$

The maximum possible "enough" value of available resource R_e would be the sum of buffer size and allocation channels, E_e would be to be a safety margin of available resource, R_n would be set to be a fraction of available resource, and E_w and NE_w are provided to tolerate the change of traffic.

Based on chosen input variables and their terms, the fuzzy admission threshold estimator has $M = 4$ fuzzy IF-THEN rules. The fuzzy rules have the following form:

Rule m: IF $P_H(t)$ is X_{1m} and $Q(t)$ is X_{2m}, THEN $h_m(t+1)$,

where X_{im} is the term of the i-th linguistic variable used in rule m, and $h_m(t+1)$ is the output function of rule m for the time period $(t, t+1]$, $1 \le m \le M$.

Here, *Sugeno's position-gradient type reasoning method* [17] is applied to derive $h_m(t+1)$ effectively and then to obtain the admission threshold $H(t+1)$. According to Sugeno's method, $h_m(t+1)$ can be expressed as

$$h_m(t+1) = \beta \cdot H(t) + a_m(t+1), \tag{6}$$

where β, $0 < \beta < 1$, is a forgetting constant that can maintain the estimator stability, $H(t)$ is the admission threshold during last time period $(t-1, t]$, and $a_m(t+1)$ is an adjustment parameter for $h_m(t+1)$. Then $H(t+1)$ is given by

$$H(t+1) = \sum_{m=1}^{M} h_m(t+1)w_m(t+1) = \beta \cdot H(t) + \sum_{m=1}^{M} a_m(t+1)w_m(t+1), \tag{7}$$

where $w_m(t+1)$ is the weighting factor for the output variable of rule m, defined as $w_m(t+1) = \mu_{X_{1m}}(P_H(t)) \cdot \mu_{X_{2m}}(Q(t))$. Since membership functions are set to be symmetrical, $\sum_{m=1}^{M} w_m(t+1) = 1$.

The adjustment parameter $a_m(t+1)$ is obtained by the gradient decent method, where an error function at time t is defined as

$$E(t) = \frac{1}{2}(P_H(t) - P_H^*)^2. \tag{8}$$

Note that $E(t)$ is given in the sense that $P_H(t)$ needs to be controlled around P_H^*. Then $a_m(t+1)$ is given by

$$a_m(t+1) = a_m(t) - \eta \cdot \frac{\partial E(t)}{\partial a_m(t)}, \tag{9}$$

$$= a_m(t) - \eta \cdot (P_H(t) - P_H^*) \cdot \frac{\partial P_H(t)}{\partial a_m(t)}.$$

where η is an adaptation gain which must be properly chosen.

Because the admission threshold $H(t+1)$ could be regarded as an entry barrier to regulate new calls coming to the system, and the change of $P_H(t+1)$ during $(t, t+1]$, denoted by $\Delta P_H(t+1)$, would be varied in accordance with $H(t+1)$ so that $P_H(t+1)$ can be kept at around P_H^* to fulfill QoS requirement, $\Delta P_H(t+1)$ is approximated heuristically as a first-order polynomial of $H(t+1)$.

Then $P_H(t+1)$ can be expressed as

$$P_H(t+1) = P_H(t) + \Delta P_H(t+1) \approx P_H(t) - \sigma \cdot H(t+1) + c, \tag{10}$$

where σ is an experience value and c is a constant value. For example, $\Delta P_H(t+1) \approx -0.2\% \cdot H(t+1) + 0.1\%$ means that $\Delta P_H(t+1)$ in the range of $[0.1\%, \ -0.1\%]$ is inversely varied with respect to $H(t+1)$ in the range of $[0, \ 1]$. Since $\Delta P_H(t+1)$ is designed to change gradually at a rate of $0.1 \cdot P_H^*$, the value of σ could be chosen to be 0.002. Also, in order to attain a better adjustment of $H(t+1)$ to fulfill QoS requirement in acute traffic fluctuation, it is possible to change σ dynamically according to the number of handoff calls during $(t, t+1]$.

Fuzzy Channel Allocator Five input linguistic variables are chosen for fuzzy channel allocator: the channel utilization in macrocell 0 ($U_0(t)$), the channel utilization in microcell i ($U_i(t)$), the available resource in macrocell 0 ($Q_0(t)$), the available resource in microcell i ($Q_i(t)$), and the mobile speed (v). $U_0(t)$ and $U_i(t)$ can show the traffic load intensity and the load balancing among cells; $Q_0(t)$ and $Q_i(t)$ can implicate the remaining capacity; and v can show the handoff rate. The term set for both $U_0(t)$ and $U_i(t)$ is defined as $T(U_0(t)) = T(U_i(t)) = \{Low, \ High\} = \{L, \ H\}$, the term set for both $Q_0(t)$ and $Q_i(t)$ is $T(Q_0(t)) = T(Q_i(t)) = \{Enough, \ Not \ Enough\} = \{E, \ NE\}$, and the term set for v is $T(v) = \{Slow, \ Fast\} = \{S, \ F\}$. Let $\mu_L(U_0(t))$ ($\mu_L(U_i(t))$) and $\mu_H(U_0(t))$ ($\mu_H(U_i(t))$) denote the membership functions of terms L and H in $T(U_0(t))$ $(T(U_i(t)))$, respectively, and let $\mu_L(U_0(t))$, $\mu_H(U_0(t))$, $\mu_L(U_i(t))$, and $\mu_H(U_i(t))$ be

$$\mu_L(U_0(t)) = g(U_0(t); 0, L_e, 0, L_w), \tag{11}$$

$$\mu_H(U_0(t)) = g(U_0(t); 1, 1, H_w, 0), \tag{12}$$

$$\mu_L(U_i(t)) = g(U_i(t); 0, L_e, 0, L_w), \tag{13}$$

$$\mu_H(U_i(t)) = g(U_i(t); 1, 1, H_w, 0). \tag{14}$$

Here L_e would be a fraction of channel utilization, and L_w and H_w would be a change rate of channel utilization provided to tolerate the dynamic behavior of $U_0(t)$ and $U_i(t)$, respectively. The membership functions for terms of E and NE of $Q_0(t)$ and $Q_i(t)$ have the same definition as in Eq. (4) and Eq. (5). The membership functions for terms S and F in v, denoted by $\mu_S(v)$ and $\mu_F(v)$, are given by

$$\mu_S(v) = g(v; 0, S_e, 0, S_w) \tag{15}$$

$$\mu_F(v) = g(v; F_e, F_h, F_w, 0), \tag{16}$$

where S_e (F_e) would be a fraction of slow (fast) speed of mobile user, S_w (F_w) is provided to tolerate the change of slow (fast) speed, and F_h would be the fastest speed.

There are two output variables, O_1 and O_2, in the fuzzy channel allocator. The output variable O_1 represents whether the call is accepted or rejected, and O_2 indicates with which channel in either macrocell or microcell the call is allocated. In order to provide a soft channel allocation decision, not only "accept" and "reject" but also "weak accept" and "weak reject" are employed to describe the allocation decision. Thus, the term set for the output linguistic variable O_1 is defined as $T(O_1) = \{$*Accept* (A), *Weak Accept* (WA), *Weak Reject* (WR), *Reject* $(R)\}$, and the term set for O_2 is defined as $T(O_2) = \{$*Macrocell* (M_a), *Microcell* $(M_i)\}$.

A delta function, $\delta(x - x_0)$, is defined for the membership functions of the output linguistic variables, where $\delta(x - x_0)$ is characterized as $\int_{-\infty}^{\infty} \delta(x - x_0)dx = 1$ and $\delta(x - x_0) = \infty$, $x = x_0$; $\delta(x - x_0) = 0$, $x \neq x_0$. The membership functions for terms A, WA, WR, and R in $T(O_1)$, denoted by μ_A, μ_{WA}, μ_{WR}, and μ_R, respectively, are given by

$$\mu_A = \delta(O_1 - A_c), \tag{17}$$

$$\mu_{WA} = \delta(O_1 - WA_c), \tag{18}$$

$$\mu_{WR} = \delta(O_1 - WR_c), \tag{19}$$

$$\mu_R = \delta(O_1 - R_c). \tag{20}$$

Without loss of generality, the values of parameters are: $R_c = 0$, $A_c = 1$, $WR_c = (R_c + H(t+1))/2$, and $WA_c = (A_c + H(t+1))/2$. A call can be

Rule	IF		THEN	
	$U_0(t)$	$Q_0(t)$	O_1	O_2
1	H	E	A	M_a
2	H	NE	R	M_a
3	L	E	A	M_a
4	L	NE	WR	M_a

Table 1. The inference rules for macrocell only region

allocated with channel if O_1 is greater than the admission threshold $H(t+1)$. And the membership functions for terms M_a and M_i in $T(O_2)$, denoted by μ_{M_i} and μ_{M_a}, respectively, are given by

$$\mu_{M_a} = \delta(O_2 - M_{ac}), \tag{21}$$

$$\mu_{M_i} = \delta(O_2 - M_{ic}). \tag{22}$$

M_{ac} and M_{ic} are set to be positive one and negative one, respectively. If O_2 is greater than zero, the call is allocated with a macrocell channel; otherwise, with a microcell channel.

There are different call types in hierarchical cellular systems. For calls that can use only macrocell's channel, only input linguistic variables of $U_0(t)$ and $Q_0(t)$ are enabled. For calls that could use channels in either macrocell or microcell, input linguistic variables of $U_0(t)$, $U_i(t)$, $Q_0(t)$, $Q_i(t)$, and v are enabled. The fuzzy rule base with dimension: $|T(U_0(t))| \times |T(Q_0(t))|$ is shown in Table 1 for the former, and the one with dimension $|T(v)| \times |T(U_0(t))| \times |T(U_i(t))| \times |T(Q_0(t))| \times |T(Q_i(t))|$ is shown in Table 2 for the latter, where $|T(\cdot)|$ denotes the number of terms in $T(\cdot)$.

The *max-min* inference method [18] is adopted for the fuzzy channel allocator. It first applies the *min* operator on membership values of the terms of all input linguistic variable for each rule. Assume that a call is originated in microcell i and the inference rules in Table 2 are applied. We denote the minimal result for rule i to be m_i, $1 \leq i \leq 32$, and obtain m_{16}, for example, by

$$m_{16} = min(\mu_F(v), \mu_L(U_0(t)), \mu_L(U_i(t)), \mu_{NE}(Q_0(t)), \mu_{NE}(Q_i(t))). \tag{23}$$

Then the method applies the *max* operator to yield the overall membership value. For the output variable O_1, there are 4 rules for term R in Table 2, which are rule 4, 8, 20, 28. Then the overall membership value for the term R, denoted by m_R, is given by

$$m_R = max(m_4, m_8, m_{20}, m_{28}). \tag{24}$$

Similarly, m_A, m_{WA} and m_{WR} are yielded as

$$m_A = max(m_1, m_2, m_5, m_6, m_9, m_{10},$$

Rule	IF					THEN		Rule	IF					THEN	
	v	$U_0(t)$	$U_i(t)$	$Q_0(t)$	$Q_i(t)$	O_1	O_2		v	$U_0(t)$	$U_i(t)$	$Q_0(t)$	$Q_i(t)$	O_1	O_2
1	F	H	H	E	E	A	M_a	17	S	H	H	E	E	A	M_i
2	F	H	H	E	NE	A	M_a	18	S	H	H	E	NE	WA	M_a
3	F	H	H	NE	E	WA	M_i	19	S	H	H	NE	E	A	M_i
4	F	H	H	NE	NE	R	M_a	20	S	H	H	NE	NE	R	M_i
5	F	H	L	E	E	A	M_i	21	S	H	L	E	E	A	M_i
6	F	H	L	E	NE	A	M_a	22	S	H	L	E	NE	WA	M_a
7	F	H	L	NE	E	WA	M_i	23	S	H	L	NE	E	A	M_i
8	F	H	L	NE	NE	R	M_i	24	S	H	L	NE	NE	WR	M_i
9	F	L	H	E	E	A	M_a	25	S	L	H	E	E	A	M_a
10	F	L	H	E	NE	A	M_a	26	S	L	H	E	NE	WA	M_a
11	F	L	H	NE	E	WA	M_i	27	S	L	H	NE	E	A	M_i
12	F	L	H	NE	NE	WR	M_a	28	S	L	H	NE	NE	R	M_a
13	F	L	L	E	E	A	M_a	29	S	L	L	E	E	A	M_i
14	F	L	L	E	NE	A	M_a	30	S	L	L	E	NE	WA	M_a
15	F	L	L	NE	E	WA	M_i	31	S	L	L	NE	E	A	M_i
16	F	L	L	NE	NE	WR	M_a	32	S	L	L	NE	NE	WR	M_i

Table 2. The inference rules for overlay region

$$m_{13}, m_{14}, m_{17}, m_{19}, m_{21}, m_{23}, m_{25}, m_{27}, m_{29}, m_{31}), \tag{25}$$

$$m_{WA} = max(m_3, m_7, m_{11}, m_{15}, m_{18}, m_{22}, m_{26}, m_{30}), \tag{26}$$

$$m_{WR} = max(m_{12}, m_{16}, m_{24}, m_{32}). \tag{27}$$

Afterwards, the *center-of-area* defuzzification method [19] is employed to derive the defuzzification value. The defuzzification value, denoted by $\tilde{O}_1$, is given by

$$\tilde{O}_1 = \frac{m_A \times A_c + m_{WA} \times WA_c + m_{WR} \times WR_c + m_R \times R_c}{m_A + m_{WA} + m_{WR} + m_R}. \tag{28}$$

Then the output variable O_1 is obtained by

$$O_1 = \begin{cases} 0, & \text{for } \tilde{O}_1 \geq H(t+1), \\ 1, & \text{for } \tilde{O}_1 < H(t+1). \end{cases} \tag{29}$$

Note that $H(t+1) = H_0(t+1)$ if the call is a new call and is originated in macrocell-only region, and $H(t+1) = min(H_0(t+1),\ H_i(t+1))$ if the call is a new call and is originated in microcell i. On the other hand, in order to give a good protection for handoff calls, $H(t+1) = 0$ if the call is a handoff.

Similarly, the *max-min* inference method and the *center-of-area* defuzzification method are adopted for output variable O_2, which are not further described here.

3 Neural Fuzzy Resource Manager (NFRM)

3.1 System Architecture

Fig. 3 shows the functional blocks of our proposed NFRM for hierarchical cellular systems [14]. The NFRM contains functional blocks such as *base-station interface module* (BIM), *resource estimator*, *performance evaluator*, and *neural fuzzy channel allocation processor* (NFCAP). Note that BIM, resource estimator, and performance estimator in the NFRM are the same as those in the FCAC, which are described in Sec. 2.1. NFCAP performs the channel allocation using neural fuzzy logic control to attain QoS guaranteed, high channel utilization, and good user satisfaction. In the neural fuzzy logic control, a reinforcement learning is designed to adjust the mean and the variance of the membership functions to cope with the input traffic fluctuation. The detailed design of NFCAP is described in the next section.

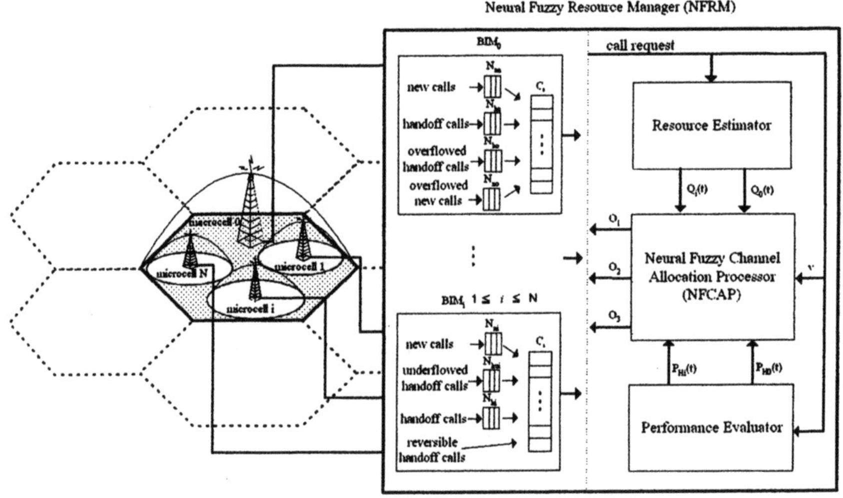

Fig. 3. The neural fuzzy resource manager (NFRM) for hierarchical cellular systems

The service profiles of the hierarchical cellular systems are briefly described here. The new call request generated in each MS is modelled as a Poisson process with mean rate λ, in which the arrival rate of the new voice calls is $\lambda_{nv} = \gamma\lambda$, and the arrival rate of the new data calls is $\lambda_{nd} = (1-\gamma)\lambda$, $0 \leq \gamma \leq 1$. The call durations for the voice and data calls are assumed to be exponentially distributed with averages equal to $\frac{1}{\mu_v}$ and $\frac{1}{\mu_d}$, respectively. And all the data (voice) call requests have identical required capacity and desired capacity, denoted by R_{rd} and R_{dd} (R_{rv} and R_{dv}), respectively.

3.2 The Design of NFCAP

NFCAP contains two functional blocks: FCS in the first layer and NFCRC in the second layer, as shown in Fig. 4.

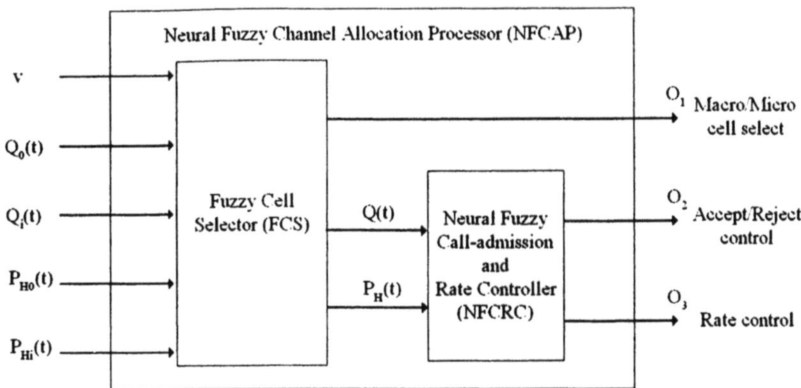

Fig. 4. The block diagram of NFCAP

Fuzzy Cell Selector (FCS) NFCAP chooses five input linguistic variables for FCS: available resources in macrocell 0 ($Q_0(t)$) and in microcell i ($Q_i(t)$), handoff failure probabilities in macrocell 0 ($P_{H0}(t)$) and in microcell i ($P_{Hi}(t)$), and mobile speed (v), and has one output linguistic variable for FCS: the selection of macrocell or microcell (O_1). The available resource of cells can indicate the remaining capacity, the handoff failure probability can show the QoS, and the mobile speed can implicate the handoff rate. Term sets for both $Q_0(t)$ and $Q_i(t)$ are $T(Q_0(t)) = T(Q_i(t)) = T\{More\ Enough,\ Slightly\ Enough,\ Not\ Enough\} = T\{ME, SE, NE\}$, term sets for both $P_{H0}(t)$ and $P_{Hi}(t)$ are $T(P_{H0}(t)) = T(P_{Hi}(t)) = T\{Low, Medium, High\} = T\{L, M, H\}$, and the term set for v is $T(v) = T\{Slow,\ Fast\} = T\{S, F\}$. The trapezoidal function $g(x; x_0, x_1, a_0, a_1)$ in Eq. (1) is chosen to implement the membership function once again.

Denote $\mu_{ME}(Q_0(t))$ ($\mu_{ME}(Q_i(t))$), $\mu_{SE}(Q_0(t))$ ($\mu_{SE}(Q_i(t))$), and $\mu_{NE}(Q_0(t))$ ($\mu_{NE}(Q_i(t))$) to be the membership functions for ME, SE, and NE in $T(Q_0(t))$ ($T(Q_i(t))$), respectively, and define $\mu_{ME}(Q_0(t))$, $\mu_{SE}(Q_0(t))$, $\mu_{NE}(Q_0(t))$, $\mu_{ME}(Q_i(t))$, $\mu_{SE}(Q_i(t))$, and $\mu_{NE}(Q_i(t))$ as

$$\mu_{ME}(Q_0(t)) = g(Q_0(t); E_{m0}, R_{m0}, E_{w0}, 0), \tag{30}$$

$$\mu_{SE}(Q_0(t)) = g(Q_0(t); S_0, S_0, S_{lw0}, S_{rw0}), \tag{31}$$

$$\mu_{NE}(Q_0(t)) = g(Q_0(t); 0, R_{n0}, 0, NE_{w0}), \tag{32}$$

$$\mu_{ME}(Q_i(t)) = g(Q_i(t); E_{mi}, R_{mi}, E_{wi}, 0), \tag{33}$$

$$\mu_{SE}(Q_i(t)) = g(Q_i(t); S_i, S_i, S_{lwi}, S_{rwi}), \tag{34}$$

$$\mu_{NE}(Q_i(t)) = g(Q_i(t); 0, R_{ni}, 0, NE_{wi}). \tag{35}$$

The maximum possible "*More Enough*" value of available resource R_{m0} (R_{mi}) would be the sum of buffer size and allocation channels, E_{m0} (E_{mi}) would be a safety margin of available resource in macrocell (microcells) in QoS requirement and traffic fluctuation, R_{n0} (R_{ni}) would be set to be a fraction of available resource in macrocell (microcells), $S_0 = \frac{1}{2}(E_{m0} + R_{n0})$ ($S_i = \frac{1}{2}(E_{mi} + R_{ni})$), and $E_{w0} = S_{rw0} = (E_{m0} - S_0)$ ($E_{wi} = S_{rwi} = (E_{mi} - S_i)$) and $NE_{w0} = S_{lw0} = (S_0 - R_{n0})$ ($NE_{wi} = S_{lwi} = (S_i - R_{ni})$) are provided to tolerate the change of traffic in macrocell (microcells).

Denote $\mu_L(P_{H0}(t))$, $\mu_M(P_{H0}(t))$, and $\mu_H(P_{H0}(t))$ ($\mu_L(P_{Hi}(t))$, $\mu_M(P_{Hi}(t))$, and $\mu_H(P_{Hi}(t))$) to be the membership functions for L, M, and H in $T(P_{H0}(t))$ ($T(P_{Hi}(t))$), respectively, and define $\mu_L(P_{H0}(t))$, $\mu_M(P_{H0}(t))$, $\mu_H(P_{H0}(t))$, $\mu_L(P_{Hi}(t))$, $\mu_M(P_{Hi}(t))$, and $\mu_H(P_{Hi}(t))$ as

$$\mu_L(P_{H0}(t)) = g(P_{H0}(t); 0, L_{e0}, 0, L_{w0}), \tag{36}$$

$$\mu_M(P_{H0}(t)) = g(P_{H0}(t); M_{e0}, M_{e0}, M_{lw0}, M_{rw0}), \tag{37}$$

$$\mu_H(P_{H0}(t)) = g(P_{H0}(t); H_{e0}, 1, H_{w0}, 0), \tag{38}$$

$$\mu_L(P_{Hi}(t)) = g(P_{Hi}(t); 0, L_{ei}, 0, L_{wi}), \tag{39}$$

$$\mu_M(P_{Hi}(t)) = g(P_{Hi}(t); M_{ei}, M_{ei}, M_{lwi}, M_{rwi}), \tag{40}$$

$$\mu_H(P_{Hi}(t)) = g(P_{Hi}(t); H_{ei}, 1, H_{wi}, 0). \tag{41}$$

H_{e0} (H_{ei}) would be set to be P_H^* provided to guarantee the QoS requirement in macrocell (microcells), L_{e0} (L_{ei}) would be set to be a safety margin of the handoff failure probability in QoS requirement in macrocell (microcells), $M_{e0} = \frac{1}{2}(H_{e0} + L_{e0})$ ($M_{ei} = \frac{1}{2}(H_{ei} + L_{ei})$), and $L_{w0} = M_{lw0} = M_{e0} - L_{e0}$ ($L_{wi} = M_{lwi} = M_{ei} - L_{ei}$) and $H_{w0} = M_{rw0} = H_{e0} - M_{e0}$ ($H_{wi} = M_{rwi} = H_{ei} - M_{ei}$) are provided to tolerate the dynamic behavior of the handoff failure probability in macrocell (microcells).

The membership functions for terms S and F in v, denoted by $\mu_S(v)$ and $\mu_F(v)$, are given by

$$\mu_S(v) = g(v; 0, S_e, 0, S_w), \tag{42}$$

$$\mu_F(v) = g(v; F_e, F_h, F_w, 0), \tag{43}$$

where S_e (F_e) would be a fraction of slow (fast) speed of mobile user, S_w (F_w) is provided to tolerate the change of slow (fast) speed, and F_h would be the fastest speed.

There are different call types in hierarchical cellular systems. For calls that can use only macrocell channels, FCS has to choose the macrocell, and send $P_{H0}(t)$ and $Q_0(t)$ to NFCRC. For calls that could use channels either

IF					THEN	IF					THEN
$Q_O(t)$	$Q_i(t)$	$P_{HO}(t)$	$P_{Hi}(t)$	v	O_1	$Q_O(t)$	$Q_i(t)$	$P_{HO}(t)$	$P_{Hi}(t)$	v	O_1
ME	SE	–	–	–	M_a	SE	ME	–	–	–	M_i
ME	NE	–	–	–	M_a	NE	ME	–	–	–	M_i
SE	NE	–	–	–	M_a	NE	SE	–	–	–	M_i
ME	ME	L	M	–	M_a	ME	ME	M	L	–	M_i
ME	ME	L	H	–	M_a	ME	ME	H	L	–	M_i
ME	ME	M	H	–	M_a	ME	ME	H	M	–	M_i
SE	SE	L	M	–	M_a	SE	SE	M	L	–	M_i
SE	SE	L	H	–	M_a	SE	SE	H	L	–	M_i
SE	SE	M	H	–	M_a	SE	SE	H	M	–	M_i
NE	NE	L	M	–	M_a	NE	NE	M	L	–	M_i
NE	NE	L	H	–	M_a	NE	NE	H	L	–	M_i
NE	NE	M	H	–	M_a	NE	NE	H	M	–	M_i
ME	ME	L	L	F	M_a	ME	ME	L	L	S	M_i
ME	ME	M	M	F	M_a	ME	ME	M	M	S	M_i
ME	ME	H	H	F	M_a	ME	ME	H	H	S	M_i

Table 3. THE INFERENCE RULES FOR THE OVERLAY REGION

in macrocell or microcell, FCS determines the serving cell according to input linguistic variables of $Q_O(t)$, $Q_i(t)$, $P_{HO}(t)$, $P_{Hi}(t)$, and v. The output linguistic variable $O_1 = M_a$ if the macrocell is assigned, and $O_1 = M_i$ if the microcell is allocated. $T(O_1) = \{M_a, M_i\}$. The fuzzy rule base with dimension: $|T(Q_O(t))| \times |T(Q_i(t))| \times |T(P_{HO}(t))| \times |T(P_{Hi}(t))| \times |T(v)|$ are shown in Table 3.

Membership functions for M_a and M_i in $T(O_1)$ are defined as

$$\mu_{Ma} = g(O_1; 0, 0, 0, 0), \tag{44}$$

$$\mu_{Mi} = g(O_1; 1, 1, 0, 0). \tag{45}$$

The *max-min* inference method and apply the *center-of-area* defuzzification method are adopted for output variable O_1. There are $P_H(t)$ and $Q(t)$ output to NFCRC determined by O_1: if the call is with channels in the macrocell $O_1 = 0$, $P_H(t) = P_{HO}(t)$ and $Q(t) = Q_O(t)$; otherwise the call is with channels in the microcell $O_1 = 1$, and $P_H(t) = P_{Hi}(t)$, $Q(t) = Q_i(t)$.

Neural Fuzzy Call-Admission and Rate Controller (NFCRC) The NFCRC takes the handoff failure probability $P_H(t)$ and available resource $Q(t)$ as input linguistic variables. The handoff failure probability shows the QoS, and the available resource implicates the traffic load intensity. This is a feedback control system that the handoff failure probability acts as a QoS index feedback to indicate how effectively the NFCRC is managing the radio resource.

A five-layer neural fuzzy controller is adopted to design the NFCRC. The best structure of NFCRC is obtained via structure learning which measures

Rule	IF		THEN	
	$P_H(t)$	$Q(t)$	O_2	O_3
1	H	ME	WA	BR
2	H	SE	WR	BR
3	H	NE	R	BR
4	M	ME	A	HM
5	M	SE	A	LM
6	M	NE	WR	BR
7	L	ME	A	HR
8	L	SE	A	HM
9	L	NE	WR	BR

Table 4. The Inference Rules for NFCRC

the degree of fuzzy similarity and decides the size of the fuzzy partition of the linguistic [20], [21]. Usually, a hybrid learning algorithm is applied to construct the NFCRC. The algorithm is a two-phase learning scheme. In phase I, a self-organized learning scheme is used to construct the presence of rules and to locate the initial membership functions; in phase II, a reinforcement learning scheme is used to optimally adjust the membership functions for desired outputs. To initiate the learning process, the size of the term set for each input/output linguistic variable, the fuzzy control rules, and training data must be provided. In the self-organized training phase, the initial structure of the controller could be constructed via Kohonen's feature-maps algorithm and the *N-nearest-neighbors* scheme [15] to provide a rough estimate of the structure, if the controller is not provided with an initial knowledge base. Here, based on the domain knowledge obtained from FCAC scheme, an initial form of the controller is constructed. Only a slight modification for the structure is needed in the self-organized training phase.

The term set used to describe the handoff failure probability is defined as $T(P_H(t)) = \{Low\ (L),\ Medium\ (M),\ High\ (H)\}$. And the term set for the available resource is defined as $T(Q(t)) = \{More\ Enough\ (ME), Slightly\ Enough\ (SE), Not\ Enough\ (NE)\}$. In order to provide a soft admission decision, the term set of the output linguistic variable O_2 is defined as $T(O_2) = \{Reject\ (R),\ Weakly\ Reject\ (WR),\ Weakly\ Accept\ (WA),\ Accept\ (A)\}$. Similarly, the term set of the output linguistic variable O_3 is defined as $T(O_3) = \{Basic\ Rate\ (BR), Low\ Medium\ Rate\ (LM),\ High\ Medium\ Rate\ (HM),\ High\ Rate(HR)\}$. The rule structure for NFCRC is shown in Table 4.

The connectionist structure of the NFCRC is constructed in Fig. 5. The NFCRC has the nodes in layer 1 as input linguistic nodes. It has two pairs of nodes in layer 5, where each pair of output nodes have two kinds of linguistic nodes. One is for feeding training data (desired output) into the net and the other is for pumping decision signals (actual output) out of the net. The nodes in layer 2 and layer 4 are term nodes which act as membership functions of the respective linguistic variables. The nodes in layer 3 are rule nodes; each node represents one fuzzy rule and all nodes form a fuzzy rule base. The links in layer 3 and layer 4 function as an inference engine; the links in layer 3 define preconditions of the rule nodes and the links in layer 4 define consequences of the rule nodes. The links in layer 2 and layer 5 are fully connected between the linguistic nodes and their corresponding term nodes.

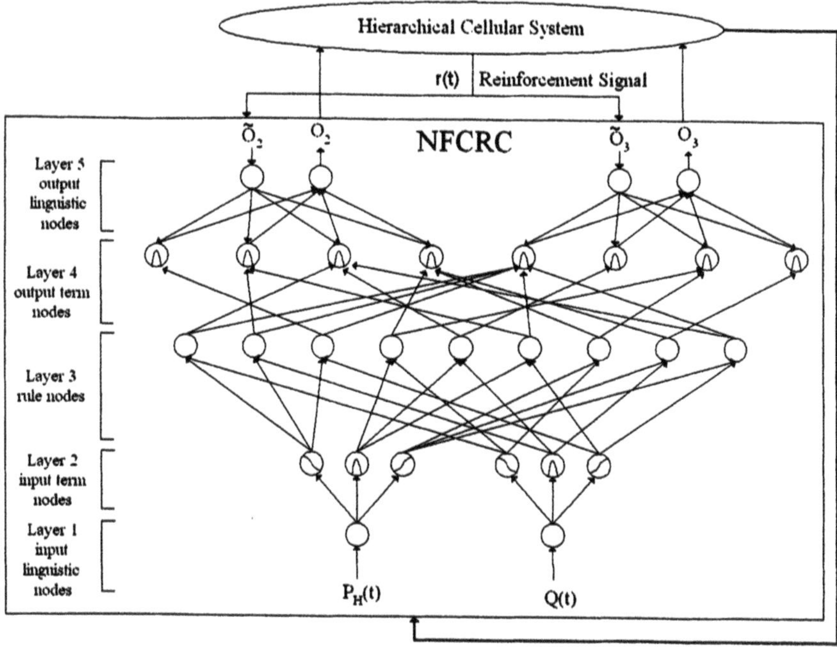

Fig. 5. The structure of the NFCRC controller

NFCRC has a net input function $f_i^{(k)}(u_{ij}^{(k)})$ and an activation output function $a_i^{(k)}(f_i^{(k)})$ for node i in layer k, where $u_{ij}^{(k)}$ denotes the input to node i in layer k from node j in layer $(k-1)$. The layers are described in the following.

Layer 1: In this layer, there are two input nodes with the respective input linguistic variables $P_H(t)$ and $Q(t)$. Define

$$f_i^{(1)}(u_i^{(1)}) = u_i^{(1)} \text{ and } a_i^{(1)} = f_i^{(1)}, \; 1 \le i \le 2, \tag{46}$$

where $u_1^{(1)} = P_H(t)$ and $u_2^{(1)} = Q(t)$.

Layer 2: The nodes in this layer are used as the fuzzifier. Each node performs a bell-shaped function defined as

$$f_i^{(2)}(u_{ij}^{(2)}) = -\frac{(u_{ij}^{(2)} - m_{jn}^{(I)})^2}{\sigma_{jn}^{(I)^2}} \text{ and } a_i^{(2)} = e^{f_i^{(2)}}, \ 1 \le i \le 6, \quad (47)$$

where $u_{ij}^{(2)} = a_j^{(1)}$, $j = \lfloor \frac{i+2}{3} \rfloor$, and $m_{jn}^{(I)}$ and $\sigma_{jn}^{(I)}$ are the mean and the standard deviation of the n-th term of the input linguistic variable from node j in input layer, respectively, $n = i$ if $i \le 3$ and $n = i - 3$ if $i > 3$.

Layer 3: In this layer, the links perform precondition matching of fuzzy control rules. Each rule node performs the fuzzy AND operation defined as

$$f_i^{(3)}(u_{ij}^{(3)}) = min(u_{ij}^{(3)}; \forall j \in P_i) \text{ and } a_i^{(3)} = f_i^{(3)}, \ 1 \le i \le 9, \quad (48)$$

where $u_{ij}^{(3)} = a_j^{(2)}$ and $P_i = \{j|$ all j that are precondition nodes of the i-th rule$\}$.

Layer 4: There are two groups of output in this layer: one for the output of admission control O_2 and the other for the output rate control O_3. Also, nodes in this layer have two operating modes: *down-up* and *up-down*. In the down-up operating mode, the links perform consequence matching of fuzzy control rules. Each node performs a fuzzy OR operation which integrates the fired strength of rules that have the same consequence and is defined as

$$f_i^{(4)}(u_{ij}^{(4)}) = max(u_{ij}^{(4)}; \forall j \in C_i) \text{ and } a_i^{(4)} = f_i^{(4)}, \ 1 \le i \le 8, \quad (49)$$

where $u_{ij}^{(4)} = a_j^{(3)}$ and $C_i = \{j|$ all j that have the same consequence of the i-th term in the term set of O_2 and $O_3\}$. The up-down operating mode is used during learning periods, which will be described later.

Layer 5: There are two pairs of nodes in this layer. One node in each pair performs the down-up operation for the decision signals O_2 and O_3. The node and its links act as the defuzzifier. The function used to simulate a center-of-area defuzzification method for O_2 signal is approximated by

$$f_i^{(5)}(u_{ij}^{(5)}) = \sum_{j=1}^{4} m_j^{(O)} \sigma_j^{(O)} u_{ij}^{(5)}$$

$$\text{and } a_i^{(5)} = U\left(\frac{f_i^{(5)}}{\sum_{j=1}^{4} \sigma_j^{(O)} u_{ij}^{(5)}} - \theta\right), \ i = 1 \quad (50)$$

where $u_{ij}^{(5)} = a_j^{(4)}$, θ is the decision threshold, and

$$U(x) = \begin{cases} 1 & \text{if } x \geq 0, \\ 0 & \text{otherwise.} \end{cases} \tag{51}$$

Clearly, $O_2 = a_1^{(5)}$ and a new connection will be accepted only if $O_2 = 1$. Similarly, the O_3 signal is also to simulate a center-of-area defuzzification method approximated by

$$f_i^{(5)}(u_{ij}^{(5)}) = \sum_{j=5}^{8} m_j^{(O)} \sigma_j^{(O)} u_{ij}^{(5)}$$

$$\text{and } a_i^{(5)} = \frac{f_i^{(5)}}{\sum_{j=5}^{8} \sigma_j^{(O)} u_{ij}^{(5)}} \times R_{dx},\ i = 2, \tag{52}$$

where $u_{ij}^{(5)} = a_j^{(4)}$, and R_{dx} is the number of desired channels for a call; $x = v$ denotes the voice call; $x = d$ denotes the data call. Clearly, $O_3 = \lceil a_2^{(5)} \rceil$ and a new connection is assigned to use a number of O_3 channels. The other node performs the up-down operation during the learning period.

The parameter initialization of membership functions is described below. To locate the mean m_i of the i-th membership function for linguistic variable x, $1 \leq i \leq M$, given a set of training data x_j for x, a statistical clustering technique called Kohonen's feature-maps algorithm [15] is employed.

[Obtain m_i by using Kohonen's feature-maps algorithm]

Step 1: Set initial values of m_i for all membership functions, $1 \leq i \leq M$, such that

$$\min_{1 \leq i \leq N} x_j \leq m_i \leq \max_{1 \leq i \leq N} x_j.$$

Set an initial learning rate α $(0 < \alpha < 1)$.

Step 2: Set $j = 1$.

Step 3: Present training data x_j and compute the distance $d_i = |x_j - m_i|$, $1 \leq i \leq M$.

Step 4: Determine the kth membership function which has the minimum distance d_k $(d_k = \min_{1 \leq i \leq M} d_i)$.

Update m_k by

$$m_k = m_k + \alpha(x_j - m_k).$$

Step 5: If $j < N$, $j = j + 1$, Goto **Step 3**

ELSE

Decrease α and Goto **Step 2**.

EndIf

The above procedure will stop until $\alpha \leq 0$. The determination of which d_i is minimum at **Step 4** can be quickly accomplished in constant time via a winner-take-all circuit [15]. The adaptive algorithm can be independently performed to obtain m_i for each input and output linguistic variables.

As for the corresponding standard deviations σ_i of the ith membership function of x, since m_i and σ_i will be finely tuned in the reinforcement learning phase, a first-nearest-neighbor heuristic is used to estimate σ_i, which is given by

$$\sigma_i = \frac{|m_i - m^*|}{\gamma}, \tag{53}$$

where

$$m^* = \begin{cases} m_{i-1}, & \text{for } |m_i - m_{i-1}| < |m_i - m_{i+1}|; \\ m_{i+1}, & \text{otherwise,} \end{cases} \tag{54}$$

and γ is called an overlap parameter used to describe the degree of overlapping with two membership functions.

Reinforcement Learning Algorithm Since there are no measurable output values feedback to instruct the NFCRC to learn, a reinforcement learning algorithm is adopted and an evaluative handoff failure probability is used as a reinforcement signal. Fig. 5 also shows the diagram of the reinforcement learning for NFCRC, where the hierarchical cellular system provides the reinforcement signal $r(t)$ as a desired output to NFCRC and it receives the call admission control value O_2 and rate control value O_3 from NFCRC. The reinforcement signal is the difference between the desired handoff failure probability and measured handoff failure probability at time t, which is express as

$$r(t) = P_H^* - P_H(t). \tag{55}$$

The reinforcement learning is applied to adjust parameters of input and output membership functions optimally, according to the input training data, the reinforcement signal, and the fuzzy logic rules. It derives updating rules for the mean and the standard deviation of the bell-shaped membership functions so as to minimize the error function, defined as

$$E(t) = \frac{1}{2}r^2(t) = \frac{1}{2}(P_H^* - P_H(t))^2. \tag{56}$$

For each training data set, starting at the input nodes, the *down* − *up* operation can compute to obtain the actual outputs of call admission control O_2 and rate control O_3, and consequently $P_H(t)$ is measured. On the other hand, from the output node, the *up* − *down* operation is used to compute $\frac{\partial E(t)}{\partial w(t)}$ for all hidden nodes, where $w(t)$ is the adjustable parameters such as the mean and the standard deviation for the input and output bell-shaped membership functions. the general learning rule is adopted, which is given by

$$w(t+1) = w(t) + \eta \cdot (-\frac{\partial E(t)}{\partial w(t)}), \tag{57}$$

where η is the learning rate.

4 Simulation Results and Discussions

4.1 Hierarchical Cellular System Model

In the simulations, a Manhattan-street-type hierarchical cellular system with N = 9 microcells constructed is assumed. To demonstrate the traffic flow within the hierarchical cellular system, the handoff behavior of users is characterized by a teletraffic flow matrix [3], defined as

$$A = \begin{bmatrix} a_{00} & a_{01} & a_{02} & \cdots & a_{0N} & a_{0d} \\ a_{10} & a_{11} & a_{12} & \cdots & a_{1N} & a_{1d} \\ a_{20} & a_{21} & a_{22} & \cdots & a_{2N} & a_{2d} \\ \cdots & \cdots & \cdots & \cdots & \cdots & \cdots \\ a_{N0} & a_{N1} & a_{N2} & \cdots & a_{NN} & a_{Nd} \end{bmatrix}$$

$$= \begin{bmatrix} 0.0 & 0.1 & 0.1 & 0.1 & 0.1 & 0.1 & 0.1 & 0.1 & 0.1 & 0.1 & 0.1 \\ 0.1 & 0.0 & 0.3 & 0.0 & 0.3 & 0.0 & 0.0 & 0.0 & 0.0 & 0.0 & 0.3 \\ 0.1 & 0.2 & 0.0 & 0.2 & 0.0 & 0.2 & 0.0 & 0.0 & 0.0 & 0.0 & 0.3 \\ 0.1 & 0.0 & 0.3 & 0.0 & 0.0 & 0.0 & 0.3 & 0.0 & 0.0 & 0.0 & 0.3 \\ 0.1 & 0.2 & 0.0 & 0.0 & 0.0 & 0.2 & 0.0 & 0.2 & 0.0 & 0.0 & 0.3 \\ 0.0 & 0.0 & 0.25 & 0.0 & 0.25 & 0.0 & 0.25 & 0.0 & 0.25 & 0.0 & 0.0 \\ 0.1 & 0.0 & 0.0 & 0.2 & 0.0 & 0.2 & 0.0 & 0.0 & 0.0 & 0.2 & 0.3 \\ 0.1 & 0.0 & 0.0 & 0.0 & 0.3 & 0.0 & 0.0 & 0.0 & 0.3 & 0.0 & 0.3 \\ 0.1 & 0.0 & 0.0 & 0.0 & 0.0 & 0.2 & 0.0 & 0.2 & 0.0 & 0.2 & 0.3 \\ 0.1 & 0.0 & 0.0 & 0.0 & 0.0 & 0.0 & 0.3 & 0.0 & 0.3 & 0.0 & 0.3 \end{bmatrix},$$

where a_{ij}, $i \neq j$, represents the probability of a handoff call originated in cell i and directed to cell j, $1 \leq j \leq N$, and a_{id} denotes the probability of this handoff call directed to the adjacent macrocell, $0 \leq i \leq N$. $\sum_{j=0} a_{ij} = 1$ and a_{ii} would be zero.

In each cell, the number of mobile stations is assumed to be 550, and λ_{nv}=0.8λ, λ_{nd}=0.2λ. Suppose R_{rv}=1 and R_{dv}=1 for voice calls, and R_{rd}=1 and R_{dd}=4 for data calls. Low- and high-mobility users are generated in a ratio of 7:3, and the cell dwell time is exponentially distributed with mean 180 sec. (18 sec.) and 1440 sec. (144 sec.) for high-mobility users and low-mobility users in macrocell (microcells), respectively. And the speed of mobile users is assumed to be uniformly distributed in the range of 0 km to 40 km (40 km to 80 km) for low- (high-) mobility users. It is also assumed that the mean unencumbered session duration is $\frac{1}{\mu_v}$=100 seconds for voice call and $\frac{1}{\mu_d}$=60 seconds for data call, and the patience time for queued voice (data) calls is in the range of 5 to 20 seconds.

As to the channel partition in this simulation, 150 channels are fixedly allocated to macrocell and microcells with a pattern of $(C_0, C_1, \cdots, C_N) = (42, 12, \cdots, 12)$. For the OCA and CCA schemes, a number of channels C_{ri} are reserved as guard channels for handoff calls in cell i, $0 \leq i \leq N$, which are denoted by $(C_{r0}, C_{r1}, \cdots, C_{rN})$. According to our simulation experiments, the

appropriate values are $(C_{r0}, C_{r1}, \cdots, C_{rN}) = (8, 4, \cdots, 4)$ for OCA scheme and $(C_{r0}, C_{r1}, \cdots,$
$C_{rN}) = (3, 2, \cdots, 2)$ for CCA schemes at $\lambda = 5 \times 10^{-4}$. Since the reneging (dropping) process is considered, it is not necessary to provide a large buffer size for new and handoff calls [5], [12]; all buffer sizes in macrocell and microcells are assumed to 3. Note that in the following performance comparison, OCA scheme provides no buffer and CCA and FCAC schemes support the same buffering scheme and capacity as NFRM does.

4.2 Parameter Settings

In this subsection, the values of the parameters for each function blocks in the FCAC and NFRM are given. These values are determined based on the QoS requirements and the knowledge of conventional CCA scheme.

For FCAC, in the fuzzy admission threshold estimator, parameters of membership functions of input linguistic variables are selected as follows: $L_e = 0.01$, $H_e = 0.02$, and $L_w = H_w = 0.01$ for $\mu_L(P_H(t))$ and $\mu_H(P_H(t))$ in (2) and (3); $E_e = E_w = NE_w = 17$, $R_e = 45$, and $R_n = 0$ for $\mu_E(Q(t))$ and $\mu_{NE}(Q(t))$ with macrocell in (4) and (5); $E_e = E_w = NE_w = 7$, $R_e = 15$, and $R_n = 0$ for $\mu_E(Q(t))$ and $\mu_{NE}(Q(t))$ with microcells in (4) and (5). In the fuzzy channel allocator, parameters of membership functions of input linguistic variables are selected as follows: $L_e = L_w = H_w = 0.5$ for $\mu_L(U_0(t))$, $\mu_H(U_0(t))$, $\mu_L(U_i(t))$, and $\mu_H(U_i(t))$ in (11)-(14); $E_e = E_w = NE_w = 17$, $R_e = 45$, and $R_n = 0$ for $\mu_E(Q(t))$ and $\mu_{NE}(Q(t))$ with macrocell in (4) and (5); $E_e = E_w = NE_w = 7$, $R_e = 15$, and $R_n = 0$ for $\mu_E(Q(t))$ and $\mu_{NE}(Q(t))$ with microcells in (4) and (5); and $S_w = F_w = 40$ km, $S_e = 20$ km, $F_e = 60$ km, and $F_h = 80$ km for $\mu_S(v))$ and $\mu_L(v)$ in (15) and (16). And constant parameters are set to be: $\eta = 0.01$ and $\beta = \gamma = 0.9$.

As to NFRM, parameters of membership functions for input linguistic variables in the FCS are selected as follows: $L_{e0} = L_{ei} = 0$, $L_{w0} = L_{wi} = M_{e0} = M_{ei} = M_{lw0} = M_{lwi} = 0.01$, $H_{e0} = H_{ei} = 0.02$, and $M_{rw0} = M_{rwi} = H_{w0} = H_{wi} = 0.01$ for $\mu_L(P_{H0}(t))$, $\mu_M(P_{H0}(t))$, $\mu_H(P_{H0}(t))$, $\mu_L(P_{Hi}(t))$, $\mu_M(P_{Hi}(t))$, and $\mu_H(P_{Hi}(t))$ in (36)-(41); $E_{m0} = 12$, $R_{m0} = 45$, $E_{w0} = S_0 = S_{lw0} = S_{rw0} = NE_{w0} = 6$, and $R_{n0} = 0$, for $\mu_{ME}(Q_0(t))$, $\mu_{SE}(Q_0(t))$, and $\mu_{NE}(Q_0(t))$ in (30)-(32); $E_{mi} = 10$, $R_{mi} = 15$, $E_{wi} = S_0 = S_{lwi} = S_{rwi} = NE_{wi} = 5$, and $R_{ni} = 0$, for $\mu_{ME}(Q_i(t))$, $\mu_{SE}(Q_i(t))$, and $\mu_{NE}(Q_i(t))$ in (33)-(35).

In NFCAP, the initial values of membership functions of term sets for $P_H(t)$ are chosen according to QoS requirement and then properly adjusted via the learning algorithm. Thus, the mean value $m_{11}^{(I)}$ ($m_{12}^{(I)}$, $m_{13}^{(I)}$) in membership function of H (M, L) of $P_H(t)$ is set to be 0.05 (0.02, 0), and let $\sigma_{11}^{(I)} = \frac{1}{2} \cdot (m_{11}^{(I)} - m_{12}^{(I)})$ ($\sigma_{12}^{(I)} = \sigma_{13}^{(I)} = \frac{1}{2} \cdot (m_{12}^{(I)} - m_{13}^{(I)})$). In order to utilize the resource as much as possible and to guarantee the QoS requirement, the initial values of membership functions of ME, SE, and NE for $Q(t)$ is set to be $m_{21}^{(I)} = 9$, $m_{22}^{(I)} = 3$, and $m_{23}^{(I)} = 0$ ($m_{21}^{(I)} = 7$, $m_{22}^{(I)} = 3$, and $m_{23}^{(I)} = 0$) if the call is assigned to use the channels in macrocell (microcell), and let $\sigma_{21}^{(I)} = \frac{1}{2} \cdot (m_{21}^{(I)} - m_{22}^{(I)})$ and $\sigma_{22}^{(I)} = \sigma_{23}^{(I)} = \frac{1}{2} \cdot (m_{22}^{(I)} - m_{23}^{(I)})$.

The initial membership functions of the mean $m_j^{(O)}$ of the term set O_2 (O_3) are set to be equally spaced in the range of $[0, 1]$, and let $\sigma_j^{(O)} = 0.01$. The decision threshold θ in (23) is set to be $\theta = 0$ for handoff call and $\theta = 0.5$ for new call because handoffs are given priority higher than new calls. The use of different η may drastically reduce the training time required in the learning phase. As for $P_H(t)$ and $Q(t)$, their initial membership functions were heuristically set and required further optimization in the learning phase. Thus, $\eta = 0.01$ was used.

4.3 Simulation Results

Five performance measures such as the system utilization, the new-call blocking probability, the handoff failure probability, the forced termination probability, and the handoff rate are observed. The system utilization at time t, denoted by $U(t)$, is defined as

$$U(t) = \frac{K_0(t) + \sum_{i=1}^{N} K_i(t)}{C_0 + \sum_{i=1}^{N} C_i}, \tag{58}$$

where $K_0(t)$ ($K_i(t)$) is the average number of busy channels in macrocell 0 (microcell i) at time t and C_0 (C_i) is the channel capacity for macrocell 0 (microcell i). The new-call blocking probability at time t, denoted by $P_N(t)$, is defined as

$$P_N(t) = \frac{\sum_{i=0}^{N}(NB_i(t) + NR_i(t))}{\sum_{i=0}^{N} NN_i(t)}, \tag{59}$$

where $NB_i(t)$ ($NR_i(t)$) is the number of blocked (reneging) new calls in cell i and $NN_i(t)$ ($NN_0(t)$) is the number of new calls originating in microcell i (macrocell-only region), at time t. Similarly, the handoff failure probability at time t, denoted by $P_H(t)$, is given by

$$P_H(t) = \frac{\sum_{i=0}^{N}(HB_i(t) + HR_i(t))}{\sum_{i=0}^{N} NH_i(t)}. \tag{60}$$

A call will be forced termination if it is corrupted due to a handoff failure during its conversation time. The forced termination probability at time t, denoted by $P_F(t)$, is defined as

$$P_F(t) = \frac{\sum_{i=0}^{N}(HB_i(t) + HR_i(t))}{\sum_{i=0}^{N} NS_i(t)}, \tag{61}$$

where $HB_i(t)$ ($HR_i(t)$) is the number of blocked (dropped) handoff calls in cell i and $NS_i(t)$ is the number of admitted new calls originated in cell i, at time t. The handoff rate at time t, denoted by $R_H(t)$, is defined as

$$R_H(t) = \frac{\sum_{i=0}^{N} NH_i(t)}{\sum_{i=0}^{N} NS_i(t)}. \tag{62}$$

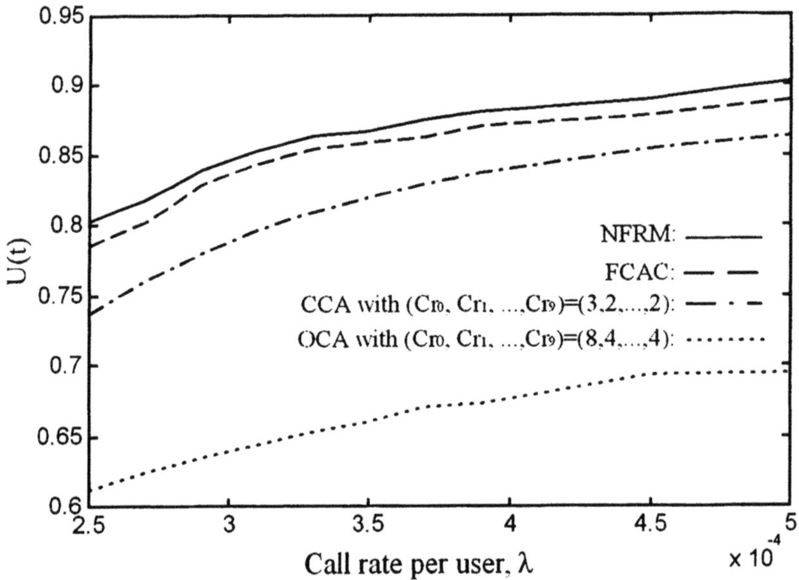

Fig. 6. $U(t)$ for NFRM, FCAC, OCA and CCA schemes

Fig. 6 shows the system utilization $U(t)$ versus the calling rate per user λ for schemes of NFRM, FCAC, OCA and CCA at time $t = 10^8$. It reveals that the system utilization of NFRM gains 31.1%, 6.3%, and 1.4% improvement over the OCA, CCA, and FCAC methods, respectively. The superior performance of NFRM is because FCS in NFRM refers much more effective information than other conventional controllers, and it adopts fuzzy logic theory to balance traffic load between macrocell and microcells and provide a soft and accurate control during traffic fluctuation, and NFCRC in NFRM possesses the learning capability of neural networks to reduce the decision error and the fuzzy logic theory to qualitatively represent control rules naturally in the neural network to overcome some uncertainty and imprecision.

Fig. 7 shows the new-call blocking probability $P_N(t)$ and the handoff failure probability $P_H(t)$ for schemes of NFRM, FCAC, OCA, and CCA versus the calling rate per user λ at time $t = 10^8$. It can be seen that, as λ varies,

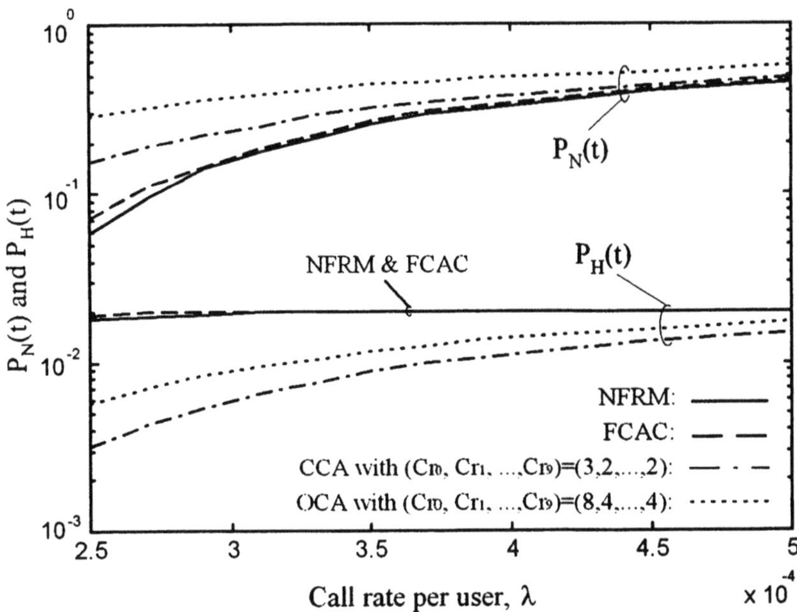

Fig. 7. $P_H(t)$ and $P_N(t)$ for NFRM, FCAC, OCA and CCA schemes

$P_H(t)$ of NFRM and FCAC remains constant at around $P_H^* = 2\%$. Also, compared to FCAC, OCA, and CCA, the $P_N(t)$ of NFRM is minimal. The simulation result indicates that NFRM achieves highest system utilization while fulfilling the system QoS requirement. This is because NFRM uses neural fuzzy control to the allocation of channels. Neural networks have merits of ability to learn from examples and to cope with incomplete input data. Fuzzy logic is a soft logic which is appropriate to represent in determining if a given requirement constraint is complied or violated. This in effect removes the imposition of worse case assumption from the decision making of channel selection. And the neural networks used in fuzzy call admission control and rate manager can effectively estimate the optimal call admission and appropriately allocate a number of channels for each call. On the contrary, the other schemes maintain a fixed number of guard channels regardless that the traffic load is fluctuating and the changing λ is unpredictable. Fig. 8 shows the forced termination probability $P_F(t)$ versus the calling rate per user λ for schemes of NFRM, FCAC, OCA and CCA at time $t = 10^8$. It is found that $P_F(t)$ of NFRM has flat curve under 2%. The reason is that NFRM obtains the unchanged $P_H(t)$, as shown in Fig. 7.

Fig. 9 shows the handoff rate $R_H(t)$ versus the calling rate per user λ for schemes of NFRM, FCAC, OCA, and CCA at time $t = 10^8$. It reveals that NFRM has more handoff rate than OCA by an amount of 2%. The reason is

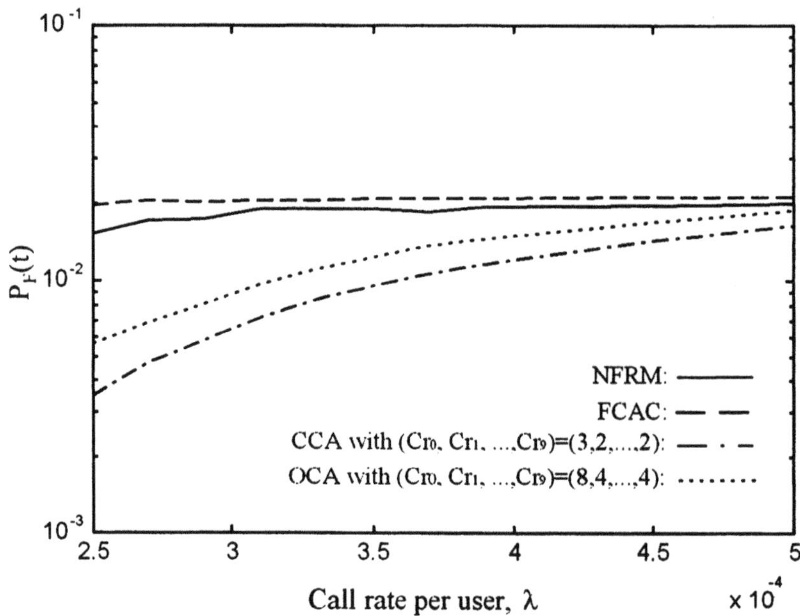

Fig. 8. $P_F(t)$ for NFRM, FCAC, OCA and CCA schemes

that the design of NFRM is based on the knowledge of FCAC and CCA which combines overflow, reversible, and underflow. Fortunately, the signaling overheads for these handoffs might not cost so much as those for conventional handoffs between macrocells since most of these handoffs are occurred in the same macrocell. It also reveals that NFRM achieves less handoff rate than CCA and FCAC by an amount of 14.9% and 6.8%, respectively. It is not only because of more information such as the speed of mobile station considered in NFRM but also because of the neural fuzzy logic control that can provide decision support and expert system with powerful reasoning and learning capabilities.

5 Concluding Remarks

In this chapter, two intelligent channel assignment schemes: fuzzy channel allocation controller (FCAC) and neural fuzzy resource manager (NFRM), are introduced for hierarchical cellular system providing multimedia services. The FCAC employs the concept of fuzzy logic control system, while the NFRM employs the synergy of fuzzy logic system and neural network. The design of FCAC is based on the domain knowledge obtained from our previous work on

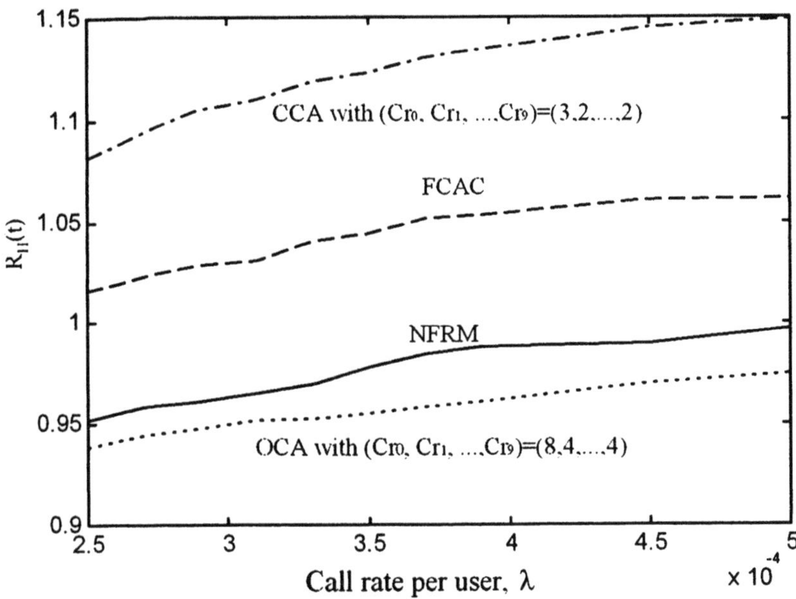

Fig. 9. $R_H(t)$ for NFRM, FCAC, OCA and CCA schemes

CCA scheme [5]. The FCAC consists of two fuzzy function blocks: fuzzy admission threshold estimator and fuzzy channel allocator. In the fuzzy admission threshold estimator, *Sugeno's position-gradient type reasoning method* is adopted to adjust the admission threshold adaptively. In the fuzzy channel allocator, the *max-min* inference method is adopted to balance the utilization between macrocell and microcells to achieve a higher overall system utilization.

Based on the domain knowledge obtained from FCAC and CCA, the NFRM improves furtherly the system performance with rate control. In the NFRM, a five-layer neural fuzzy controller is used to design the neural fuzzy call-admission and rate controller (NFCRC). The best structure of NFCRC is obtained via structure learning which measures the degree of fuzzy similarity and decides the size of the fuzzy partition of the linguistic variables. To initialize the parameters of the membership functions for linguistic variable, Kohonen's feature-maps algorithm is employed for statistical clustering from a set of training data. Moreover, without a obvious decision target for the NFCRC to learn, a reinforcement learning algorithm is adopted and an evaluative handoff failure probability is used as a reinforcement signal.

Simulation results show that the proposed NFRM improves the overall channel utilization by an amount of 31.1% higher than the OCA scheme, 6.3% better than the CCA scheme, and 1.4% larger than the FCAC scheme, while maintaining the QoS requirement. And it still reduces the handoff rate by an amount of 14.9% under the CCA mechanism and 6.8% lower than the FCAC scheme, but increases the handoff rate by an amount of 2% over the OCA mechanism.

For well-experienced engineers, soft computing techniques are very powerful since they can achieve tractability, robust, and low cost solution at exploiting the tolerance for imprecision , uncertainty, and partial truth. The core of the soft computing techniques is domain knowledge which is usually obtained from simulation or design experience. Sometimes, it is not easy to obtain these domain knowledge, especially for a new, under-developing telecommunication system. Therefore, some researchers turn to combine model-based methodology and soft computing techniques, such as: neuro-dynamic programming (NDP) [22]. The NDP provides a mathematical foundation for system modelling and improve the system performance through on-line learning. This is an interesting approach for future study.

References

1. Kolavennu VR, Rappaport SS (1986) Traffic performance characterisation of a personal radiocommunication system, IEE, pt. F, **133**, pp. 550–561
2. I CL, Greenstein LJ, Gitlin RD (1993) A microcell/macrocell cellular architecture for low- and high-mobility wireless users, IEEE J. Selected Areas Commun. **11**, pp. 885–891
3. Rappaport SS, Hu LR (1994) Microcellular communication systems with hierarchical microcell overlays: traffic performance models and analysis, Proceedings of the IEEE **82**, pp. 1383–1397
4. ——— (1995) Personal communication systems using multiple hierarchical cellular overlays, IEEE J. Selected Areas in Commun. **13**, pp. 406–415
5. Lo KR, Chang CJ, Chang C, Shung B (1998) A combined channel assignment strategy for hierarchical cellular systems, ICUPC 1997, pp. 128–132, and Computer Commun. **21**, pp. 1143–1152
6. Chang CJ, Su TT, Chiang YY (1994) Analysis of a cutoff priority cellular radio system with finite queueing and reneging/dropping, IEEE/ACM Trans. Networking **2**, pp. 166–175
7. Dahlman E, Beming P, Knutsson J, Ovesiö F, Persson M, Roobol C (1998) WCDMA—The radio interface for future mobile multimedia communications, IEEE Trans. Veh. Technol. **VT-47**, pp. 1105–1117
8. Samukic A (1998) UMTS universal mobile telecommunications system: development of standards for the third generation, IEEE Trans. Veh. Technol. **VT-47**, pp. 1099–1104
9. Shum KW, Sung CW (1999) Fuzzy layer selection method in hierarchical cellular systems, IEEE Trans. Veh. Technol. **VT-48**, pp. 1840–1849
10. Chang CJ, Chen BW, Liu TY, Ren FC (2000) Fuzzy/Neural congestion control for integrated voice and data DS-CDMA/FRMA cellular networks, IEEE J. Selected Areas Commun. **18**, pp. 283–293
11. Abdul-Haleem M, Cheung KF, Chuang J (1994) Fuzzy logic based dynamic channel assignment, ICCS 1994, pp. 773–777

12. Lo KR, Chang CJ, Chang C, Shung B (2000) A QoS-guaranteed fuzzy channel allocation controller for hierarchical cellular systems, IEEE Trans. Veh. Technol. **VT-49**, pp. 1588–1598
13. Chang PR, Wang BC (1996) Adaptive fuzzy power control for CDMA mobile radio systems, IEEE Trans. Veh. Technol. **VT-45**, pp. 225–236
14. Lo KR, Chang CJ, Chang C, Shung B A neural fuzzy resource manager for hierarchical Cellular systems supporting multimedia services, submitted to IEEE Trans. Veh. Technol.
15. Lin CT, Lee CSG (1996) Neural Fuzzy Systems, Englewood Cliffs, NJ: Prentice-Hall
16. Sinha MK, Gupta MM (1999) Soft Computing & Intelligent Systems, Academic Press
17. Nguyen HT, Sugeno M, Tong R, Yager RR (1995) Theoretical aspects of fuzzy control, John Wiley & SONS, Inc.
18. Zimmermann HJ (1991) Fuzzy Set Theory and Its Applications, 2nd edition, Kluwer Academic Publishers
19. Yager RR, Zadeh LA (1992) An Introduction To Fuzzy Logic Applications in Intelligent Systems, Kluwer Academic Publishers
20. Lin CT, Lee CSG (1991) Neural-network-based fuzzy logic control and decision system, IEEE Trans. Computers **40**, pp. 1320–1336
21. ——— (1994) Reinforcement structure/parameter learning for Neural-network-based fuzzy logic control systems, IEEE Trans. Fuzzy Systems **2**, pp. 46–63
22. Bertsekas DP, Rsitsiklis JN (1996) Neuro-Dynamic Programming, Athena Scientific

A New Mobility Prediction System using Neuro-Fuzzy Model and Its Application to Restoration of Mobility Databases

Joon-Min Gil[1] and Chan Yeol Park[2]

[1] Institute of Basic Science, Korea University
1-5, Anam-Dong, Sungbuk-Gu, Seoul 136-701, KOREA
[2] Supercomputing Center, Korea Institute of Science & Technology Information
P.O. Box 122, Yusong, Taejon 305-600, KOREA

Abstract. We introduce a new system of predicting future location of users in mobile networks, and apply the system to the failure restoration of mobility databases. The prediction system is based on a neuro-fuzzy inference system that we developed. The system is adaptable in the sense that it "learns" and "predicts" (in a sense that is formally defined) from users' past movement patterns in order to obtain future locations. We also report on an experimental evaluation of our system. The evaluation uses the mobility data, which are derived from the mobility model based on the traveling demand model. The performance of our system shows that our prediction system has high accuracy ratio regardless of the various mobility of users. Also, we show that our restoration scheme can reduce the cost needed to restore the location records of lost users after a failure when compared to the checkpointing scheme.

Keywords: Mobile Networks, Neuro-Fuzzy Model, Mobility Learning and Prediction, Mobility Databases, Failure Restoration

1 Introduction

Mobile Networks [1] enable users to communicate with each other at any time from any location. Thus, users no longer need to remain at a fixed location to receive messages. However, due to the intrinsic characteristics of mobile networks in themselves, there are some problems, such as communication resource deficiency and frequently disconnection. To solve these problems, users (or mobile terminals) should execute mobile applications without an effect of their location. So, users can maintain the communications that do not depend so extremely on locality, the capacity of holding resources, and so on.

Users' mobility is a result of the location change of users, and also plays an important role in maintaining continuous connectivity to mobile systems regardless of the geographical location of users. Unfortunately, continuous connectivity is not always possible. For example, when users move rapidly between cells, the data or services that they are served in a previous cell are frequently forwarded to a new cell. These users have a short residual time in each cell and so will encounter many periodic breaks in the course of services. These breaks could bring about unacceptable Quality of Service (QoS) for some multimedia mobile services such as video conferencing, image transfer, web browsing, and so on. If these breaks are frequently, the data or services cannot be served to users, and the QoS is not guaranteed. Thus, it is required that mobile networks aggressively consider user mobility. This is achieved through mobility prediction. In other words, mobile systems should predict the future location where users will move, so that users' location change is known to mobile systems before users move into a new cell. Then, the data and services are reserved by the systems in advance, and users can immediately receive mobile data or services with the same degree of efficiency as at the prior location. As a result, a mobility prediction scheme is positively necessary to provide users with a high quality of services.

Users' location information is usually maintained in mobility databases, permanently or temporarily. When there are calls for users, the calls are delivered using the mobility databases. However, if the mobility databases fail due to their malfunction, the location records stored in the databases are lost and incoming calls to users may be rejected. This results in a large service delay and a serious deterioration in the performance of mobile systems. Thus, there is a need for an explicit restoration procedure for mobility databases in order to guarantee continuous service availability to users.

Without an explicit restoration procedure, the delay in restoring location information after a failure depends on the length of the users' silence period. In IS-41 [2], there is no explicit expedient to restore mobility databases. After a failure, mobility databases incrementally reconstruct the location information of a user each time the user sends a registration message. Before the users originate any message, all incoming calls to them are rejected. In GSM [3], mobility databases are backed up periodically; i.e., a checkpointing procedure is performed for the recovery of failed mobility databases. After a failure, the databases are immediately restored from a stable storage. However, some backup records may be obsolete. If the interval between the checkpointing time and the failure time of mobility databases is relatively long, the users' latest location information may not be updated, in which case, location records during this period may be rendered obsolete.

Users' mobility plays an important role in locating lost users after a failure. To restore lost location information, lost users should be paged, starting from the area recorded on a backup. Unfortunately, unsuccessful paging occurs when users move far away from the area recorded on the backup, in which case entire cells in a RA should be paged. This results in deteriorated system performance and unacceptable QoS. If it would can predict the probable location of users after a failure, it would cut the paging cost tremendously. Thus, mobility prediction would be useful in enhancing the performance of mobile systems during the restoration of failed mobility databases.

In this paper, we introduce a new method for mobility prediction which is based on a Neuro-Fuzzy Inference System (NFIS). Our mobility prediction scheme predicts a future location from movement patterns using users' location history. The NFIS plays a role as a coordinator for smooth communication and adaptive deals with variously changeable mobility. In order to show an applicability of our prediction scheme, we apply our prediction scheme to the failure restoration of mobility databases. Whenever users move into a new location, their movement patterns are learned by the NFIS. When a failure occurs in mobility databases, the probable location of users may be predicted by the NFIS. The predicted locations are used to find the users' current location. Contrary to other approaches using checkpoint, our restoration scheme does not need any backup process. Thus, it has less costs during restoration process. In addition, there's no need for additional storage space to store checkpoint information and users experience only slight delays in service.

This paper is organized as follows: Sect. 2 provides an overview of related works on the mobility prediction and the failure restoration of mobility databases in mobile networks. In Sect. 3, we look at the system model used in this paper. Sect. 4 introduces the mobility learning and prediction scheme using an NFIS. In Sect. 5, we propose a restoration scheme based on mobility learning and prediction. Procedures for the restoration of mobility databases are also presented in this section. Sect. 6 shows the performance evaluation of our prediction and restoration schemes with simulations. Finally, the conclusion of this paper is given in Sect. 7.

2 Related Works

Although mobile networks has been studied extensively, most of the recent researches focuse on network architectures or protocols. There have been a few research efforts for the aspect of mobility prediction.

A widely used mobility model is fluid flow model [4] and Hong and Rappaport's model [5]. They have been used to analyze the performance of mobile networks such as the impact of a hand-off function on the assumption that users have a mean movement velocity and direction in each cell. However, they are not suitable for predicting users' future location with different movement property at each cell. Another method proposed for modeling the location information of users is introduced in [6]. The method subdivides users' movements into two patterns: regular and irregular. For modeling these patterns, the movement circle/track model and Markov chain model are used, respectively. All the movement patterns are stored in the Itinerary-Pattern Base (IPB), which is a database containing them. Then, movement patterns for a current location are retrieved from the IPB by a pattern matching method. This model has a possibility to predict the future location of users. However, it is not adaptive and robust in case that users' current location does not match any patterns stored in the IPB. In [7], when users are in a cell boundary, the next cell prediction using only current location information is proposed. However, this model employs only current location information during prediction process. Thus, it may incur incorrect prediction because past location information is not used for prediction.

When a failure occurs in mobility databases, location records stored in them are lost. So, the exact location of users may not be identified, and the latest location information of users may not be updated. In this situation, a connection may not be established for incoming calls to users. Typically, without an explicit restoration procedure, the location information can be restored after a failure by means of one of the following events: autonomous location registration, call origination, and registration boundary crossing. These three events are used to restore mobility databases in the absence of an explicit restoration procedure. However, the delay and cost in restoring mobility databases depend on the occurrence of three events by users. If any of these three events does not occur after a failure, the location information of users are permanently not restored. After a long time passes, if one of these events occurs, services may not be served to users for the failure period.

The existing two standards for mobile networks are IS-41 and GSM. Unlike the IS-41, GSM have provided an explicit restoration procedure [3] to recover mobility databases after a failure occurs. During failure-free operations, the location information recorded in the mobility databases is periodically checkpointed into a stable storage co-located in the mobility databases. After a failure occurs, the location information is immediately reconstructed from the backup. However, backup records are not always exact; some backup records may be obsolete, because the location information may have changed during the interval between checkpoint time and failure time. In such instances, mobility databases are restored from incorrect location information. Then, incoming calls to users may be lost due to the absence of location information for the users. Furthermore, users often experience delays in service using this checkpointing scheme, not to mention it takes up additional storage space. The bigger the number of users, the larger the amount of records needed to store the location information of users.

Several restoration schemes have been proposed in [8,9]. These works have focused on deriving the optimal checkpoint interval to balance the checkpointing cost against the paging cost. However, it is not adequate to apply the interval to all users due to a different mobility of each user. The movement behavior of users is closely related to the cost and delay required to locate users after a failure. For example, fast-moving users may quickly move out of the last checkpointed location. In this case, systems should page entire cells within a RA due to the first unsuccessful paging in the last checkpointed location. Moreover, if users move far away from the last checkpointed location, more successive paging must be executed, resulting in redundant paging.

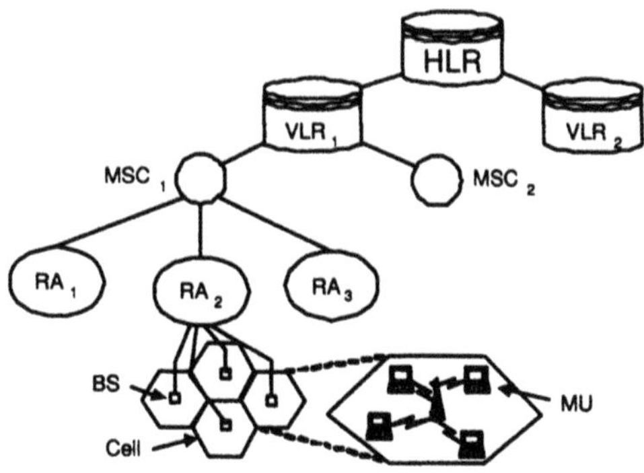

Fig. 1. Mobility Database Architecture

Thus, the checkpointing of mobility databases does not seem to be an appropriate approach for improving the efficiency of a restoration procedure.

3 System Model

The network architecture of mobile networks is composed of two sets of entities: one is the fixed network and the other is mobile units (or users). At the end of the fixed network, a Base Station (BS) is augmented to provide users with an interface. Users can only communicate with a BS through a wireless link. The service area is partitioned into cells. A cell is the geographical area covered by a BS. In each cell, a BS periodically broadcasts location information to its own area. Users in each cell are able to identify their current location by listening to the location information. Several cells compose a Registration Area (RA). Each RA is connected to a Mobile Switching Center (MSC) through a wired network. An MSC typically provides switching functions which coordinate location registration and call delivery. The MSC has access to the mobility databases in the networks. These mobility databases are used to store the location and service information for each registered user in mobile networks. The architecture of mobility databases is typically based on a two-tier hierarchy structure, Home Location Register (HLR) and Visitor Location Register (VLR). The HLR is a global database which stores information about all users registered in mobile networks. A VLR is a local database usually associated with the MSC. It stores information about users visiting the MSC's RA. Whenever a handover occurs during the wireless connection of users, the RA maintains information about the current cell location of users. Figure 1 shows the systems architecture of mobile networks.

4 Mobility Learning and Prediction System

In this section, we present a mobility learning and prediction system that plays an important part in this paper. The motivation to use a Neuro-Fuzzy

Inference System (NFIS) arises from the simple heuristic of existing prediction schemes. Since these schemes are based on direct observation in the current mobile situation, they lead to poor prediction performance, particularly if the current location of users does not match any location information known to mobile systems. On the other hand, our scheme derives an appropriate prediction for a future location from the previously known mobile information using the NFIS, even if the location history never occur in any of the mobile information.

4.1 Neuro-fuzzy inference model

In this paper, we use the simplified fuzzy inference model form among various fuzzy models. The simplified fuzzy inference model, referred to as the zero-order Sugeno' model in [10], is based on the fuzzy IF-THEN rules whose consequence is a real number. Thus, this model provides the inference structure that reduces the computation complexity of defuzzification in an inference procedure. The form of fuzzy IF-THEN rules is as follows:

$$\text{Rule } i: \ \texttt{IF } x_1 \texttt{ is } A_1^i \text{ and}, \ldots, \text{ and } x_d \texttt{ is } A_d^i, \texttt{ THEN } y \texttt{ is } \omega_i \tag{1}$$

where, i is a fuzzy rule number; $x_1, \ldots, x_d$ are input variables; A_j^i is a fuzzy set for input variable x_j in the ith fuzzy rule; ω_i is a real number for output variable y in the ith fuzzy rule ($i = 1, 2, \ldots, n$, $j = 1, 2, \ldots, d$). Given the real-valued input vector $\boldsymbol{x} = [x_1, x_2, \ldots, x_d]$, the real-valued output of the fuzzy model is inferred as follows:

$$f(\boldsymbol{x}) = \frac{\sum_{i=1}^{n} \mu_i \cdot \omega_i}{\sum_{i=1}^{n} \mu_i}, \quad \mu_i = \prod_{j=1}^{d} \mu_{A_j^i}(x_j) \tag{2}$$

Here, $\mu_{A_j^i}$ is a fuzzy membership function of the fuzzy set A_j^i and μ_i is a membership value of ith fuzzy rule. This mapping is performed using singleton fuzzifier as fuzzification, product-inference operator as fuzzy inference procedure, and weighted-average technique as defuzzification. The fuzzy membership function can employ various forms such as triangular, trapezoid, and Gaussian according to application problems. Because of its smoothness and concise notation, Gaussian function is becoming increasingly popular for specifying fuzzy sets. Moreover, since the function has a continuous and differentiable property, it suitably applies to the learning rules. Thus, we employ the Gaussian function as the fuzzy membership function.

$$\mu_{A_j^i}(x_j) = \exp\left[-\frac{1}{2} \cdot \left(\frac{x_j - c_j^i}{\sigma_j^i}\right)^2\right] \tag{3}$$

Here, c_j^i is the central value of the fuzzy membership function and σ_j^i is the variance of the central value of the fuzzy membership function.

The parameters c_j^i , σ_j^i in (3) and the parameter ω_i in (2) are iteratively adjusted by back-propagation learning algorithms [10] in order to minimize a learning error ($e = \frac{1}{2}\sum_p (y_p - \hat{y}_p)^2$) and produce a better output. The learning rules for these parameters are

$$\begin{aligned} c_j^i(t+1) &= c_j^i(t) - \eta \cdot \frac{\partial e}{\partial c_j^i} = c_j^i(t) - \eta \cdot \frac{\partial e}{\partial \mu_i} \cdot \frac{\partial \mu_i}{\partial \mu_{A_j^i}} \cdot \frac{\partial \mu_{A_j^i}}{\partial c_j^i} \\ &= c_j^i(t) - \eta \cdot (y_p - \hat{y}_p) \cdot \left(\frac{(\omega_i(t) - \hat{y}_p)}{\sum_{k=1}^{n} \mu_k} \right) \cdot \mu_i \cdot \left(\frac{x_j - c_j^i(t)}{(\sigma_j^i(t))^2} \right) \end{aligned} \quad (4)$$

$$\begin{aligned} \sigma_j^i(t+1) &= \sigma_j^i(t) - \eta \cdot \frac{\partial e}{\partial \sigma_j^i} = \sigma_j^i(t) - \eta \cdot \frac{\partial e}{\partial \mu_i} \cdot \frac{\partial \mu_i}{\partial \mu_{A_j^i}} \cdot \frac{\partial \mu_{A_j^i}}{\partial \sigma_j^i} \\ &= \sigma_j^i(t) - \eta \cdot (y_p - \hat{y}_p) \cdot \left(\frac{(\omega_i(t) - \hat{y}_p)}{\sum_{k=1}^{n} \mu_k} \right) \cdot \mu_i \cdot \left(\frac{(x_j - c_j^i(t))^2}{(\sigma_j^i(t))^3} \right) \end{aligned} \quad (5)$$

$$\begin{aligned} \omega_i(t+1) &= \omega_i(t) - \eta \cdot \frac{\partial e}{\partial \omega_i} \\ &= \omega_i(t) - \eta \cdot \left(\frac{\mu_i}{\sum_{k=1}^{n} \mu_k} \right) \cdot (y_p - \hat{y}_p) \end{aligned} \quad (6)$$

where, y_p is the actual output for the pth input-output data, $\hat{y}_p$ is the output of the neuro-fuzzy inference system for the pth input-output data, and η is a learning rate.

4.2 Mobility learning and prediction

We have devised a mobility learning and prediction system to predict the probable location of users. The system expresses user mobility as the movement velocity and direction. It predicts a future location from the movement factors of current and past locations. Let S_k^t be the state k at time t of the movement of users. S_k^t represents the movement factors that determine the next location to which users will move. In this paper, $S_k^t = (v_k^t, \theta_k^t)$ such that $k \leq K$ and $t \in T$. Here, K is the number of states in the state space, T is the set of time, v_k^t is the kth movement velocity at time t, and θ_k^t is the kth movement direction at time t. We assume that a movement state can be changed when a period of time τ elapses.

Using the movement states defined above, the movement function which maps current and past movement states into a future movement state can be denoted by $\text{NFIS}(S_k^t, S_k^{t-1}, \ldots, S_k^{t-(h-1)}) \mapsto S_k^{t+1}$. Here, NFIS and h are a neuro-fuzzy inference system and the number of current and past movement states, respectively. The movement function uses the current and past movement states $S_k^t, S_k^{t-1}, \ldots, S_k^{t-(h-1)}$ to predict the future movement state S_k^{t+1}. $S_k^t, S_k^{t-1}, \ldots, S_k^{t-(h-1)}$ represents the location history which includes the movement patterns accumulated in the previous days or months. They are used for the input of the NFIS. The output of the movement function is the future movement state inferred by the NFIS. The probable location of users is obtained from the velocity and direction of the future movement state S_k^{t+1}.

In order to construct fuzzy rulebase from movement states for several days or months, we use clustering-based techniques [10] that can automatically generate fuzzy rules according to the degree of similarity of movement states. The clustering-based techniques partition movement states into some clusters so that the similarity within a cluster is larger than that within others. Each cluster has a cluster center. The cluster center is used as the central value of the fuzzy membership function in the antecedent part of a fuzzy rule. Let $\boldsymbol{R}_i (= [R_i^1, R_i^2, \ldots, R_i^h])$ be the ith cluster center among M clusters and $\boldsymbol{S}_k^t (= [S_k^t, S_k^{t-1}, \ldots, S_k^{t-(h-1)}])$ be the kth movement state vector at time t. In order to measure a similarity degree between $\boldsymbol{R_i}$ and $\boldsymbol{S_k^t}$, a distance function can be defined by

$$D_i = |\boldsymbol{S}_k^t - \boldsymbol{R}_i| \tag{7}$$

where, D_i is a distance between $\boldsymbol{R_i}$ and $\boldsymbol{S_k^t}$. In (7), as $\boldsymbol{R_i}$ and $\boldsymbol{S_k^t}$ become more similar, D_i grows smaller. Then, it is necessary to determine a degree of similarity between $\boldsymbol{R_i}$ and $\boldsymbol{S_k^t}$. To determine this similarity, we define a radius r. By using (7) and the radius r, a degree of similarity between a movement state vector and each cluster is defined as follows: If $D_i \leq r$, the similarity is high, otherwise it is low.

Based on this criterion, the movement state vector of users can be subdivided into three classes:

- Not defining a movement state vector as a cluster (fuzzy rule).
- Defining a movement state vector as a cluster.
- Defining as a movement state vector as a cluster, but considering the movement state vector as an unnecessary cluster.

The first class indicates that a movement state vector is added to fuzzy rulebase as a new cluster (fuzzy rule) because the vector is not correspondent to any cluster among predefined ones. In this case, the antecedent and consequence of a new fuzzy rule are organized as follows:

$$\mu_{A_j^i}(x_j) = \exp\left[-\frac{1}{2} \cdot \left(\frac{x_j - S_k^{t-(j-1)}}{r}\right)^2\right], \quad \omega_i = S_k^{t+1} \tag{8}$$

where, $S_k^{t-(j-1)}$ and r correspond to the central value (c_j^i) and the variance (σ_j^i) of a fuzzy membership function, respectively. S_k^{t+1} corresponds to the real number (ω_i) of consequence part.

The second class indicates that a movement state vector has a high degree of similarity to several clusters among predefined ones. In this case, by increasing the importance of the clusters, the NFIS results in a more exact prediction when a similar movement state vector with predefined movements states appears in the movement path of users. In order that the movement state can affect the real value of the consequence part in fuzzy rules, ω_i is updated as follows:

$$\omega_i = \frac{\sum_{k=1}^{m_i} S_k^{t+1}}{m_i} \tag{9}$$

where, m_i is the number of movement state vectors within the ith cluster. Using (9), we get the updated weight which corresponds to the mean of future movement states within a cluster.

The third class indicates that a movement state vector had been previously defined as a fuzzy rule, but is currently considered unnecessary. This means the wrong reflection of the latest movement state in fuzzy rulebase. This situation happens when the past movement of users differs from their ordinary movement patterns. The cluster (fuzzy rule) made from this situation causes incorrect predictions. So, the unnecessary fuzzy rule should be eliminated from fuzzy rulebase according to an appropriate mobile situation. In other words, those fuzzy rules that suitably reflect the latest movement state should be stored in the fuzzy rulebase. Thus, it is necessary for a criterion to determine whether or not there exists unnecessary fuzzy rules in the fuzzy rulebase. Such a criterion is given as

$$L(age_i) = \exp[-\alpha \cdot age_i] \ , \quad \text{Rule } i = \begin{cases} young & \text{if } L(age_i) \leq \beta \\ old & \text{if } L(age_i) > \beta \end{cases} \tag{10}$$

where, age_i is the age of the ith fuzzy rule, α is the parameter which controls the age of a fuzzy rule, and β is the parameter which determines the elimination of a fuzzy rule.

The age of a fuzzy rule is updated as follows: If a distance between a current movement state vector and a cluster center is less than the radius, the age of a fuzzy rule becomes zero. Otherwise, the age of a fuzzy rule is increased by one. To determine how much the movement state of users affects fuzzy rules, we introduce the parameter β. According to the value of β, fuzzy rules are subdivided into two groups: young fuzzy rules and old fuzzy rules. Since the young fuzzy rules suitably reflect the latest movement state, it is desirable that they be used as fuzzy rules. The old fuzzy rules are built from the movement state which is found just once in the movement path of users during β. There is a very strong possibility that the old fuzzy rules bring about incorrect prediction. Thus, the old fuzzy rules should be eliminated from the fuzzy rulebase. In our system, if the age of a fuzzy rule is more than β (the old fuzzy rules), this fuzzy rule is eliminated as an unnecessary fuzzy rule.

Here, we describe an overall algorithm for mobility learning and prediction. The algorithm performs a learning procedure for the movement patterns of users in order to construct more an elaborate fuzzy rulebase. It also performs a prediction procedure to predict the future location. The various steps are described as follows.

Algorithm 1: Mobility Learning and Prediction

Let $\boldsymbol{S}_k^t (= [S_k^t, S_k^{t-1}, \ldots, S_k^{t-(h-1)}])$ be the movement state vector containing the current and past movement information at time t ($k = 1, 2, \ldots, K$, $t = 1, 2, \ldots, T$). We predict the future movement state S_k^{t+1} from $\boldsymbol{S}_k^t$ at time t by the NFIS.

Step 1: Starting with the first movement state vector $\boldsymbol{S}_1^t$ and the first future movement state S_1^{t+1}, establish a cluster center $\boldsymbol{R}_1$ at $\boldsymbol{S}_1^t$, set $\omega_1 = S_1^{t+1}$, and assign zero to the age of the cluster age_1. Select a radius r.

Step 2: Suppose that when we consider the kth movement state vector $\boldsymbol{S}_k^t$, there are M clusters with centers at $\boldsymbol{R}_1, \boldsymbol{R}_2, \ldots, \boldsymbol{R}_M$. Compute the similarity of $\boldsymbol{S}_k^t$ to these M cluster centers by (7) and let the highest similarity be $D_{k'}$, that is, the nearest cluster to $\boldsymbol{S}_k^t$ is $\boldsymbol{R}_{k'}$ ($k' = 1, 2, \ldots, M$).

a) If $D_{k'} > r$, establish $\boldsymbol{S}_k^t$ as a new cluster center $\boldsymbol{R}_{M+1} = \boldsymbol{S}_k^t$, set $\omega_{M+1} = S_k^{t+1}$, and assign zero to the age of the new cluster age_{M+1}.

b) If $D_{k'} \le r$, update $\omega_{k'}$ by (9) and set $age_l = age_l + 1$ for $l = 1, 2, \ldots, M$ with $l \ne k'$.

Step 3: Eliminate unnecessary clusters by (10).

Step 4: Compute the future movement state $\hat{S}_k^{t+1}$ at the kth movement state vector $\boldsymbol{S}_k^t$ by (2).

Step 5: Adjust the parameters c_j^i ,σ_j^i, and ω_i as much as the number of learning iterations by (4), (5), and (6).

Step 6: Execute from Step 2 to Step 5 iteratively.

So far, we have described the mobility learning and prediction system based on the NFIS for modeling and predicting the movement patterns of users. The next section proposes an restoration scheme with the mobility learning and prediction.

5 Restoration Scheme with Mobility Learning and Prediction

In this section, we describe an restoration scheme to provide service availability to users even after the failure of mobility databases. Our restoration scheme consists of two primary operations: failure-free operations and failure recovery operations.

5.1 Failure-free operations

A single fault model is assumed in our restoration scheme. The mobility databases are assumed to be fail-stop and so the lost location records are restored immediately after a failure by the mobility learning and prediction described in Sect. 4.2.

During failure-free operations, our restoration scheme learns the users' moving trajectory. We consider the current movement direction and velocity as user mobility parameters. Basically, our restoration scheme assumes that users inform mobility databases of their current location every time interval τ_r. Also, mobility databases update their location records when users originate a call, when a handover occurs, and so on. These assumptions reduce wireless link consumption. The movement patterns are gathered at the MSC and periodically transferred to the HLR via the VLR. Then, the NFIS learns the movement patterns using the latest location information. The failure-free operations are same to the mobility learning and prediction algorithm described in Sect. 4.2

5.2 Failure recovery operations

After a failure, mobility databases must restore the same location information as before. Our approach to restore failed mobility databases is based on the mobility learning and prediction. When a failure occurs, the mobility learning and prediction system is initiated to predict a user's probable location. Let t_f be the time at which a failure occurs in mobility databases, and $\boldsymbol{S}_k^{t_f}$ be the kth movement state vector at time t_f. After time t_f, the movement state vector at future time t cannot employ current and past movement information any more ($t_f < t$). Thus, in our scheme, to predict the kth movement state at current time t_p ($t < t_p$), future movement state vectors are recursively constructed with the location information predicted by NFIS up to time t_p. The kth movement state vector at time t, $\boldsymbol{S}_k^t$, is organized as follows:

$$\boldsymbol{S}_k^t = \begin{cases} \left[\hat{S}_k^t, \hat{S}_k^{t-1}, \ldots, \hat{S}_k^{t_f+1}, S_k^{t_f}, S_k^{t_f-1}, \ldots, S_k^{t-(h-1)}\right] & \text{if } t_f < t < t_f + h \\ \left[\hat{S}_k^t, \hat{S}_k^{t-1}, \ldots, \hat{S}_k^{t-(h-1)}\right] & \text{if } t \geq t_f + h \end{cases} \tag{11}$$

where, S_k^* and $\hat{S}_k^*$ represent the movement states included with $\boldsymbol{S}_k^{t_f}$ and the predicted movement states by NFIS, respectively.

Using (11), we can get different movement state vectors whether or not S_k^* is included with the movement state vector at time t. $\boldsymbol{S}_k^t$ is used for predicting the movement state $\hat{S}_k^{t+1}$ at time $t+1$, and $\hat{S}_k^{t+1}$ is used for organizing the movement state vector S_k^{t+1} at time $t+1$. This process is repeated up to time t_p by NFIS in order to predict the movement state $\hat{S}_k^{t_p}$ at time t_p. The prediction information derived in this way is used as the first paging cell to locate lost users. This prediction information represents the probable cell of users at the failure time of the mobility databases. The accuracy of predicting the probable cell location of users depends on the time interval between t_f and t_p. In general, users may move farther and farther away from their location at time t_f, as the time interval grows longer. However, if the mobility learning and prediction system correctly learns user mobility during failure-free operations, it can predict the probable cell of users at time t_p, with a high rate of accuracy. In this case, only one paging is needed to locate lost users after a failure. So, as the rate of accuracy in predicting location increases, paging cost goes down.

Our restoration scheme consists of HLR failure restoration procedure and VLR failure restoration procedure. Figure 2(a) and (b) shows the overall flow of the failure restoration of two databases, the HLR and the VLR, respectively. Algorithms 2 and 3 also show the restoration step of two mobility databases after a failure.

Algorithm 2: HLR Failure Restoration by Mobility Prediction

Step 1: The failed HLR sends a restoration initiation message to all VLRs in order to gather the lost location information of users.

Step 2: Each VLR that receives the restoration initiation message is queried to search the lost location information of users.

Step 3: The failed HLR receives an exact location information of users from all VLRs. The HLR is reconstructed using the location information.

Algorithm 3: VLR Failure Restoration by Mobility Prediction

Step 1: The failed VLR sends a restoration initiation message to the HLR in order to gather the lost location information of users.

Step 2: The HLR that receives the restoration initiation message predicts the location of the requested users.

Step 3: The failed VLR receives the predicted location from the HLR.

Step 4: The predicted location is paged to locate users.

- **Step 4.1:** A cell corresponding to the predicted location is paged. If any response from the paged user is returned, the location information for the user is reconstructed.
- **Step 4.2:** If there is no response within a specific time, the adjacent cells of the predicted location are paged.
- **Step 4.3:** If there's still no response from the adjacent cell paging, entire cells within a RA are paged.

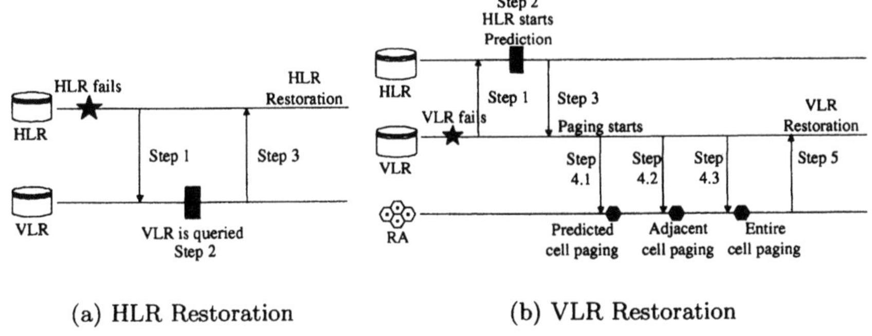

(a) HLR Restoration (b) VLR Restoration

Fig. 2. Mobility Database Restoration by Mobility Prediction

Step 5: The failed VLR receives an exact location information of the paged users. The VLR is reconstructed using the location information. If any response from the paged users is not returned after three successive pagings, the failed VLR waits until the users originate a call.

The checkpointing scheme is a passive restoration strategy in which the location information of users is backed up into a stable storage during failure-free operations. Because failed databases are restored from storage, restored location information may be obsolete. So, restoration efficiency under the checkpointing scheme is very sensitive to user mobility. On the other hand, our scheme is an aggressive restoration strategy in which the lost location information is recovered by predicting the users' movements at the failure time. It predicts the probable location of users after a failure and uses this the predicted location information as a starting paging cell. Since the predicted location information is based on the movement patterns of users during failure-free operations, we expect it to be more accurate and therefore entail less number of paging.

6 Performance Evaluation

In this section, we evaluate the performance of our mobility prediction and restoration schemes through simulations. Our simulations are performed with the mobility data, which are approximated to users' real movements. So, we first describe how to generate mobility data. Then, we show an efficiency of two schemes. In particular, we compare our restoration scheme with a restoration scheme that uses a periodic checkpointing scheme [9].

6.1 Mobility Data Generation

In general, mobility data are too complicated to be collected directly from the real mobile environments. Our procedure to generate mobility data consists of a user mobility model and user mobility data generation. The user mobility model presents a method of modeling users' movement behavior in terms of the velocity and direction. Based on the user mobility model, mobility data are generated.

Users have their own movement behavior. The movement behavior is defined as the act of moving from one given geographical location to another, over a spatial dimension. According to their movement behavior, users have specific movement patterns. Most previous works [4,5] have used some version of a random-walk mobility model, which fails to take into account certain aspects of movement behavior, such as users' movement patterns. In [11], the traveling demand model is proposed to represent observed travel patterns by travel mode. This is based on transportation planning and traffic engineering models that describe and predict travel demand, typically concentrating on urban areas [12]. In this model, it is assumed that individual travelers make travel choices which are "best" for them. Based on the traveling demand model, we extract the mobility data in terms of users' movement velocity and direction.

The velocity of users can be interpreted in two ways: using mean velocity and an arbitrary velocity variation. In a real situation, when users move at a mean velocity, the velocity variation would be nearly zero. If users move without any periodicity for a velocity per period of time, they move at a random velocity in the range of a mean velocity. Thus, at time $t+1$, users have the following velocity:

$$v_{t+1} = \bar{v_t} + \tilde{v_t} \tag{12}$$

where, $\bar{v_t}$ is a mean velocity at time t; $\tilde{v_t}$ is the variation of a mean velocity at time t and it is selected using a uniform distribution in the velocity variation range $[-\tilde{v_c}, \tilde{v_c}]$. In (12), as the velocity variation range increases, the velocity grows more random.

On the assumption that users move toward a destination along their route, users are differentiated using their target direction toward the destination and a variation of the target direction. Thus, the movement direction at time $t+1$, θ_{t+1} can be defined by

$$\theta_{t+1} = \bar{\theta_t} + \frac{2\pi}{d} \cdot k \tag{13}$$

where, $\bar{\theta_t}$ is a current direction toward a destination at time t; d is the number of movement directions; k indicates the movement direction index and is selected with the probability p_k, where $k \in \{0, 1, \ldots, d-1\}$. In (13), if the probability of the direction k toward a destination, p_k, is larger than that of other directions, users would move into a destination without abrupt direction

Table 1. User Classes and Parameter Values

Velocity			*Direction*		
Class	Meaning	Value ($\tilde{v_c}$)	Class	Meaning	Value ($\boldsymbol{p}$)
A	regular	(5, 25)	X	almost straight	[0.0 0.1 0.8 0.1 0.0]
B	moderate	(10, 40)	Y	moderate	[0.1 0.2 0.4 0.2 0.1]
C	fluctuating	(20, 60)	Z	staggering	[0.2 0.2 0.2 0.2 0.2]

changes. On the other hand, if all probabilities are equal, users would move into a destination with a random direction.

Through the mobility model described above, users are categorized into several movement types. In this paper, we will make the mobility data per each movement types and employ them in evaluating the performance of our restoration scheme.

6.2 Simulation Environments

When carrying out the simulations, we assume that users move to any cell in the hexagon cell environments along unknown movement paths according to their movement properties. The assumed total number of cells is 256. Each VLR has the same number of cells. It is also assumed that the radius of a cell is 500 m. Within these geographical parameters, the location information of users is measured at every 0.0835 hours for 30 days.

In order to extract mobility data, we classify users into nine user types by the combination of three kinds of velocities (A, B, and C) and three kinds of directions (X, Y, and Z). Table 1 shows the user classes for velocity and direction. Among nine user types, the users of the type AX always move toward a destination, keeping a mean velocity and direction. On the other hand, the users of the type CZ display a random velocity and direction in their movement paths.

For the sake of simplicity, we assume that $\bar{v_t} = 5km/h$ and $d = 5$ for all time t. Thus, users have different movement paths according to the velocity variation range ($\tilde{v_c}$) and the direction probability vector ($\boldsymbol{p} = [p_0, p_1, \ldots, p_{d-1}]$ such that $\sum_{k=1}^{d-1} p_k = 1$). Parameters values for each user class are also shown in Table 2.

Our mobility prediction system gets the location history of users at current location whenever users move into any location. Then, the future location of users is obtained by the NFIS. The predicted future location is the cell to be entered by users. The performance of our mobility prediction system is determined as the hit degree between the predicted cell and the real cell where users visit. Thus, in order to analyze the performance of our mobility prediction system, we use the prediction ratio which is defined as a ratio of correctly predicted cells out of the total cells in a movement path. We will also compare our prediction system with a simple prediction scheme named Average Prediction (AP). This scheme uses the mean velocity and direction

Table 2. Simulation Parameters

Parameter Description	Values
The number of mobility data (*days*)	30
The number of current and past movement states (h)	4
The change period of movement states (τ)	0.0835
Radius (r)	0.3
Learning rate (η)	0.05
The number of learning iterations	50
The control parameter for the age of fuzzy rules (α)	0.1
The remove parameter for unnecessary fuzzy rules (β)	0.005

of current and past locations to derive the next location from the location history.

With mobility data for each user types, we evaluated the performance of our restoration scheme. During failure-free operations, NFIS learns the movement patterns of users with the mobility data. For the sake of simplicity, it was assumed that the failure of mobility databases happens after it passes 30 days. Then, a hit ratio between a predicted location and actual location was collected. These results were compared to those gathered using the periodic checkpointing scheme.

The performance of the restoration scheme with periodic checkpointing is shown in [9]. After failed mobility databases are restored from a stable storage, lost users are paged to find their exact locations. If lost users are not found by the paging, entire paging within a RA should be performed. This renders the checkpointed location information obsolete because users have moved away from the checkpointed location. Thus, the performance of the restoration scheme with periodic checkpointing can be determined by a hit ratio between a restored location and the probable location of lost users after a failure. Accordingly, it is shown in [9] that users' residence time in a RA follows an exponential distribution with mean $\frac{1}{\lambda_m}$, where λ_m is a mean probability that users move out of the RA. Under this observation, the probability P_{T_V} which users move in a RA at a time between 0 and T_V before checkpointing and move out of the RA at the time T_V after a failure, is derived as

$$P_{T_V} = \frac{1}{T_V} \int_0^{T_V} e^{-\lambda_m (T_V - s)} ds = \frac{1 - e^{-\lambda_m \cdot T_V}}{T_V \cdot \lambda_m} \quad (14)$$

where, T_V is an average checkpoint interval. From (14), we can deduce that users do not move out of the RA of a checkpointed location after a failure occurs at mobility databases. It should be noted that, as the probability becomes lower, the paging area to locate lost users grows wider. In our simulations, (14) was used to compare the performance of our restoration scheme with that of a restoration scheme with periodic checkpointing. For demonstration purposes, we assume that $\lambda_m = 0.4$.

The performance of our restoration scheme is determined by the degree of accuracy of actual location vis-à-vis predicted location by the NFIS. If our scheme can predicts the probable location of users after a failure, only one paging is needed to locate the users. This dramatically decreases paging cost. In order to analyze the performance of our restoration scheme, we evaluate how many location records are exactly restored after the failure of mobility databases elapses. As the criterion of the performance, we use an Exact Cell Hit Ratio (ECHR) and an Adjacent Cell Hit Ratio (ACHR). The ECHR is defined as a ratio of the correctly restored location information by a single paging to the total number of lost location information at a future time after a failure. The ACHR is defined as a ratio of correctly restored location information by the paging of adjacent cells around predicted cell to the total number of lost location information at a future time after a failure.

6.3 Simulation Results

Figure 3 shows the prediction ratios of our scheme and the AP scheme for all user types. As shown in this figure, our prediction system has a higher prediction ratio than the AP scheme. We can see that the mean prediction ratio of the AP scheme sharply decreases as randomness for the movement patterns of users increases. On the other hand, our prediction system has an uniformly high prediction ratio regardless of a variety of user types. It is also shown that as randomness for the movement patterns of users increases, the prediction ratio range of the AP scheme become wider than that of our system. In addition, our system has prediction ratios in the range of 88 to 100% for all user types, while the AP scheme has prediction ratios in the range of 0 to 95%. Because the AP scheme uses the mean of current and past locations to predict the future location, all location information is not included during the learning phase of the NFIS. Thus, it cannot express any periodic characteristics in users' movement patterns and so it is heavily affected by the randomness of the movement patterns.

Figure 4 shows the average hit ratio of all user types as a time after the failure of mobility databases elapses. In this figure, the solid line indicates the average hit ratio of the checkpointing scheme. The dotted line and the dashed line indicate the average ECHR and ACHR of our restoration scheme, respectively. The checkpointing scheme shows a steep descent as a time elapses. Besides, we can see that the curve sharply goes down at the initial stage of the failure. On the other hand, our restoration scheme maintains a high average hit ratio for all user types until a time after a failure elapses in the range from 50 to 70 minutes. Even though the ECHR are somewhat low, the ACHR which is a hit ratio for the adjacent cells of a predicted cell is considerably high. Thus, this figure shows that the performance of the checkpointing scheme is very sensitive to the time elapsed after a failure. On the

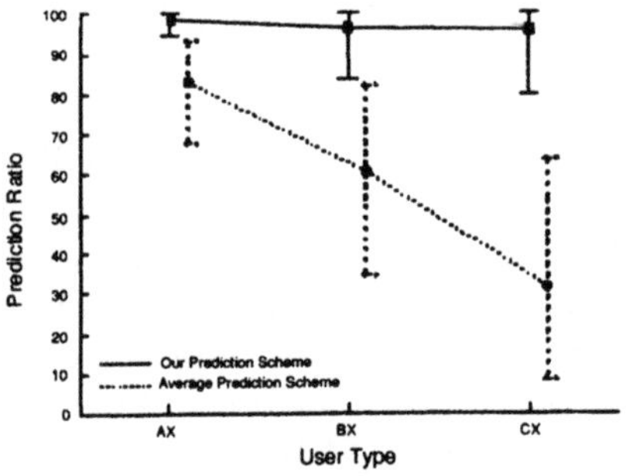

(a) Almost Straight Direction

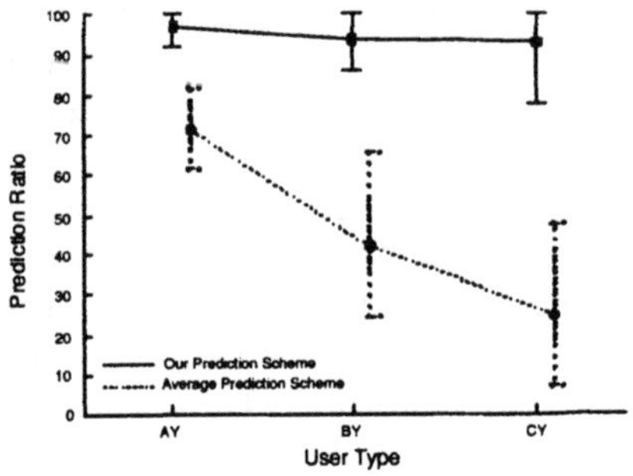

(b) Moderate Direction

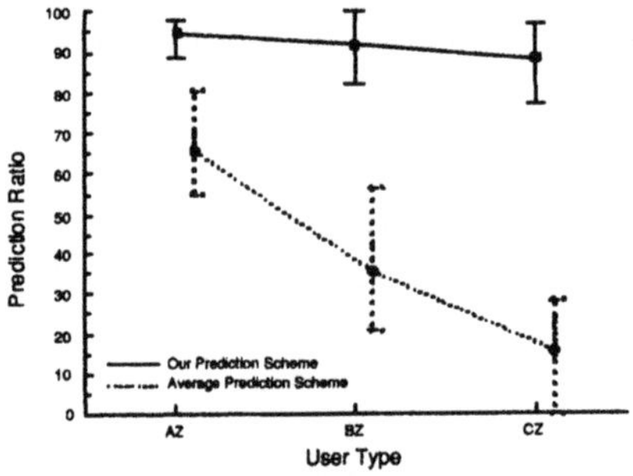

(c) Staggering Direction

Fig. 3. ECHR for All Users

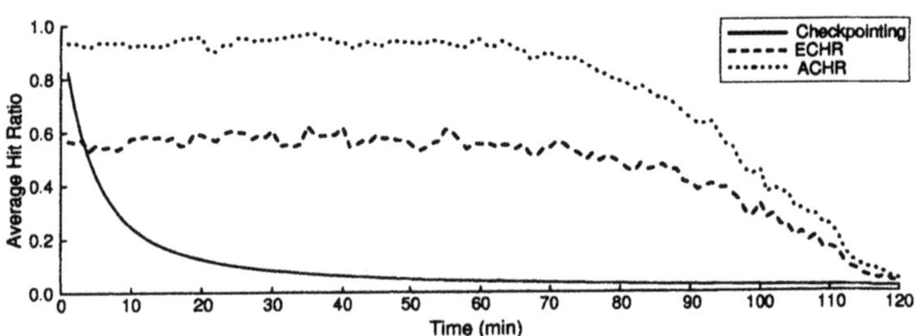

Fig. 4. Average Hit Ratio as a Time after a Failure Elapses

other hand, the performance of our restoration scheme does not depend on the elapsed time. Even though a lot of time has elapsed after a failure, our restoration scheme maintains a high average hit ratio.

From the results of Fig. 4, we observe that our restoration scheme can reduce the cost needed to restore location records in the presence of the failure of mobility databases. This reduction indicates that our restoration scheme is able to predicts the probable location of lost users after a failure. In addition, our restoration scheme can sufficiently locate lost users using less number of paging, resulting in a dramatic decrease in paging cost. On the other hand, the checkpointing scheme can locate lost users by paging an entire RA after the first unsuccessful paging. Because a RA in practical systems consists of dozens or hundreds of cells, checkpointing scheme should page a lot of cells to locate lost users.

We now examine the effect of the users' random movement on the ECHR. Figure 5 shows the ECHR of nine user types as the elapsed time after a failure. It shows the probability that users exactly reside on the predicted cell by the NFIS after a failure occurs. In Fig. 5, the users of the type AX have highest hit ratio from among nine user types because they have fairly regular movement patterns. The ECHR of the users is in the range of 80 and 90% up to a failure duration time of roughly 110 minutes. On the other hand, the users of the type CZ who have somewhat random movement patterns have the lowest hit ratio. Even though the ECHR of this type is not high as that of the other types, there is no steep slope during an initial stage after a failure. We can see in Fig. 5 that the users of this type maintain a uniform ECHR up to a failure duration time of roughly 60 minutes.

Upon analyzing the ECHR of Fig. 5, we can find that as the variation of velocity and direction increases, the ECHR decreases. Nevertheless, for all user types, there is no steep slope during the initial stage of a failure. This means that the location records of lost users can be restored with uniform paging costs in the initial state of a failure regardless of users' random movements.

7 Conclusions

We have proposed an approach based on mobility learning and prediction to predict the future locations of users. The main issue addressed is the prediction of user mobility in order to improve the QoS for mobile services. The predicted location information by the NFIS could be used for supporting the preconnection of services and the preallocation of resources. Thus, users can maintain continuous connectivity to mobile systems, regardless of the new location to which they move. They can get mobile services or data with the same efficiency as at the previous location.

We also showed the applicability of the prediction system to the failure restoration of mobility databases. Unlike previous approaches using the checkpointing scheme, our restoration scheme can reduce the cost needed to restore the location records of lost users after a failure by using a lesser-number of paging. The simulations showed that our scheme provides con-

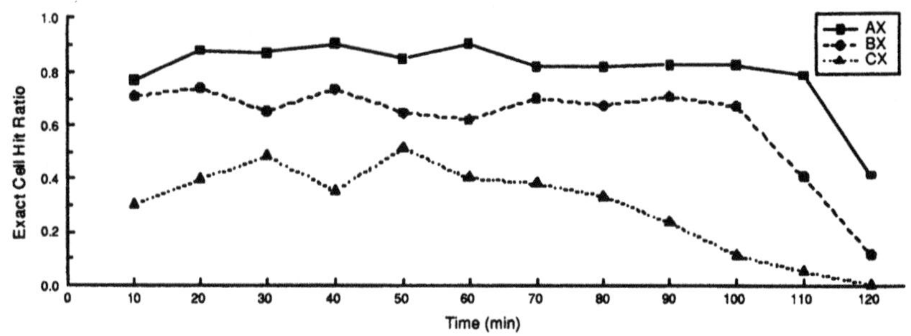

(a) Almost Straight Direction

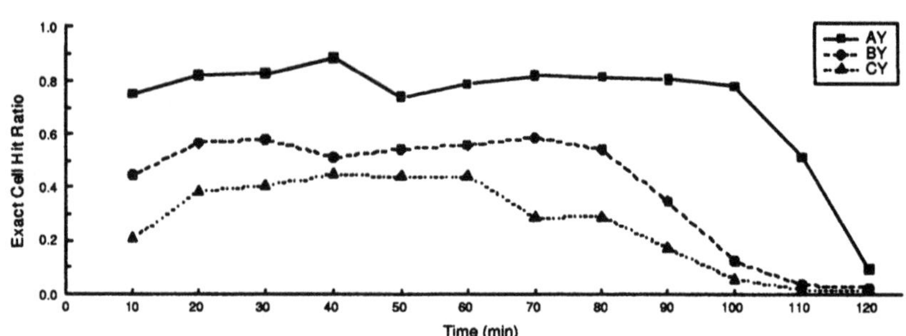

(b) Moderate Direction

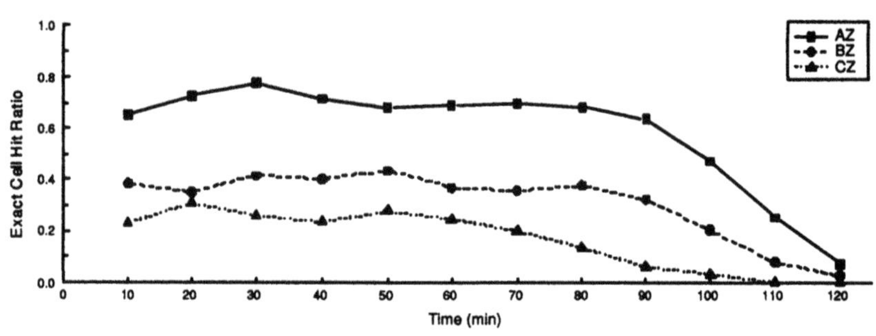

(c) Staggering Direction

Fig. 5. ECHR for All Users

siderably high restoration accuracy. While the checkpointing scheme has becomes less accurate as the checkpoint interval becomes larger, our restoration scheme with mobility learning and prediction capability maintains a considerably high restoration rate regardless of the time elapsed after the failure of mobility databases. In our restoration scheme, the NFIS replaces the role of the checkpoint as mobility learning and prediction. Thus, our restoration scheme has no backup process and so fewer failure-free operation costs. In addition, it needs no storage space to store the checkpoints. We conclude that our scheme produces more accurate restoration rate than the checkpointing scheme, at less delay and lower cost.

References

1. Pitoura E, Samaras G (1998) Data management for mobile computing. Kluwer Academic Publishers
2. EIA/TIA (1991) Cellular radio-telecommunications intersystems operations: Automatic roaming. Technical Report IS-41.3-B, EIA/TIA
3. ETSI/TC (1993) Restoration procedures, version 4.2.0. Technical Report Recommendation GSM 03.07, ETSI/TC
4. Thomas R, Gilbert H, Mazziotto G (1988) Influence of the movement of mobile station on the performance of the radio cellular network. Proc. of 3rd Nordic Seminar
5. Hong D, Rappaport S (1986) Traffic model and performance analysis for cellular mobile radio telephone systems with prioritized and non-prioritized handoff procedure. IEEE Trans. on Vehicular Technology 35: 77-91
6. Liu G, Maguire G (1996) A class of mobile motion prediction algorithms for wireless mobile computing and communications. Mobile Networks and Applications 1: 113-121
7. Liu T, Bahl P, Chlamtac I (1998) Mobility modeling, location tracking, and trajectory prediction in wireless ATM networks. IEEE Journal on Selected Areas in Communications 16: 922-936
8. Lin Y-B (1995) Failure restoration of mobility databases for persoanl communication networks. ACM-Baltzer Journal of Wireless Networks 1: 365-372
9. Wang T-P, Tseng C-C, Chou W-K (1997) An aggressive approach to failure restoration of PCS mobility databases. Mobile Computing and Communication Review 1: 21-28
10. Jang R, Sun C-T, Mizutani E (1997) Neuro-fuzzy and soft computing. Prentice-Hall
11. Scourias J, Kunz T (1997) A dynamic individualized location management algorithm. Proc. of the 8th IEEE Int. Symp. on Personal, Indoor, and Mobile Radio Communications, pp. 1004-1008
12. Oppenheim N (1994) Urban travel modeling. A Wiley-Interscience Publication

Voice Over Wireless LAN Using Intelligent Control

Maria C. Yuang and Po-Lung Tien

Department of Computer Science and Information Engineering
National Chiao Tung University
Hsinchu, Taiwan 30050
mcyuang@csie.nctu.edu.tw
tbl@csie.nctu.edu.tw

Abstract. In this chapter, we present a network framework in which Variable Bit Rate (VBR) voice is transported over a wireless Local Area Network (LAN) using two novel intelligent control systems. First, the Intelligent Multiple Access Control System (IMACS) at the base station governs the medium access of the uplink channel. In particular, IMACS supports both Available Bit Rate (ABR) data traffic and in-band signaling control (SCR) traffic for VBR voice based on a TDM-based contention access scheme. Significantly, dynamic allocation of contention bandwidth between ABR and SCR traffic is facilitated through predicting ABR self-similar traffic characteristics based on a neural-fuzzy approach. As a result, IMACS offers various QoS guarantees and a maximum of network throughput irrelevant to traffic variation. With such guarantees, the second system- Intelligent Voice Smoother (IVoS) at the application layer of each mobile terminal facilitates intramedia synchronization of voice data streams. The traffic predictor of IVoS predicts three traffic characteristics of every newly encountered talkspurt period. Based on the predicted characteristics, IVoS determines the corresponding buffering delay to be imposed on the first frame. All subsequent frames of the talkspurt can be playout in a quasi-Constant-Bit-Rate (CBR) manner. Finally, we demonstrate via experimental results that with such intelligent control, the playout Quality of Service (QoS) of VBR voice can be guaranteed irrespective of any traffic and load variation.

Keywords. Wireless Local Area Network (WLAN), Asynchronous Transfer Mode (ATM), Medium Access Control (MAC), intramedia synchronization, Quality of Service (QoS), self-similar traffic, neural-fuzzy method

1. Introduction

With the rapid proliferation of personal communication services provided to multimedia portable computers, wireless access to existing networks has

This work was supported in part by the MOE Program of Excellence Research, Taiwan, R.O.C., under Contract 89-E-FA04-1-4, and in part by Institute for Information Industry (III), MOEA, Taiwan, R.O.C., under Contract 91-0238.

emerged as a significant concern [1]. In particular, a wireless Asynchronous Transfer Mode (ATM) Local Area Network (LAN) [2,3] is a wireless LAN operating the ATM data transport technology and protocol [4] that has been successfully employed in broadband ISDN networks. Essentially, wireless ATM LAN has been envisioned as a potential framework for next-generation wireless local/access networks due to two critical strengths. They are the facilitation of seamless internetworking between LAN and Wide Area Network (WAN) technology, and the support of integrated multimedia services with a wide range of service rates and different Quality-of-Service (QoS) requirements.

Expected supported services include Constant Bit Rate (CBR), Variable Bit Rate (VBR), Available Bit Rate (ABR), and in-band Signaling Control (SCR) for making CBR/VBR bandwidth reservation. Examples of QoS requirements for CBR/VBR, ABR, and SCR traffic are bounded jitter/delay, minimum cell rate, and minimum non-blocking probability, respectively. Owing to being stream-oriented, CBR and VBR traffic has been shown efficiently governed by reservation-based access [5,6]. On the other hand, ABR and SCR traffic that is bursty in nature is most suitably directed by contention-based access [2,5]. In our line of work, we employ the hybrid access, namely the combination of reservation and contention access. Two challenging tasks pertaining to such network are the design of Multiple Access Control (MAC) and dynamic bandwidth allocation for all four traffic types. Our first objective is to present an intelligent system facilitating a hybrid-access MAC scheme with dynamic bandwidth allocation based on a neural-fuzzy traffic prediction technique.

Furthermore, for making more efficient use of bandwidth, voice traffic is considered to be better transported by VBR, instead of CBR, using speech activity detection [7] at the sending terminal. The success of VBR voice then hinges on the resulted playout quality in terms of Playout Delay (*PD*) and Distortion of Talkspurt (*DOT*). While interactive voice applications are more sensitive to *PD*, voice distribution services are more susceptible to *DOT*. Furthermore, these two qualities are mostly affected by end-to-end packet delay, packet loss, and delay variation, or so-called jitter, incurred from the network. Owing to the local environment and guaranteed bandwidth provision for voice, we only consider the jitter problem [8,9].

To remove jitter, intramedia synchronization methods [8] have been proposed. Existing prevailing schemes can be categorized as: static delay-based, dynamic feedback-based, and dynamic delay-based. First, the static delay-based method preserves playout continuity by buffering massive packets at receiving systems [10]. The method is simple but at the expense of a drastic increase in *PD*. Besides, the method requires immense buffer at the receiving system, which is unviable for mobile terminals in the wireless LAN environment. Second, the dynamic feedback-based method [11] adjusts the source generation rate by means of sending feedbacks from receiving systems. The method is unviable for most live-source applications. Finally, the dynamic delay-based method [12,13] performs adaptive buffering of frames that have been time-stamped at sources. The method is effective but requires much processing and framing overhead from frequent time stamping. It is thus unviable for the mobile terminal case. Our second objective is that, by taking advantage of on-line traffic prediction, we propose a simple, dynamic delay-based intramedia synchronization scheme [8] resulting in minimal *DOT* and *PD*.

In this chapter, we present a network framework in which VBR voice is transported over a wireless ATM LAN with two novel intelligent control systems, as shown in **Fig. 1**. They are the Intelligent Multiple Access Control System (IMACS) [2], and Intelligent Voice Smoother (IVoS) [8]. Significantly, both IMACS and IVoS make use of neural-fuzzy-based traffic prediction during their operations. First of all, the network consists of a Base Station (BS) serving a number of Mobile Terminals (MT's). The medium bandwidth is divided into two separate channels- uplink and downlink. IMACS, which operates in the BS, governs the medium access of the uplink channel based on a hybrid-TDM-based MAC protocol. In addition, augmented with on-line neural-fuzzy-based prediction of ABR self-similar traffic characteristics, IMACS offers dynamic allocation of contention bandwidth between ABR and SCR traffic.

Furthermore, IVoS at the application layer of each mobile terminal performs intramedia synchronization of voice data streams. Facilitated with on-line traffic prediction, IVoS first predicts three traffic characteristics of every newly encountered talkspurt period. On the basis of the predicted characteristics for each talkspurt, IVoS adaptively determines the corresponding buffering delay to be imposed on the first frame. With such delay, all subsequent frames of the talkspurt can be playout in a quasi-CBR manner. Finally, we demonstrate via experimental results that with such intelligent control, the playout QoS of voice traffic can be guaranteed irrespective of any traffic and load variation.

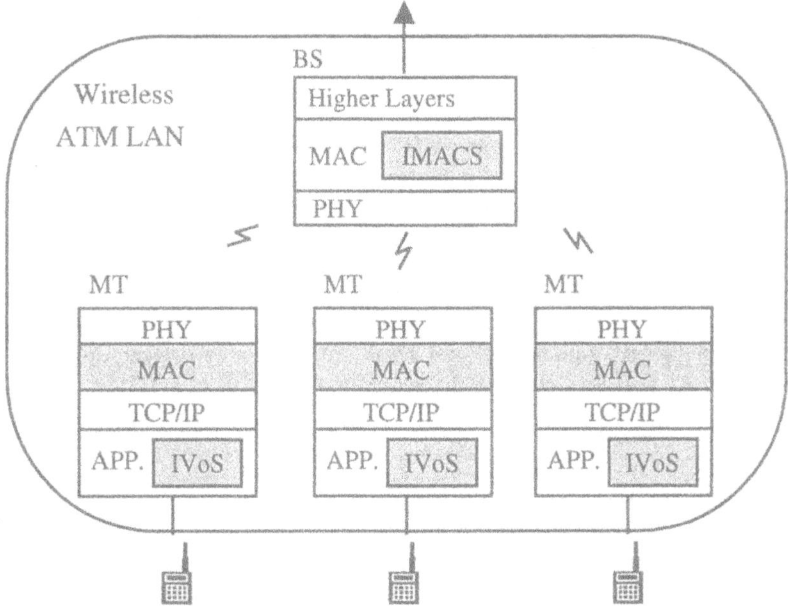

Legend: BS = Base Station;
MT= Mobile Terminal;
IMACS= Intelligent Multiple Access Control System;
IVoS= Intelligent Voice Smoother;

Fig. 1. Voice over Wireless ATM LAN

The remainder of this chapter is organized as follows. In Section 2, we present the architecture and design of IMACS. In Section 3, we introduce the neural-fuzzy traffic prediction technique employed in IMACS. In Section 4, we describe the design of IVoS and demonstrate experimental results. Finally, concluding remarks are given in Section 5.

2. Intelligent Multiple Access Control System (IMACS)

2.1 Background

The wireless LAN architecture considered in this work is infrastructure-based, namely the classical cell with a Base Station (BS) serving a finite set of Mobile Terminals (MT's) by means of a shared radio medium. The sharing of medium bandwidth among connected stations is governed by multiple access control schemes with the bandwidth divided into accessible units. In general, existing multiple access schemes fall into one of three categories [14,15]: Frequency-Division Multiple Access (FDMA), Time-Division Multiple Access (TDMA), and Code-Division Multiple Access (CDMA). While FDMA schemes divide the bandwidth into portions of frequency spectrum, TDMA schemes divide the bandwidth into time slots. CDMA divides the bandwidth into a set of codes through which different stations are allowed to operate at the same frequency and time.

In our line of work, we focus on TDMA-based schemes. The schemes support both uplink and downlink traffic carrying information sent from mobile stations to the BS and vise versa, respectively, in wireless LAN networks. To this end, TDMA-based schemes can be further categorized as either Frequency-Division-Duplex (FDD) [14], or Time-Division-Duplex (TDD) [15]. In FDD-based systems, uplink and downlink traffic are carried by two distinct carrier frequencies. On the contrary in TDD-based systems, while using one common carrier frequency, uplink and downlink traffic are carried at different time intervals. As will be shown in the sequel that IMACS is a TDMA FDD-based system.

2.2 Architecture

IMACS operates in the BS of an infrastructure-based wireless ATM LAN. On the basis of FDD, the medium bandwidth is divided into two separate channels: uplink and downlink. The uplink channel transfers information from MT's to the BS, based on a new hybrid TDMA-based MAC scheme (described later). The downlink channel typically broadcasts information and acknowledges previous transmissions made on the uplink channel. This operation is beyond the scope of this paper. Furthermore, time on the uplink channel is divided into a contiguous sequence of fixed-size TDMA frames (see **Fig. 2**).

Each frame is further subdivided into a fixed number of slots to be dynamically allocated to four ATM-traffic classes: CBR, VBR, ABR, and SCR. As was mentioned, while CBR and VBR traffic are governed by reservation access using Reservation (R)-type bandwidth, ABR and SCR traffic are controlled by contention access using Contention (C)-type bandwidth. Each slot contains a data packet, or more specifically, an ATM cell, other than guard times, sync, and other control fields [3]. Notice that, with guard times provided, the propagation delay between the BS and MT's can be ignored. This in turn allows acknowledgements for all packet transmissions made in the current slot to be available to all MT's prior to the beginning of the next slot.

Most significantly, the network is assumed to use Phase-Shift Keying (PSK)-based encoding equipped with simple CDMA capability [16], namely pseudo-code sequence generation. Specifically, all MT's with ABR packets in their buffers are required to inform the BS through placing different code sequences at the last slot of each frame, called the Common Notification Field (CNF). Due to orthogonality and phase differences [16] of CDMA, the BS is able to identify the total number of different codes, which corresponds to the total number of MT's with backlogged ABR traffic during the last frame. This information is made available by IMACS for on-line traffic estimation and prediction described later.

IMACS is composed of three major components (see **Fig. 3**): Multiple Access Controller (MACER), Traffic Estimator/Predictor (TEP), and Intelligent Bandwidth Allocator (IBA). It supports CBR, VBR, ABR, and in-band SCR traffic. IMACS has been designed to satisfy delay and throughput guarantees for CBR/VBR traffic while offering minimal access delay and blocking probability for ABR and SCR traffic, respectively. In general, MACER employs a reservation-based access protocol for CBR and VBR traffic. It uses fixed amounts of R_C-type bandwidth for CBR, and R_V-type bandwidth for VBR (R_C+R_V= R) (in slots). By contrast, MACER adopts a contention-based access protocol using C_S-type bandwidth for SCR and C_A-type bandwidth for ABR (C_S+C_A= C) (in slots). In particular, to achieve an acceptable non-blocking probability, MACER governs SCR traffic by contention access using the

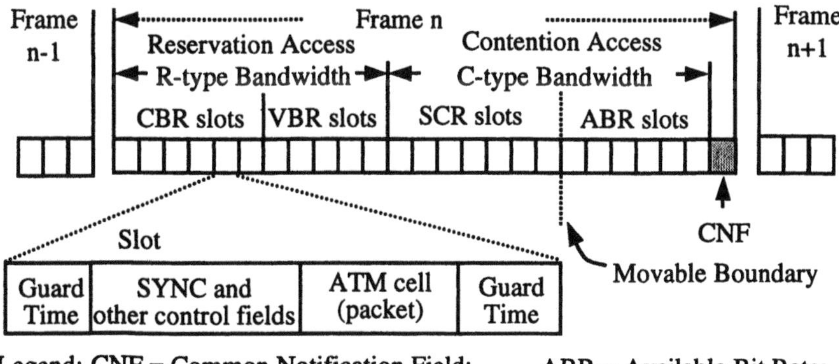

Fig. 2. Frame and slot structures

Dynamic Tree Splitting (DTS) collision-resolution algorithm parameterized by the optimal Splitting Depth (SD), denoted as DTS-*d*, if SD=*d*.

IBA is responsible for allocating R-type bandwidth on a call basis and allocating C-type bandwidth on a frame basis. On behalf of IBA, TEP performs periodic estimation and on-line prediction of characteristics of ABR traffic that is assumed self-similar in nature. Specifically, TEP estimates the Hurst (H) parameter of self-similar traffic based on wavelet analysis [17,18]. In addition, with CNF values of the past frames, TEP predicts the short-term mean and variance of the subsequent frame using an on-line Neural-Fuzzy-based Traffic Prediction (NFTP) technique [2,19] described in the next section. Given with predicted ABR traffic characteristics and the SCR blocking probability requirement, IBA determines the optimal SD prior to every subsequent frame. Once the optimal SD is identified, C_S bandwidth is determined. The remaining bandwidth (C_A) is then allocated to ABR traffic.

In the sequel, we describe each component in detail except for the NFTP technique used in TEP. The basic design and experimental results of NFTP are separately presented in the next section.

2.3 Multiple Access Controller (MACER)

2.3.1 Reservation Access

Recall that, CBR and VBR traffic are statically allocated with fixed amounts of bandwidth (R_C and R_V) for an entire call, satisfying the duty cycle and maximum end-to-end delay requirements, respectively. Due to the allocation simplicity for CBR traffic, we focus on the determination of R_V for VBR traffic (VBR voice in our case) subject to satisfying a given end-to-end delay requirement.

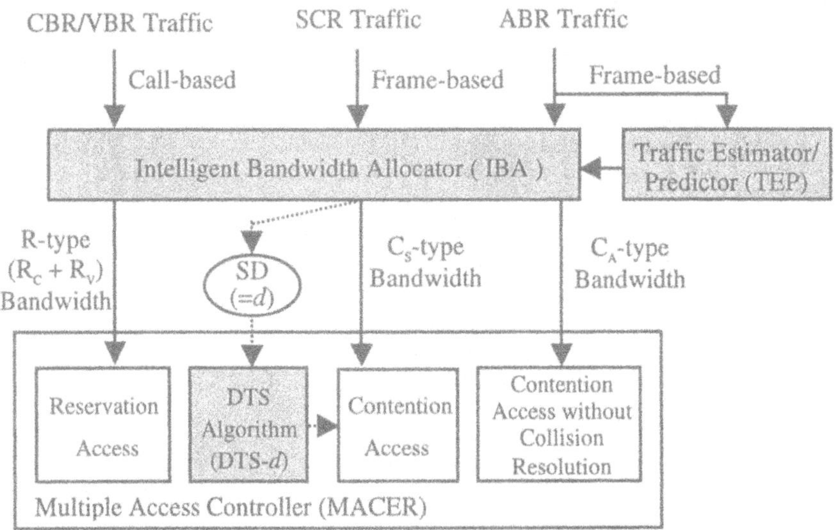

Fig. 3. IMACS architecture

VBR traffic is assumed regulated through a leaky-bucket (ρ, σ) controller [20], where ρ is the mean leaky rate, and σ is the maximum bucket size. Accordingly, a VBR traffic source can be characterized by three parameters (ρ, σ, D_{max}), where D_{max} is the maximum tolerable end-to-end delay. We now derive the minimum bandwidth R_V for VBR traffic satisfying a given end-to-end delay bound, D_{max}. Let $A(s,t)$ denote the total number of packets arriving in a time interval $(s,t]$, and $S(u,v)$ the packets served within the time interval $(u,v]$. Since arriving packets must conform to the (ρ, σ) regulator, $A(0,t) \le \lceil \rho \cdot t + \sigma \rceil$, where $\lceil\ \rceil$ denotes the ceiling function. Let $\bar{G}$ denote the maximum signaling delay for the establishment of a VBR voice connection. First, D_{max} can be given as

$$D_{max} = \sup_t \{d \ge 0 : A(0,t) = S(\bar{G}, t+d)\}. \quad (1)$$

Notice that $S(u,v) = (v-u) \times R_V$, Equation (1) becomes

$$D_{max} = \sup_t \{d : 0 \le d \le \frac{\lceil \rho \cdot t + \sigma \rceil}{R_V} + \bar{G} - t\}. \quad (2)$$

Since condition $\rho \le R_V$ must be satisfied, it follows that $D_{max} = \frac{\lceil \sigma \rceil}{R_V} + \bar{G}$, which results in the fixed bandwidth to be allocated to VBR:

$$R_V = \max(\frac{\lceil \sigma \rceil}{D_{max} - \bar{G}}, \rho). \quad (3)$$

The BS then makes R_V and slot access information available to each active connection via the downlink channel prior to data transfer of the connection.

2.3.2 *Contention Access*

To support voice signaling, i.e., SCR traffic, MACER employs a contention access protocol augmented with a DTS collision-resolution algorithm. The DTS-*d* algorithm (if SD=*d*) performs as follows. In each frame, SCR is initially allocated with the basic allocation (in slots). Slots from the basic allocation are randomly accessed. Should collisions occur and the number of splitting is less than *d*, twice as many as the number of collided slots are allocated at the next splitting level. This process repeats until either there is no collision or the number of splitting levels has reached *d*. All unresolved transmissions then back off in the next frame. It is worth noticing that SCR call requests are blocked when the number of frame backoffs exceeds a predefined threshold, called the Retry Count (RC).

2.4 Intelligent Bandwidth Allocator (IBA)- Optimal SD Determination

The bandwidth allocation problem can be elucidated by the following dilemma. We observed that, greater SD values yield appealing SCR blocking probability but at the expense of reduced ABR throughput. Nevertheless, smaller SD values still render unfavorable ABR throughput and aggregate throughput despite the price of increasing SCR blocking probability paid. Therefore, the objective of

IBA has been the determination of the optimal SD per every frame, aiming at satisfying SCR blocking probability and ABR throughput requirements, while retaining maximal aggregate network throughput. In short, IBA has been designed to provide optimal allocation between C_S and C_A types of bandwidth.

To this end, we performed simulation-based throughput analysis. In the analysis, SCR traffic is assumed Poisson distributed, and ABR traffic is modeled as self-similar. The generated throughput results postulate the optimal SD's under various traffic conditions. These results are off-line trained and constructed using a Back Propagation Neural Network (BPNN) [9], which is used on-line by IBA.

For modeling and generation of self-similar traffic, we adopted the fractional Gaussian noise (FGN) process [21], and a fast-generation algorithm [22]. Particularly for traffic generation, we considered a set of ten slots each time for generating a non-negative number of cell arrivals. Given a mean arrival, we first randomly generated a number, which represents the total number of arriving cells, in each group of ten slots. The exact arriving epochs of these cells were then uniformly distributed in ten slots.

As was previously mentioned, SCR and ABR cells are handled differently with respect to the backoff policy. Collided ABR cells back off in the next frame. Collided SCR cells (calls) back off at most d times within a frame provided that SD=d. Failed calls keep retrying the next frame until the RC is reached. Calls are then considered blocked. Notice that the RC is inferred from the maximum tolerable call-set-up delay. In our simulation, the RC was set to five (frames) corresponding to a maximum call-set-up delay of 50 milliseconds. Moreover, it is required to impose limits on the basic-allocation size and maximum SD value so that the total amount of C_S bandwidth never exceeds the frame size. In the simulation, for a frame of 300 slots in length, the size of the basic allocation ranged from 10 to 50 slots, and the maximum eligible SD value was set as 6.

I. Satisfaction of ABR QoS– ABR Throughput

We experimented on ABR throughput under a variety of SCR traffic loads based on the DTS-d (d = 0, 1, 2, 3) and the Exhaustive Binary-Tree-Splitting (EBTS) collision-resolution algorithms. Notice that EBTS corresponds to DTS-4 in this case, resolving collisions up to the entire bandwidth of a frame. Simulation results are plotted in **Fig. 4**. As shown in the figure under light ABR loads, DTS-2 outperforms other DTS and EBTS algorithms. As the SCR load increases reaching a turning point, which is located distinctively under different algorithms, ABR throughput starts declining. Among all approaches, DTS-3 and EBTS undergo the most deteriorating performance. On the other hand, under heavy ABR loads, ABR throughput invariably declines with increasing SCR load in all algorithms. The greater the SD is, the poorer the throughput. Specifically, DTS-0 achieves the best performance due to the provision of a fixed amount of bandwidth to ABR traffic despite an increase in the SCR load.

II. Satisfaction of SCR QoS– blocking probability

We then investigate the sensitivity of SCR blocking probability with respect to the SD value and basic allocation, under light and heavy SCR traffic loads. As was expected, blocking probability declines with increasing basic allocation and

SD value. Specifically, heavier loads demand greater SD values to achieve the same grade of blocking probability. For example, to achieve strictly non-blocking, SD=4 and SD=5 are required under light and heavy SCR loads, respectively.

III. Maximization of aggregate throughput

We finally examine the aggregate throughput under various traffic and SD values. In **Fig. 5**, we depict the aggregate throughput as a function of SCR load under light and heavy ABR loads, using four variants of the DTS algorithm. Initially starting from a light SCR and ABR load in part (a) of the figure, greater SD values unsurprisingly achieve better throughput. However, as the SCR load increases, greater SD values can no longer benefit the aggregate throughput resulting from substantial unresolved collision. Greater SD values yield more bandwidth waste, leading to poorer throughput. The turning point again is located differently under different variants of the DTS algorithm. Under a heavy

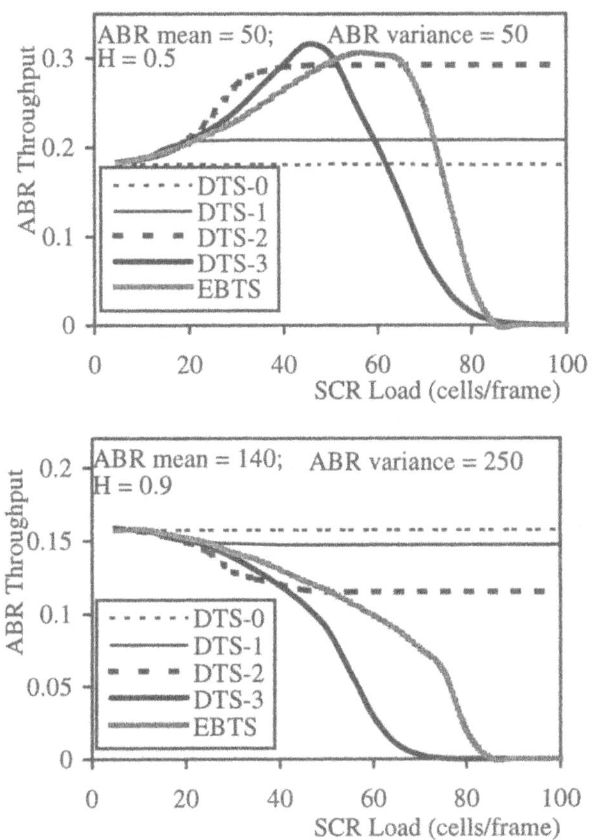

Fig. 4. ABR throughput versus SCR load. (**a**) Light-load, low-H ABR traffic. (**b**) Heavy-load, high-H ABR traffic

ABR traffic load shown in part (b), we observed consistent plots, which however exhibit earlier turning points owing to the contribution of the heavy ABR load.

Accordingly, the optimal SD is dependent on four traffic characteristics: ABR mean load, variance, the Hurst parameter, and the SCR mean load. As was previously stated, these results are then off-line trained and constructed via a BPNN, which can be effectively accessed on-line by IBA offering optimal bandwidth allocation.

3. Neural-Fuzzy-Based Traffic Prediction (NFTP)

NFTP performs on-line traffic prediction based on a self-constructing neural-fuzzy inference network [19]. It is involved in two learning phases: structure and

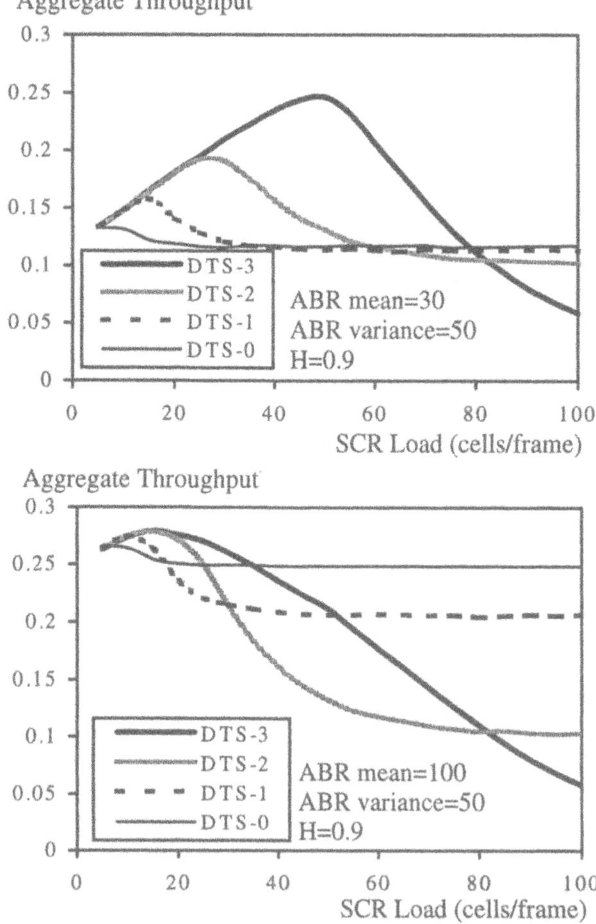

Fig. 5. Aggregate network throughput. **(a)** Light ABR load. **(b)** Heavy ABR load

parameter learning. The structure-learning phase determines the structure of fuzzy if-then rules, and the parameter-learning phase tunes the coefficients of the rules adapting to the input traffic dynamics. Unlike existing neural-fuzzy models using sequential learning, NFTP performs the structure and parameter learning in parallel. This makes NFTP advantageous for fast on-line prediction.

NFTP is a six-layer network taking on a number of input nodes and one output node, as shown in **Fig. 6**. Initially, there are no rules in the network other than input nodes (layer 1) and an output node (layer 6). Upon receiving on-line training data, the structure-learning process proceeds by dynamically self-constructing fuzzy if-then rules (layer 3) according to an input-output clustering-based space-partitioning algorithm [19]. Once a new rule is generated, the centers and widths of the corresponding set of Gaussian membership functions (layer 2 and layer 5) are assigned. The output of a layer-3 node corresponds to the firing strength of the corresponding fuzzy rule, which is in turn normalized in layer 4. Consequently, the predicted output value, y, is given as:

$$y=\sum_i y_i,\ i=\text{fuzzy rule index, and}$$

$$\text{Fuzzy rule } i\text{: } \underline{\text{If}}\ x_1 \text{ is } A_{i1}\ \underline{\text{and}} \ldots \underline{\text{and}}\ x_n \text{ is } A_{in},\ \underline{\text{Then}}\ y_i = f_i m_i. \tag{4}$$

where y_i is the contribution of fuzzy rule i to the predicted output value, x_j is the j_{th} input value, A_{ij} is the j_{th} membership function of fuzzy rule i, f_i is the normalized firing strength of fuzzy rule i, and m_i is the center of the membership function in layer 5 connected to fuzzy rule i. Meanwhile, in the parameter-learning process, the centers and widths of input membership functions (layer 2) are dynamically adjusted based on the Least Mean Squares (LMS) algorithm [19], whereas those of output membership functions (layer 5) are tuned using the Back Propagation algorithm.

Fig. 6 illustrates an NFTP network with three inputs. This network predicts

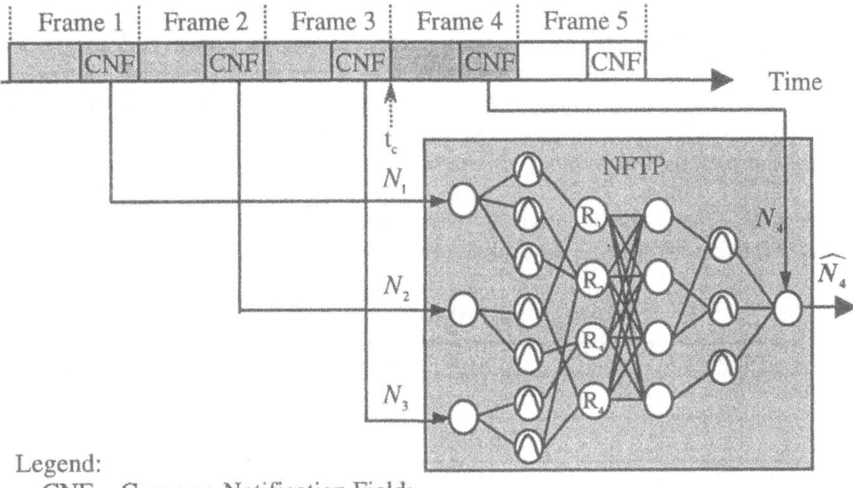

Fig. 6. NFTP architecture

the future CNF value ($\hat{N}_4$), which corresponds to *the mean number of active MT's in the subsequent frame*, based on three input values taken from three most-recent CNF values (denoted as N_i, i=1 to 3). At the end of each frame, in addition to predicting the CNF value of the next frame, NFTP also performs the learning operation described above. This is indicated in the figure by the arrowed link pointing from the CNF of Frame 4 to the NFTP output node.

We experimented on two different NFTP structures using different types of inputs, respectively, via simulation. In the first structure, called *CNF-based NFTP*, the inputs are taken directly from a set of different numbers of past CNF values (N_i), ranging from 4 to 24, similar to what is shown in **Fig. 6**. In the second structure, referred to as *CNF-correlation-based NFTP*, we adopted *exponential-averaging k-lag correlation* of CNF values as inputs. Specifically, taking an example of NFTP with four inputs x_k, k=1 to 4, at the end of the i_{th} frame, x_k will be set as the k-lag correlation $\hat{C}_i$ defined as: $\hat{C}_i = \lambda C_i + (1-\lambda)\hat{C}_{i-1}$, where $C_i = N_i \times N_{i-k}$, and λ is the smoothing constant ($0<\lambda<1$). With this structure, we also used 4 to 24 different numbers of inputs. In this simulation, we on-line predicted a set of 200 frames, using both structures of NFTP. All parameters used in the simulation are summarized in **Table 1**. In addition, the performance of NFTP is evaluated in terms of its prediction precision (error rate), time complexity, and space complexity. The error rate was computed as the normalized average deviation between the actual and predicated CNF values. The space complexity was given in terms of the total number of fuzzy rules generated at the end of 200-frame prediction. Notice that, since such inference network can be implemented in hardware, we thus disregarded its time complexity. Simulation results are displayed in **Table 2**.

We observed during the experiment that the prediction error rate using either structure is irrelevant to the Hurst parameter (H), but highly sensitive to the variance. This can be perceived by the fact that by and large, H manifests only long-term behavior, whereas variance greatly reflects short-term fluctuation. In essence, as shown in **Table 2** under traffic H=0.8, the error rate greatly increases with the variance. Furthermore, compared to CNF-based NFTP, CNF-correlation-based NFTP achieves greater precision (lower error rate) and lower space complexity (less number of fuzzy rules). We finally discovered in the table that NFTP (either structure) with 12 inputs invariably exhibits better performance under both variances. Namely, small or large numbers of inputs yield inferior performance for on-line prediction.

Table 1. NFTP parameter used in simulation

Variable	Definition	Value
$\bar{F}_{in}$	Input clustering threshold	0.1
$\bar{F}_{out}$	Output clustering threshold	0.99
$\rho(t)$	Membership function threshold	0.7
β	Initial width of Gaussion function	0.25
η	Back-propagation learning constant	0.08
λ	Smoothing constant	0.5

4. Intelligent Voice System (IVoS)

4.1 Concept

Voice data are sampled and encoded as fixed-size frames. These frames with silence suppression are in turn sent over the network. The arrivals of frames at IVoS are assumed MMBP (described later) distributed with unknown probabilistic parameters. IVoS mainly determines the time at which frames are transferred from the IVoS Smoother Buffer to the decoder buffer from which frames are playbacked. An end-to-end flow scenario is illustrated in the time-space diagram shown in **Fig. 7**. In the figure, the conversation between the sender and receiver (with IVoS) is conducted in an alternating manner between an active period and an inactive period. IVoS is in an active period during receiving frames; otherwise, it is in an inactive period. Furthermore, being in an active period as shown in **Fig. 7** (b), IVoS alternates between the talkspurt state, with frames intermittently received (during the busy state), and the pause state during which no frames appear. Moreover, the first frame in any talkspurt is tagged (marked "x" as shown in the figure) before being transmitted.

Accordingly, the goal of IVoS is the enforcement of CBR playout during the active period by dynamically adjusting the duration of pauses in an effort to compensate for jitter within talkspurts. The rationale of how IVoS achieves the goal is shown in **Fig. 7** (a). The time axis in IVoS is slottized by the processing of a single frame from the adjacent lower layer, namely the transport layer. We assume that, disregarding the framing overhead, voice frames are generated (playbacked) at the encoder (decoder) of the sending (receiving) end system at a rate of one-third of the processing rate of the transport layer [9].

Table 2. Performance of NFTP using two different structures

H = 0.8	CNF-based NFTP			CNF-correlation-based NFTP		
	Number of inputs	Number of fuzzy rules	Error rate	Number of inputs	Number of fuzzy rules	Error rate
Mean = 50 Variance = 20	4	27	6.1	4	6	5.5
	8	33	6.0	8	12	5.6
	12	42	5.9	12	23	5.4
	16	47	6.6	16	20	5.5
	20	54	6.7	20	28	5.3
	24	55	7.3	24	23	5.6
Mean = 50 Variance = 60	4	27	10.9	4	11	9.7
	8	33	10.7	8	20	9.5
	12	42	10.6	12	17	9.3
	16	47	11.8	16	20	9.8
	20	54	12.1	20	20	9.7
	24	55	13.1	24	23	9.9

Define $\mathcal{F}$ as the ratio of the generation or playout of a frame, referred to as the frame time, to the processing of a single frame at the transport layer, referred to as the slot time, i.e., $\mathcal{F}$=Frame time/slot time. For the example given in **Fig. 7** (a), $\mathcal{F} = 3$.

Frames are finally received at IVoS at the receiving end system. Ideally in a jitter-free network, the frame inter-departure times from the sender's application are the same as the frame inter-arrival times at IVoS. In this case, frames are playbacked intelligibly at the maximum rate, namely one frame per every three time-slots, during talkspurts. Unfortunately in reality, owing to delay jitter induced in the network, different frames yield different end-to-end delays causing speech unintelligibility. Denote t_i^s and t_i^p as the sample and playout time of frame i, respectively. In addition, denote D_i as the end-to-end transfer delay of frame i. Accordingly, playout discontinuity is quantified by Distortion of

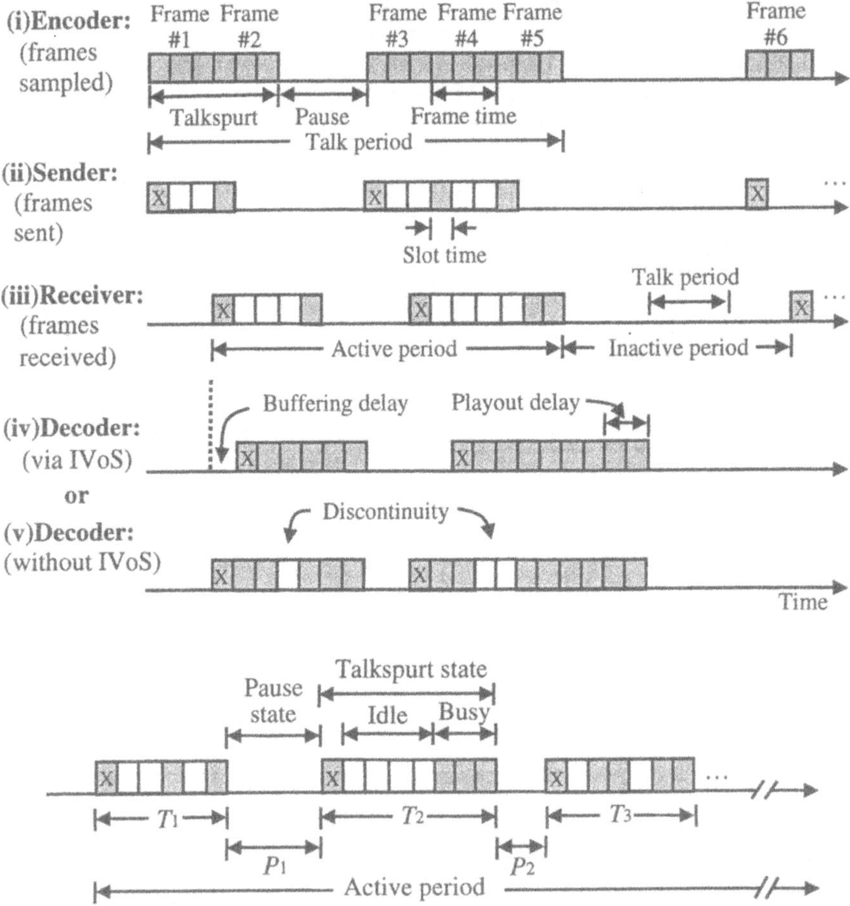

Fig. 7. Concept of IVoS. **(a)** An end-to-end voice flow scenario. **(b)** IVoS states

Talkspurt (*DOT*), where the *DOT* for talkspurt *j*, denoted as DOT_j, is defined as

$$DOT_j = \frac{\sum_{i=R(j)+1}^{L(j)} \left[(t_i^p - t_{i-1}^p) - (t_i^s - t_{i-1}^s)\right]}{F_j - 1} = \frac{\sum_{i=R(j)+1}^{L(j)} [D_i - D_{i-1}]}{F_j - 1}. \tag{5}$$

where $R(j)$ and $L(j)$ are the ordinal numbers of the first and last frame in talkspurt *j*, and F_j is the total number of frames in talkspurt *j*. Moreover, playout discontinuity can be reduced at the expense of an increase in Playout Delay (*PD*). Let PD_i denote the *PD* of talkspurt *i*, defined as the elapsed time between the fastest possible departure and real departure of the last frame in talkspurt *i*.

Consequently, IVoS aims at achieving minimal $E[D\tilde{O}T]$ and $V[D\tilde{O}T]$ (zero in the case of CBR playout) while sustaining minimal $E[P\tilde{D}]$. Two issues have been considered in the design of IVoS: (1) how and what characteristics of future traffic to be predicated; and (2) how to determine an adaptive buffering delay imposed on each talkspurt aiming at achieving a quasi-CBR playout during talkspurts. Before proposing solutions to these two issues, we first present the inbound traffic model in the next subsection.

4.2 Inbound Traffic Model

The inbound traffic to IVoS is modeled by a generic discrete-time Markov Modulated Bernoulli Process (MMBP) [23], as shown in **Fig. 8**. The process alternates between the pause state and the talkspurt state. Within the talkspurt state, the process switches between the busy state in which one frame always arrives (with probability =1) per slot time, and the idle state in which no frame is generated. The transition probabilities between states are given in the figure. The duration staying in any state is assumed geometrically distributed. The steady-state probabilities of being at the three states, denoted as Π_{pause}, Π_{busy}, and Π_{idle}, can be computed using $\Pi = \Pi P$ [24], where $\Pi = (\Pi_{pause}, \Pi_{busy}, \Pi_{idle})$, and *P* is the state transition probability matrix of the MMBP. As a result,

$$\Pi_{pause} = \frac{\alpha\delta}{\delta\beta + \gamma\beta + \alpha\delta}; \Pi_{busy} = \frac{\delta\beta}{\delta\beta + \gamma\beta + \alpha\delta}, \Pi_{idle} = \frac{\gamma\beta}{\delta\beta + \gamma\beta + \alpha\delta}. \tag{6}$$

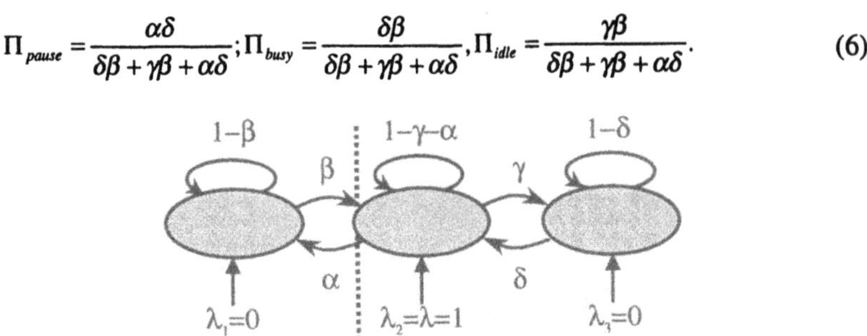

Fig. 8. Inbound traffic model

Moreover, the mean frame rate (*MFR*) and mean burst length (*MBL*) can be directly expressed as functions of α, β, γ, and δ:

$$MFR = \Pi_{busy} \times \lambda = \frac{\delta\beta \times \lambda}{\delta\beta + \gamma\beta + \alpha\delta} = \frac{\delta\beta}{\delta\beta + \gamma\beta + \alpha\delta}, \text{ and } MBL = \frac{1}{\alpha + \gamma}. \quad (7)$$

4.3 IVoS System Architecture

IVoS is composed of three major components (see **Fig. 9**): Smoother Buffer, Neural-Fuzzy Traffic Predictor, and CBR Enforcer. Newly arriving frames are first placed in the Smoother Buffer in an FCFS fashion. Each time, the reception of a marked frame, which corresponds to the initiation of a new talkspurt, triggers the NF Traffic Predictor to perform the prediction of three traffic characteristics of the upcoming talkspurt *i*. They are the talkspurt length (T_i), frame count (F_i), and the last burst length (B_i). Based on the three predicted characteristics, the CBR Enforcer then determines a dynamic delay to be imposed on the first frame, and regulates instant playout for all subsequent frames of the talkspurt. The same process repeats for the next talkspurt until the end of the talk.

Notice that the architecture of the NF Traffic Predictor is similar to the NFTP in IMACS except that there are three predicted outputs. The description is thus omitted. In the sequel, we focus on the design of the CBR Enforcer.

4.4 CBR Enforcer

Based on the three predicted characteristics, the CBR Enforcer mainly determines an Adaptive Buffering Delay (*ABD*) to be imposed on the first (marked) frame initiating the talkspurt. Let $A\tilde{B}D$ and ABD_i denote the random variable of *ABD*, and *ABD* for talkspurt *i*, respectively. The Enforcer aims at the achievement of quasi-CBR departures for all subsequent frames belonging to the same talkspurt. In principle, the buffering delay of a talkspurt should be large

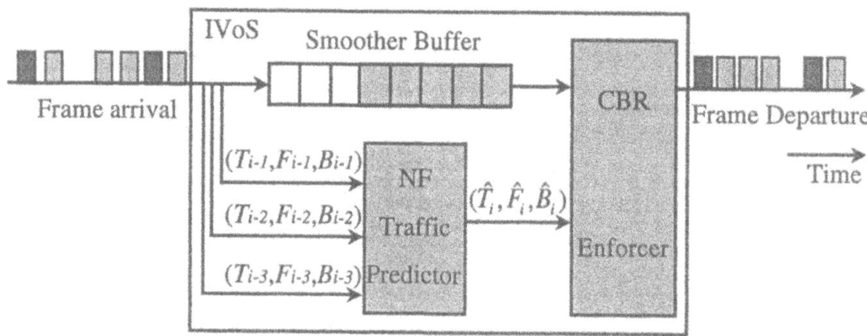

Fig. 9. System Architecture of IVoS

enough in compensation for the total number of time slots lacking the frame playout in the talkspurt.

Should the three traffic characteristics, T_i, F_i, B_i, of talkspurt i be known, let us consider the best case, namely that incurring the minimum *ABD*. In this case, the remaining frames (frames not belonging to the last burst) have arrived back to back at the beginning of the talkspurt. First, the entire playout duration, defined as the interval from the beginning of the talkspurt to the end of playout of the last frame, is clearly the sum of the length of talkspurt i (T_i) and the additional duration required to playback frames of the last burst $[B_i \times (\mathcal{F}-1)]$. Moreover, the total elapsed time required for CBR playout, subject to a total number of F_i frames in talkspurt i, is $F_i \times \mathcal{F}$. Therefore, the ABD_i can be given as the difference of the entire playout duration and the elapsed time for CBR playout, i.e., $[T_i + B_i \times (\mathcal{F}-1)] - F_i \times \mathcal{F}$.

Now, consider the real case in which the remaining frames have arrived at different locations between the beginning and the last burst. Taking the localities of the remaining frames into account, we introduce a so-called locality parameter, denoted as θ ($0 < \theta \le 1$). Thus, we attain the theoretical buffering delay, denoted as ABD_i^{TH}, yielding CBR playout, as

$$ABD_i^{TH} = \left[T_i + \left\lfloor \frac{1}{\theta} \times B_i \times (\mathcal{F}-1) \right\rfloor - F_i \times \mathcal{F} \right]^+ . \tag{8}$$

It is worth noticing that a θ being unity corresponds to the best case given previously. The smaller the θ, the later and the more widely spread frames have arrived. Furthermore, since T_i, F_i, and B_i are not known in advance, replacing them by $\hat{T}_i$, $\hat{F}_i$, and $\hat{B}_i$, predicted by the NF Predictor, we can formulate the actual buffering delay as

$$ABD_i = \left[\hat{T}_i + \left\lfloor \frac{1}{\theta} \times \hat{B}_i \times (\mathcal{F}-1) \right\rfloor - \hat{F}_i \times \mathcal{F} \right]^+ . \tag{9}$$

With such delay imposed, the CBR enforcer then regulates the departure of frames in a rate of $1/\mathcal{F}$ frames/slot until the next marked frame initiating the subsequent talkspurt has been encountered.

4.5 Experimental Results

To demonstrate the viability of IVoS, we drew comparisons in terms of mean *DOT* and mean *PD*, between IVoS and two other playout approaches via simulation. The two approaches are: instant playout and pre-buffering playout. In the instant playout approach, frames were queued and playbacked at a rate of $1/\mathcal{F}$ and below. It is worth noting that the instant playout approach differs from IVoS in the lack of buffering delay imposed on the first frame of each talkspurt. In the pre-buffering approach, a 100-slot delay is imposed on the first frame of each talkspurt. Other assumptions and parameters used for simulation are summarized as follows:

- Voice data rate during talkspurts: 64 kbps;
- Talk duration: 6.25 minutes (corresponding to 30,000 slots long);
- Slot size: 100 bytes (corresponding to 12.5 ms/slot of voice);
- $\mathcal{F}$: 3 (corresponding to a mean load of 1/3 frame/slot during talkspurts);
- Buffer size for all approaches: 1200 slots (large enough to assure loss free);

In **Fig. 10**, we plot mean *DOT* and mean *PD* under complete sets of *MFR's* and *MBL's*. It is particularly worth noting that pre-buffering playout achieves

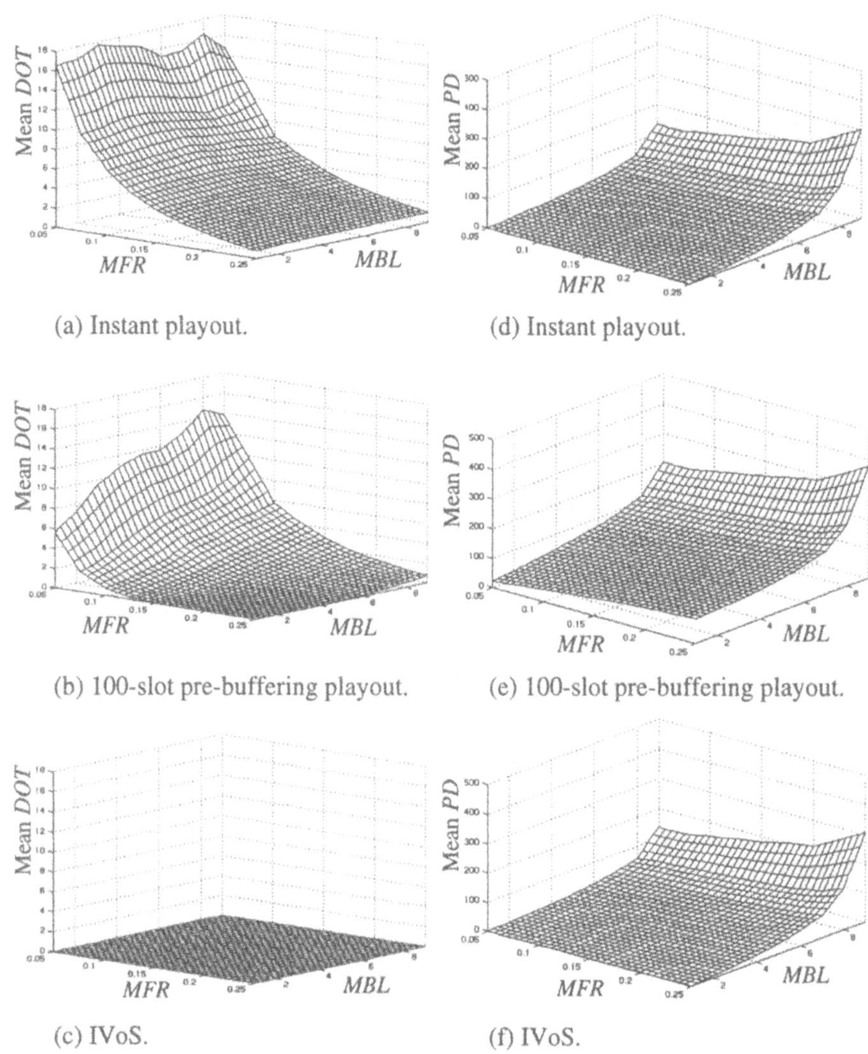

(a) Instant playout. (d) Instant playout.

(b) 100-slot pre-buffering playout. (e) 100-slot pre-buffering playout.

(c) IVoS. (f) IVoS.

Fig. 10. Mean *DOT* and *PD* performance of three playout approaches

only bearable mean *DOT*, however, at the expense of a drastic deterioration in mean *PD*. We observe that IVoS outperforms two other approaches in its invariably low *DOT* irrespective any traffic (*MBL*) and load (*MFR*) variation. As for mean *PD*, IVoS retains compatibly short delay to the optimal one, i.e., instant playout. In addition, the figure clearly reveals a trade-off problem between *DOT* and *PD* using both the instant playout and pre-buffering playout approaches. By contrast, IVoS is free from the trade-off problem and achieves near-optimal performance.

5. Conclusions

In this chapter, we have presented two intelligent control systems, IMACS and IVoS, for supporting VBR voice over wireless ATM LANs. Both systems employ on-line neural-fuzzy-based traffic prediction achiveing optimal QoS performance under any load and burstiness conditions of networks. Significantly, efficient processing and an insignificant-buffer requirement by IVoS at mobile terminals, IVoS is particurly suitable for the wireless LAN environment.

References

1. Cox D (1995) Wireless Personal Communications: What Is It? IEEE Pers Comm, **2(2)**, 20-35
2. Yuang M, Tien P (2000) Multiple Access Control with Intelligent Bandwidth Allocation for Wireless ATM Networks. IEEE J Select Areas Commun, **18(9)**, 1658-1669
3. Raychaudhuri D, French L, Siracusa R, Biswas S, Yuan R, Narasimhan P, Johnston C (1997) WATMnet: A Prototype Wireless ATM System for Multimedia Personal Communication. IEEE J Select Areas Commun, **15(1)**, 83-95
4. Prycker M (1995), Asynchronous Transfer Mode- Solution for Broadband ISDN. Prentice Hall, Third Edition
5. Levine D, Akyildiz I, Naghshineh M (1997) A Resource Estimation and Call Admission Algorithm for Wireless Multimedia Networks Using the Shadow Cluster Concept. IEEE J Select Areas Commun, **5(1)**, 1-12
6. Listanti M, Mascitelli F, Mobilia A (1998) D^2MA: A Distributed Access Protocol for Wireless ATM Networks. Proc IEEE INFOCOM, 315-321
7. Saito H (1994), Teletraffic Technologies in ATM Networks. Norwood, MA: Artech House
8. Tien P, Yuang M (1999) Intelligent Voice Smoother for Silence-Suppressed Voice over Internet. IEEE J Select Areas Commun, **17(1)**, 29-41
9. Yuang M, Tien T, Liang S (1997) Intelligent Video Smoother for Multimedia Communicaitons. IEEE J Select Areas Commun, **15(2)**, 136-146
10. Ramjee R, Kurose J, Towsley D (1994) Adaptive Playout Mechanisms for frameized Audio Applications in Wide-Area Networks. Proc IEEE INFOCOM, 680-688
11. Little T, Ghafoor A (1991) Multimedia Synchronization Protocols for Broadband Integrated Services. IEEE J Select Areas Commun, **9(9)**, 1368-1382

12. Xie Y, Liu C, Lee M, Saadawi and T (1996) Adaptive Multimedia Synchronization in a Teleconference System. Proc IEEE ICC, 1355-1359
13. Ishibashi Y, Tasaka S, Tsuji A (1996) Measured Performance of a Live Media Synchronization Mechanism in an ATM Network. Proc IEEE ICC, 1348-1354
14. Akyildiz I, et al (1999) Medium Access Control Protocols for Multimedia Traffic in Wireless Networks. IEEE Network, 39-47
15. Passas N, et al (1997) Quality-of-Service-Oriented Medium Access Control for Wireless ATM Networks. IEEE Comm Mag, **35(11)**, 42-50
16. Lehnert J, Pursley M (1987) Error Probabilities for Binary Direct-Sequence Spread-Spectrum Communications with Random Signature Sequences. IEEE Trans Comm, COM-**35(1)**, 87-98
17. Abry P, Veitch D (1998) Wavelet Analysis of Long-Range Dependent Traffic. IEEE Trans Inform Theory, **44(1)**, 2-15
18. Giordano S, Miduri S, Pagano M, Russo F, Tartarelli S (1997) A Wavelet-based Approach to the Estimation of the Hurst Parameter for Self-similar Data. Proc DSP, 479-482
19. Jung C, Lin C (1998) An On-line Self-Constructing Neural Fuzzy Inference Network and Its Applications. IEEE Trans Fuzzy Systems, **6(1)**, 12-32
20. Cruz R (1991) A Calculus for Network Delay, Part I: Network Elements in Isolation. IEEE Trans Inform Theory, **37(1)**, 114-131
21. Beran J, Sherman R, Taqqa M, Willinger W (1995) Long-Range Dependence in Variable Bit Rate Video Traffic. IEEE Trans Comm, **43(2/3/4)**, 1566-1579
22. Paxson V (1997) Fast, Approxmate Synthesis of Fractional Gaussian Noise for Generating Self-Similar Network Traffic. Proc ACM/SIGCOMM, 5-18
23. Heffes H, Lucantoni D (1986) A Markov Modulated Characterization of Packetized Voice and Data Traffic and Related Statistical Multiplexer Performance. IEEE J Select Areas Commun, **4(6)**, 856-868
24. Daigle J (1992) Queueing Theory for Telecommunications. Addison-Wesley

Least Squares Kernel Methods and Applications

Anthony Kuh
Department of Electrical Engineering, University of Hawaii
Honolulu, HI 96822

Abstract This chapter discusses kernel methods and applications with a focus on the least squares support vector machines (LS-SVM). We give an introduction to kernel methods and the LS-SVM solution in the primal and dual spaces. The focus of this chapter is on solving binary classification problems. The LS-SVM has nice properties in that the algorithm implements nonlinear decision regions, is of moderate complexity, converges to minimum mean squared error solutions, and can be implemented adaptively. We compare the performance of the LS-SVM with the performance of the SVM and the Bayesian optimal solution. We also formulate a windowed recursive least squares implementation of the LS-SVM and present an example to recover CDMA signals.

Keywords Kernel methods, Support Vector Machines (SVM), Least Squares SVM, adaptive kernel CDMA receivers.

1 Introduction

In the last several years support vector machines (SVM) and kernel methods [3, 6, 5, 25, 30] have generated considerable interest and have been used in many applications from image processing to optical character recognition to analyzing DNA data [5, 25, 19]. Efficient algorithms have been found to find good solutions to realize SVM and other kernel methods. Kernel methods and SVM are also based on principles from statistical learning theory and their performance can be theoretically analyzed. This chapter focuses on a recently developed kernel algorithm called the least squares (LS)-SVM which is based on using a quadratic error function with equality constraints, [26]. The LS-SVM have nice properties in that their solution can be found by solving a set of linear equations making the algorithm amenable for adaptive on-line application. The chapter discusses advantages of using kernel methods (focus on LS-SVM), discusses the performance of the LS-SVM, and explores using the LS-SVM algorithm for recovering Code Division Multiple Access (CDMA) signals.

SVM or optimal margin classifiers are adaptive learning systems that receive labeled training data (e.g. pattern classification or regression problems) and transform these problems into optimization problems, [30]. Unlike many machine learning algorithms where solutions are ill posed, the solution to the SVM is well defined. SVM are usually solved by finding solutions to quadratic programming problems. SVM were originally applied to binary pattern classification problems where data was linearly separable, but the algorithm has been extended to handle data that is not separable by introducing slack variables [6] and to implement nonlinear decision regions via kernel functions [3]. By working with suitable kernel functions the solution to SVM can be found by solving the quadratic programming problem in the dual obervation space rather than the primal feature space thereby reducing overall computations. Additional extensions (e.g. to solving regression problems) and efficient methods to solve the quadratic programming problems (e.g. the sequential minimization optimization (SMO) algorithm [22]) have made the SVM a popular tool to use for many practical applications [5, 25].

SVM are based on principles of learning theory and structural risk minimization [5, 25, 30]. The generalization error can be defined for both classification and regression problems and is bounded by a weighted sum of the training error and the model error which depends on the Vapnik Chervonenkis dimension or the fat shattering dimension of the SVM [5]. For many of these kernels the dimension is high or even infinite resulting in loose bounds.

In this chapter we focus on a variation to the SVM called the least squares (LS)-SVM. The LS-SVM was originally introduced by Suykens, Lukas, and Vandewalle, [26], and is closely related to Kernel Fisher Discriminant Analysis (KFDA) introduced in [1, 18]. The key difference between SVM and LS-SVM is that inequality constraints are replaced by equality constraints and LS-SVM uses a quadratic error criterion. The LS-SVM solution involves solving a set of linear equations making the solution easier to implement than SVM which involves solving a quadrating programming problem with linear inequality constraints. The LS-SVM solution is not only simpler to implement than the SVM solution, but it is easily implemented adaptively. SVM solutions are usually found via batch algorithms where as LS-SVM can be found using on-line adaptive algorithms. This allows for relatively simple implementation of the LS-SVM for adaptive signal processing and communication applications.

Both the SVM and LS-SVM capabilities are controlled by the choice of the kernel function. This allows the user to solve complex pattern recognition and regression problems by choosing the appropriate kernel (e.g. linear, gaussian, sigmoid). A nice feature about both the SVM and LS-SVM solution is that the solution can be found by working in the dual observation space where the complexity of the problem depends on the number of the observation examples and not the dimension of the input feature space, [5, 25].

The LS-SVM is easy to analyze as the solution can be expressed analytically in terms of the parameters and sample first and second order statistics. In this chapter we will conduct a Bayesian analysis of the model where we consider a simple two hypothesis detection model. There are priors associated with each of the hypothesis and costs for making decisions. We study how the the LS-SVM solution compares to the optimal Bayesian solution and the SVM solution. Our focus is on the minimum error probability criterion.

The chapter is organized as follows. Section 2 discusses the basic Bayesian two hypotheses detection problem, the optimal Bayesian solution, and notation used in this chapter. Section 3 discusses optimal margin classifiers and the Support Vector Machine (SVM). We also briefly discuss algorithms used to find the SVM solution in the dual space. In Section 4 we present the LS-SVM problem and find the solution in both the primal and dual space. Section 5 compares the SVM and the LS-SVM in terms of performance and complexity of each algorithm. We compare the performance to the optimal Bayesian solution and state results about convergence of the SVM and the LS-SVM. Section 6 introduces an adaptive windowed recursive least squares algorithm that can be implemented on-line. Section 7 introduces a classification application of the SVM and LS-SVM. We use SVM and LS-SVM detectors to recover Code Division Multiple Access (CDMA) signals. Finally, Section 8 summarizes contributions of this chapter and suggests directions for further research.

2 Bayesian Two Hypothesis Detection Problem

The Bayesian detection problem [29] can be modeled with two random variables: $X \in \mathcal{R}^n$, describes the input and $Y \in \{-1, 1\}$, describes the output. In the detection problem we see a sample (or several samples) drawn from X and then make a decision $\hat{Y}$ on what Y is. We are typically given the likelihood conditional density function $f_{X|Y}(x|y)$ and the prior probability distribution, $P(Y = 1) = p$ and the $P(Y = -1) = 1 - p$.

There are also costs associated with making decisions where $C_{i,j}$ is the cost for deciding that the output is i given that it is j. The goal of Bayesian detection is to minimize the expected cost averaged over the input distribution X. The decision that minimizes the expected cost is called the Bayesian optimal solution. Here we will consider the minimum error probability problem where

$$C_{1,1} = C_{-1,-1} = 0, \quad C_{1,-1} = C_{-1,1} = 1.$$

The optimal decision depends on maximizing the posterior probability, $f_{Y|X}(y|x)$ or by considering the likelihood function defined as

$$L(x) = f_{X|Y}(x|1)/f_{X|Y}(x|-1).$$

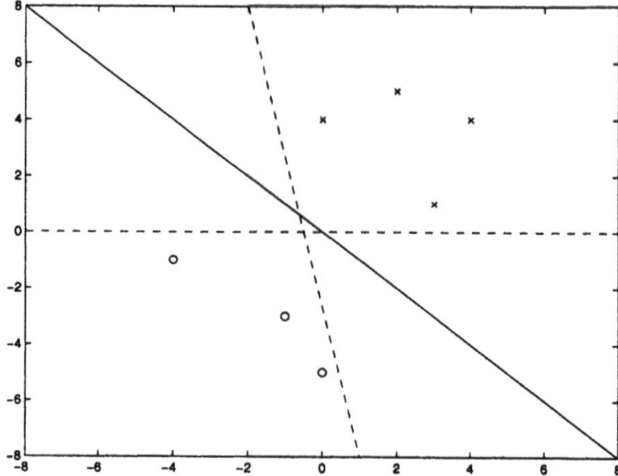

Figure 1: Labeled points that are linearly separable. All hyperplanes achieve separation, but bold hyperplane with solid line is the maximum margin hyperplane.

If $L(x) \geq (1-p)/p$ decide that $\hat{Y} = 1$ otherwise decide that $\hat{Y} = -1$. The decision region can be a simple linear threshold function or a more complex decision function depending on the two conditional likelihood densities.

If we do not have the conditional density functions or the posterior probabilities we must estimate the appropriate sufficient statistic. A sufficient statistics for the binary Bayesian detection problem is the log-likelihood function, $l(x) = \log(L(x))$ and the posterior probability, $P(x) = f_{Y|X}(1|x)$. Here we make our estimates based on observing labeled training data. We would like to estimate either $l(x), P(x)$ or the optimal Bayesian decision rule, $\mathrm{sgn}(P(x) - .5)$. In this chapter we compare the optimal Bayesian decision rule with the SVM and the LS-SVM solution.

3 Support Vector Machine

In this section we discuss the standard SVM for binary classification problems. The SVM is sometimes referred to as an optimum margin classifier. Consider a binary two hypothesis detection problem where l samples are drawn with l_1 positively labeled samples and $l - l_1$ negatively labeled samples. In Fig. 1 we label the positively labeled samples with an **x** and the negatively labeled samples with an **o**. If all the positively labeled samples can be separated from the negatively labeled samples by a linear hyperplane as in Fig. 1, then the set of labeled points is said to be linearly separable.

Mathematically, let $(x_i, y_i), 1 \leq i \leq l$ be a set of l labeled samples with $x_i \in \mathcal{R}^n$ and $y_i \in \{-1, 1\}$ for $1 \leq i \leq l$. The training points are linearly

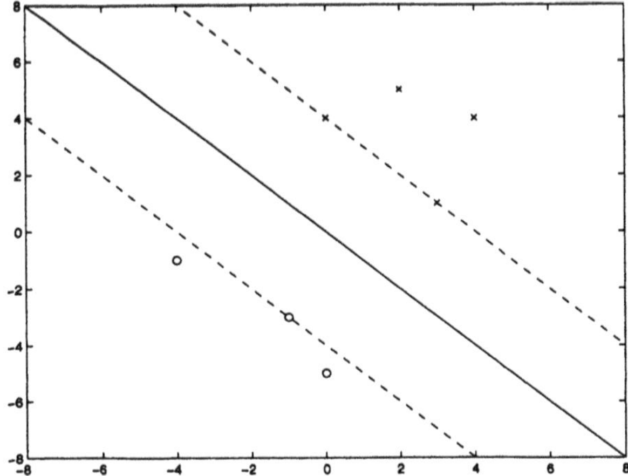

Figure 2: Optimum margin hyperplane shown by solid line with margin hyperplanes shown by dashed lines.

separable if there exists a weight vector, $w \in \mathcal{R}^n$ and a threshold value, $b \in \mathcal{R}$ such that $y_i = \text{sgn}(w^T x_i + b))$ for all $1 \leq i \leq l$ where

$$\text{sgn(s)} = \begin{cases} 1, & \text{if } s \geq 0 \\ -1, & \text{if } s < 0. \end{cases}$$

The w and b can be found via the Perceptron learning algorithm [24, 5]. If the samples are linearly separable, there are an infinite number of hyperplanes that will separate the data. The solution to finding a separating hyperplane is therefore an ill posed problem. The optimum margin classifier finds the hyperplane that maximizes the minimum distance between the hyperplane and the set of l samples.

Among all hyperplanes separating the data, there exists an optimal one yielding the maximum margin of separation between the classes,

$$\max_{w,b} \ \min(||x - x_i|| : x \in \mathcal{R}^n, (w^T x) + b = 0, i = 1, \ldots, l)$$

subject to $y_i((w^T x_i) + b) > 0, \ \ i = 1, \ldots, l.$

We can also define the margin hyperplanes as those hyperplanes that are parallel to the optimum margin hyperplane and include samples that are of minimum distance to the optimum margin hyperplane (support vectors). These margin hyperplanes are shown by dashed lines in Fig. 2. We normalize the weights so that the two margin hyperplanes describe equations $w^T x_i + b = 1$ and $w^T x_i + b = -1$. Then the margins (the minimum distance from sample to optimum margin hyperplane is given by $1/||w||$. Maximizing the margin is therefore equivalent to minimizing the magnitude of the weights or minimizing the magnitude squared value. The problem is then transformed into a quadratic programming problem [25],

$$\min \frac{1}{2}||w||^2 \tag{1}$$

subject to $y_i((w^T x_i) + b) \geq 1, \ i = 1, \dots, l$.

The SVM are made more powerful as they can be easily extended

- to construct nonlinear decision functions [3] by introducing kernel functions that satisfy Mercers Theorem [25]. Here the x_i are replaced with the functions $\phi(x_i)$ and inner products are replaced with kernel functions $K(x_i, x_j) = \phi(x_i)^T \phi(x_j)$.
- to handle data that is not separable [6] by introducing slack variables to the quadratic programming problem. Data points are penalized if they are misclassified.

The SVM with these modifications can then be described by

$$\min \frac{1}{2}||w||^2 + C\sum_{i=1}^{l} \xi_i \tag{2}$$

subject to $y_i(w^T \phi(x_i) + b) \geq 1 - \xi_i, \xi_i \geq 0 \ i = 1, \dots, l, \ C > 0$.

The quadratic programming problem can be solved by introducing Lagrangian multipliers and noting that the problem is convex. We let α_i denote the Lagrangian multipliers for constraints on (x_i, y_i). The solution satisfies the Karush-Kuhn-Tucker (KKT) conditions [14]. The weight w is then expressed as

$$w = \sum_{i=1}^{l} y_i \alpha_i \phi(x_i) \tag{3}$$

From this equation we see that only the inputs associated with nonzero Lagrangian multipliers α_i contribute to the weight vector w. These inputs are called support vectors and lie on the margins of the decision region (i.e. lie on or on the wrong side of one of the two margin hyperplanes that satisfy $w^T x + b = \pm 1$, see Fig. 2). These support vectors are the critical vectors in determining the optimal margin classifier.

The quadratic programming problem is solved by considering the dual problem where

$$\max_{\alpha} \sum_{i=1}^{l} \alpha_i - \frac{1}{2} \sum_{i,j=1}^{l} \alpha_i \alpha_j y_i y_j K(x_i, x_j). \tag{4}$$

subject to $0 \leq \alpha_i \leq C, \ \ i = 1, \dots, l$, and $\sum_{i=1}^{l} \alpha_i y_i = 0$. The decision function can then be written as

$$f(x) = \text{sign}\left(\sum_{i=1}^{l} \alpha_i y_i K(x, x_i) + b\right). \tag{5}$$

There have been many methods proposed to solve the SVM quadratic programming problem. Some methods deal with breaking the quadratic programming problem into smaller subproblems where the smaller subproblems are solved using numerical techniques (e.g. chunking [30, 20]). Another method, the Sequential Minimization Optimization (SMO) algorithm also breaks up the quadratic programming problem into subproblems, but solves the subproblems by analytical methods [22]. The SMO algorithm uses minimal storage space by successively solving a quadratic optimization problem involving two Lagrangian multipliers using analytical methods. An inner loop solves the quadratic programming problems and an outer loop searches for Lagrangian multipliers that violate the KKT complementarity conditions. The algorithm continues to operate until KKT conditions are satisfied for all Lagrangian multipliers. In recent years there have been more methods developed to solve large SVM problems. These methods decompose the quadratic programming problem into smaller problems or find approximate solutions to the quadratic programming problem by reducing its size [11, 4, 15].

4 Least Squares Support Vector Machines

4.1 LS-SVM Solution

The standard SVM are solved using quadratic programming methods, however these methods are often time consuming and are difficult to implement adaptively [25, 5]. Research has been undertaken to use a quadratic error criterion instead of the $\mathcal{L}_1$ norm used for the SVM. Ridge regression methods using a quadratic error criterion were developed for classification problems [5] and recently [17] use a quadratic error criterion to find an iterative solution to their Lagrangian SVM networks. These methods still have inequality constraints, however [26] formulated a modified least squares SVM (LS-SVM) based on using a quadratic error criterion with equality constraints. The LS-SVM is a simple modification of equation (2) with

$$\min J(w,b) = \min_{w,b} \frac{1}{2}||w||^2 + C\sum_{i=1}^{l} {e_i}^2 \tag{6}$$

subject to $y_i(w\cdot\phi(x_i)+b) = 1-e_i,\ \ i=1,\ldots,l$. This problem is easily solved by setting the partial derivatives $\partial J(w,b)/\partial(w) = 0$ and $\partial J(w,b)/\partial(b) = 0$ and solving for w and b. We get that

$$(R(l) + I/(lC))w = P(l) \tag{7}$$

and

$$m_X(l)^T w + b = m_Y(l) \tag{8}$$

where $m_X(l) = (1/l)\sum_{i=1}^{l}\phi(x_i)$ and $m_Y(l) = (1/l)\sum_{i=1}^{l} y_i$ are the sample first order statistics. The second order statistics are given by

$$R(l) = (1/l)\sum_{i=1}^{l}\phi(x_i)\phi(x_i)^T - m_X(l)m_X(l)^T$$

and

$$P(l) = (1/l)\sum_{i=1}^{l} y_i\phi(x_i) - m_Y(l)m_X(l)$$

If $(R(l) + I/(lC))$ is nonsingular we then have that

$$w = (R(l) + I/(lC))^{-1}P(l)$$

and $b = m_Y(l) - m_X(l)^T(R(l) + I/(lC))^{-1}P(l)$. As an example consider the points in Figures 1 and 2. The SVM solution depends on the support vectors and results in $w = (.25, .25)^T$ and $b = 0$. The LS-SVM solution depends on all samples and results in $w = (.1648, .1882)^T$ and $b = -.0856$. The LS-SVM solution depends on the first and second order statistics of $\phi(x(i))$ and $y(i)$. As the number of samples, l grows large we can state the following result.

Theorem: Under the assumptions of Section 2 where samples are independently drawn from the same distribution let ϕ be a Mercer kernel with $\phi(X)$ and Y being second order random variables. Let $R = \mathbf{E}(\phi(X)\phi(X)^T)$ be positive definite, $P = \mathbf{E}(Y\phi(X))$, $m_x = \mathbf{E}(\phi(X))$, and $m_Y = \mathbf{E}(Y)$. Then the LS-SVM solution converges almost surely to the kernel minimum mean squared error solution defined by weight vector, $w = R^{-1}P$ and threshold value, $b = m_Y - {m_X}^T R^{-1}P$.

Proof: A random variable $\phi(X)$ is a second order random variable if $\mathbf{E}(|\phi(X)|^2)$ is finite [32]. From the assumptions and the Law of Large Numbers [8], we therefore have that the time averages all converge almost surely to ensemble averages. We then use the fact that if two sequences of random variables converge almost surely then their products will also converge almost surely [8]. Then from equations (7-8) and that R is positive definite we have that the time averages will converge almost surely to ensemble averages with $w = R^{-1}P$ and $b = m_Y - {m_X}^T R^{-1}P$.

Remark 1: For the binary hypothesis detection problem let $m_i = \mathbf{E}(\phi(X)|Y = i)$ and $R_i = \mathbf{E}(\phi(X)\phi(X)^T|Y = i)$ $i = -1, 1$, then under assumptions of Section 2 and the Theorem above

$$w = (pR_1 + (1-p)R_{-1})^{-1}(pm_1 - (1-p)m_{-1}) \; , \tag{9}$$

$$b = 2p-1-(pm_1+(1-p)m_{-1})^T(pR_1+(1-p)R_{-1})^{-1}(pm_1-(1-p)m_{-1}) . \tag{10}$$

In general, this solution will be different from the minimum probability of error solution which depends on the likelihood ratio $L(x)$. In certain cases where the conditional densities under each hypothesis are symmetric and the conditional densities are translations of one another the LS-SVM asymptotic solution could be the same as the minimum probability of error solution.

Remark 2: Computation of the weight vector w and threshold value b depends on the dimensionality of the vector, $\phi(x) \in \mathcal{R}^d$. For polynomial kernels this can be quite high and for Gaussian kernels the dimensionality is infinite. This is a reason that we use the kernel trick to work in the dual observation space when $l < d$. The formulation in the dual observation space is given below.

4.2 LS-SVM Solution in Dual Observation Space

The dual problem can be implemented [26] by considering equation (6) and augmenting it with the equality constraints and lagrange multipliers, α_i to get

$$\min J(w,b,\alpha,e) = \min_{w,b} \frac{1}{2}||w||^2 + C\sum_{i=1}^{l} {e_i}^2 - \sum_{i=1}^{l} \alpha_i(y_i(w^T\phi(x_i)+b) - 1 + e_i)$$

Using the KKT conditions we set the partial derivatives equal to zero [26] to get

$$\frac{\partial J}{\partial w} = 0 \ \rightarrow \ w = \sum_{i=1}^{l} \alpha_i y_i \phi(x_i)$$

$$\frac{\partial J}{\partial b} = 0 \ \rightarrow \ \sum_{i=1}^{l} \alpha_i y_i = 0$$

$$\frac{\partial J}{\partial e_i} = 0 \ \rightarrow \ \alpha_i = Ce_i$$

$$\frac{\partial J}{\partial \alpha_i} = 0 \ \rightarrow \ y_i(w^T\phi(x_i)+b) - 1 + e_i = 0.$$

We can express these equations in more compact form to get

$$\alpha^T y = 0 \tag{11}$$

where $y = (y_1 \dots y_l)^T$, $\alpha = (\alpha_1 \dots \alpha_l)^T$, and

$$yb + (\Omega + C^{-1}I)\alpha = \mathbf{1}_l \tag{12}$$

where $\mathbf{1}_l$ is a one vector with l components and

$$\Omega(i,j) = y_i y_j K(x_i, x_j), \ \ 1 \le i,j \le l \tag{13}$$

The formulation and solution is presented in [26] and because the constraints are all equality constraints the solution involves solving a set of linear equations. Here C is a regularization factor. If C is large, then the solution approaches the standard minimum mean squared error estimate. If C is made small, then the regularization term, $||w||^2$ becomes more important and outlier points are deemphasized.

4.3 Comments about LS-SVM solution

Remark 3: The algorithm complexity depends on d and l. Assuming R and K are invertible, the algorithm involves inverting either R or K. If $l < d$ the complexity is $\mathcal{O}(l)^3$ using direct methods and if $d < l$ the complexity is $\mathcal{O}(d)^3$ using direct methods. If linear machines are used or the number of training examples is low or moderate, the LS-SVM algorithm complexity will be relatively low and have advantages over standard SVM.

Remark 4: The algorithm involves solving a set of d linear equations in the primal space or solving a set of $l+1$ equations in the dual observation space. On-line least squares forms of the algorithm can easily be implemented using adaptive recursive methods. A windowed recursive least squares formulation is given in Section 6.

Remark 5: The LS-SVM algorithm is very similar to many other kernel algorithms including Kernel Fisher Discriminant Analysis (KFDA) discussed in [1, 18, 19]. The LS-SVM algorithm is easily transformed into KFDA. In KFDA the goal in the kernel space is to choose a projection of the data to maximize the separation of the means of the two different data sets while minimizing the variances

5 Comparison of SVM and LS-SVM with Optimal Bayesian Detector

This section is divided into three subsections. In the first subsection we study the behavior of the SVM and the LS-SVM if we restrict the kernel to be linear. In the second subsection we discuss some results from [16] where the kernel space is rich enough to approximate a sufficient statistic of the Bayesian detection problem. Results have been developed for the SVM and the LS-SVM. In the last subsection we discuss some computational comparisons between SVM and LS-SVM.

In this section we assume examples are drawn independently at each time for the Bayesian detection model described in Section 2. With probability p the example is a positive example and with probability $1-p$ the example is a negative example.

5.1 Linear Machines

We start by examining linear machines to get some understanding of the differences between three detectors: the linear SVM, the linear LS-SVM, and the optimal linear Bayesian solution. In general each detector gives different solutions. Under certain circumstances where we have symmetries in the likelihood densities and when priors are equal all three detectors could give the same solution. For specific distributions with large amounts of examples we can numerically compare the performance of simple linear SVM and LS-SVM.

In Section 4 we saw that the LS-SVM depends on the first and second order statistics of the data and the regularization parameter C. We found that the LS-SVM solution under mild assumptions converges to a minimum mean squared error solution as we increase the number of training samples.

The SVM solution is more complicated and depends on the conditional distribution functions and various conditional expectations. The solution depends on Ω and equation (4) can be stated compactly as

$$\max_{\alpha} \alpha^T \mathbf{1}_l - .5\alpha^T \Omega \alpha \tag{14}$$

subject to $\alpha^T y = 0$ and $0 \leq \alpha_i \leq C, \quad i = 1, \ldots, l$. We cannot come up with a closed form solution in the general case. However, for the linear SVM in one dimension the SVM solution can be found by ordering the positively labeled points and the negatively labeled points. Find all points that violate margins, then continue adding pairs of positive and negative points until the dual objective function is maximized. In the limit as l gets large the solution will depend on distribution functions and the conditional mean function. Here we will assume that $\mathbf{E}(X|Y = -1) \leq \mathbf{E}(X|Y = 1)$ with equal priors and that both conditional random variables have infinite support. Define

$$G(x, i) = 1 - F(x, i) = \int_x^\infty f_{X|Y}(x|i)dx$$

as conditional distribution functions and

$$M(x, i) = \int_{-\infty}^x x f_{X|Y}(x|i)dx$$

as conditional mean functions. Find function

$$u = G^{-1}(F(x, 1), -1)$$

and then find x and u such that

$$M(x, 1) = \mathbf{E}(X|Y = -1) - M(u, -1).$$

Then we have that

$$w = 2/(x - u), \quad b = -(x + u)/(x - u). \tag{15}$$

These solutions are both different from the Bayesian optimal solution which depends on the conditional density functions or the likelihood ratios.

In general, the linear optimal Bayesian solution, the LS-SVM solution, and the SVM solution all given different detectors. Consider a simple example where all detectors give different solutions. Assume lots of examples are drawn, priors are the same, and we consider a minimum error probability problem. When $Y = 1$ assume X is Gaussian with mean 1 and variance 4 and when $Y = -1$ assume X is Gaussian with mean -2 and variance 1. For this case the best linear detector has a threshold at $x = -.582$ with an error probability of .2928. The minimum mean squared error detector (LS-SVM with C large) has a threshold of $x = -.5$ with an error probability of .2934. Finally, the SVM detector has a threshold of $x = -.65$ with an error probability of .2937.

5.2 Kernel Machines

Despite the fact that the SVM and the LS-SVM give different solutions when we restrict the kernel to be linear, we have observed that both machines give similar good performance on applications such as the recovery of Code Division Multiple Access (CDMA) signals [7, 12] which is discussed in Section 7. In the Bayesian framework that we have constructed what is the performance of the SVM and the LS-SVM if we allow the SVM to use a richer set of kernel functions?

The SVM problem is cast as a regularization problem in a reproducing kernel Hilbert space (RKHS) [5, 23]. A reproducing kernel over $\mathcal{R}^d$ is a positive definite function on $\mathcal{R}^d \times \mathcal{R}^d$. If we choose a reproducing kernel function $K(t, s)$ in a RKHS denoted by H_K, [16] shows that if functions in the RKHS can approximate the sufficient statistics $P(x)$, then the SVM solution and the LS-SVM solution will converge to the optimal Bayesian solution. A proof is given for the case where the $\mathcal{L}_2$ norm is used and simulations are conducted for Sobolev Hilbert space kernels and Gaussian kernels for the $\mathcal{L}_1$ norm.

An intuitive reason is that if we use proper kernel functions then the input space is transformed into feature space. The transformed conditional density functions will be altered such that each conditional density under each hypothesis will look similar (in many cases resembling Gaussian distributions) except for a translation shift. In these cases the solution for each algorithm can coincide.

The SVM and LS-SVM algorithm are based on receiving a finite set of labeled samples. When proper kernels are chosen the version space [21] (space where training samples are correctly classified) will also include the Bayesian solution. In order to get good solutions for the SVM and LS-SVM the parameters of the algorithm and model order must be chosen properly (e.g. for Gaussian kernels widths σ and C are parameters to be chosen) . This can be done on an ad hoc basis by trying different sets of parameters and model orders. Another method is to use the Bayesian evidence framework discussed for multi-layer perceptrons in [2], for SVM in [13], and for LS-SVM in [28].

5.3 Complexity of SVM and LS-SVM

A big advantage of the LS-SVM problem is that it has equality constraints versus inequality constraints for the SVM problem. We can therefore solve the LS-SVM by solving a set of linear equations where as we must use more complicated quadratic programming methods to solve the SVM. The LS-SVM is therefore more amenable to solution via on-line adaptive methods where as the SVM is usually solved by off-line batch methods. The memory requirements for LS-SVM are generally less than SVM as we only need to store the current training example and a representation of previous examples where as the SVM using a batch algorithm generally require storage of all examples in the batch. There has been a significant amount of new research at finding quick solutions to standard SVM problems using decomposition methods [11, 4, 15]. These methods establish analytical methods for decomposing quadratic programming problems, reduce the size of the quadratic programming problem, and find additional methods to reduce overall computation time. These methods may also be amenable to on-line implementation.

In the next section we implement a recursive least squares algorithm in the dual space. The algorithm takes $\mathcal{O}(N^2)$ operations per update where N is a finite sized window of data. We can then apply this algorithm to many signal processing and communications problems.

One disadvantage is that the Lagrangian multipliers for the LS-SVM tend to be all nonzero where as for the SVM case most of the multipliers are zero (only support vectors are nonzero). The Lagrangian multipliers for the LS-SVM also have physical meaning in that they are directly proportional to the error. Variations can be found for the LS-SVM to eliminate many of the least relevant support vectors and obtain a sparse representation as discussed in [27].

6 Adaptive LS-SVM Algorithm

In this section we formulate an adaptive on-line solution for the LS-SVM based on equations (11,12). We use a window size of length N. The training data are described by the inputs $x(k) = [x_k| \dots |x_{k+N-1}]$ and targets $y(k) = (y_k \dots y_{k+N-1})^T$. Let $U(k) = \Omega_k + C^{-1}I$ and

$$\Omega_k(i,j) = y_{i+k}y_{j+k}K(x_{i+k}, x_{j+k}), \;\; 0 \leq i,j \leq N-1 \tag{16}$$

The parameters of the LS SVM algorithm at time k are described by the threshold value $b(k)$ and the Lagrangian multipliers $\alpha(k) = (\alpha_k \dots \alpha_{k+N-1})^T$. We then get the matrix equation described by

$$\begin{bmatrix} 0 & y(k)^T \\ y(k) & U(k) \end{bmatrix} \begin{bmatrix} b(k) \\ \alpha(k) \end{bmatrix} = \begin{bmatrix} 0 \\ \mathbf{1}_N \end{bmatrix} \tag{17}$$

We then use the following simple matrix identity,

$$\begin{bmatrix} A & B \\ B^T & d \end{bmatrix}^{-1} = \begin{bmatrix} A^{-1} & 0 \\ 0 & 0 \end{bmatrix} + vv^T c \tag{18}$$

where A is a nonsingular square matrix, d is a constant, $v = (B^T A^{-1}, -1)^T$, and $c = (d - B^T A^{-1} B)^{-1}$.

Assuming $U(k)$ is nonsingular when equation (18) is applied to equation (17) we get that

$$b(k) = g(k) = (y(k)^T P(k) \mathbf{1}_N)/(y(k)^T P(k) y(k)) \tag{19}$$

where $P(k) = U(k)^{-1}$ and

$$\alpha(k) = P(k)(\mathbf{1}_N - g(k)y(k)). \tag{20}$$

A key to this algorithm is the computation of $P(k) = U(k)^{-1}$. Note that we can partition $U(K)$ as follows:

$$U(k) = \begin{bmatrix} l(k) & L(k)^T \\ L(k) & D(k) \end{bmatrix}$$

where $l(k)$ is a scalar and $D(k)$ is a square matrix of size $N - 1 \times N - 1$. At time $k + 1$, old data $(L(k), l(k))$ will be shifted out to produce

$$U(k+1) = \begin{bmatrix} D(k) & R(k+1) \\ R(k+1)^T & r(k+1) \end{bmatrix}$$

where $r(k + 1)$ is a scalar and the new data at time $k + 1$ is given by $(R(k + 1), r(k + 1))$.

We can then use the matrix identity from equation (18) to compute $P(k)$ to get that

$$P(k) = \begin{bmatrix} 0 & 0 \\ 0 & Q(k) \end{bmatrix} + v_l(k) v_l(k)^T g_l(k) \tag{21}$$

where $Q(k) = D(k)^{-1}$, $v_l(k) = (-1, L(k)^T Q(k))^T$, and $g_l(k) = l(k) - L(k)^T Q(k) L(k)$. We can also use the matrix identity from equation (18) to compute $P(k + 1)$ to get that

$$P(k+1) = \begin{bmatrix} Q(k) & 0 \\ 0 & 0 \end{bmatrix} + v_r(k) v_r(k)^T g_r(k) \tag{22}$$

where $v_r(k) = (R(k)^T Q(k), -1)^T$, and $g_r(k) = r(k) - R(k)^T Q(k) R(k)$. Note that both $P(k)$ and $P(k + 1)$ contain $Q(k)$. The right side of equation (22) is broken up into two terms: the first term containing old data and $Q(k)$ and the second term contains the new information from update $k + 1$. If we are given $P(k)$ we can easily determine $Q(k)$ from equation (21). We can express $P(k)$ as

$$P(k) = \begin{bmatrix} g_l(k) & p_1(k)^T \\ p_1(k) & P_2(k) \end{bmatrix}. \tag{23}$$

Then we have that

$$Q(k) = P_2(k) - p_1(k)p_1(k)^T / g_l(k). \tag{24}$$

The algorithm for finding threshold value $b(k)$ and $\alpha(k)$ can then be summarized by

1. Initialization: $k = 1$,
2. Get data $(x(k), y(k))$ and compute $P(k)$ (from $Q(k-1)$ equation (22) or initialization).
3. Compute $b(k)$ from equation (19) and $\alpha(k)$ from equation (20).
4. Compute $Q(k)$ from $P(k)$ from equation (24).
5. $k \leftarrow k + 1$ go to 2.

Each update of algorithm runs in $\mathcal{O}(N^2)$ time. The parameter N will depend on the application used and could include factors such as the degree of stationarity of the data.

Using other windowing formats we can get different adaptive least squares algorithms. We can easily reformat the algorithm to have a weighted least squares SVM. Examples include an exponential weighting factor as commonly used for the Recursive Least Squares (RLS) algorithm [9].

7 LS-SVM Application

7.1 CDMA Signal Model

Code Division Multiple Access (CDMA) systems are multiple access communication systems where multiple users transmit different signals over the same channel. The transmitted signals are mixed in time and frequency (unlike Time Division Multiple Access (TDMA) and Frequency Division Multiple Access (FDMA) sustems) and are encoded using a signature spreading signal. Each original message signal is recovered by having some knowledge of the signature signal and / or applying signal processing methods. The simplest receiver is to use the signature sequence as a matched filter to despread the desired signal. This system is the most commonly used type of receiver, but does not work well when interference from other users called multiple access interference (MAI) is high [31].

CDMA systems have received an increasing amount of attention because of performance benefits when compared to alternate TDMA and FDMA options and the possibilities of using sophisticated signal processing solutions for interference suppression. Current CDMA systems (e.g. IS-95) use limited signal processing capabilities, but many third-generation wireless networks will use CDMA and will require more complex signal processing in order to meet the increasing user requirements for bandwidth and mobility. There has been substantial research in using signal processing to improve the performance of CDMA systems by reducing the effects of MAI. When all signature sequences are available optimal receivers can be designed to minimize the effects of MAI and additive noise, but these receivers often have high complexity [31]. Tradeoffs can be obtained by reducing the complexity of receivers by using linear receivers such as the decorrelator receiver or the linear minimum mean squared error receiver. These receivers are superior to matched filters in reducing MAI and have lower complexity than optimal nonlinear receivers [31].

These performance gains via signal processing are more fully realized with CDMA systems with short spreading codes and a small number of users. These systems are assumed here. Adaptive algorithms are also attractive for short spreading codes because of their relative low complexity, minimal information required for the algorithm, and the ability of the algorithms to adapt to changing environments. A tutorial giving a comprehensive description of adaptive CDMA receivers is given in [10].

For CDMA systems users are distinguished by their signature sequences. Given K active users, the received signal is the sum of the spread signals contaminated by additive white Gaussian noise described by

$$x(t) = \sum_{k=1}^{K} \sum_{m=1}^{M} b_{k,m} u_k(t - mT - d_k) + \sigma n(t) \tag{25}$$

where $b_{k,m} \in \{-1, +1\}$ is the bit transmitted by the kth user at the mth time, $u_k(t)$ contains the spreading signal, T is duration of the information sampling interval, and d_k is the delay of user k. The noise is described by a constant factor σ and $n(t)$ which is Gaussian white noise with unit power spectral density. The spreading signal is contained in $u(t)$ and is described by

$$u(t) = \int_0^T A_k s_k(\tau) h(t, \tau) d\tau. \tag{26}$$

The spreading signal of the kth user is described by $s_k(t)$, A_k is the amplitude of the signal, and $h(t, \tau)$ describes the dispersive channel impulse response. The spreading signal is sampled every T_s seconds with the spreading gain given by $L = T/T_s$.

The focus here is on downlink receivers where we do not have information about spreading signals. We assume that the receiver is given a training sequence to learn and is interested in finding on-line adaptive learning algorithms. We will assume s synchronous model where the delays $d_k = 0$. Then the CDMA model of equation (25) is easily converted to a discrete time model which can be described in a compact matrix form by

$$\mathbf{x} = \mathbf{U}\mathbf{b} + \sigma\mathbf{n} \tag{27}$$

where $\mathbf{x}$ are the outputs received by the downlink, $\mathbf{U}$ contains the spreading signals and amplitudes when passed through the channel, $\mathbf{b}$ contains the information bits, and $\mathbf{n}$ is the additive white noise.

7.2 LS SVM CDMA Receivers

In previous work [7], we implemented the SVM to recover CDMA signals. To solve the quadratic programming problem we implemented the Sequential Minimization Optimization (SMO) algorithm developed in [22]. We simulated the algorithm on several CDMA configurations and found that the linear SVM performed similarly to the linear minimum mean squared error receiver and that SVM with Gaussian kernels performed better than linear receivers and approached the performance of optimal receivers.

Here we would like to implement an adaptive nonlinear receiver at the mobile station that has moderate complexity. The adaptive receiver is trained to recover the first message signal $b_{1,m}$. In training mode the adaptive receiver adjusts its weights and threshold value as it receives the CDMA output signal, $\mathbf{x}$ (from equation (27)) and the information bits $b_{1,m}$. The receiver does not require signature spreading sequences. During testing mode the the weights and threshold value of the receiver are fixed as the receiver gets the CDMA output signal, $\mathbf{x}$ and makes estimates $\hat{b}_{1,m}$ on the information bits of the first message signal.

Other nonlinear adaptive receivers such as decision feedback equalizers and interference cancellers are implemented at the basestation [31], as these receivers usually require signature sequences of other users. The SVM receiver is also a nonlinear receiver, however it is more difficult to develop a simple low complexity on-line adaptive implementation to solve the quadratic programming problem. This has led us to examine the LS-SVM which can construct nonlinear classifiers that are implemented in an on-line manner with moderate complexity (see Section 6).

To examine the performance of LS-SVM we simulated a simple CDMA system with five users and no multipath fading. We considered different situations where the power of the desired user (user 1) was small compared to the other users ($A_1/A_i = .2, 2 \leq i \leq 5$) and where the power of all users were the same. We also considered a situation where the MAI was high (the correlation between different signature spreading sequences was high (.429)). We used 400 training examples to train a linear SVM and an SVM with Gaussian kernels. Once the receiver was trained, we tested the algorithm

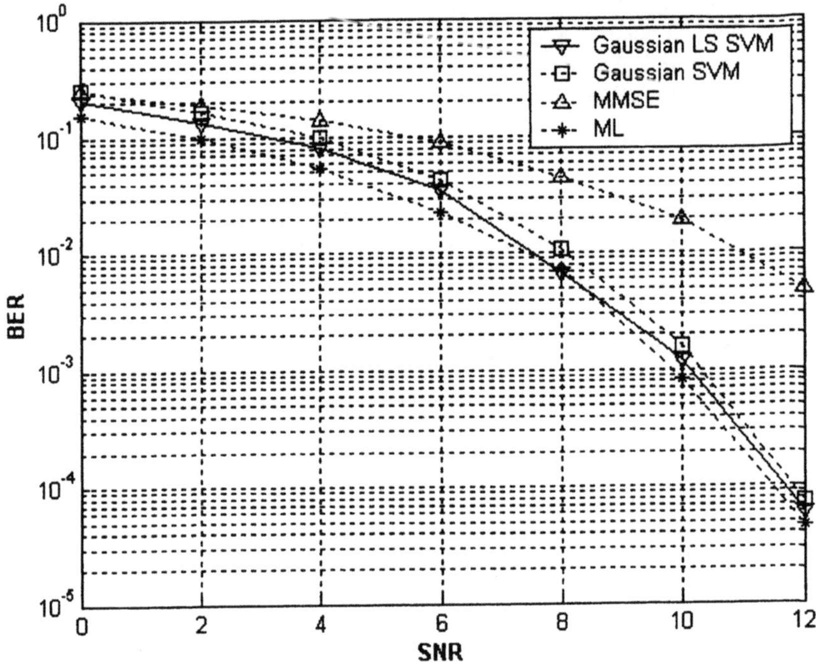

Figure 3: Five unequal energy users. High MAI with correlation between users given by ρ= 2/7.

on 10,000 test examples. We also compared to the linear minimum mean squared error (LMMSE) receiver and the optimal minimum probability of error receiver. We trained both the standard SVM algorithm (using the SMO algorithm [22]) and the LS-SVM algorithm.

Fig. 3 shows the case where the power of the desired user is much less than other users and Fig. 4 shows the case where the power of all users is the same. For both figures, the SNR refers to the desired signal power divided by noise power. For both simulations the LMMSE receiver performed poorly and this is due to the high multiple access interference (MAI). The linear SVM and linear LS-SVM had similar performance to the LMMSE receiver. We used Gaussian kernels for both the SVM and LS-SVM and tuned the parameters C and the Gaussian widths σ by running several simulations with different values for the parameters. Both the SVM and LS-SVM detectors have similar performance that is close to the optimal minimum probability of error receiver. For the unequal power case, the SVM and LS-SVM are about .2dB away from the optimal receiver and for the equal power case, the SVM and LS-SVM are about .6dB away from the optimal receiver.

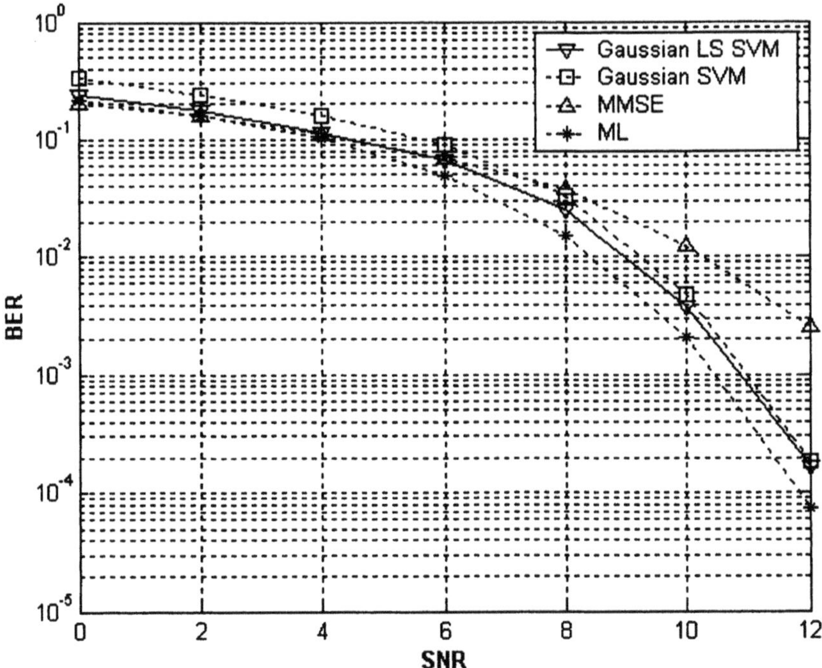

Figure 4: Five equal energy users. High MAI with correlation between users given by $\rho = 2/7$.

We also tested the adaptive windowed recursive least squares algorithm discussed in the last section. Fig. 5 shows the performance of the adaptive recursive LS-SVM algorithm in a time varying environment. Each epoch refers to 200 training samples. For the first 400 epochs there are a total of five users. Initially the Signal to Interference Ratio (SIR), is low, but the LS-SVM algorithm converges to the desired detector within a few hundred epochs. At 400 epochs we dropped two users. The LS-SVM algorithm can again adapt to the desired detector within another couple of hundred epochs.

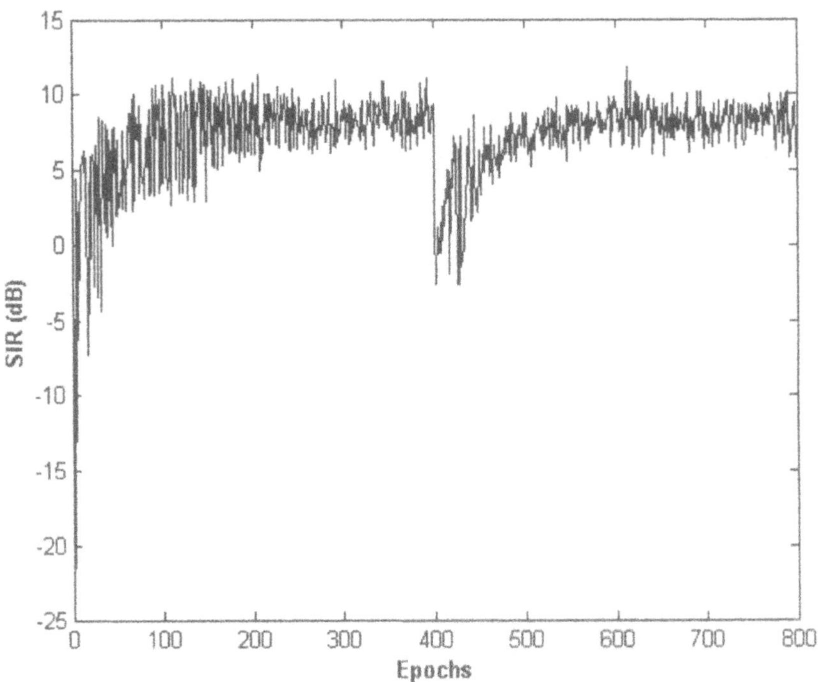

Figure 5: Five unequal energy users. High MAI with correlation between users given by ρ= 2/7. Adaptive RLS-SVM implemented with window size of200. At time 400 two users drop out.

8 Summary and Discussion

This chapter discusses kernel methods with a focus on the least squares support vector machine (LS-SVM) for binary classification problems. The chapter first introduces optimum margin classifiers and the support vector machine (SVM). We then discuss the LS-SVM algorithm and give solutions in the primal and dual space. We show that the algorithm converges to the minimum mean squared error kernel solution.

The chapter then compares the LS-SVM algorithm to the SVM algorithm and Bayesian optimal solution. In general solutions are different as each algorithm is based on different criteria (LS-SVM solutions are based on second order statistics where as SVM solutions depend on the conditional distributions). Under suitable assumptions (including use of proper kernels), the SVM and LS-SVM closely approximate the Bayesian optimal solution. We also discuss the complexity and memory requirements of SVM and LS-SVM and note that the LS-SVM solution has computational advantages over the SVM solution. The LS-SVM can be implemented via adaptive on-line algorithms and a windowed recursive least squares (RLS) algorithm is presented. Finally, we discussed recovery of CDMA signals with SVM and LS-SVM detectors. We find that the SVM and LS-SVM algorithm using Gaussian kernel can achieve performance close to the minimum probability of error detector when MAI is high. We also implemented the windowed RLS-SVM algorithm and found that it can adapt to changing environments.

There are many directions for further research. More analysis can be conducted comparing SVM and LS-SVM to optimal Bayesian solutions. For proper kernels and tuning of parameters we can show the optimality of the LS-SVM algorithm for recovery of CDMA signals.

Recently LS-SVM have also been applied to unsupervised learning for kernel principal component analysis. There are also are many applications in signal processing and communications where LS-SVM could be successfully applied (e.g. image processing, channel equalization).

Acknowledgements

Special thanks go to Xin Zhao and Xiaohong Gong who helped with many of the simulations. Thanks also goes to Dr. Johan Suykens for many fruitful discussions and his introducing me to LS-SVM.

References

[1] Baudat G, Anouar F (2000) Generalized discriminant analysis using a kernel approach. *Neural Computation*, vol. 12, 10, pp 2385-2404.

[2] Bishop C (1995) *Neural Networks for Pattern Recognition*, Oxford University Press.

[3] Boser B, Guyon I, Vapnik V (1992) A training algorithm for optimal margin classifiers. In D. Haussler, editor, *Proceedings of the 5th Annual ACM Workshop on Computational Learning Theory*, ACM Press, pp 144–152.

[4] Chang C-C, Hsu C-W, Lin C-J (2000) The analysis of decomposition methods for support vector machines. *IEEE Trans. on Neural Networks*, Vol. 11, #4, pp 1003-1008.

[5] Cristianini N, Shawe-Taylor J (2000) *An Introduction to Support Vector Machines.* Cambridge University Press, Cambridge, U.K.

[6] Cortes C, Vapnik V (1995) Support vector networks. *Machine Learning*, 20, pp 273–297.

[7] Gong X, Kuh A (1999) Support Vector Machine for Multiuser Detection. *33rd Asilomar Conference on Signals, Systems, and Computers*, Monterey CA, Vol. 1, pp 685-689.

[8] Grimmett G, Stirzaker D (2001) *Probability and Random Processes*, 3rd Ed., New York, Oxford Science.

[9] Haykin S (1996) *Adaptive filter theory, 3rd Ed.*, Prentice Hall, Englewood Cliffs, NJ.

[10] Honig M, Tsatsanis M (2000) Adaptive Techniques for Multiuser CDMA Receivers. In *IEEE Signal Processing Magazine*, Vol. 17, #3, pp 49-61.

[11] Joachims T (1999) "Making large-scale SVM learning practical". *Advances in Kernel Methods - Support Vector Learning* B. Schölkopf, C. Burges, and A. Smola Eds, Cambridge, MA, MIT Press, pp 169-184.

[12] Kuh A (2001) "Adaptive Kernel Methods for CDMA Systems", IJCNN'01, Washington D.C., pp 1404–1409.

[13] Kwok J, (2000) "The evidence framework applied to support vector machines", *IEEE Trans. on Neural Networks*, Vol. 11, pp 1162-1173.

[14] Kuhn H, and Tucker A (1951) Nonlinear programming. In *Proceedings of the 2nd Berkeley Symposium on Mathematical Statistics and Probabilistics*, University of California Press, pp 481–492.

[15] Lee Y-J, Mangasarian O (2001) "RSVM: Reduced support vector machines". *Proceedings First SIAM International Conference on Data Mining.*

[16] Lin Y, (1999) "Support Vector Machines and the Bayes Rule in Classification", Technical Report 1014, Dept. of Statistics, University of Wisconsin, Madison WI.

[17] Magasarian O, Musicant D (2001) Lagrangian Support Vector Machines. in *Journal of Machine Learning Research*, #I, pp 161-177.

[18] Mika S, Ratsch G, Weston J, Schölkopf B, Müller K (1999) "Fisher discriminant analysis with kernels", *Neural Networks for Signal Processing*, IX, Y.-H. Hu, J. Larsen, E. Wilson, and S. Douglas Eds. IEEE, pp 41-48.

[19] Müller K, Mika S, Ratsch G, Tsuda K, Schölkopf B (2001) An Introduction to Kernel-Based Learning Algorithms. *IEEE Trans. on Neural Networks*, Vol. 12, #2, pp 181-202.

[20] Ossuna E, Freund R, Girosi F (1999) "An improved training algorithm for support vector machines", *Neural Networks for Signal Processing*, VII, J. Principe, L. Giles, N. Morgan, and E. Wilson Eds. IEEE, pp 276-285.

[21] Opper M, Haussler D (1991) Generalization performance of Bayes optimal classification algorithm for learning a perceptron. *Physical Review Letters.* vol. 66, pp 2677.

[22] Platt J (1999) "Sequential minimal optimization: a fast algorithm for training support vector machines". *Advances in Kernel Methods - Support Vector Learning* B. Schölkopf, C. Burges, and A. Smola Eds, Cambridge, MA, MIT Press, pp 185-208.

[23] Poggio T, Girosi F (1998) A Sparse Representation for Function Approximation. *Neural Computation*, Vol. 10 # 6, pp 1445-1454.

[24] Rosenblatt F (1962) *Principles of Neurodynamics.* Spartan, New York.

[25] Schölkopf B, Burges C, Smola A (1999) *Advances in kernel methods support vector learning.* MIT Press, Cambridge, MA.

[26] Suykens J, Lukas L, and Vandewalle J (1999) Least squares support vector machine classifiers. *Neural Processing Letters*, Vol. 9, #3, pp 293-300.

[27] Suykens J, Lukas L, Vandewalle J (2000) Sparse approximation using least squares support vector machines. *ISCAS 2000*, Geneva, Switzerland, Vol. II, pp 757-760.

[28] Van Gestel T, Suykens J, Lanckriet G, Lambrechts A, De Moor B, Vandewalle J (2002) A Bayesian Framework for Least Squares Support Vector Machine Classifiers. *Neural Computation*, vol 15, 5, pp 1115-1148.

[29] Van Trees H (1968) *Detection, Estimation, and Modulation Theory Part I* John Wiley.

[30] Vapnik V (1998) *Statistical learning theory.* John Wiley, New York City, NY.

[31] Verdu S (1998) *Multiuser Detection.* Cambridge University Press, Cambridge, United Kingdom.

[32] Wong E, Hajek B (1985) *Stochastic Processes for Engineering Systems.* New York, Springer Verlag.

Least Bit Error Rate Adaptive Multiuser Detection

Sheng Chen

Department of Electronics and Computer Science, the University of Southampton, Highfield, Southampton SO17 1BJ, U.K.

Abstract. Linear detector required at direct-sequence code division multiple access (DS-CDMA) communication systems is classically designed based on the minimum mean squares error (MMSE) criterion, which can efficiently be implemented using the standard adaptive algorithms, such as the least mean square (LMS) algorithm. As the probability distribution of the linear detector's soft output is generally non-Gaussian, the MMSE solution can be far away from the optimal minimum bit error rate (MBER) solution. Adopting a non-Gaussian approach naturally leads to the MBER linear detector. Based on the approach of Parzen window or kernel density estimation for approximating the probability density function (p.d.f.), a stochastic gradient adaptive MBER algorithm, called the least bit error rate (LBER), is developed for training a linear multiuser detector. A simplified or approximate LBER (ALBER) algorithm is particularly promising, as it has a computational complexity similar to that of the classical LMS algorithm. Furthermore, this ALBER algorithm can be extended to the nonlinear multiuser detection.

Keywords: CDMA, linear detector, mean square error, bit error rate, probability density function, adaptive algorithm, least mean square algorithm, least bit error rate algorithm, nonlinear detector.

1 Introduction

DS-CDMA constitutes an attractive multiuser scheme that allows users to transmit at the same carrier frequency. However, this creates multiuser interference (MUI) which, if not controlled, can seriously degrade the quality of reception. Multiuser detection provides an effective means to combating MUI [1],[2]. In a DS-CDMA communication system, the objective of the receiver is to detect the transmitted information bits of one (at mobile-end) or many (at base station) users. The first case is concerned with communication from base station to mobile and is commonly called downlink. This chapter considers multiuser detection at mobile. In the downlink scenario, a mobile has a very limited computing power, and computational complexity is a critical issue. Typically, a linear detector is employed at the receiver to meet the strict real-time computational constraint.

A variety of linear multiuser detectors have been proposed [2]–[8]. A very simple linear detector is the matched filter (MF) with the detector weight vector being set to the user spreading code. However, multipath distortions, which are often encountered in DS-CDMA systems, can seriously degrade the performance of the MF. Another linear detector structure is called the zero-forcing (ZF) detector. As this ZF solution ignores the noise, it suffers from a serious noise enhancing problem. Moreover, other interfering user codes required in calculating the ZF solution may not be available to the detector, and it is not known how to adaptively implement the ZF solution. The most popular linear multiuser detector is the MMSE detector, as it usually provides good performance and can readily be implemented using standard adaptive filter techniques such as the LMS algorithm.

The ultimate performance criterion of a detector is its bit error rate (BER). Although for the linear detector a small mean square error (MSE) is associated with a small BER, the MMSE solution is generally not the MBER one. This is obvious, since in order for the MMSE solution to be the MBER solution the detector's soft output should be Gaussian distributed. The conditional p.d.f. of the linear detector output is however a sum of Gaussian distributions and therefore non-Gaussian. In the situation where only a few strong interfering users exist, the MMSE solution can be considerably inferior to the MBER one. A non-adaptive MBER linear multiuser detector is considered in [9] based on gradient optimization for narrow-band Gaussian CDMA channels which do not introduce intersymbol interference (ISI). There are a few adaptive MBER linear multiuser detectors in the literature [10]–[12].

The adaptive MBER algorithm given in [10] uses a difference approximation to estimate the gradient of one-sample error probability. Its main drawback is very slow convergence, particularly in the situation where the error rate is very low. Furthermore, the computational complexity of the algorithm is high and is in the order $O(M^2)$, M being the detector length. The adaptive MBER algorithm reported in [11], called the approximate MBER (AMBER) detector, is appealing due to its computational simplicity. It is a stochastic gradient algorithm that is identical to the signed-error LMS algorithm [13], except in the vicinity of the decision boundary where it is modified to continue updating the weights when the signed-error LMS algorithm would not. The AMBER algorithm therefore can continuously update when the detector weight vector has reached the regions of very low error rate.

This chapter considers the adaptive MBER algorithm based on a density approximation approach [12], called the LBER algorithm. Previous studies have shown that this LBER algorithm outperforms the AMBER algorithm in terms of convergence speed and steady-state BER misadjustment, in both the linear multiuser detection application [12] and the single-user equalization application [14],[15]. Although the computational requirement of this LBER linear detector is considerably higher than the AMBER linear detector, its complexity is still in the order $O(M)$. Furthermore, as will be shown in this chapter, a simplified LBER algorithm called the ALBER has a similar performance to the full LBER algorithm and yet has a complexity similar to the very simple LMS algorithm. An added advantage of this ALBER algorithm is that it can be extended to the nonlinear multiuser detector.

A basic assumption for a linear detector to work adequately is that the two classes of signal states related to the transmitted bit being +1 and −1, respectively, are linearly separable. Multipath distortions however may result in linearly inseparable situation. In the nonlinear separable case, a nonlinear multiuser detector is required to achieve good performance. Classically, training a nonlinear detector is based on the LMS algorithm. Previous work [16],[17] has shown that a nonlinear detector trained by the ALBER algorithm can closely match the theoretical optimal performance.

2 System Model

The discrete-time baseband model of the synchronous DS-CDMA downlink system supporting N users and transmitting M ($> N$) chips per bit is depicted in Fig. 1, where $b_i(k) \in \{\pm 1\}$ denotes the k-th bit of user i, the unit-length signature sequence for user i is $\bar{\mathbf{s}}_i = [\bar{s}_{i,1} \ \cdots \ \bar{s}_{i,M}]^T$ and the transfer function associated with the channel impulse response (CIR) at chip rate is

$$H(z) = \sum_{i=0}^{n_h - 1} h_i z^{-i}. \tag{1}$$

The bit vector of N users at bit instant k is $\mathbf{b}(k) = [b_1(k) \ \cdots \ b_N(k)]^T$, and the received signal vector obtained by sampling at chip rate is $\mathbf{r}(k) = [r_1(k) \ \cdots \ r_M(k)]^T$. It can be shown that the baseband model for $\mathbf{r}(k)$ is

$$\mathbf{r}(k) = \mathbf{P} \begin{bmatrix} \mathbf{b}(k) \\ \mathbf{b}(k-1) \\ \vdots \\ \mathbf{b}(k-L+1) \end{bmatrix} + \mathbf{n}(k) = \bar{\mathbf{r}}(k) + \mathbf{n}(k) \tag{2}$$

where L is the channel ISI span, $\bar{\mathbf{r}}(k)$ denotes the noise-free received signal vector, the white Gaussian noise vector $\mathbf{n}(k) = [n_1(k) \ \cdots \ n_M(k)]^T$ with

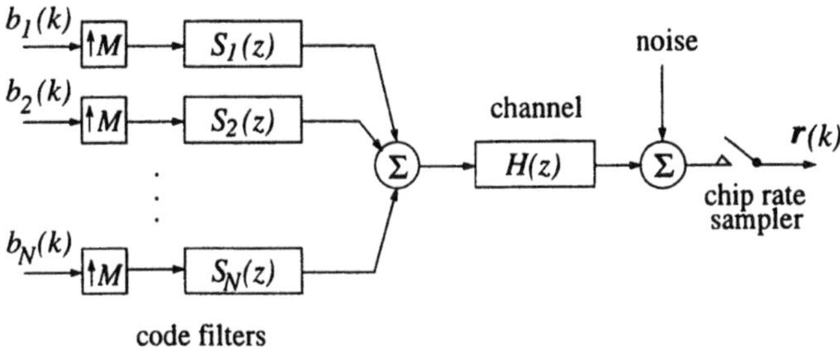

Fig. 1. Discrete-time model of synchronous CDMA downlink.

$E[\mathbf{n}(k)\mathbf{n}^T(k)] = \sigma_n^2\mathbf{I}$, and the $M \times LN$ system matrix $\mathbf{P}$ is given by

$$\mathbf{P} = \mathbf{H}\begin{bmatrix} \bar{\mathbf{S}}\mathbf{A} & \mathbf{0} & \cdots & \mathbf{0} \\ \mathbf{0} & \bar{\mathbf{S}}\mathbf{A} & \ddots & \vdots \\ \vdots & \ddots & \ddots & \mathbf{0} \\ \mathbf{0} & \cdots & \mathbf{0} & \bar{\mathbf{S}}\mathbf{A} \end{bmatrix} \tag{3}$$

with the user signature matrix $\bar{\mathbf{S}} = [\bar{\mathbf{s}}_1 \ \cdots \ \bar{\mathbf{s}}_N]$, the diagonal user signal amplitude matrix $\mathbf{A} = \text{diag}\{A_1 \ \cdots \ A_N\}$ and the $M \times LM$ CIR matrix

$$\mathbf{H} = \begin{bmatrix} h_0 & h_1 & \cdots & h_{n_h-1} & 0 & \cdots & 0 & 0 & 0 & 0 \\ 0 & h_0 & h_1 & \cdots & h_{n_h-1} & 0 & \cdots & 0 & 0 & 0 \\ \vdots & \ddots & \ddots & \ddots & \cdots & \ddots & \ddots & & \ddots & \vdots \\ 0 & \cdots & 0 & h_0 & h_1 & \cdots & h_{n_h-1} & 0 & \cdots & 0 \end{bmatrix}. \tag{4}$$

The channel ISI span L depends on the length of the CIR, n_h, related to the length of the chip sequence, M. For $n_h = 1$, $L = 1$; for $1 < n_h \leq M$, $L = 2$; for $M < n_h \leq 2M$, $L = 3$; and so on.

Let the $N_b = 2^{LN}$ possible combinations of $[\mathbf{b}^T(k) \ \cdots \ \mathbf{b}^T(k-L+1)]^T$ be

$$\mathbf{b}^{(j)} = \begin{bmatrix} \mathbf{b}^{(j)}(k) \\ \mathbf{b}^{(j)}(k-1) \\ \vdots \\ \mathbf{b}^{(j)}(k-L+1) \end{bmatrix}, \ 1 \leq j \leq N_b, \tag{5}$$

and $b_i^{(j)}$ the i-th element of $\mathbf{b}^{(j)}(k)$. Clearly,

$$\bar{\mathbf{r}}(k) \in \mathcal{R} \triangleq \{\bar{\mathbf{r}}_j = \mathbf{P}\mathbf{b}^{(j)}, \ 1 \leq j \leq N_b\}. \tag{6}$$

$\mathcal{R}$ is called the set of noise-free received signal states. For user i, it can be divided into two subsets depending on the value of $b_i(k)$

$$\mathcal{R}^{(\pm)} \triangleq \{\bar{\mathbf{r}}_j^{(\pm)} \in \mathcal{R} : \ b_i(k) = \pm 1\}. \tag{7}$$

Consider the linear detector for user i which takes the form

$$y(k) = \mathbf{w}^T\mathbf{r}(k) = \mathbf{w}^T(\bar{\mathbf{r}}(k) + \mathbf{n}(k)) = \bar{y}(k) + e(k) \tag{8}$$

where $\mathbf{w} = [w_1 \ \cdots \ w_M]^T$ is the detector weight vector and $e(k)$ is Gaussian with zero mean and variance $\mathbf{w}^T\mathbf{w}\sigma_n^2$. The estimated $b_i(k)$ is given by

$$\hat{b}_i(k) = \text{sgn}(y(k)) = \begin{cases} +1, & y(k) \geq 0, \\ -1, & y(k) < 0. \end{cases} \tag{9}$$

Obviously, the scalar $\bar{y}(k)$ can only take values from the set

$$\mathcal{Y} \triangleq \{\bar{y}_j = \mathbf{w}^T\bar{\mathbf{r}}_j, \ 1 \leq j \leq N_b\} \tag{10}$$

which can be divided into two subsets depending on the value of $b_i(k)$

$$\mathcal{Y}^{(\pm)} \triangleq \{\bar{y}_j^{(\pm)} \in \mathcal{Y} : \; b_i(k) = \pm 1\}. \tag{11}$$

For the linear detector (8) to perform adequately, $\mathcal{Y}^{(+)}$ and $\mathcal{Y}^{(-)}$ must be linearly separable. Otherwise, a nonlinear detector should be used. The simple MF detector is given by $\mathbf{w}_{\mathrm{MF}} = \bar{\mathbf{s}}_i$. The most popular solution for the linear detector (8) is the MMSE one given by

$$\mathbf{w}_{\mathrm{MMSE}} = \left(\sigma_n^2 \mathbf{I} + \mathbf{P}\mathbf{P}^T\right)^{-1} \mathbf{p}_i \tag{12}$$

with $\mathbf{p}_i$ denoting the i-th column of $\mathbf{P}$. Although user i may not know the other user codes and therefore may be unable to compute $\mathbf{w}_{\mathrm{MMSE}}$ directly, adaptive implementation using the LMS does not require to know the other user codes. The ZF solution $\mathbf{w}_{\mathrm{ZF}}$ is obtained by setting $\sigma_n^2 = 0$ in (12).

3 The MBER Linear Multiuser Detector

The conditional p.d.f. of $y(k)$ given $b_i(k) = +1$ is

$$p_y(y| + 1) = \frac{1}{N_{sb}\sqrt{2\pi}\sigma_n\sqrt{\mathbf{w}^T\mathbf{w}}} \sum_{j=1}^{N_{sb}} \exp\left(-\frac{\left(y - \bar{y}_j^{(+)}\right)^2}{2\sigma_n^2 \mathbf{w}^T\mathbf{w}}\right) \tag{13}$$

where $N_{sb} = N_b/2$ is the number of points in $\mathcal{Y}^{(+)}$ and $\bar{y}_j^{(+)} \in \mathcal{Y}^{(+)}$. Thus, the conditional BER of the linear detector given $b_i(k) = +1$ is

$$P_{E,+}(\mathbf{w}) = \int_{-\infty}^{0} p_y(y| + 1)\, dy = \frac{1}{N_{sb}} \sum_{j=1}^{N_{sb}} Q\left(g_{j,+}(\mathbf{w})\right) \tag{14}$$

where

$$g_{j,+}(\mathbf{w}) = \frac{\bar{y}_j^{(+)}}{\sigma_n\sqrt{\mathbf{w}^T\mathbf{w}}} = \frac{\mathbf{w}^T\bar{\mathbf{r}}_j^{(+)}}{\sigma_n\sqrt{\mathbf{w}^T\mathbf{w}}} = \frac{\mathrm{sgn}(b_i^{(j)})\bar{y}_j^{(+)}}{\sigma_n\sqrt{\mathbf{w}^T\mathbf{w}}} \tag{15}$$

and

$$Q(x) = \frac{1}{\sqrt{2\pi}} \int_x^{\infty} \exp\left(-\frac{u^2}{2}\right) du. \tag{16}$$

Similarly, the conditional p.d.f. of $y(k)$ given $b_i(k) = -1$ is

$$p_y(y| - 1) = \frac{1}{N_{sb}\sqrt{2\pi}\sigma_n\sqrt{\mathbf{w}^T\mathbf{w}}} \sum_{j=1}^{N_{sb}} \exp\left(-\frac{\left(y - \bar{y}_j^{(-)}\right)^2}{2\sigma_n^2 \mathbf{w}^T\mathbf{w}}\right) \tag{17}$$

where $\bar{y}_j^{(-)} \in \mathcal{Y}^{(-)}$, and the conditional BER given $b_i(k) = -1$ is

$$P_{E,-}(\mathbf{w}) = \int_0^{\infty} p_y(y|-1)\, d\, y = \frac{1}{N_{sb}} \sum_{j=1}^{N_{sb}} Q\left(g_{j,-}(\mathbf{w})\right) \tag{18}$$

where

$$g_{j,-}(\mathbf{w}) = -\frac{\bar{y}_j^{(-)}}{\sigma_n \sqrt{\mathbf{w}^T \mathbf{w}}} = \frac{\operatorname{sgn}(b_i^{(j)}) \bar{y}_j^{(-)}}{\sigma_n \sqrt{\mathbf{w}^T \mathbf{w}}}. \tag{19}$$

Because of the symmetric distribution of $\mathcal{Y}$, $P_{E,-}(\mathbf{w}) = P_{E,+}(\mathbf{w})$. Thus, the BER of the linear detector with the weight vector $\mathbf{w}$ is

$$P_E(\mathbf{w}) = P_{E,+}(\mathbf{w}) = \frac{1}{N_{sb}} \sum_{j=1}^{N_{sb}} Q\left(g_{j,+}(\mathbf{w})\right). \tag{20}$$

It is seen that the BER can be evaluated based on a single subset $\mathcal{Y}^{(+)}$ (or $\mathcal{Y}^{(-)}$). Also the BER is invariant to a positive scaling of the weight vector, that is, the BER depends on the direction of $\mathbf{w}$ only.

The MBER solution is defined as

$$\mathbf{w}_{\text{MBER}} = \arg\min_{\mathbf{w}} P_E(\mathbf{w}). \tag{21}$$

The gradient of $P_E(\mathbf{w})$ with respect to $\mathbf{w}$ is

$$\nabla P_E(\mathbf{w}) = \frac{1}{N_{sb}\sqrt{2\pi}\sigma_n} \sum_{j=1}^{N_{sb}} \exp\left(-\frac{\left(\bar{y}_j^{(+)}\right)^2}{2\sigma_n^2 \mathbf{w}^T \mathbf{w}}\right) \times$$

$$\operatorname{sgn}(b_i^{(j)}) \left(\frac{\bar{y}_j^{(+)} \mathbf{w}}{\left(\mathbf{w}^T \mathbf{w}\right)^{3/2}} - \frac{\bar{\mathbf{r}}_j^{(+)}}{\sqrt{\mathbf{w}^T \mathbf{w}}}\right). \tag{22}$$

With the gradient, the optimization problem (21) can be solved for iteratively using a conjugated gradient algorithm [18],[12] with a resetting of the search direction to the negative gradient $-\nabla P_E(\mathbf{w})$ every J iterations. It is computationally advantageous to normalize $\mathbf{w}$ to a unit-length after every iteration, so that the gradient can be simplified as

$$\nabla P_E(\mathbf{w}) = \frac{1}{N_{sb}\sqrt{2\pi}\sigma_n} \sum_{j=1}^{N_{sb}} \exp\left(-\frac{\left(\bar{y}_j^{(+)}\right)^2}{2\sigma_n^2}\right) \operatorname{sgn}(b_i^{(j)}) \left(\bar{y}_j^{(+)} \mathbf{w} - \bar{\mathbf{r}}_j^{(+)}\right). \tag{23}$$

It is obvious that, if $\mathcal{Y}^{(-)}$ is used for the BER evaluation, one only needs to substitute $\bar{y}_j^{(+)}$ and $\bar{\mathbf{r}}_j^{(+)}$ by $\bar{y}_j^{(-)}$ and $\bar{\mathbf{r}}_j^{(-)}$ in the gradient formula.

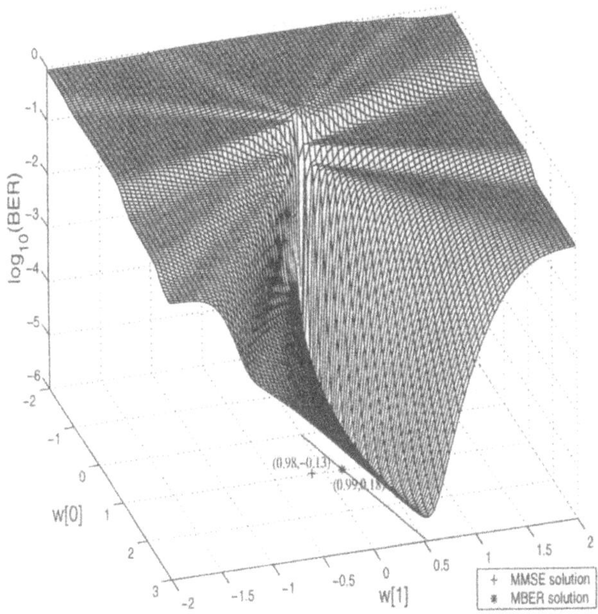

Fig. 2. Bit error rate surface of user-one detector for a simple two-user system with two chips per bit and SNR_1=25 dB given in Example 1.

Unlike the MMSE solution (12), there exists no close-form solution for $\mathbf{w}_{\mathrm{MBER}}$. For $\mathbf{w}_{\mathrm{MMSE}}$ to achieve the MBER, the p.d.f. (13) or (17) must be Gaussian. If the number of users is large, this p.d.f. will be close to Gaussian, and the BER difference between $\mathbf{w}_{\mathrm{MMSE}}$ and $\mathbf{w}_{\mathrm{MBER}}$ is expected to be small, if any gap exists. If however, there exist only a few interferers, it is likely that the BER difference between $\mathbf{w}_{\mathrm{MMSE}}$ and $\mathbf{w}_{\mathrm{MBER}}$ will be large. Whether the MMSE solution can achieve a BER close to the MBER also depends on the CIR. Before turning to the adaptive MBER algorithm, consider the following simple example, which provides some insights to the MBER solutions.

Example 1. This is a simplest system with two equal-power users and two chips per bit. The two chip codes are $(+1,+1)$ and $(+1,-1)$, respectively, and the transfer function of the CIR at chip rate is $H(z) = 1 + 0.8z^{-1} + 0.6z^{-2}$. The signal to noise ratio for user 1 is $\mathrm{SNR}_1 = 25$ dB. The $\log_{10}(\mathrm{BER})$ surface for user 1 is plotted in Fig. 2. Some general observations can be drawn from Fig. 2. The MBER solutions form a half line in the weight space, with one end of the line approaching the original and the other end approaching infinity. Any point in this half line is a global MBER solution. The point marked in the MBER solution half line is the unit-length one. The origin of the weight space is the singular (discontinuity) point of the BER surface. The MMSE solution for this example is also depicted in Fig. 2. For the MMSE solution, $\log_{10}(\mathrm{BER}) = -3.88$, while for a MBER one, $\log_{10}(\mathrm{BER}) = -5.56$.

4 Adaptive MBER Algorithms

The p.d.f. of $y(k)$ is explicitly given by

$$p_y(y) = \frac{1}{N_b\sqrt{2\pi}\sigma_n\sqrt{\mathbf{w}^T\mathbf{w}}}\sum_{j=1}^{N_b}\exp\left(-\frac{(y-\bar{y}_j)^2}{2\sigma_n^2\mathbf{w}^T\mathbf{w}}\right) \tag{24}$$

and the BER can alternatively be expressed as

$$P_E(\mathbf{w}) = \frac{1}{N_b}\sum_{j=1}^{N_b} Q\left(g_j(\mathbf{w})\right) \tag{25}$$

where $\bar{y}_j \in \mathcal{Y}$ and

$$g_j(\mathbf{w}) = \frac{\mathrm{sgn}(b_i^{(j)})\bar{y}_j}{\sigma_n\sqrt{\mathbf{w}^T\mathbf{w}}}. \tag{26}$$

In reality, the p.d.f. of $y(k)$ is unknown. A widely used approach to approximate a p.d.f. is known as the kernel density estimate [19]–[21].

Given a block of K training samples $\{\mathbf{r}(k), b_i(k)\}$, a kernel density or Parzen window estimate of the p.d.f. (24) is given by:

$$\hat{p}_y(y) = \frac{1}{K\sqrt{2\pi}\rho_n\sqrt{\mathbf{w}^T\mathbf{w}}}\sum_{k=1}^{K}\exp\left(-\frac{(y-y(k))^2}{2\rho_n^2\mathbf{w}^T\mathbf{w}}\right) \tag{27}$$

where the kernel width ρ_n is related to the noise standard deviation σ_n [20]. From this estimated p.d.f., the estimated BER is given by

$$\hat{P}_E(\mathbf{w}) = \frac{1}{K}\sum_{k=1}^{K} Q\left(\hat{g}_k(\mathbf{w})\right) \tag{28}$$

with

$$\hat{g}_k(\mathbf{w}) = \frac{\mathrm{sgn}(b_i(k))y(k)}{\rho_n\sqrt{\mathbf{w}^T\mathbf{w}}}. \tag{29}$$

The gradient of $\hat{P}_E(\mathbf{w})$ is

$$\nabla\hat{P}_E(\mathbf{w}) = \frac{1}{K\sqrt{2\pi}\rho_n}\sum_{k=1}^{K}\exp\left(-\frac{y^2(k)}{2\rho_n^2\mathbf{w}^T\mathbf{w}}\right)\times$$

$$\mathrm{sgn}(b_i(k))\left(\frac{y(k)\mathbf{w}}{(\mathbf{w}^T\mathbf{w})^{3/2}} - \frac{\mathbf{r}(k)}{\sqrt{\mathbf{w}^T\mathbf{w}}}\right). \tag{30}$$

By substituting $\nabla P_E(\mathbf{w})$ with $\nabla\hat{P}_E(\mathbf{w})$ in the conjugate gradient updating mechanism, a block-data adaptive algorithm can readily be obtained [12].

4.1 Least Bit Error Rate Algorithm

To derive a sample-by-sample adaptive algorithm, consider a single-sample estimate of $p_y(y)$

$$\hat{p}_y(y,k) = \frac{1}{\sqrt{2\pi}\rho_n\sqrt{\mathbf{w}^T\mathbf{w}}} \exp\left(-\frac{(y-y(k))^2}{2\rho_n^2\mathbf{w}^T\mathbf{w}}\right). \quad (31)$$

Using the instantaneous stochastic gradient

$$\nabla\hat{P}_E(\mathbf{w},k) = \frac{\mathrm{sgn}(b_i(k))}{\sqrt{2\pi}\rho_n} \exp\left(-\frac{y^2(k)}{2\rho_n^2\mathbf{w}^T\mathbf{w}}\right)\left(\frac{y(k)\mathbf{w}}{(\mathbf{w}^T\mathbf{w})^{3/2}} - \frac{\mathbf{r}(k)}{\sqrt{\mathbf{w}^T\mathbf{w}}}\right), \quad (32)$$

with a re-scaling to ensure $\mathbf{w}^T\mathbf{w} = 1$, gives rise to the LBER algorithm:

$$\mathbf{w}(k+1) = \mathbf{w}(k) + \mu\frac{\mathrm{sgn}(b_i(k))}{\sqrt{2\pi}\rho_n} \exp\left(-\frac{y^2(k)}{2\rho_n^2}\right)(\mathbf{r}(k) - y(k)\mathbf{w}(k)), \quad (33)$$

$$\mathbf{w}(k+1) = \frac{\mathbf{w}(k+1)}{\sqrt{\mathbf{w}^T(k+1)\mathbf{w}(k+1)}}, \quad (34)$$

where the adaptive gain μ and the kernel width ρ_n are the two algorithm parameters that need to be set appropriately.

4.2 Approximate Least Bit Error Rate Algorithm

In the kernel density estimate (27), a variable width $\rho_n\sqrt{\mathbf{w}^T\mathbf{w}}$ is used. This is because the true standard deviation of $y(k)$ is $\sigma_n\sqrt{\mathbf{w}^T\mathbf{w}}$, which depends on the detector weight vector. If an approximation is made by using a constant width ρ_n in a kernel density estimate, computational complexity can be reduced considerably. Formally, this is to use the kernel density estimate

$$\tilde{p}_y(y) = \frac{1}{K\sqrt{2\pi}\rho_n}\sum_{k=1}^{K} \exp\left(-\frac{(y-y(k))^2}{2\rho_n^2}\right) \quad (35)$$

as an approximation of the true density (24), and to use

$$\tilde{P}_E(\mathbf{w}) = \frac{1}{K}\sum_{k=1}^{K} Q\left(\tilde{g}_k(\mathbf{w})\right) \quad (36)$$

with

$$\tilde{g}_k(\mathbf{w}) = \frac{\mathrm{sgn}(b_i(k))y(k)}{\rho_n} \quad (37)$$

as a BER estimate. The gradient of $\tilde{P}_E(\mathbf{w})$ has a much simpler form.

Adopting this approach, an approximate LBER algorithm is obtained:

$$\mathbf{w}(k+1) = \mathbf{w}(k) + \mu\frac{\mathrm{sgn}(b_i(k))}{\sqrt{2\pi}\rho_n} \exp\left(-\frac{y^2(k)}{2\rho_n^2}\right)\mathbf{r}(k). \quad (38)$$

For this ALBER algorithm, there is no need to normalize $\mathbf{w}$ after each update, and the algorithm has a similar complexity to the LMS algorithm.

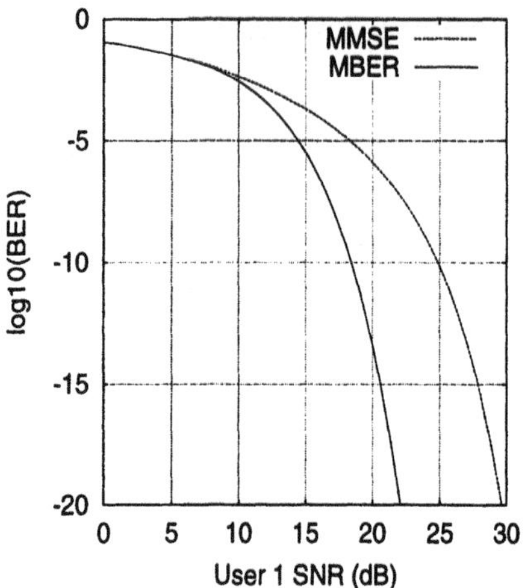

Fig. 3. Linear detector bit error rates for user 1 of Example 2. SNR_i, $1 \leq i \leq 4$, are identical.

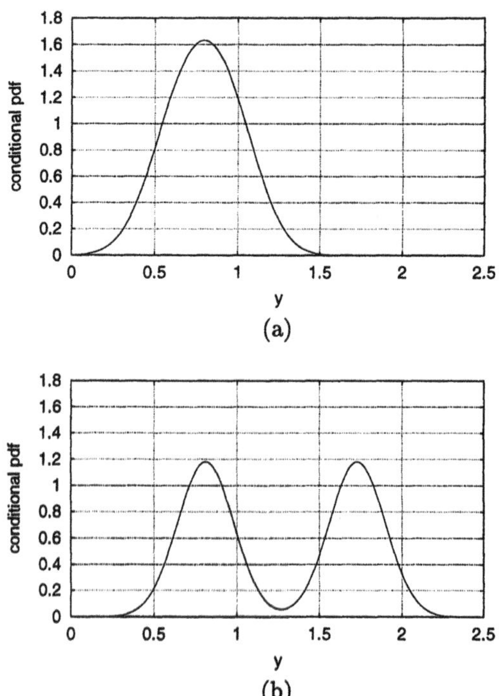

Fig. 4. Conditional probability density function $p_y(y|+1)$: (a) the MMSE detector and (b) the MBER detector, for user 1 of Example 2. $\text{SNR}_i = 16$ dB, $1 \leq i \leq 4$, and the weight vector is normalized to a unit-length.

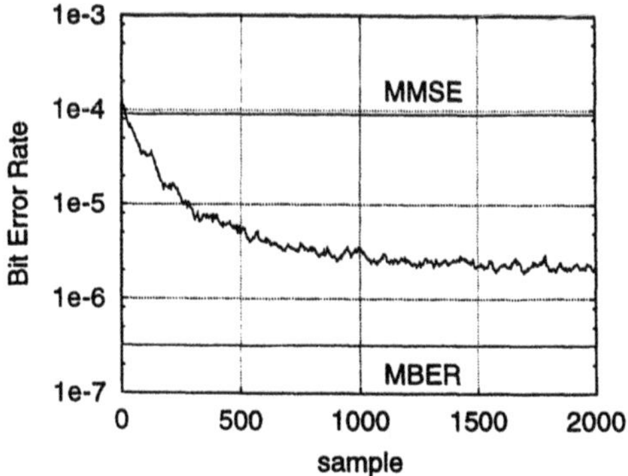

Fig. 5. Learning curves (solid: training and dashed: decision directed adaptation) of the LBER algorithm for user 1 of Example 2. $\text{SNR}_i = 16$ dB, $1 \le i \le 4$, $\mathbf{w}(0) = \mathbf{w}_{\text{MMSE}}$, adaptive gain $\mu = 0.05$ and width $\rho_n^2 = 4\sigma_n^2$. Two curves are indistinguishable.

5 Simulation Study

The adaptive MBER algorithms discussed in the previous section are investigated using computer simulation.

Example 2. This is a 4-user system with 8 chips per bit. The four user code sequences are $(+1, +1, +1, +1, -1, -1, -1, -1)$, $(+1, -1, +1, -1, -1, +1, -1, +1)$, $(+1, +1, -1, -1, -1, -1, +1, +1)$ and $(+1, -1, -1, +1, -1, +1, +1, -1)$, respectively. The four users have equal signal power and the transfer function of the CIR at chip rate is

$$H(z) = 0.4 + 0.7z^{-1} + 0.4z^{-2}. \tag{39}$$

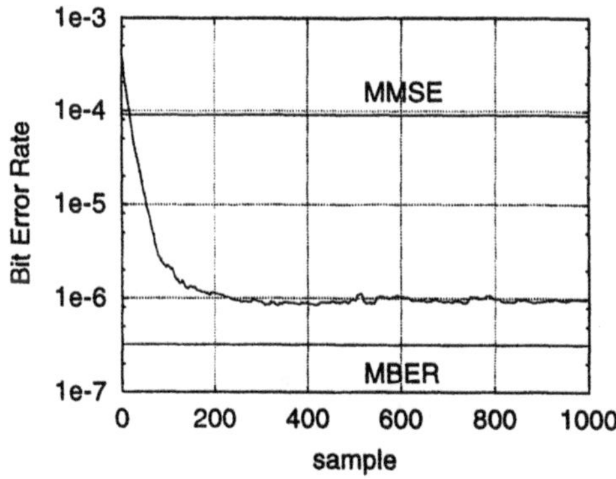

Fig. 6. Learning curves (solid: training and dashed: decision directed adaptation) of the LBER algorithm for user 1 of Example 2. $\text{SNR}_i = 16$ dB, $1 \le i \le 4$, $\mathbf{w}(0) = [0.01\ 0.01\ 0.01\ -0.01\ -0.01\ -0.01\ -0.01\ 0.01]^T$, adaptive gain $\mu = 0.05$ and width $\rho_n^2 = 4\sigma_n^2$. Two curves are indistinguishable.

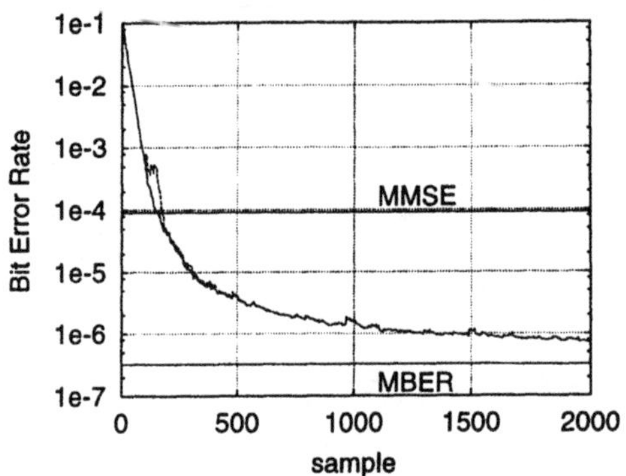

Fig. 7. Learning curves (solid: training and dashed: first 80-sample training followed by decision directed adaptation) of the LBER algorithm for user 1 of Example 2. $\mathrm{SNR}_i = 16$ dB, $1 \leq i \leq 4$, $\mathbf{w}(0) = \mathbf{w}_{\mathrm{MF}}$, adaptive gain $\mu = 0.05$ and width $\rho_n^2 = 2\sigma_n^2$.

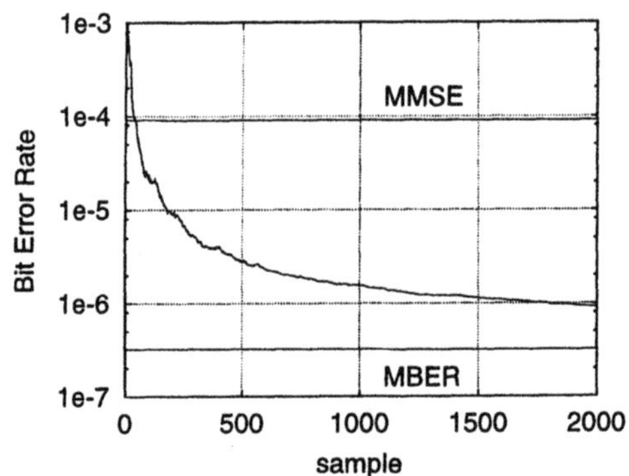

Fig. 8. Learning curves (solid: training and dashed: decision directed adaptation) of the ALBER algorithm for user 1 of Example 2. $\mathrm{SNR}_i = 16$ dB, $1 \leq i \leq 4$, $\mathbf{w}(0) = \mathbf{w}_{\mathrm{MMSE}}$, adaptive gain $\mu = 0.2$ and width $\rho_n^2 = 25\sigma_n^2$. Two curves are indistinguishable.

The linear detector for user 1 is considered. Fig. 3 compares the BER performance of the MMSE detector with that of the MBER detector. The MBER solution is obtained using the conjugate gradient algorithm with a periodic resetting of search direction. It can be seen that for this case the BERs of the two detectors are considerably different. Given the user 1 SNR to be $\mathrm{SNR}_1 = 16$ dB, Fig. 4 depicts the conditional p.d.f.s, $p_y(y|+1)$, for the MMSE and MBER detectors. It can be see that the conditional p.d.f. of the MMSE detector resembles a Gaussian. This is not surprising. The conditional p.d.f. $p_y(y|+1)$ is a sum of Gaussian distributions. The MMSE solution can be viewed to set the parameter vector $\mathbf{w}$ of $p_y(y|+1)$ in such a way that

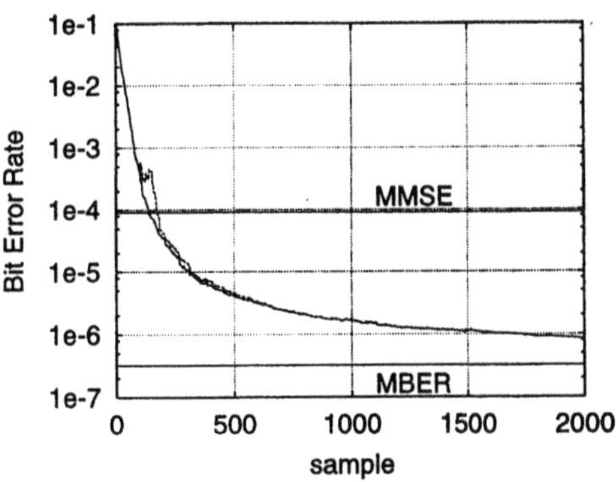

Fig. 9. Learning curves (solid: training and dashed: first 80-sample training followed by decision directed adaptation) of the ALBER algorithm for user 1 of Example 2. $\text{SNR}_i = 16$ dB, $1 \leq i \leq 4$, $\mathbf{w}(0) = \mathbf{w}_{\text{MF}}$, adaptive gain $\mu = 0.5$ and width $\rho_n^2 = 25\sigma_n^2$.

this non-Gaussian distribution looks like a Gaussian distribution. The non-Gaussian nature of $p_y(y|+1)$ is clearly demonstrated in the case of the MBER solution. The two adaptive MBER algorithms, the LBER and ALBER, are next studied. In the investigation, all the results are averaged over 100 runs.

Fig. 5 shows the learning curves of the LBER algorithm, given $\text{SNR}_1 =$ 16 dB and $\mathbf{w}(0) = \mathbf{w}_{\text{MMSE}}$, where the adaptive gain $\mu = 0.05$ and the width parameter $\rho_n^2 = 4\sigma_n^2 \approx 0.1$. There are in fact two indistinguishable learning curves in Fig. 5, the solid one indicates the training performance, and the dashed one is the decision-directed performance with $b_1(k)$ being substituted by the detector's decisions $\hat{b}_1(k)$. It is well known that the BER surface is highly complicated and may contain local minima. Our experience has suggested that initializing $\mathbf{w}(0)$ to the MMSE solution is generally a bad choice for the LBER algorithm, as the convergence speed is generally slow and the steady-state BER misadjustment is relatively large for this initial condition. The training and decision-directed learning curves for the LBER algorithm given $\mathbf{w}(0) = [0.01\ 0.01\ 0.01\ -0.01\ -0.01\ -0.01\ -0.01\ 0.01]^T$ are shown in Fig. 6. It can be seen that for this choice of $\mathbf{w}(0)$ the algorithm has a much faster convergence rate and a smaller steady-state BER misadjustment. Unlike the MMSE solution, the MF is known to a detector and can conveniently be used as $\mathbf{w}(0)$. With $\mathbf{w}(0) = \mathbf{w}_{\text{MF}}$, the two learning curves of the LBER algorithm are illustrated in Fig. 7. It can be seen that the algorithm performs better with the MF solution as $\mathbf{w}(0)$ than with the MMSE solution as $\mathbf{w}(0)$.

The learning curves of the ALBER algorithm with $\mathbf{w}(0)$ set to the MMSE solution and MF one are depicted in Figs. 8 and 9, respectively. Comparing Figs. 8 and 9 with Figs. 5 and 7, it can be seen that this approximate LBER algorithm does not seem to result in performance degradation. In fact, the ALBER algorithm appears to be less sensitive to the choice of initial condition, and the algorithm with $\mathbf{w}(0)$ set to the MMSE solution performs better than the LBER algorithm with the same initial condition. The ALBER algorithm has an added advantage of simpler computational requirements.

Example 3. This example investigate the near-far effect to the adaptive MBER algorithm. The system has two users with the two user chip codes given by $(+1, +1, -1, -1)$ and $(+1, -1, -1, +1)$, respectively. The transfer function of the CIR at chip rate is given by

$$H(z) = 1.0 + 0.25z^{-1} + 0.5z^{-3}. \tag{40}$$

The linear detector for user 1 is considered. With a fixed user 1 signal power A_1 and a fixed $\mathrm{SNR}_1 = 16$ dB, the interfering user 2 power A_2 is varied to provide different desired signal to interference ratios (SIR). Fig. 10 summarizes the performance of various detectors.

Example 4. The settings of this example are identical to Example 2, except that the three-path channel

$$H(z) = h_0 + h_1 z^{-1} + h_2 z^{-2} \tag{41}$$

is a Rayleigh fading one with the normalized Doppler frequency $f_d = 7.69 \times 10^{-7}$. Transmission is organized into frames, and a frame consists of 37 training bits and 112 data bits. Decision-directed adaptation is employed during data transmission. The channel is assumed to be frame faded, that is, the channel is kept constant within a frame. A typical set of the channel taps

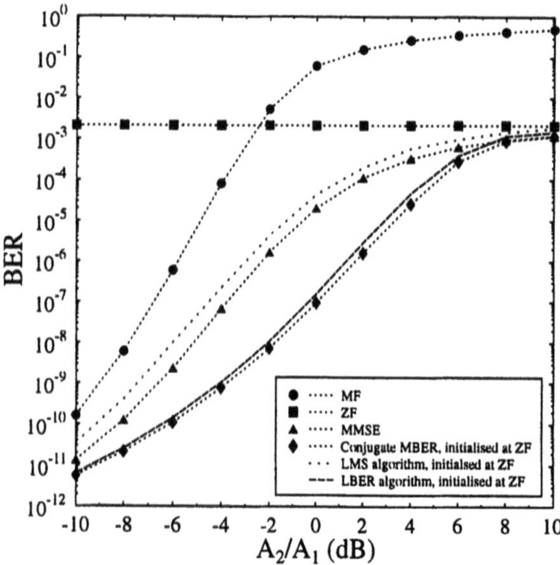

Fig. 10. Comparison of bit error rates of various linear detectors for user 1 of Example 3. $\mathrm{SNR}_1 = 16$ dB with varying SIR.

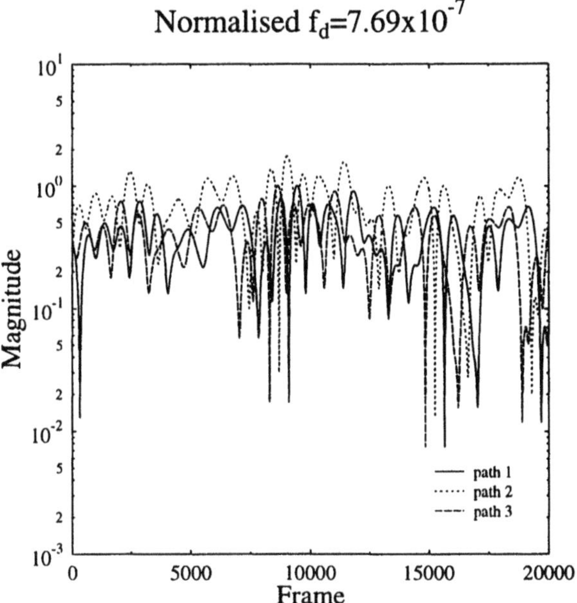

Fig. 11. A typical set of channel fading paths for Example 4.

is shown in Fig. 11. Linear detector for user 1 is considered. Fig. 12 depicts the performance of the two adaptive algorithms, the LBER and LMS, in comparison with the benchmark of the theoretical MMSE performance.

6 Extension to the nonlinear multiuser detector

If we are not restricting to the class of linear detector, then the optimal detector for the system model described in Section 2 can easily be shown to be the following Bayesian detector [22]

$$y_B(k) = f_B(\mathbf{r}(k)) = \sum_{j=1}^{N_b} \beta_j b_i^{(j)} \exp\left(-\frac{\|\mathbf{r}(k) - \bar{\mathbf{r}}_j\|^2}{2\sigma_n^2}\right) \tag{42}$$

where $\bar{\mathbf{r}}_j \in \mathcal{R}$ and β_j is a positive constant incorporating ***a priori*** probability of $\bar{\mathbf{r}}_j$. Since all the states in $\mathcal{R}$ are equiprobable, all the β_j have the same value. The hard decision is made by quantizing $y_B(k)$ according to the rule (9). Although this Bayesian detector provides the best performance, it is computationally very expensive. Furthermore, the set of the channel states $\bar{\mathbf{r}}_j$ are unknown and have to be learned by some means.

Consider the general nonlinear detector for user i, which has the form

$$y(k) = f(\mathbf{r}(k); \mathbf{w}) \tag{43}$$

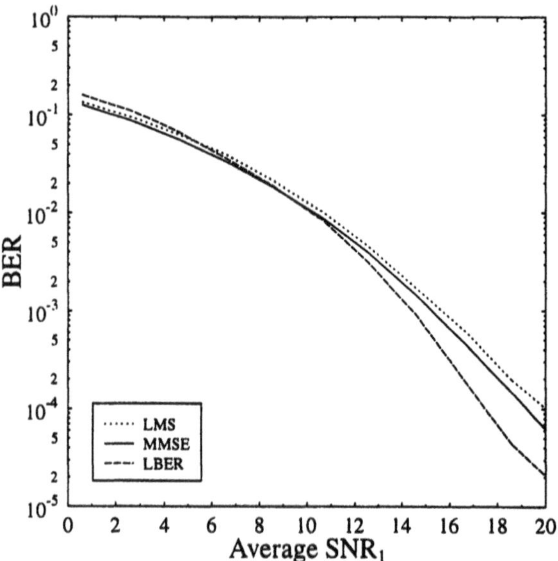

Fig. 12. Performance comparison of three user-1 detectors for Example 4. For the two adaptive algorithms, $\mathbf{w}(0) = \mathbf{w}_{\mathrm{MF}}$.

where the detector map f is realized for example by a neural network and the vector $\mathbf{w}$ consists of all the adjustable parameters of the detector. Classically, adaptive training of such a nonlinear structure is based on the MMSE principle and typically implemented using the LMS algorithm

$$\left.\begin{array}{l} y(k) = f(\mathbf{r}(k); \mathbf{w}(k-1)) \\ \mathbf{w}(k) = \mathbf{w}(k-1) + \mu(b_i(k) - y(k)) \frac{\partial f(\mathbf{r}(k); \mathbf{w}(k-1))}{\partial \mathbf{w}} \end{array}\right\} \tag{44}$$

However, the true performance criterion should be the BER, and we consider how to construct an adaptive MBER algorithm.

By linearizing the detector (43) around $\bar{\mathbf{r}}(k)$, it can be approximated as

$$y(k) \approx \bar{y}(k) + e(k) \tag{45}$$

where

$$\bar{y}(k) = f(\bar{\mathbf{r}}(k); \mathbf{w}) \tag{46}$$

can only take the value from the set

$$\mathcal{Y} \triangleq \{\bar{y}_j = f(\bar{\mathbf{r}}_j; \mathbf{w}),\ 1 \leq j \leq N_b\} \tag{47}$$

and

$$e(k) = \left[\frac{\partial f(\bar{\mathbf{r}}(k); \mathbf{w})}{\partial \mathbf{r}}\right]^T \mathbf{n}(k) \tag{48}$$

is Gaussian with zero mean and variance

$$\rho_n^2(\mathbf{w}) = \frac{\sigma_n^2}{N_b} \sum_{j=1}^{N_b} \left[\frac{\partial f(\bar{\mathbf{r}}_j; \mathbf{w})}{\partial \mathbf{r}} \right]^T \frac{\partial f(\bar{\mathbf{r}}_j; \mathbf{w})}{\partial \mathbf{r}} \tag{49}$$

The p.d.f. of $y(k)$ can then be approximated by

$$p_y(y) \approx \frac{1}{N_b \sqrt{2\pi \rho_n^2(\mathbf{w})}} \sum_{j=1}^{N_b} \exp\left(-\frac{(y - \bar{y}_j)^2}{2\rho_n^2(\mathbf{w})} \right) \tag{50}$$

and the BER of the detector is approximately

$$P_E(\mathbf{w}) \approx \frac{1}{N_b} \sum_{j=1}^{N_b} Q(g_j(\mathbf{w})) \tag{51}$$

where

$$g_j(\mathbf{w}) = \frac{\mathrm{sgn}(b_i^{(j)}) \bar{y}_j}{\rho_n(\mathbf{w})} = \frac{\mathrm{sgn}(b_i^{(j)}) f(\bar{\mathbf{r}}_j; \mathbf{w})}{\rho_n(\mathbf{w})}. \tag{52}$$

Using the kernel density estimate (35) with a constant ρ_n^2 to approximate the density (50) naturally leads to the ALBER algorithm

$$\left. \begin{array}{l} y(k) = f(\mathbf{r}(k); \mathbf{w}(k-1)) \\ \mathbf{w}(k) = \mathbf{w}(k-1) + \mu \frac{\mathrm{sgn}(b_i(k))}{\sqrt{2\pi}\rho_n} \exp\left(-\frac{y^2(k)}{2\rho_n^2} \right) \frac{\partial f(\mathbf{r}(k); \mathbf{w}(k-1))}{\partial \mathbf{w}} \end{array} \right\} \tag{53}$$

for training the nonlinear detector (43). The derivative $\frac{\partial f}{\partial \mathbf{w}}$ depends on the particular detector map employed. For example, consider the radial basis function (RBF) detector of the form

$$y(k) = f_{RBF}(\mathbf{r}(k); \mathbf{w}) = \sum_{j=1}^{n_c} \alpha_j \exp\left(-\frac{\|\mathbf{r}(k) - \mathbf{c}_j\|^2}{\tilde{\sigma}_j} \right). \tag{54}$$

The parameter vector $\mathbf{w}$ contains all the RBF weights α_j, widths $\tilde{\sigma}_j$ and centers $\mathbf{c}_j$. The dimension of $\mathbf{w}$ is therefore $n_c \times (M+2)$. The derivatives $\frac{\partial f_{RBF}}{\partial \mathbf{w}}$ are given in the forms

$$\left. \begin{array}{l} \frac{\partial f_{RBF}}{\partial \alpha_j} = \exp\left(-\frac{\|\mathbf{r}(k) - \mathbf{c}_j\|^2}{\tilde{\sigma}_j} \right) \\ \frac{\partial f_{RBF}}{\partial \tilde{\sigma}_j} = \alpha_j \exp\left(-\frac{\|\mathbf{r}(k) - \mathbf{c}_j\|^2}{\tilde{\sigma}_j} \right) \frac{\|\mathbf{r}(k) - \mathbf{c}_j\|^2}{\tilde{\sigma}_j^2} \\ \frac{\partial f_{RBF}}{\partial \mathbf{c}_j} = 2\alpha_j \exp\left(-\frac{\|\mathbf{r}(k) - \mathbf{c}_j\|^2}{\tilde{\sigma}_j} \right) \frac{\mathbf{r}(k) - \mathbf{c}_j}{\tilde{\sigma}_j} \end{array} \right\} \quad 1 \le j \le n_c. \tag{55}$$

Example 5. The settings of this example are identical to Example 2, except that this time the detector for user 2 is considered. For user 2, $\mathcal{R}^{(+)}$ and

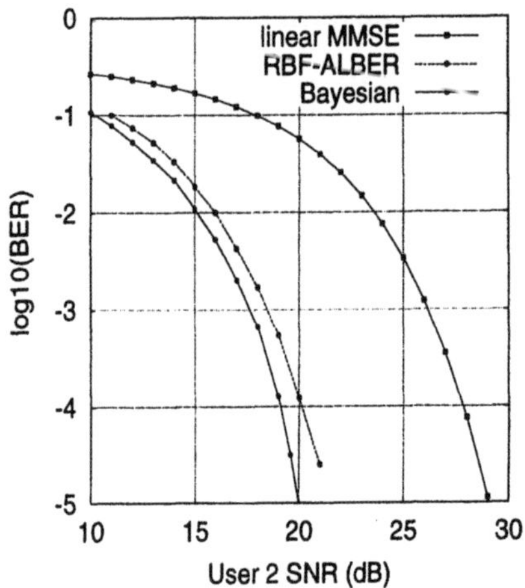

Fig. 13. Linear and nonlinear detector bit error rates for user 2 of Example 5. SNR_i, $1 \leq i \leq 4$, are identical. The RBF detector has 64 centers.

$\mathcal{R}^{(-)}$ are almost linearly inseparable, and a linear detector has poor BER performance, as is shown in Fig. 13. For this example, the linear MMSE solution and the linear MBER one produce the same BER. The BERs of the optimal Bayesian detector is also shown in Fig. 13. Note that in this example the number of channel states $N_b = 256$, and the Bayesian detector is highly complex. The performance of the 64-center RBF detector trained by the ALBER algorithm (53) is depicted in Fig. 13. It can be seen that the performance of this ALBER RBF detector is very close to the full optimal Bayesian performance. Interestingly, in the simulation it is observed that the same 64-center RBF detector under the identical conditions but trained by the LMS algorithm (44), although converged well in the MSE, often results in a BER near 0.5. This confirms with the results given in [16],[17].

7 Conclusions

Adaptive multiuser detection has been considered based on the principle of minimizing the BER. It has clearly been demonstrated that, even for the linear detector, the MBER solution can be considerably better than the classical MMSE solution at least for certain situations. A fully adaptive MBER approach has been developed for training the linear detector. In particular, the ALBER algorithm has a computational complexity similar to that of the very simple LMS algorithm. Furthermore, it has been shown how to extend the adaptive MBER approach to the nonlinear multiuser detector.

Acknowledgement

Figs. 2, 10, 11 and 12 were provided by Mr. A.K. Samingan.

References

1. Prasad, R. (1996) *CDMA for Wireless Personal Communications.* Artech House, Inc.
2. Verdú, S. (1998) *Multiuser Detection.* Cambridge University Press
3. Xie, Z., Short, R.T., Rushforth, C.K. (1990) A family of suboptimum detectors for coherent multiuser communications. *IEEE J. Selected Areas in Communications*, **8**, 683–690
4. Madhow, U., Honig, M.L. (1994) MMSE interference suppression for direct-sequence spread-spectrum CDMA. *IEEE Trans. Communications*, **42**, 3178–3188
5. Miller, S.L. (1995) An adaptive direct-sequence code-division multiple-access receiver for multiuser interference rejection. *IEEE Trans. Communications*, **43**, 1746–1755
6. Moshavi, S. (1996) Multi-user detection for DS-CDMA communications. *IEEE Communications Magazine*, **34**, 124–136
7. Poor, H.V., Verdú, S. (1997) Probability of error in MMSE multiuser detection. *IEEE Trans. Information Theory*, **43**, 858–871
8. Woodward, G., Vucetic, B.S. (1998) Adaptive detection for DS-CDMA. *Proc. IEEE*, **86**, 1413–1434
9. Mandayam, N.B., Aazhang, B. (1997) Gradient estimation for sensitivity analysis and adaptive multiuser interference rejection in code-division multi-access systems. *IEEE Trans. Communications*, **45**, 848–858
10. Psaromiligkos, I.N., Batalama, S.N., Pados, D.A. (1999) On adaptive minimum probability of error linear filter receivers for DS-CDMA channels. *IEEE Trans. Communications*, **47**, 1092–1102
11. Yeh, C.C., Lopes, R.R., Barry, J.R. (1998) Approximate minimum bit-error rate multiuser detection. In: *Proc. Globecom'98* (Sydney, Australia), Nov. 1998, 3590–3595
12. Chen, S., Samingan, A.K., Mulgrew, M., Hanzo, L. (2001) Adaptive minimum-BER linear multiuser detection for DS-CDMA signals in multipath channels. *IEEE Trans. Signal Processing*, **49**, 1240–1247
13. Sharma, R., Sethares, W.A., Bucklew, J.A. (1996) Asymptotic analysis of stochastic gradient-based adaptive filtering algorithms with general cost functions. *IEEE Trans. Signal Processing*, **44**, 2186–2194
14. Mulgrew, B., Chen, S. (2000) Stochastic gradient minimum-BER decision feedback equalisers. In: *Proc. IEEE Symp. Adaptive Systems for Signal Processing, Communication and Control* (Lake Louise, Alberta, Canada), Oct.1-4, 2000, 93–98
15. Mulgrew, B., Chen, S. (2001) Adaptive minimum-BER decision feedback equalisers for binary signalling. *Signal Processing*, **81**, 1479–1489
16. Chen, S., Samingan, A.K., Hanzo, L. (2002) Adaptive near minimum error rate training for neural networks with application to multiuser detection in CDMA communication systems. submitted to *Neural Networks*
17. Chen, S., Mulgrew, B., Hanzo, L. (2001) Adaptive least error rate algorithm for neural network classifier. In: *Proc. 2001 IEEE Workshop Neural Networks for Signal Processing* (Falmouth, MA, USA), Sept.10-12, 2001, 223–232
18. Bazaraa, M.S., Sherali, H.D., Shetty, C.M. (1993) *Nonlinear Programming: Theory and Algorithms.* New York: John Wiley

19. Parzen, E. (1962) On estimation of a probability density function and mode. *The Annals of Mathematical Statistics*, **33**, 1066–1076
20. Silverman, B.W. (1996) *Density Estimation.* London: Chapman Hall
21. Bowman, A.W., Azzalini, A. (1997) *Applied Smoothing Techniques for Data Analysis.* Oxford: Oxford University Press
22. Chen, S., Samingan, A.K., Hanzo, L. (2001) Support vector machine multiuser receiver for DS-CDMA signals in multipath channels. *IEEE Trans. Neural Networks*, **12**, 604-611

www.ingramcontent.com/pod-product-compliance
Ingram Content Group UK Ltd.
Pitfield, Milton Keynes, MK11 3LW, UK
UKHW021859190726
13853UKWH00003B/1331
* 9 7 8 3 6 4 2 5 3 6 2 2 9 *